Instructor Solutions Manual

for Olmsted/Williams

Chemistry

THE MOLECULAR SCIENCE

Larry Peck
and
Frank Kolar
Texas A & M University

Mosby
St. Louis Baltimore Berlin Boston Carlsbad Chicago London Madrid
Naples New York Philadelphia Sydney Tokyo Toronto

Dedicated to Publishing Excellence

Printed in the United States of America

Mosby–Year Book, Inc.
11830 Westline Industrial Drive
St. Louis, Missouri 63146

ISBN 0-8151-6502-1
24074

PREFACE

This solutions manual has been prepared as a supplement to CHEMISTRY-THE MOLECULAR SCIENCE by John Olmsted III and Gregory M. Williams.

The textbook by Olmsted and Williams presents an excellent, modern approach to general chemistry. To thoroughly understand the molecular world as it is presented in the textbook, a student needs to develop the ability to solve problems and answer questions related to the principles and concepts presented. We recognize that many students have difficulty developing the necessary problem-solving ability to best utilize the knowledge gained from material presented in the textbook. Through the use of this book and the textbook, the authors hope that students will improve their problem-solving skills while learning a considerable amount of chemistry. Students are strongly encouraged to attempt to solve each problem and to consult the textbook before referring to the solutions in this manual. It is highly likely that the knowledge gained and the problem-solving skills acquired in a general chemistry course will prove valuable in subsequent science and technical courses as well as assisting students to prepare for a variety of professions.

Many problems can be solved in more than one way. We encourage students to develop methods that they understand. For consistency we have striven to follow the methods used in the textbook. Where practical, we have tried to include dimensions and to show the correct number of significant figures; in some problems an extra digit is carried in intermediate steps. Remember, authors can make mistakes, too.

We wish to thank our friends and family for their support in the production of this book. Special thanks are extended to Sandra Peck, typist.

Larry Peck and Frank Kolar
Chemistry Department
Texas A&M University
College Station, TX 77843

CONTENTS

CHAPTER 1: THE SCIENCE OF CHEMISTRY

1.1 You may think of other examples to elaborate on, but a few are:
a) Setting any pollution (environmental) standards
b) Setting any food, drug or cosmetics purity standards
c) Passing any drug control laws

1.2 There are other examples, but a few ideas include:
a) Cooking b) Cleaning c) Repairing (painting and plumbing)

1.3 A pharmacist must know nomenclature for chemicals (especially complex organic molecules) in order to identify chemically equivalent commercial drugs, must know how to handle and protect the potency of drugs, must be able to weigh out drugs (in some cases small amounts) and thoroughly mix them, and must know enough about chemical reactions so that he does not mix chemically incompatible drugs.

1.4 The chemist needs more confidence; the theory may be wrong. The chemist should repeat his experiment several times to see if the results are repeatable. If they are repeatable, then he should check his interpretation of the data. If he finds nothing wrong with this experiment or interpretation, he should announce his results so that others may test his results. If their results confirm his results, the theory will need to be changed.

1.5 a) hydrogen, H b) helium, He c) hafnium, Hf d) nitrogen, N
e) neon, Ne f) niobium, Nb

1.6 a) potassium, K b) platinum, Pt c) plutonium, Pu d) lead, Pb
e) palladium, Pd

1.7 a) As, arsenic b) Ar, argon c) Al, aluminum d) Am, americium
e) Ag, silver f) Au, gold g) At, astatine h) Ac, actinium

1.8 a) Br, bromine b) Be, beryllium c) B, boron d) Bk, berkelium
e) Ba, barium f) Bi, bismuth

1.9 CCl_4

1.10 C_5H_{12}

1.11 a) Br_2 b)HCl c) C_2H_5I d) PCl_3 e) SF_4 f) N_2O_4

1.12 a) C_3H_6O b) $C_3H_6O_2$ c) $C_7H_5N_3O_6$ d) C_3H_8O e) H_2S f) $C_8H_9NO_2$

1.13 Cs, cesium

1.14 Ne, neon

1.15 O, oxygen; Cl, chlorine; Se, selenium; and P, phosphorus. Oxygen and selenium have chemical properties similar to those of sulfur.

1.16 Ge, germanium; Sb, antimony; Pb, lead; and In, indium. Germanium and lead have chemical properties similar to those of tin.

1.17 a) Li, lithium; Na, sodium; K, potassium; Rb, rubidium; and Cs, cesium are possible answers. (Select any one).
b) Be, beryllium; Mg, magnesium; Ca, calcium; Sr, strontium; Ba, barium; and Ra, radium are possible answers.

1.18 a) F, fluorine; Cl, chlorine; Br, bromine; I, iodine; and At, astatine are possible answers. (Select any three)
b) Be, beryllium; Mg, magnesium; Ca, calcium; Sr, strontium; Ba, barium; and Ra, radium are possible answers. (Select any three).
c) Select any three elements from atomic number 89 (actinium) through 102 (nobelium).
d) Select any three from: Ne, neon; Ar, argon; Kr, krypton; Xe, xenon; and Rn, radon.
e) Select any three of the 37 elements located between columns II and III on the Periodic Table.
f) Select any three elements from atomic numbers 57 through 70.
g) Select three nonmetals or metalloids found in columns I through VII of the Periodic Table.
h) Li, lithium; Na, sodium; K, potassium; Rb, rubidium; Cs, cesium; and Fr, francium are possible answers. (Select any three).

1.19 lithium, Li; beryllium, Be; boron, B; carbon, C; oxygen, O; fluorine, F; neon, Ne

1.20 phosphorus, P; arsenic, As; antimony, Sb; and bismuth, Bi

1.21 a) iron - pure substance
b) cup of coffee - solution
c) glass of milk - heterogeneous mixture
d) dust free atmosphere - solution
e) dusty atmosphere - heterogeneous mixture
f) block of wood - heterogeneous mixture

1.22 a) blood - heterogeneous mixture
b) dry ice - pure substance
c) krypton gas - pure substance
d) platinum-iridium bar - solution
e) uniodized table salt - pure substance
f) lemonade - heterogeneous mixture

1.23 a) gasoline - liquid b) molasses - liquid c) snow - solid d) chewing gum - solid

1.24 a) tree sap - liquid b) ozone- gas c) shaving cream - liquid d) water vapor - gas

1.25 a) formation of frost - physical change
b) drying of clothes - physical change
c) burning of leaves - chemical change

1.26 a) boiling water - physical change
b) coffee brewing - physical change
c) photographic film being developed - chemical change

1.27 a) lake water - mixture
b) distilled water- compound
c) mud - mixture
d) helium - element
e) rubbing alcohol - compound
f) paint - mixture

1.28 a) Earth's atmosphere - mixture b) beer - mixture c) iron magnet - element
d) ice - compound e) liquid bromine - element f) mercury - element

1.29 a) $100,000 = 1.00000 \times 10^5$ b) $10,000 \pm 100 = 1.00 \times 10^4$
c) $0.000400 = 4.00 \times 10^{-4}$ d) $0.0003 = 3 \times 10^{-4}$ e) $275.3 = 2.753 \times 10^2$

1.30 a) $175,906 = 1.75906 \times 10^5$ b) $0.0000605 = 6.05 \times 10^{-5}$
c) $2,500,000 \pm 100 = 2.5000 \times 10^6$ d) $2,500,000,000 = 2.50 \times 10^9$

1.31 a) $430 \text{ kg} = 4.30 \times 10^2 \text{ kg}$ b) $1.35\ \mu\text{m} \times \frac{1 \times 10^{-6}\ \text{m}}{\mu\text{m}} = 1.35 \times 10^{-6}\ \text{m}$

c) $624 \text{ ps} \times \frac{1 \times 10^{-12}\ \text{s}}{\text{ps}} = 6.24 \times 10^{-10}\ \text{s}$

d) $1024 \text{ ng} \times \frac{10^{-9}\ \text{g}}{\text{ng}} \times \frac{1\ \text{kg}}{10^3\ \text{g}} = 1.024 \times 10^{-9}\ \text{kg}$

e) $93,000 \text{ km} \times \frac{10^3\ \text{m}}{\text{km}} = 9.3000 \times 10^7\ \text{m}$

f) $1 \text{ day} \times \frac{24\ \text{h}}{\text{d}} \times \frac{60\ \text{min}}{\text{h}} \times \frac{60\ \text{s}}{\text{min}} = 9 \times 10^4\ \text{s}$
(assuming that one day is only one significant figure)

g) $0.0426 \text{ in} \times \frac{2.54\ \text{cm}}{\text{in}} \times \frac{10^{-2}\ \text{m}}{\text{cm}} = 1.08 \times 10^{-3}\ \text{m}$

1.32 a) $1 \text{ wk} \times \frac{7\ \text{d}}{\text{wk}} \times \frac{24\ \text{h}}{\text{d}} \times \frac{60\ \text{min}}{\text{h}} \times \frac{60\ \text{s}}{\text{min}} = 6.04800 \times 10^5\ \text{s}$
(assuming exactly one week)

b) $15 \text{ miles} \times \frac{1.6093\ \text{km}}{1\ \text{mile}} \times \frac{10^3\ \text{m}}{1\ \text{km}} = 2.4 \times 10^4\ \text{m}$

c) $4.567\ \mu\text{s} \times \frac{10^{-6}\ \text{s}}{\mu\text{s}} = 4.567 \times 10^{-6}\ \text{s}$

d) $6.45 \text{ mL} \times \frac{10^{-3}\ \text{L}}{\text{mL}} \times \frac{10^{-3}\ \text{m}^3}{\text{L}} = 6.45 \times 10^{-6}\ \text{m}^3$

e) $47\ \mu\text{g} \times \frac{10^{-6}\ \text{g}}{\mu\text{g}} \times \frac{\text{kg}}{10^3\ \text{g}} = 4.7 \times 10^{-8}\ \text{kg}$

1.33 $5.0 \times 10^{-1} \text{ carat} \times \frac{3.168 \text{ grains}}{1 \text{ carat}} \times \frac{1 \text{ g}}{15.4 \text{ grains}} \times \frac{10^{-3} \text{ kg}}{1 \text{ g}} = 1.0 \times 10^{-4} \text{ kg}$

$7.00 \text{ g of Au} \times \frac{10^{-3} \text{ kg}}{1 \text{ g}} = 7.00 \times 10^{-3} \text{ kg}$

$1.0 \times 10^{-4} \text{kg} + 7.00 \times 10^{-3} \text{ kg} = 7.10 \times 10^{-3} \text{ kg}$

1.34 $10.00 \text{ kg } H_2O + 0.0065 \text{ kg NaCl} + 0.047546 \text{ kg sugar} =$

$10.05\underline{4046} \rightarrow 10.05 \text{ kg} = 1.005 \times 10^4 \text{ g}$

1.35 $1 \text{ qt } H_2O \times \frac{1 \text{ L}}{1.057 \text{ qt}} \times \frac{10^3 \text{mL}}{1 \text{ L}} \times \frac{1 \text{ cm}^3}{1 \text{ mL}} \times \frac{1.00 \text{ g}}{1 \text{ cm}^3} = 946 \text{ g}$

(assuming exactly 1 qt of H_2O)

1.36 $1 \text{ qt Hg} \times \frac{1 \text{ L}}{1.057 \text{ qt}} \times \frac{10^3 \text{mL}}{1 \text{ L}} \times \frac{1 \text{ cm}^3}{1 \text{ mL}} \times \frac{13.55 \text{ g}}{\text{cm}^3} = 12819.3 \text{ g} \rightarrow 1.282 \times 10^4 \text{ g}$

(assuming exactly 1 qt of Hg)

1.37 $\text{Volume} = 15.5 \text{ cm} \times 4.6 \text{ cm} \times 1.75 \text{ cm} = 125 \text{ cm}^3$

$\text{density} = \frac{98.456 \text{ g}}{125 \text{ cm}^3} = 0.79 \frac{\text{g}}{\text{cm}^3}$

1.38 $\text{mass of liquid} = 0.4827 \text{ g} - 0.4763 \text{ g} = 0.0064 \text{ g}$

$\text{density} = \frac{0.0064 \text{ g}}{8.00 \times 10^{-3} \text{ cm}^3} = 0.80 \frac{\text{g}}{\text{cm}^3}$

1.39 $\text{Volume} = \frac{\text{mass}}{\text{density}} = \frac{15.4 \text{ g}}{2.70 \text{ g/cm}^3} = 5.70 \text{ cm}^3$

1.40 $\text{Volume} = \frac{0.246 \text{ g}}{2.65 \text{ g/cm}^3} = 0.0928 \text{ cm}^3$

1.41 $r = 0.875 \text{ in} \times \frac{2.54 \text{ cm}}{1 \text{ in}} = 2.2225 \text{ cm}$; $\text{height} = 4.500 \text{ in} \times \frac{2.54 \text{ cm}}{1 \text{ in}} = 11.43 \text{ cm}$

$V = \pi r^2 h = \pi(2.2225)^2(11.43) = 177.37 \text{ cm}^3$

mass of $H_2O = 270.064 \text{ g} - 93.054 \text{ g} = 177.010 \text{ g}$

$$d = \frac{177.010 \text{ g}}{177.37 \text{ cm}^3} = 0.998 \text{ g / cm}^3$$

1.42 V of Pb = $(2.00 \text{ cm})^3 = 8.00 \text{ cm}^3$;
mass = V x d = $(8.00 \text{ cm}^3)(11.34 \text{ g / cm}^3) = 90.7 \text{ g}$
V of Hg = $4\pi r^3/3 = 4\pi(1.00 \text{ cm})^3/3 = 4.18879 \text{ cm}^3$
mass = V x d = $(4.18879 \text{ cm}^3)(13.55 \text{ g/cm}^3) = 56.8 \text{ g}$
The lead cube possesses more mass.

1.43 a) 176 kg b) 2.54 s c) 73 mi hr^{-1} d) 924 kg m^2s^{-2} e) 34 J

1.44 a) $340 \text{ in}^3 \times \left(\frac{2.54 \text{ cm}}{\text{in}}\right)^3 \times \left(\frac{\text{m}}{10^2 \text{ cm}}\right)^3 = 5.57 \times 10^{-3} \text{ m}^3$

b) $\frac{35 \text{ mile}}{\text{hr}} \times \frac{5280 \text{ ft}}{\text{mile}} \times \frac{12 \text{ in}}{\text{ft}} \times \frac{2.54 \text{ cm}}{\text{in}} \times \frac{\text{m}}{10^2 \text{ cm}} \times \frac{1 \text{hr}}{60 \text{ min}} \times \frac{1 \text{ min}}{60 \text{ s}}$
$= 16 \text{ m s}^{-1}$

c) $\left(\left(6 \text{ ft} \times 12 \frac{\text{in}}{\text{ft}}\right) + 10 \text{ in}\right) \times \frac{2.54 \text{ cm}}{\text{in}} \times \frac{1 \text{ m}}{10^2 \text{ cm}} = 2.1 \text{ m}$

d) $220 \text{ lb} \times \frac{1 \text{ kg}}{2.2 \text{ lb}} = 100 \text{ kg} \rightarrow 1.0 \times 10^2 \text{ kg}$

e) $\frac{32 \text{ mi}}{\text{gal}} \times \frac{1.61 \text{ km}}{1 \text{ mi}} \times \frac{0.266 \text{ gal}}{1 \text{ L}} = 13.70432 \frac{\text{km}}{\text{L}} \rightarrow 14 \text{ km / L}$

1.45 $1 \text{ year} \times \frac{365 \text{ days}}{\text{yr}} \times \frac{24 \text{ hr}}{1 \text{ day}} \times \frac{60 \text{ min}}{1 \text{ hr}} \times \frac{60 \text{ s}}{1 \text{ min}} \times \frac{186{,}000 \text{ mi}}{1 \text{ s}}$
$\times \frac{1.61 \text{ km}}{1 \text{ mi}} = 9.44 \times 10^{12} \text{ km}$

1.46

S

O O

1.47

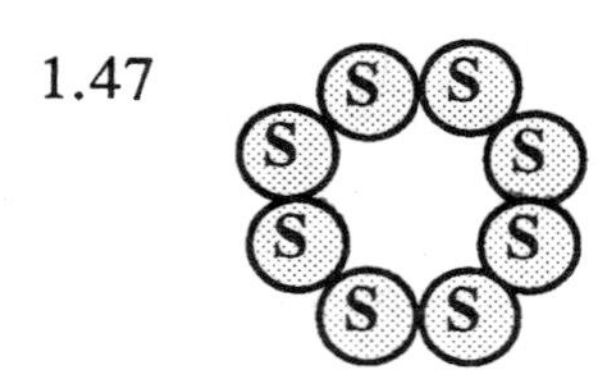

1.48 Nine elements have symbols beginning with "P." They are: Pb, lead; Pd, palladium; P, phosphorus; Pt, platinum; Pu, plutonium; Po, polonium; Pr, praseodymium; Pm, promethium; and Pa, protactinium.

1.49 Eight elements have symbols beginning with "T." They are: Ta, tantalum; Tc, technetium; Te, tellurium; Tb, terbium; Tl, thallium; Th, thorium; Tm, thulium; and Ti, titanium.

1.50 $$\text{Volume} = \frac{\text{mass}}{\text{density}} = \frac{36.5\text{ g}}{3.12\text{ g / mL}} = 11.7\text{ mL}$$

1.51 Many of these metals were found in pure or nearly pure form and were used without chemical treatment. Others were available as ores that could be easily reduced to the metal.

1.52 Sodium, Na

1.53 Lead, Pb

1.54 a) 39.15 b) 4 c) 37.5 d) 183 e) 0.03

1.55 $$100\text{ years} \times \frac{365.24\text{ day}}{\text{year}} \times \frac{24\text{ hr}}{\text{day}} \times \frac{60\text{ min}}{\text{hr}} \times \frac{60\text{ s}}{\text{min}} = 3.1557 \times 10^9\text{ s}$$

1.56

metals	nonmetals
good conductor of heat	poor conductor of heat
good conductor of electricity	poor conductor of electricity
malleable and ductile	brittle
shiny	dull appearing
form cations	form anions

1.57 silver and copper

1.58 $$\text{Area} = 3.00\text{ in} \times 3.00\text{ in} \times \frac{2.54\text{ cm}}{\text{in}} \times \frac{2.54\text{ cm}}{\text{in}} = 58.06\text{ cm}^2$$

Weight = 0.255 g; Density = 2.70 g cm^{-3} (Table 1 - 5)

$$\text{Thickness} = 0.255\text{ g} \times \frac{1\text{ cm}^3}{2.70\text{ g}} \times \frac{1}{58.06\text{ cm}^2} \times \frac{\text{pm}}{10^{-10}\text{ cm}} = 1.63 \times 10^7\text{ pm}$$

1.59 Some examples are given. You may be able to find other examples. Be and Ba, Pd and Pt, Ag and Au, S and Se, and Pr and Pa.

1.60 Yes, many other sciences have areas which are based on chemistry.

1.61 Try by starting with things that you are interested in. What happens chemically when we taste? What happens chemically when we smell? What chemical materials can serve as superconductors? Are there chemical substances that can stimulate the immune system? Develop a method of storing radioactive wastes. If you are interested in food preparation, you might include examples associated with (1) the preserving of vitamins during cooking, (2) better microwave ovens, (3) new food additives that are natural substances, (4) genetically engineered plants, (5) the effects of metal ions in foods, (6) or what essential nutrients really are.

1.62 $\frac{65\text{ miles}}{\text{hr}} \times \frac{1.61\text{ km}}{\text{mile}} = 1.0 \times 10^2\text{ km hr}^{-1}$

$\frac{65\text{ miles}}{\text{hr}} \times \frac{1.61\text{ km}}{\text{mile}} \times \frac{10^3\text{ m}}{\text{km}} \times \frac{1\text{ hr}}{60\text{ min}} \times \frac{1\text{ min}}{60\text{ s}} = 29\text{ m s}^{-1}$

1.63 $-11.5°\text{C} + 273.15 = 261.7\text{ K}$

1.64 a) $2.37 \times 10^5\text{ mile} \times 1.61\text{ km mile}^{-1} = 3.82 \times 10^5\text{ km}$

b) $2.37 \times 10^5\text{ mile} \times \frac{1.61\text{ km}}{\text{mile}} \times \frac{10^3\text{ m}}{\text{km}} \times \frac{1\text{ s}}{3.00 \times 10^8\text{ m}} = 1.27\text{ s}$

1.65 a) 4.52×10^{34} b) 1×10^{-4}

1.66 a) $CHClF_2$ b) CH_2O_2 c) BrF_3 d) C_4H_{10}

1.67 $5.00\text{ lb} \times \frac{454\text{ g}}{1\text{ lb}} \times \frac{1\text{ cm}^3}{8.92\text{ g}} = 254\text{ cm}^3$;

$\text{Diameter} = 0.0508\text{ in} \times \frac{2.54\text{ cm}}{1\text{ in}} = 0.129\text{ cm}$

$V = \pi r^2 h = 254\text{ cm}^3 = \pi(0.129\text{ cm}/2)^2 h$;

$h = \frac{254\text{ cm}^3}{\pi(0.129\text{ cm}/2)^2} = 1.94 \times 10^4\text{ cm} \times \frac{10^{-2}\text{ m}}{\text{cm}} = 1.94 \times 10^2\text{ m}$

1.68 a) germanium, arsenic, antimony
b) lanthanum (La) – ytterbium (Yb)
c) sodium, potassium, cesium
d) oxygen, sulfur, nitrogen
e) chlorine, fluorine, oxygen
f) helium, neon, argon, etc.

1.69 3 min 57 s = 237 s; 1 mile = 1.61×10^3 m;

$1500\text{ m} \times \frac{237\text{ s}}{1.61 \times 10^3\text{ m}} = 221\text{ s}$

1.70 a) intensive b) intensive c) extensive d) intensive e) extensive

1.71 $12 \text{ ft} \times 9.5 \text{ ft} \times 10.5 \text{ ft} \times \left(\frac{12 \text{ in}}{\text{ft}}\right)^3 \times \left(\frac{2.54 \text{ cm}}{\text{in}}\right)^3 \times \left(\frac{1 \text{ m}}{10^2 \text{ cm}}\right)^3 = 33.9 \text{ m}^3$

$\frac{5.5 \text{ mg}}{1.000 \text{ m}^3} \times 33.9 \text{ m}^3 \times \frac{1 \text{ g}}{10^3 \text{ mg}} = 0.19 \text{ g}$

1.72 $1.0 \text{ in} \times \frac{2.54 \text{ cm}}{\text{in}} \times \frac{1 \text{ m}}{10^2 \text{ cm}} \times \frac{1 \text{ atom}}{200 \text{ pm}} \times \frac{1 \text{ pm}}{10^{-12} \text{ m}} = 1.3 \times 10^8 \text{ atoms}$

1.73 $((7 \text{ ft} \times 12 \text{ in/ft}) + 1 \text{ in}) \times \frac{2.54 \text{ cm}}{\text{in}} \times \frac{10 \text{ mm}}{\text{cm}} = 2160. \text{ mm}$

$\pm 1/4 \text{ in} \times 2.54 \text{ cm in}^{-1} \times 10 \text{ mm cm}^{-1} = \pm 6.35 \text{ mm}$

Answer = 2160 ± 6 mm

1.74 a) Sr b) In c) Br or O d) Ge e) Co f) Ne g) Pu

1.75 $9.3 \times 10^7 \text{ miles} \times \frac{1.61 \text{ km}}{\text{mile}} \times \frac{1000 \text{ m}}{\text{km}} \times \frac{1 \text{ s}}{3.00 \times 10^8 \text{ m}} \times \frac{1 \text{ min}}{60 \text{ s}} = 8.3 \text{ min}$

1.76 Use Figures 1-4 and 1-5 to guide you in answering this question.

1.77 Follow the notations used in Figures 1-4 and 1-5.

1.78 a) Vinegar. Vinegar is mostly water and oil floats on water.
b) Table salt. Salt dropped into water sinks before it dissolves.
c) Iron. An iron pan is heavier than an aluminum pan of the same size.

1.79 a) $8.97 \times 10^5 \text{ acre ft} \times \frac{43{,}560 \text{ ft}^3}{\text{acre ft}} \times (12 \text{ in/ft})^3 \times (2.54 \text{ cm/m})^3$

$\times (1 \text{ m}/10^2 \text{ cm})^3 \times \frac{1 \text{ L}}{10^{-3} \text{ m}^3} = 1.11 \times 10^{12} \text{ L}$

b) $8.97 \times 10^5 \text{ acre ft} \times \frac{43{,}560 \text{ ft}^3}{\text{acre ft}} = 3.91 \times 10^{10} \text{ ft}^3$

c) $8.97 \times 10^5 \times 43{,}560 \text{ ft}^3 \times (12 \text{ in/ft})^3 \times (2.54 \text{ cm/in})^3 \times (1 \text{ m}/10^2 \text{ cm})^3$

$= 1.11 \times 10^9 \text{ m}^3$

1.80 $\frac{2.65 \text{ kg}}{145 \text{ cm}^3} \times \frac{1000 \text{ g}}{1 \text{ kg}} = 18.3 \text{ g/cm}^3$

Density of Au = 19.3 g/cm^3 (Section 1.6)

The crown was not pure gold.

CHAPTER 2: THE ATOMIC NATURE OF MATTER

2.1 a)

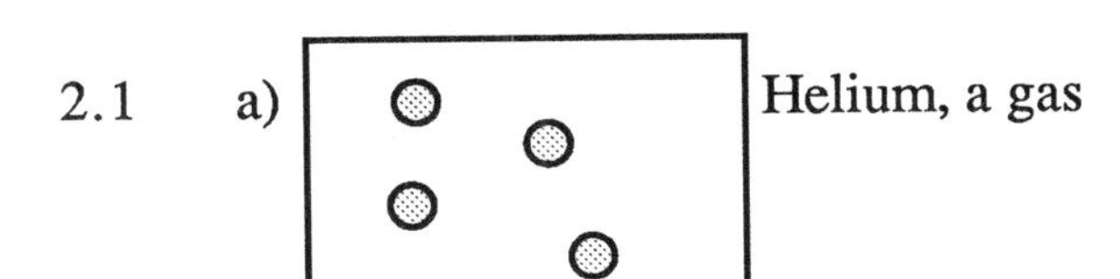

Helium, a gas

b)

Tungsten, a solid

c)

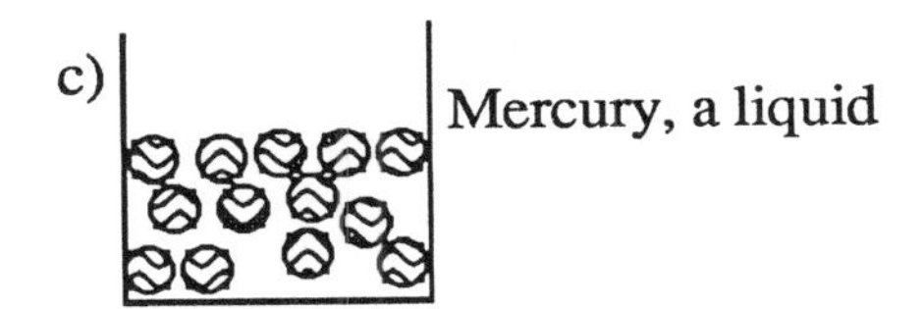

Mercury, a liquid

2.2

Solid, diatomic iodine

2.3 a) Oxygen gas contains oxygen atoms.

b) Oxygen atoms combine into diatomic oxygen molecules.

c) The diatomic oxygen molecules look and act differently than the solid carbon made from carbon atoms.

d) The molecules of oxygen and carbon react to give carbon monoxide; atoms of carbon combine with atoms of oxygen in a 1:1 ratio.

e) Atoms are rearranged in this chemical process, but the total number of each type of atom remains the same (4 atoms of oxygen and 13 atoms of carbon).

2.4

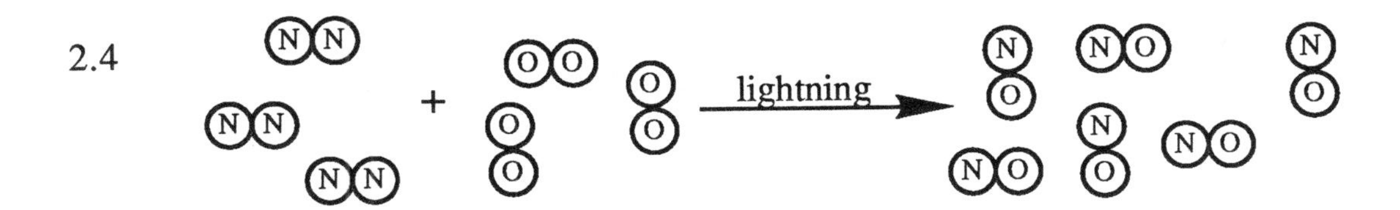

2.5 a) 25 atoms C x $\frac{\text{1 molecule } CH_4}{\text{1 atom C}}$ = 25 molecules CH_4

b) 25 atoms C x $\frac{\text{2 molecules } H_2}{\text{1 atom C}}$ = 50 molecules H_2

c) 25 atoms C x $\frac{\text{2 molecules } H_2}{\text{1 atom C}}$ x $\frac{\text{2 atoms H}}{\text{molecule } H_2}$ = 100 atoms H

2.6 HH HH HH HH HH HH

C C C ⟶ CH_4 CH_4 CH_4

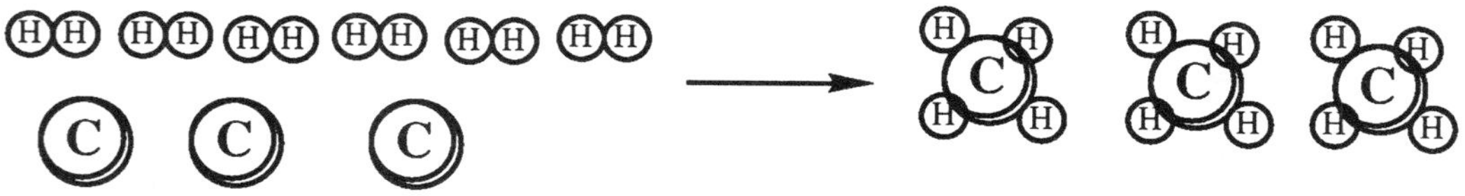

2.7 In order for the odor to reach the olfactory nerves in the nose, molecules must move from the rose through the air to the nose.

2.8 Dynamic equilibrium is established. However, the materials coming out of solution are not being deposited at the place left by those going into solution. Thus, the crystal shapes and size change over time.

2.9 Use Coulomb's Law to answer this question. Force = $k\frac{q_1q_2}{r^2}$

a) If one of the q values increases by a factor of 3, the force increases by a factor of 3.
b) If both q values increase by a factor of 3, the force increases by a factor of 9.
c) If the distance (r) doubles, the force changes by a factor of one-fourth.

2.10 a) With the increased positive charge the foil will move further away from the plate.
b) If the foil becomes negatively charged, it will be attracted to the positively charged plate.
c) As the electrons are added, the foil would approach the plate and then move away to be the original distance once the magnitude of the charge equals the original charge.

2.11 By adjusting the amount of electrical force, the downward-acting gravitational force could be exactly counterbalanced by the upward-attracting electrical force. From the electrical force required to suspend the different droplets, calculation of the charges on the droplets is possible. The calculated values of charge would be multiples of the electron charge just like Millikan determines because the positively-charged oil droplets result from the removal of one or more electron.

2.12 The alpha particles would be expected to slow down and change direction as they passed through the foil, but only by a small amount.

2.13 (See Table 2-1) -1.00×10^{-6} C x = $\frac{9.1094 \times 10^{-31}\text{ kg}}{-1.6022 \times 10^{-19}\text{ C}}$ = 5.69×10^{-18} kg

2.14 (See Table 2-1) 5.26×10^{14} p's = $\frac{(+1.6022 \times 10^{-19}\text{ C})}{\text{p}}$ = 8.43×10^{-5} C

2.15 a) two electrons

b) Total mass of particles = $(2\ e)\left(\frac{9.1094 \times 10^{-31}\ kg}{1\ e}\right)$

$$+(2\ p)\left(\frac{1.6726 \times 10^{-27}\ kg}{1\ p}\right)+(2\ n)\left(\frac{1.6749 \times 10^{-27}\ kg}{1\ n}\right)$$

$$= 6.6968 \times 10^{-27}\ kg;$$

$$\frac{(2\ e)\left(\frac{9.1094 \times 10^{-31}\ kg}{1\ e}\right)}{6.69698 \times 10^{-27}\ kg} \times 100\% = 0.027205\% \text{ or } 0.00027205;$$

Using at. mass of He: $\frac{(2\ e)\left(\frac{9.1094 \times 10^{-28}\ g}{1\ e}\right)}{\frac{4.002602\ g / mole}{6.022 \times 10\ atom^{23} / mole}} \times 100\% = 0.027411\%$

or 0.00027411

2.16 Sodium, $_{11}Na$; silicon, $_{14}Si$; sulfur, $_{16}S$; silver, $_{47}Ag$; samarium, $_{62}Sm$; strontium, $_{38}Sr$; scandium, $_{21}Sc$; selenium, $_{34}Se$

2.17 $_{12}Mg$, magnesium; $_{25}Mn$, manganese; $_{29}Cu$, copper; $_{27}Co$, cobalt; $_{15}P$, phosphorus; $_{82}Pb$, lead

2.18 26–iron, Fe; 7–nitrogen, N; 18–argon, Ar; 80–mercury, Hg; 50–tin, Sn; 19–potassium, K

2.19 a) $^{56}_{26}Fe$ b) $^{38}_{18}Ar$ c) $^{236}_{92}U$ d) $^{19}_{9}F$

2.20 a) $^{3}_{2}He$ b) $^{66}_{30}Zn$ c) $^{132}_{54}Xe$ d) $^{14}_{7}N$

2.21

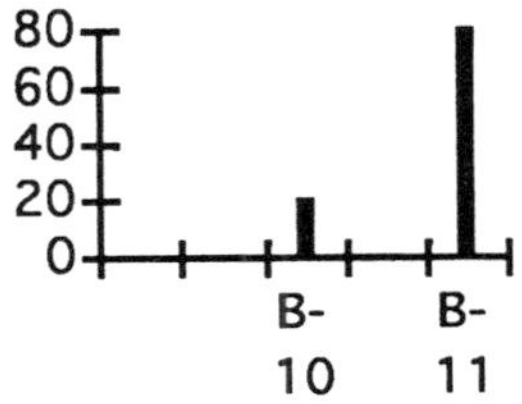

80
60
40
20
0
B-10
B-11

2.22 To answer this question, construct a pie chart like that for germanium in Figure 2-23 of text except the percentages need to be: 0.8 % (Pt-192), 32.9 % (Pt-194), 33.8 % (Pt-195), 25.3 % (Pt-196), and 7.2 % (Pt-198).

2.23 There are 3 peaks for Cl_2^+. They are 70, 72, and 74; 70 is the most intense; 74 is the least intense.

2.24 There will be 3 peaks for Br_2^+. They are 158, 160, and 162; 158 is the most intense; 162 is the least intense.

2.25 Cations: Cl_2^+, CO^+, Cr^{3+}; Anions: Cl^- and $Cr_2O_7^{2-}$; Neutral species: C, CCl_4, CO_2

2.26 a) Ar^+ b) OH^- (or HO^-) c) O_2^- d) H_3O^+

2.27 a) $OH^- + H^+ \rightarrow H_2O$ b) $Na \rightarrow Na^+ + e^-$
c) $HCl \rightarrow H^+ + Cl^-$ d) $O + 2\,e^- \rightarrow O^{2-}$

2.28

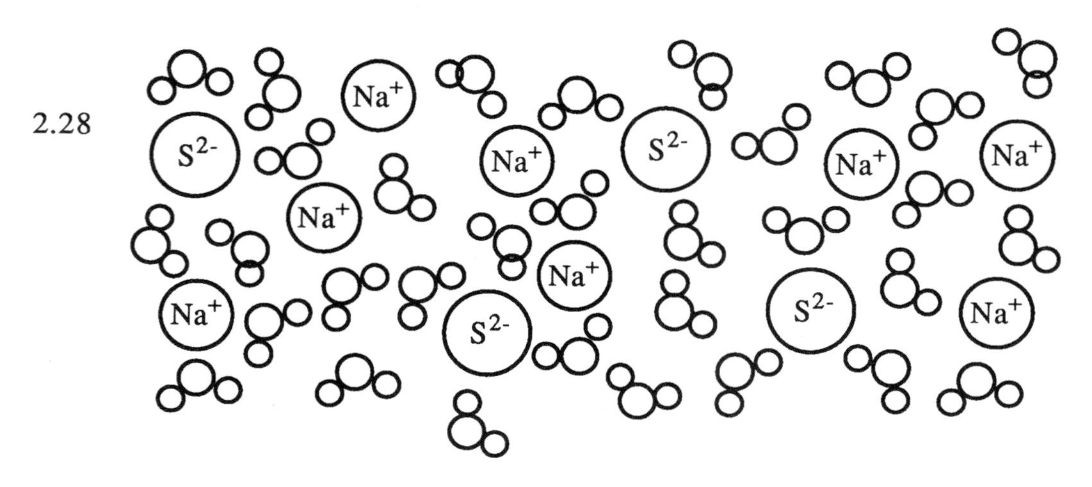

2.29 Al^{3+} and O^{2-} yield Al_2O_3 as a neutral representation.

2.30 a) cesium, Group I, Cs^+ b) strontium, Group II, Sr^{2+} c) iodine, Group VII, I^-

2.31 CsI and SrI_2

2.32 a) All chemical reactions require that the number of atoms of each chemical element be conserved.
b) All processes that involve movement of electrons neither create nor destroy electrons. Electrons are conserved. Net electrical charge is always conserved.
c) Mass is neither created nor destroyed during physical and chemical transformations.
d) Energy is neither created nor destroyed in any process, although it may be transferred from one body to another or converted from one form into another.

2.33 a) Some of the runner's fat is converted into energy for running.
b) As the apple falls, potential energy is converted to kinetic energy.
c) The kinetic energy is converted into thermal energy when the apple hits the ground.

2.34 4.55×10^5 m s^{-1} = V; 9.1094×10^{-31} kg = m. (Table 2-1); 1 kg m^{2}s^{-2} = 1 J

k.E. = 1/2 mV^2 = 1/2(9.1094×10^{-31} kg)(4.55×10^5 m s^{-1})^2

= 9.43×10^{-20} kg m^2 s^{-2} = 9.43×10^{-20} J

2.35 k.E. = 1/2 mV^2; $V = \sqrt{\dfrac{k.E.}{1/2\ m}} = \sqrt{\dfrac{3.75 \times 10^{-23}\ kg\ m^2/s^2}{1/2\left(\dfrac{4.002602 \times 10^{-3}\ kg}{6.022 \times 10^{23}}\right)}} = 106\ m/s$

2.36 a) $CH_4 \rightarrow CH_4^+ + e^-$ b) $CH_4^+ \rightarrow 3H + CH^+$

2.37 a) radiant energy becomes thermal energy b) potential energy becomes kinetic energy c) potential energy (chemical energy) becomes thermal and radiant energy

2.38 V = 7.44×10^3 km/hr; mass = 1.6749×10^{-27} kg; 1000 m = 1 km;
1 kg m^{2}s^{-1} = 1 J; 1 hr = 60 min; 1 min = 60 s

$$\frac{7.44 \times 10^3\ km}{hr} \times \frac{1000\ m}{km} \times \frac{1\ hr}{60\ min} \times \frac{1\ min}{60\ s} = 2.0667 \times 10^3\ m\ s^{-1}$$

k.E. = 1/2 mV^2 = 1/2(1.6749×10^{-27} kg)(2.0667×10^{-3} m s^{-1})^2 = 3.58×10^{-21} J

2.39 X could be one of many elements such as Be, Mg, Ca, Sr, Ba, and Ra, but they must be metals if the compound formed is to be ionic as stated in the problem. OF_2, XeF_2 and other nonmetallic compounds are not ionic.

2.40 X must be a nonmetallic element that forms an X^- ion. That limits one to the Group VII elements. Fluorine, chlorine, bromine, astatine and iodine are the obvious answers to this question.

2.41 solid liquid

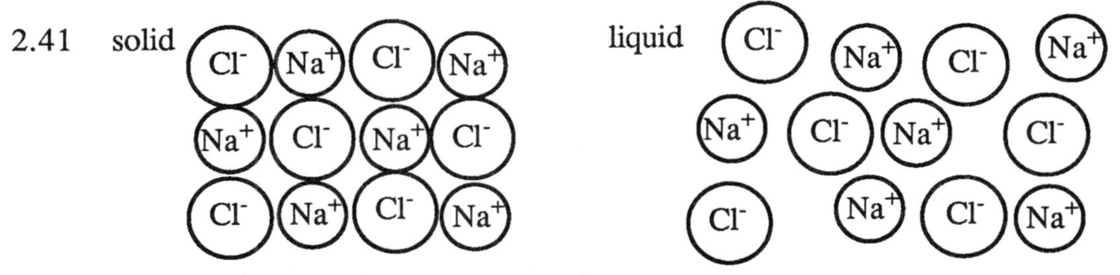

2.42 A mass spectrometer would be used; it can separate isotopes and give very precise results.

2.43 N_2^+, O_2^+, N^+ and O^+; N_2^-, O_2^-, N^- and O^-

2.44 ${}^1_1H$, ${}^2_1H$ or ${}^2_1D$, ${}^{19}_9F$

2.45 a) 8 protons, 8 neutrons and 10 electrons
b) 7 protons, 8 neutrons and 7 electrons
c) 25 protons, 30 neutrons and 22 electrons
d) 17 protons, 18 neutrons and 18 electrons
e) 17 protons, 20 neutrons and 16 electrons

2.46 Atomic number of Mg = 12
3.984×10^{-26} kg = $(12 \times 1.6726 \times 10^{-27}) + (? \times 1.6749 \times 10^{-27})$
? = ~12 neutrons (If the mass of electrons is considered, exactly 12 neutrons will be obtained in the above calculation.) Mg – 24 78.6%

4.150×10^{-26} kg = $(12 \times 1.6726 \times 10^{-27}) + (? \times 1.6749 \times 10^{-27})$
? = ~13 neutrons Mg – 25 10.1%

4.315×10^{-26} kg = $(12 \times 1.6726 \times 10^{-27}) + (? \times 1.6749 \times 10^{-27})$
? = ~14 neutrons Mg – 26 11.3%

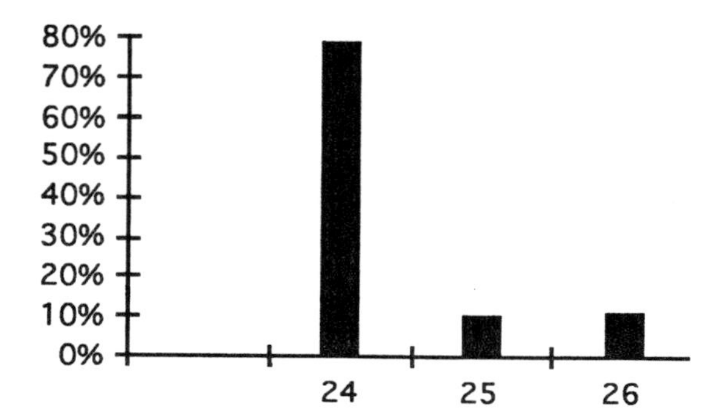

2.47 a) Co-60, 27 protons and 33 neutrons; b) C-14, 6 protons and 8 neutrons;
c) U-235, 92 protons and 143 neutrons; d) U-238, 92 protons and 146 neutrons

2.48

particles	charge	mass	location in atom
electrons	negative	9.1×10^{-31} kg	outside nucleus
protons	positive	1.7×10^{-27} kg	inside nucleus
neutrons	none	1.7×10^{-27} kg	inside nucleus

2.49 $^{14}_{7}N$, $^{14}_{6}C$

2.50 a) $H_2O \rightarrow H^+ + OH^- + 0\ e^-$ (none) b) $P_4 \rightarrow 4P^{3+} + 12\ e^-$ (twelve)
c) $C_2O_4^{2-} \rightarrow 2CO_2 + 2\ e^-$ (two)

2.51 Read the amounts shown in the pie chart and from those values construct a mass spectrum similar to that at the end of the answer to question 2.46

2.52 $C_2H_4^+$ (28), $C_2H_3^+$ (27), $C_2H_2^+$ (26), C_2H^+ (25), C_2^+ (24), CH_2^+ (14), CH^+ (13), C^+ (12), H^+ (1)

2.53 57.5 mile hr^{-1} = velocity 2250 lb = mass 1 km = 0.6215 mi

454 g = 1 lb 10^3 m = km 1000 g = kg

60 min = 1 hr 1 kg m^2 s^{-2} = 1 J 60 s = 1 min

k.E. = 1/2 mV^2

$$\frac{57.5 \text{ mi}}{\text{hr}} \text{x} \frac{1 \text{ km}}{0.6215 \text{ mi}} \text{x} \frac{10^3 \text{ m}}{\text{kg}} \text{x} \frac{1 \text{ hr}}{60 \text{ min}} \text{x} \frac{1 \text{ min}}{60 \text{ s}} = 25.7 \text{ m s}^{-1} = \text{velocity}$$

$$2250 \text{ lb x} \frac{454 \text{ g}}{\text{lb}} \text{x} \frac{\text{kg}}{10^3 \text{ g}} = 1.02 \text{ x } 10^3 \text{ kg}$$

k.E. = $(1/2)(1.02 \text{ x } 10^3 \text{ kg})(25.7 \text{ m s}^{-1})^2 = 3.37 \text{ x } 10^5$ J

2.54

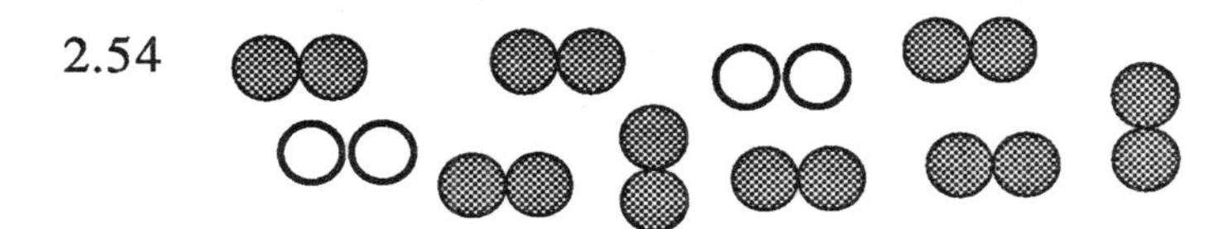

2.55 ${}^{40}_{18}Ar$, ${}^{40}_{19}K$, ${}^{40}_{20}Ca$

2.56 ${}^{2}_{1}D$, ${}^{4}_{2}He$, ${}^{6}_{3}Li$, ${}^{10}_{5}B$, ${}^{14}_{7}N$ and ${}^{16}_{8}O$

2.57

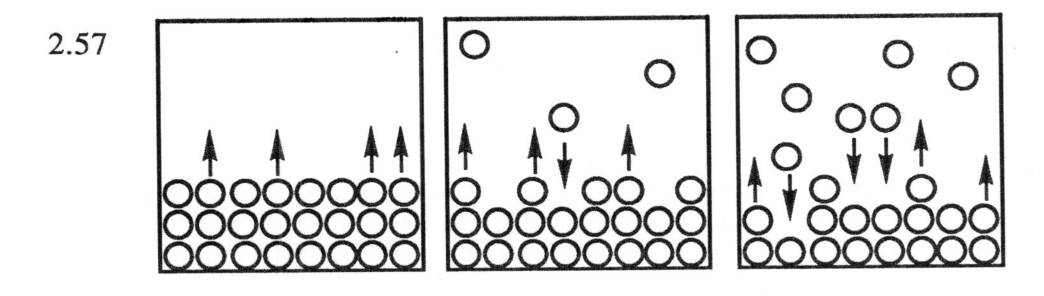

2.58 $$(+1.00 \text{ x } 10^{-6} \text{ C})\left(\frac{1.6726 \text{ x } 10^{-27} \text{ kg}}{1.6022 \text{ x } 10^{-19} \text{ C}}\right) = 1.04 \text{ x } 10^{-14} \text{ kg};$$

$$(+1.00 \text{ x } 10^{-6} \text{ C})\left(\frac{9.1094 \text{ x } 10^{-31} \text{ kg}}{1.6022 \text{ x } 10^{-19} \text{ C}}\right) = 5.69 \text{ x } 10^{-18} \text{ kg}$$

2.59 a) 30 b) 20 c) 20 N and 60 H

2.60

2.61

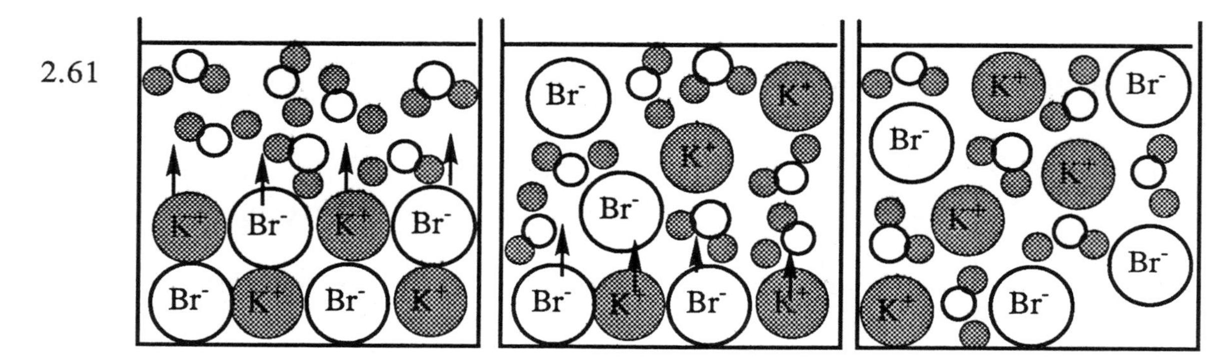

2.62 H_2O^+ mass = 3.01×10^{-26} kg
HO^+ mass = 2.84×10^{-26} kg
O^+ mass = 2.67×10^{-26} kg
H^+ mass = 1.67×10^{-27} kg

2.63 ^{98}Tc - 43 protons, 55 neutrons, 43 electrons;
^{99}Tc - 43 protons, 56 neutrons, 43 electrons

2.64 $^{45}_{20}Ca$, $^{54}_{24}Cr$, $^{63}_{28}Ni$, $^{72}_{32}Ge$, $^{81}_{36}Kr$, $^{90}_{40}Zr$

2.65 Two peaks will be obtained; one for F_2^+ and one for F^+.

2.66 Cinnabar is HgS. During heating in air, the mercury can be freed and much of the S reacts with oxygen in the air to produce the gas SO_2 which is lost to the atmosphere.

2.67

proton	electron
mass = 1.6726×10^{-27} kg	mass = 9.1094×10^{-31} kg
velocity = ?	velocity = 1.55×10^6 m s^{-1}
k.E. = $1/2\ mV^2$	k.E. = $1/2\ mV^2$

k.E.'s are equal or $1/2\ m_pV_p^2 = 1/2\ m_eV_e^2$

$(1.6726 \times 10^{-27}\ kg)V_p^2 = (9.1094 \times 10^{-31}\ kg)(1.55 \times 10^6\ m\ s^{-1})^2$
$V_p^2 = 1.308 \times 10^9\ m^2\ s^{-2}$
$V_p = 3.62 \times 10^4\ m\ s^{-1}$

2.68 a) Al^{3+} 13 p, 10 e b) Se^{2-} 34 p, 36 e c) K^+ 19 p, 18 e d) Ca^{2+} 20 p, 18 e

2.69 When magnesium metal is burned in air, the magnesium combines with oxygen from the air to produce MgO. The mass of MgO produced equals the mass of magnesium burned plus the mass of oxygen from the air that combined with it.

CHAPTER 3: THE COMPOSITION OF MOLECULES

3.1 a) methane, CH_4 b) ethylene, C_2H_4 c) dimethyl ether, C_2H_6O
d) hydrogen bromide, HBr e) phosphorus trichloride, PCl_3
f) urea, CH_4N_2O g) iodoethane, C_2H_5I h) hydrazine, N_2H_4

3.2 a) ammonia, NH_3 b) ethane, C_2H_6 c) methanol, CH_4O
d) iodine, I_2 e) hydrogen cyanide, HCN f) DMSO, C_2H_6SO
g) acetone, C_3H_6O h) glycine, $C_2H_5NO_2$

3.3

```
a)     H              b)  H        H          c)  H     O     H
       |                   \      /                 \  / \  /
   H - C - H                C == C                H - C     C - H
       |                   /      \                 /       \
       H                  H        H               H         H

                                                        O
                                                        ||
d) H - Br             e)  H - P - H            f)  H    C    H
                               |                    \  / \  /
                               H                     N     N
                                                    /       \
                                                   H         H

g)   H     H          h)  H - N - N - H
      \   /                   |   |
  H - C - C - H               H   H
      /   \
     H     I
```

3.4

```
a)  H - N - H         b)   H     H             c)  H
        |                   \   /                   \
        H               H - C - C - H            H - C - O - H
                            /   \                   /
                           H     H                 H

                                                        O
                                                        ||
d)  I - I             e)  H - C≡N              f)  H    S    H
                                                    \  / \  /
                                                H - C     C - H
                                                    /       \
                                                   H         H

          O                  H          O
          ||                  \         ||
g)   H    C    H      h)  H - N         C
      \  / \  /                \      /   \
  H - C     C - H               C         O - H
      /       \                / \
     H         H              H   H
```

3.5 a) 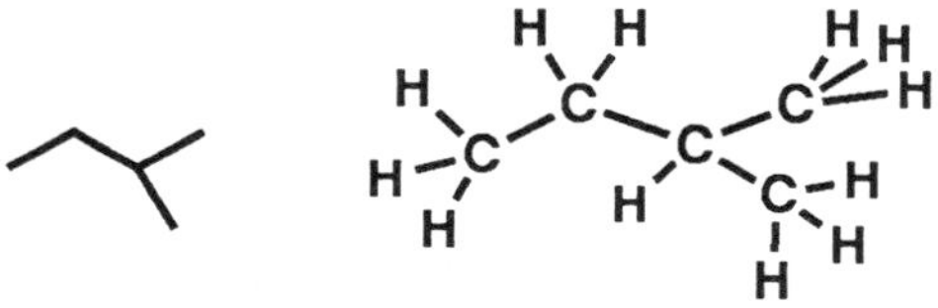 b) 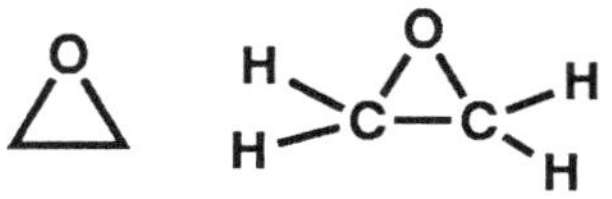

c) 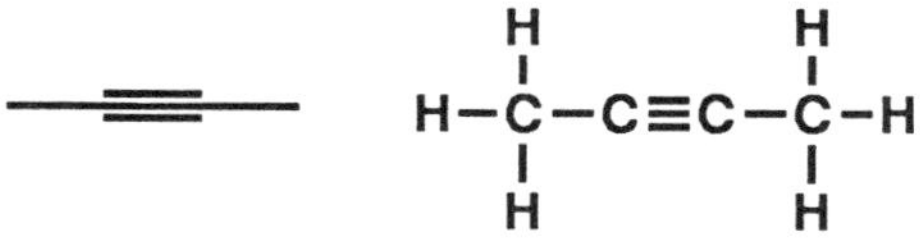 d) 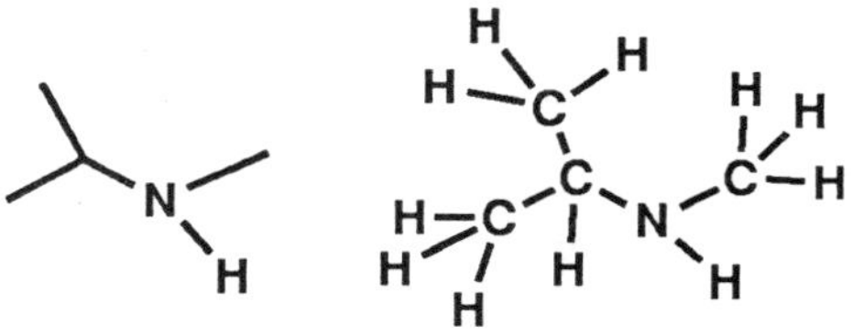

e) 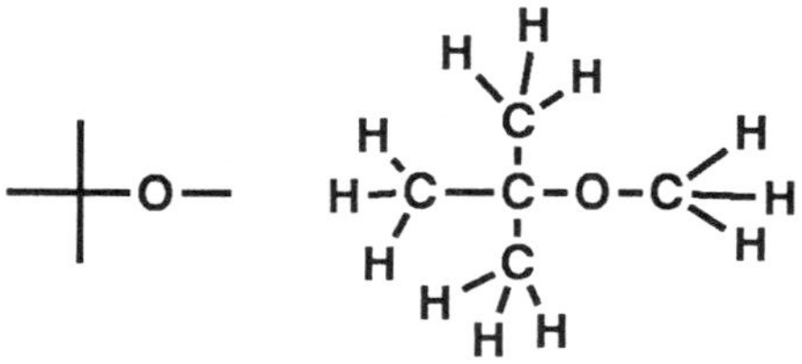 f) 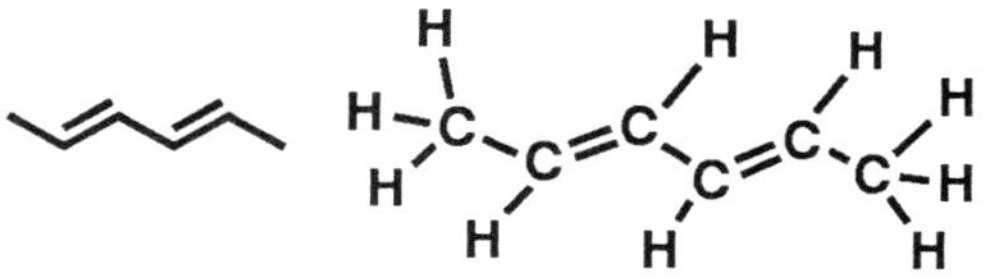

g) 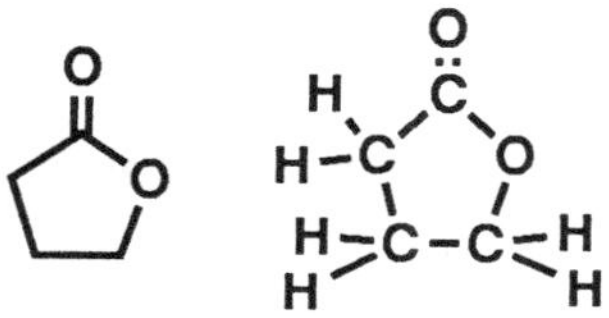 h) 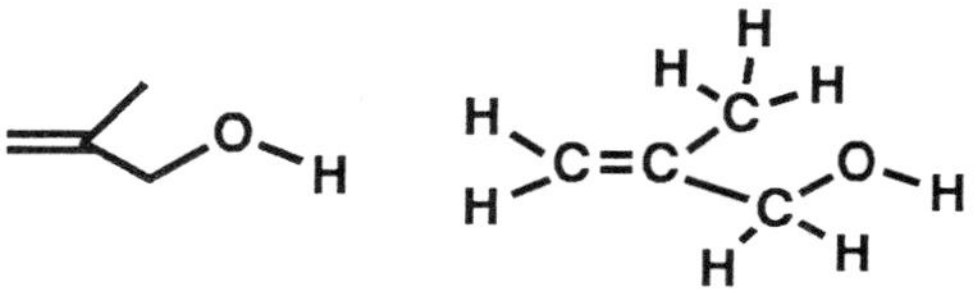

i) 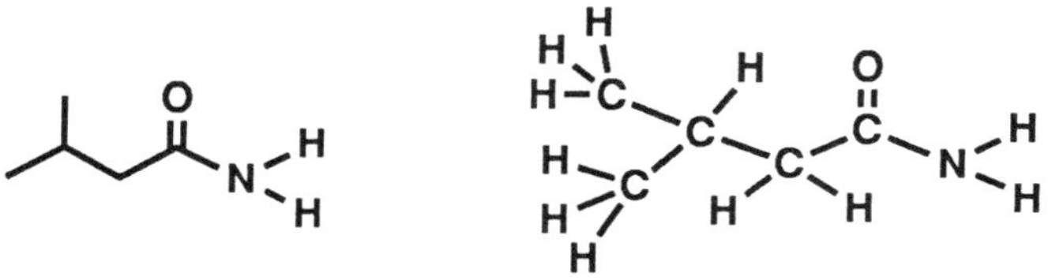

3.6 a)  b) 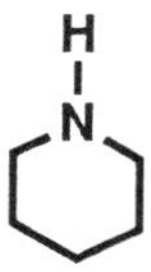 c) 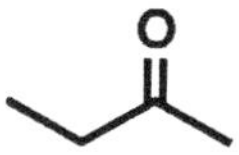

d) 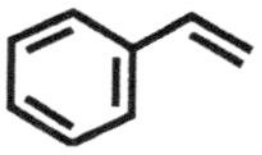 e) 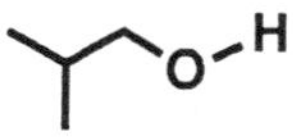 f) 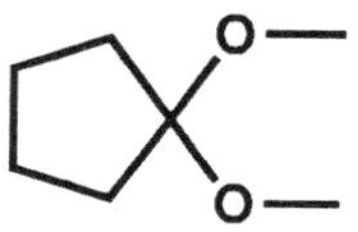

3.7 a) 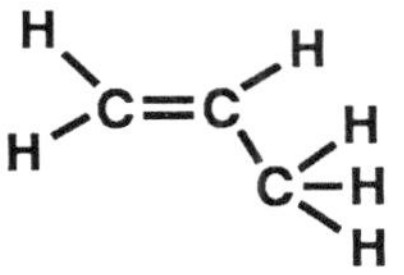 b) 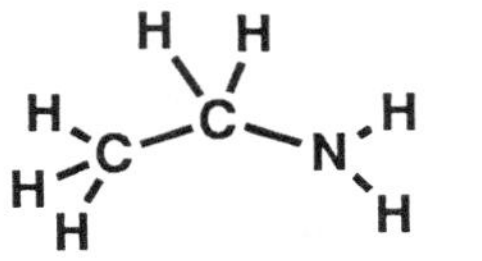 c) d) 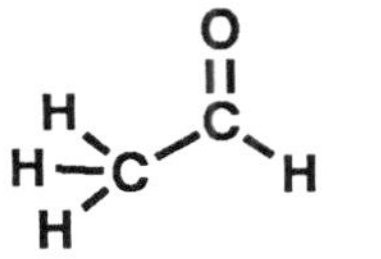 e) 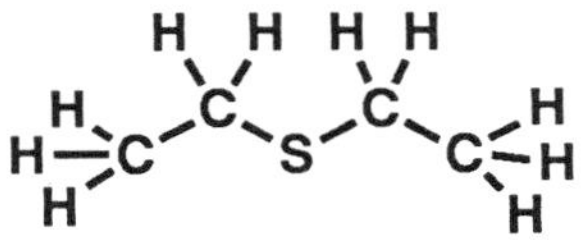 f) 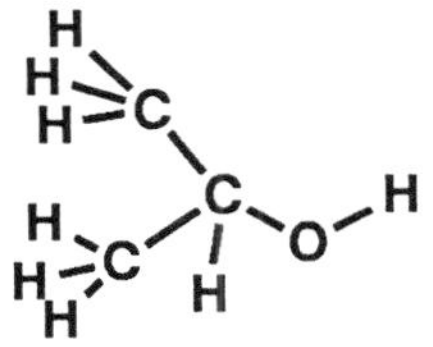 g)

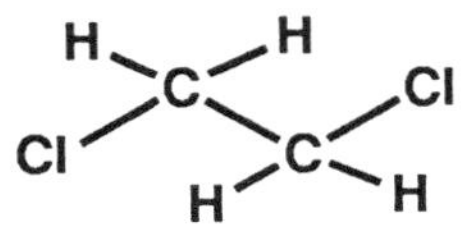

3.8 a) 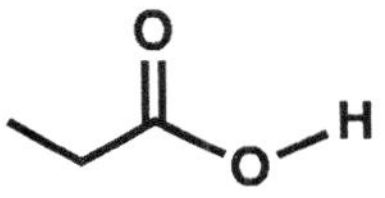 b)  c) 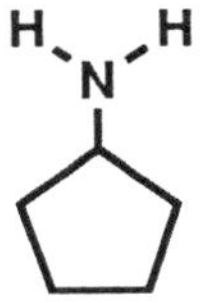 d)  e) 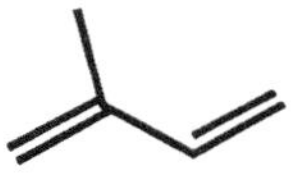 f)

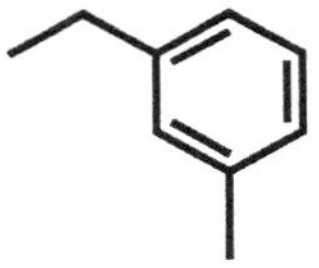

3.9 a) CO_2, carbon dioxide b) HCl, hydrogen chloride c) CCl_4, carbon tetrachloride
d) ClF_3, chlorine trifluoride e) PH_3, phosphine
f) N_2O, dinitrogen oxide (nitrous oxide)

3.10 a) CH_4, methane b) O_2, oxygen c) PCl_3, phosphorus trichloride
d) H_2S, hydrogen sulfide e) SF_4, sulfur tetrafluoride f) HF, hydrogen fluoride

3.11 a) CH_4 b) HF c) CaH_2 d) PCl_3 e) N_2O_5 f) SF_6
g) BF_3 h) $CH_3CH_2CH(OH)CH_2CH_3$

3.12 a) NH_3 b) H_2S c) $CH_3CHClCH_3$ d) SiO_2 e) N_2
f) C_2H_2 g) XeF_4 h) BrF_5

3.13 a) disulfur dichloride b) iodine heptafluoride c) hydrogen bromide
d) dinitrogen trioxide e) silicon carbide
f) methanol (other names also possible for this compound)

3.14 a) xenon difluoride b) germanium tetrachloride c) dinitrogen tertafluoride
d) lithium hydride e) selenium dioxide f) ethanol

3.15 a) HF b) CaF_2, ionic c) $Al_2(SO_4)_3$, ionic
d) $(NH_4)_2S$, ionic e) SO_2 f) CCl_4

3.16 a) Mn $(CH_3CO_2)_2$, ionic b) NaClO, ionic c) $SiCl_4$
d) $LiIO_4$, ionic e) $MgBr_2$, ionic f) H_2Se

3.17 The ionic compounds are: CaO, K_2CO_3 and Na_2HPO_4.

3.18 a) $KClO_3$ b) NH_4HCO_3 c) $Fe_3(PO_4)_2$ d) $Cu(NO_3)_2•6H_2O$
e) $AlCl_3$ f) $CdCl_2$ g) K_2O h) $NaHCO_3$ i) $LiBrO_4$

3.19 a) calcium chloride hexahydrate b) iron(II) ammonium sulfate
c) potassium carbonate d) tin(II) chloride dihydrate
e) sodium hypochlorite f) silver sulfate
g) copper(II) sulfate h) potassium dihydrogen phosphate
i) sodium nitrate j) calcium sulfite
k) potassium permanganate

3.20 a) potassium dichromate b) sodium nitrite
c) magnesium phosphate d) chromium(III) chloride
e) vanadium(III) oxide
f) potassium hydrogensulfate or potassium bisulfate
g) cesium bromide h) indium(III) nitrate pentahydrate
i) aluminum perchlorate j) tin(IV) chloride
k) tantalum(V) chloride

3.21 a) MM of Fe = 55.85 $7.85 \text{ g Fe} \times \frac{1 \text{ mol}}{55.85 \text{ g}} = 0.141 \text{ mol}$

b) $6.55 \times 10^{13} \text{ atoms} \times \frac{1 \text{ mol}}{6.022 \times 10^{23} \text{ atoms}} = 1.09 \times 10^{-10} \text{ mol}$

c) 4.68 μg of Si, MM of Si = 28.09, $1 \, \mu g = 10^{-6}$ g

$$4.68 \, \mu g \times \frac{10^{-6} \text{ g}}{\mu g} \times \frac{1 \text{ mole}}{28.09 \text{ g}} = 1.67 \times 10^{-7} \text{ mol}$$

d) 1.46 met. ton of Al 1 met. ton = 10^3 kg MM of Al = 26.98

$$1.46 \text{ met. ton} \times \frac{10^3 \text{ kg}}{\text{met. ton}} \times \frac{1 \text{ mole}}{26.98 \text{ g}} \times \frac{10^3 \text{ g}}{\text{kg}} = 5.41 \times 10^4 \text{ mol}$$

3.22 a) 3.67 kg of Ti, MM of Ti = 47.88, 10^3 g = kg

$$3.67 \text{ kg} \times \frac{10^3 \text{ g}}{\text{kg}} \times \frac{1 \text{ mol}}{47.88 \text{ g}} = 76.6 \text{ mol}$$

(continued)

(3.22 continued)

b) 7.9 mg of Ca, MM of Ca = 40.08, 10^{-3} g = mg

$$7.9 \text{ mg} \times \frac{10^{-3} \text{ g}}{\text{mg}} \times \frac{1 \text{ mol}}{40.08 \text{ g}} = 2.0 \times 10^{-4} \text{ mol}$$

c) 1.56 g of Ru, MM of Ru = 101.07; $1.56 \text{ g} \times \frac{1 \text{ mol}}{101.07 \text{ g}} = 1.54 \times 10^{-2} \text{ mol}$

d) 9.63 pg of Tc, MM of Tc = 98, 10^{-12} g = pg

$$9.63 \text{ pg} \times \frac{10^{-12} \text{ g}}{\text{pg}} \times \frac{1 \text{ mol}}{98 \text{ g}} = 9.8 \times 10^{-14} \text{ mol}$$

3.23

Ar-36:	35.96755 x 0.00337	=	0.121
Ar-38:	37.96272 x 0.00063	=	0.024
Ar-40:	39.9624 x 0.99600	=	<u>39.803</u>
	Total	=	39.948 g/mol

3.24 a) MM = 12.01 + 4(35.45) = 153.8 b) MM = 2(14.01) + 16.00 = 44.02

c) MM = 3(16.00) = 48.00 d) MM = 14.01 + 3(1.008) = 17.03

e) MM = 69.72 + 74.92 = 144.64 f) MM = 4(12.01) + 10(1.008) = 58.12

3.25 a) MM of $(NH_4)_2CO_3$ = 2(14.01) + 8(1.008) + 12.01 + 3(16.00) = 96.09

b) MM of K_2S = 2(39.10) + 32.07 = 110.27

c) MM of $CaCO_2$ = 40.08 + 12.01 + 3(16.00) = 100.09

d) MM of LiBr = 6.94 + 79.90 = 86.84

e) MM of Na_2SO_4 = 2(22.99) + 32.07 + 4(16.00) = 142.05

f) MM of $AgNO_3$ = 107.87 + 14.01 + 3(16.00) = 169.88

3.26 Biotin, $C_{10}H_{16}N_2O_3S$,
MM = 10(12.01) + 16(1.008) + 2(14.01) + 3(16.00) + 32.07 = 244.3

Nicotinamide, $C_6H_6N_2O$,
MM = 6(12.01) + 6(1.008) + 2(14.01) + 16.00 = 122.13

Pyridoxamine, $C_8H_{12}N_2O_2$,
MM = 8(12.01) + 12(1.008) + 2(14.01) + 2(16.00) = 168.20

Pantothenic acid, $C_9H_{17}NO_5$,
MM = 9(12.01) + 17(1.008) + 14.01 + 5(16.00) = 219.24

3.27 5.89 μg, 10^{-6} = 1 μg, and 6.022×10^{23} atom mol^{-1}

a) Be, MM = 9.01

$$5.86 \text{ μg} \times \frac{10^{-6} \text{ g}}{\text{μg}} \times \frac{1 \text{ mol}}{9.01 \text{ g}} \times \frac{6.022 \times 10^{23} \text{ atoms}}{\text{mol}} = 3.92 \times 10^{17} \text{ atoms}$$

(continued)

(3.27 continued)

b) P, MM = 30.97

$$5.86\ \mu g \times \frac{10^{-6}\ g}{\mu g} \times \frac{1\ mol}{30.97\ g} \times \frac{6.022 \times 10^{23}\ atoms}{mol} = 1.14 \times 10^{17}\ atoms$$

c) Zr, MM = 91.22

$$5.86\ \mu g \times \frac{10^{-6}\ g}{\mu g} \times \frac{1\ mol}{91.22\ g} \times \frac{6.022 \times 10^{23}\ atoms}{mol} = 3.87 \times 10^{16}\ atoms$$

d) U, MM = 238.03

$$5.86\ \mu g \times \frac{10^{-6}\ g}{\mu g} \times \frac{1\ mol}{238.03\ g} \times \frac{6.022 \times 10^{23}\ atoms}{mol} = 1.48 \times 10^{16}\ atoms$$

3.28 a) MM = 27(12.01) + 46(1.008) + 16 = 386.6

$$5000\ molecules \times \frac{1\ mol}{6.022 \times 10^{23}\ atoms} \times \frac{386.6\ g}{mole} = 3.210 \times 10^{-18}\ g$$

b) MM = 3(16.00) = 48.00

$$10^{15}\ molecules \times \frac{1\ mol}{6.022 \times 10^{23}\ atoms} \times \frac{48.00\ g}{mole} = 7.971 \times 10^{-8}\ g$$

c) MM =

63(12.01) + 88(1.008) + 58.93 + 14(14.01) + 14(16.00) + 30.97 = 1355.37

$$1\ molecule \times \frac{1\ mole}{6.022 \times 10^{23}\ molecules} \times \frac{1355.37\ g}{mole} = 2.251 \times 10^{-21}\ g$$

3.29 a) MM of CH_4 = 12.01 + 4(1.008) = 16.04

$$375{,}000\ molecules \times \frac{1\ mole}{6.022 \times 10^{23}\ molecules} \times \frac{16.04\ g}{mole} = 9.99 \times 10^{-18}\ g$$

b) MM = 183.2

$$2.5 \times 10^{9}\ molecules \times \frac{1\ mole}{6.022 \times 10^{23}\ molecules} \times \frac{183.2\ g}{mole} = 7.6 \times 10^{-13}\ g$$

c) MM = 893.5

$$1\ molecule \times \frac{1\ mole}{6.022 \times 10^{23}\ molecules} \times \frac{893.5\ g}{mole} = 1.484 \times 10^{-21}\ g$$

3.30 a) MM = 176.1

$$60 \text{ mg} \times \frac{10^{-3} \text{ g}}{\text{mg}} \times \frac{1 \text{ mole}}{176.1 \text{ g}} \times \frac{6.022 \times 10^{23} \text{ molecules}}{\text{mole}} = 2.1 \times 10^{20} \text{ molecules}$$

b) MM = 441.40

$$400 \ \mu\text{g} \times \frac{10^{-6} \text{ g}}{\mu\text{g}} \times \frac{1 \text{ mole}}{441.40 \text{ g}} \times \frac{6.022 \times 10^{23} \text{ molecules}}{\text{mole}} = 5.46 \times 10^{17} \text{ molecules}$$

c) MM = 286.4

$$1.5 \text{ mg} \times \frac{10^{-3} \text{ g}}{\text{mg}} \times \frac{1 \text{ mole}}{286.4 \text{ g}} \times \frac{6.022 \times 10^{23} \text{ molecules}}{\text{mole}} = 3.2 \times 10^{18} \text{ molecules}$$

d) MM = 376.4

$$1.70 \text{ mg} \times \frac{10^{-3} \text{ g}}{\text{mg}} \times \frac{1 \text{ mole}}{376.4 \text{ g}} \times \frac{6.022 \times 10^{23} \text{ molecules}}{\text{mole}} = 2.72 \times 10^{18} \text{ molecules}$$

3.31 MM of H_3PO_4 = 98.00, 454 g = 1 lb

$$2.47 \times 10^{8} \text{ lb} \times \frac{454 \text{ g}}{\text{lb}} \times \frac{1 \text{ mole}}{98.00 \text{ g}} = 1.14 \times 10^{9} \text{ moles of } H_3PO_4$$

1.14×10^{9} mole of H_3PO_4 x 35% = 4.00×10^{8} moles from P. Assume that for the H_3PO_4 made from P that one molecule P will produce one molecule of H_3PO_4. Therefore, 4.00×10^{8} moles of H_3PO_4 produced from P will utilize 4.00×10^{8} moles of P.

$$4.00 \times 10^{8} \text{ mol of P} \times \frac{30.97 \text{ g}}{\text{mole}} \times \frac{\text{kg}}{10^{3} \text{ g}} = 1.24 \times 10^{7} \text{ kg of P}$$

3.32 MM of $C_{20}H_{24}O_2$ = 296.4

$$0.035 \text{ mg} \times \frac{10^{-3} \text{ g}}{\text{mg}} \times \frac{1 \text{ mole}}{296.4 \text{ g}} = 1.2 \times 10^{-7} \text{ moles of } C_{20}H_{24}O_2$$

$$1.2 \times 10^{-7} \text{ mole} \times \frac{6.022 \times 10^{23} \text{ molecules}}{\text{mole}} = 7.2 \times 10^{16} \text{ molecules of } C_{20}H_{24}O_2$$

$$7.2 \times 10^{16} \text{ molecules of } C_{20}H_{24}O_2 \times \frac{20 \text{ C atoms}}{\text{molecule } C_{20}H_{24}O_2} = 1.4 \times 10^{18} \text{ atoms of C}$$

$$1.4 \times 10^{18} \text{ atoms C} \times \frac{1 \text{ mole C}}{6.022 \times 10^{23} \text{ atoms}} \times \frac{12.01 \text{ g}}{\text{mole C}} = 2.8 \times 10^{-5} \text{ g of C}$$

3.33 CaO,
MM = 40.08 + 16.00 = 56.08
Ca = (40.08/56.08) x 100 = 71.47%
O = (16.00/56.08) x 100 = 28.53%

SiO_2,
MM = 28.09 + 2(16.00) = 60.09
Si = (28.09/60.09) x 100 = 46.75%
O = (32.00/60.09) x 100 = 53.25%

Al_2O_3,
MM = 2(26.98) + 3(16.00) = 101.96
Al = (53.96/101.96) x 100 = 52.92%
O = (48.00/101.96) x 100 = 47.08%

Fe_2O_3,
MM = 2(55.85) + 3(16.00) = 159.70
Fe = (111.70/159.70) x 100 = 69.94%
O = (48.00/159.70) x 100 = 30.06%

3.34 CaF_2,
MM = 40.08 + 2(19.00) = 78.08
Ca = (40.08/78.08) x 100 = 51.33%
F = (38.00/78.08) x 100 = 48.67%

MnO_2,
MM = 54.94 + 2(16.00) = 86.94
Mn = (54.94/86.94) x 100 = 63.19%
O = (32.00/86.94) x 100 = 36.81%

CoO,
MM = 58.93 + 16.00 = 74.93
Co = (58.93/74.93) x 100 = 78.65%
O = (16.00/74.93) x 100 = 21.35%

Cu_2O,
MM = 2(63.55) + 16.00 = 143.10
Cu = (127.10/143.10) x 100 = 88.82%
O = (16.00/143.10) x 100 = 11.18%

3.35 $$\text{C : } 74.0 \text{ g x} \frac{1 \text{ mole}}{12.01 \text{ g}} = 6.1615 \text{ mole C}; \quad \frac{6.1615 \text{ mole C}}{1.2384} = 4.975 \text{ mole C}$$

$$\text{H : } 8.65 \text{ g x} \frac{1 \text{ mole}}{1.008 \text{ g}} = 8.5813 \text{ mole H}; \quad \frac{8.5813 \text{ mole H}}{1.2384} = 6.929 \text{ mole H}$$

$$\text{N : } 17.35 \text{ g x} \frac{1 \text{ mole}}{14.01 \text{ g}} = 1.2384 \text{ mole N}; \quad \frac{1.2384 \text{ mole N}}{1.2384} = 1.000 \text{ mole N}$$

Empirical Formula = C_5H_7N
Mass of Empirical Formula = 5(12.01) + 7(1.008) + 14.01 = 81.12
If MM is twice the mass of the Empirical Formula, then the Molecular Formula must be twice the Empirical Formula. Molecular Formula is $C_{10}H_{14}N_2$

3.36 $$\text{O : } 41.41 \text{ g x} \frac{1 \text{ mole}}{16.00 \text{ g}} = 2.588 \text{ moles O}; \quad \frac{2.588 \text{ moles O}}{0.198} = 13.07 \text{ mole O}$$

$$\text{P : } 18.50 \text{ g x} \frac{1 \text{ mole}}{30.97 \text{ g}} = 0.597 \text{ mole P}; \quad \frac{0.597 \text{ mole P}}{0.198} = 3.01 \text{ mole P}$$

$$\text{H : } 0.20 \text{ g x} \frac{1 \text{ mole}}{1.008 \text{ g}} = 0.198 \text{ mole H}; \quad \frac{0.198 \text{ mole H}}{0.198} = 1.00 \text{ mole H}$$

$$\text{Ca : } 39.89 \text{ g x} \frac{1 \text{ mole}}{40.08 \text{ g}} = 0.9953 \text{ mole Ca}; \quad \frac{0.9953 \text{ mole Ca}}{0.198} = 5.03 \text{ mole Ca}$$

Empirical Formula = $Ca_5HP_3O_{13}$

3.37 $$\% \text{ C} = \left(\left(97.46 \text{ mg } CO_2 \text{ x} \left[\frac{12.01 \text{ mg C}}{44.01 \text{ mg } CO_2} \right] \right) \Big/ 38.7 \text{ mg} \right) \text{x } 100 = 68.7 \%$$

$$\% \text{ H} = \left(\left(20.81 \text{ mg } H_2O \text{ x} \left[\frac{2 \text{ x } 1.008 \text{ mg H}}{18.02 \text{ mg } H_2O} \right] \right) \Big/ 38.7 \text{ mg} \right) \text{x } 100 = 6.02 \%$$

% N = 3.8%

% O = 100 - 68.7 - 6.02 - 3.8 = 21.5%

(continued)

(3.37 continued)

$$C: 68.7 \text{ g x} \frac{1 \text{ mole}}{12.01 \text{ g}} = 5.72 \text{ mole C}; \quad \frac{5.72 \text{ mole C}}{0.271} = 21.11 \text{ mole C}$$

$$H: 6.02 \text{ g x} \frac{1 \text{ mole}}{1.008 \text{ g}} = 5.97 \text{ mole H}; \quad \frac{5.97 \text{ mole H}}{0.271} = 22.0 \text{ mole H}$$

$$N: 3.8 \text{ g x} \frac{1 \text{ mole}}{14.01 \text{ g}} = 0.271 \text{ mole N}; \quad \frac{0.271 \text{ mole N}}{0.271} = 1.00 \text{ mole N}$$

$$O: 21.5 \text{ g x} \frac{1 \text{ mole}}{16.00 \text{ g}} = 1.34 \text{ mole O}; \quad \frac{1.34 \text{ mole O}}{0.271} = 4.94 \text{ mole O}$$

These values are very indicative of heroin.

3.38

$$\% \text{ C} = \left(\left(0.372 \text{ mg } CO_2 \text{ x} \left[\frac{12.01 \text{ mg C}}{44.01 \text{ mg } CO_2}\right]\right) \Big/ 0.137 \text{ mg}\right) \text{x } 100 = 74.1 \%$$

$$\% \text{ H} = \left(\left(0.0910 \text{ mg } H_2O \text{ x} \left[\frac{2(1.008) \text{ mg H}}{18.02 \text{ mg } H_2O}\right]\right) \Big/ 0.137 \text{ mg}\right) \text{x } 100 = 7.43 \%$$

% N = (0.0158 g/0.183) x 100 = 8.63%; % O = 100 - 74.1 - 7.43 - 8.63 = 9.84%

$$C: 74.1 \text{ g x} \frac{1 \text{ mole C}}{12.01 \text{ g}} = 6.17 \text{ mole C}; \quad \frac{6.17 \text{ mole C}}{0.616} = 10.0 \text{ mole C}$$

$$H: 7.43 \text{ g x} \frac{1 \text{ mole H}}{1.008 \text{ g}} = 7.37 \text{ mole H}; \quad \frac{7.37 \text{ mole H}}{0.616} = 12.0 \text{ mole H}$$

$$N: 8.63 \text{ g x} \frac{1 \text{ mole}}{14.01 \text{ g}} = 0.616 \text{ mole N}; \quad \frac{0.616 \text{ mole N}}{0.616} = 1.00 \text{ mole N}$$

$$O: 9.84 \text{ g x} \frac{1 \text{ mole}}{16.00 \text{ g}} = 0.615 \text{ mole O}; \quad \frac{0.615 \text{ mole O}}{0.616} = 1.00 \text{ mole O}$$

Calculated Empirical Formula = $C_{10}H_{12}NO$; mass = 162.

Calculated Molecular Formula is $C_{20}H_{24}N_2O_2$; MM = 324.

The true formula of quinine is $C_{20}H_{24}N_2O_2$.

3.39 MM of KOH = 39.10 + 16.00 + 1.008 = 56.11

$$4.75 \text{ g x} \frac{1 \text{ mole}}{56.11 \text{ g}} = 0.0847 \text{ mole}; \quad M = \frac{0.0847 \text{ mole}}{0.275 \text{ L}} = 0.308 \text{ M} = [KOH]$$

Since: $KOH \rightarrow K^+ + OH^-$; 0.308 M = $[K^+]$ = $[OH^-]$

3.40 $V_i \times M_i = V_f \times M_f$; (25 mL)(0.308 M) = (100 mL) M_f

M_f = 0.077 M = [KOH]

3.41

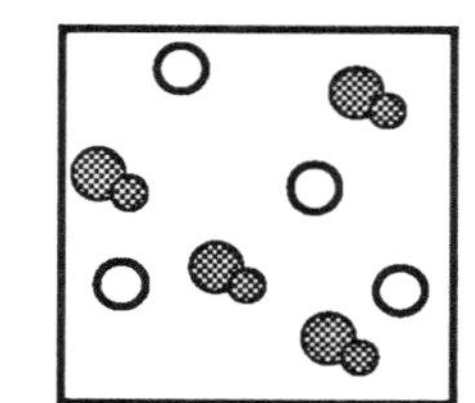

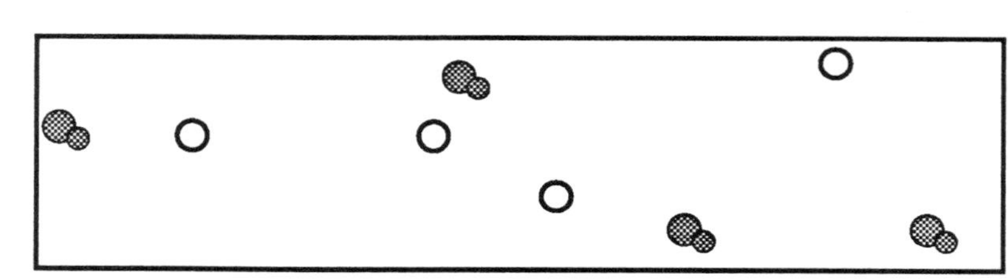

$0.308\ M = [K^+] = [OH^-]$ $\qquad$ $0.077\ M = [K^+] = [OH^-]$

$K^+ = $ O $\quad OH^- = $ [shaded pair] $\qquad$ The second solution is much more dilute.

3.42 MM of $Na_2CO_3 = 106.0$

$$3.25\ g \times \frac{1\ mole}{106.0\ g} = 0.0307\ mole$$

$$M = \frac{0.307\ mole}{0.250\ L} = 0.123\ M\ ;\quad 0.123\ M = [Na_2CO_3]$$

Since: $Na_2CO_3 \rightarrow 2Na^+ + CO_3^{2-}$ $\qquad [Na^+] = 0.246\ M$ and $[CO_3^{2-}] = 0.123\ M$

3.43

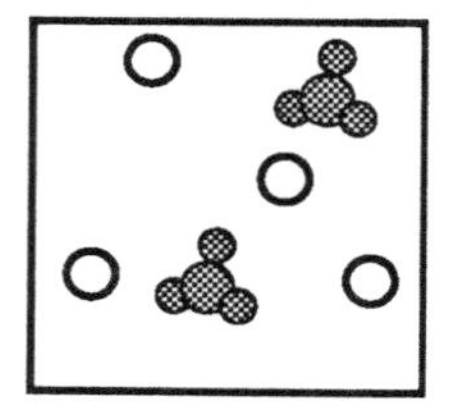

$Na^+ = $ O $\qquad CO_3^{2-} = $ [shaded cluster]

3.44 $\% H_2O = ((0.500 - 0.320)/0.500) \times 100 = 36.0\ \%$

$$\% SO_4^{2-} = \left(\left(0.280\ g \times \left[\frac{96.07\ g\ SO_4^{2-}}{233.40\ g\ BaSO_4}\right]\right) \Big/ 0.300\ g\right) \times 100 = 38.4\ \%$$

$\% Cu = 100 - 36.0 - 38.4 = 25.6\%$

$$\text{moles } H_2O:\ 36.0\ g \times \frac{1\ mole}{18.02\ g} = 1.998\ mole\ H_2O;\quad \frac{1.998\ mole\ H_2O}{0.400} = 4.995$$

$$\text{moles } SO_4^{2-}:\ 38.4\ g \times \frac{1\ mole}{96.07\ g} = 0.400\ mole\ SO_4^{2-};\quad \frac{0.400\ mole\ SO_4^{2-}}{0.400} = 1.000$$

$$\text{moles } Cu^{2+}:\ 25.6\ g \times \frac{1\ mole}{63.55\ g} = 0.403\ mole\ Cu^{2+};\quad \frac{0.403\ mole\ Cu^{2+}}{0.400} = 1.01$$

$CuSO_4 \cdot 5H_2O$

3.45 $$\% C = \left(\left(4.34 \text{ g } CO_2 \times \left[\frac{12.01 \text{ g C}}{44.01 \text{ g } CO_2}\right]\right) \Big/ 4.00 \text{ g}\right) \times 100 = 29.61\ \%$$

$$\% Cl = \left(\left(0.334 \text{ g AgCl} \times \left[\frac{35.45 \text{ g Cl}}{143.32 \text{ g AgCl}}\right]\right) \Big/ 0.125 \text{ g}\right) \times 100 = 66.09\ \%$$

% H = 100.00 - 29.61 - 66.09 = 4.30

$$C: 29.61 \text{ g} \times \frac{1 \text{ mole C}}{12.01 \text{ g}} = 2.47 \text{ mole C};\quad \frac{2.47 \text{ mole C}}{1.86} = 1.33 \text{ mole C}$$

$$Cl: 66.09 \text{ g} \times \frac{1 \text{ mole}}{35.45 \text{ g}} = 1.86 \text{ mole Cl};\quad \frac{1.86 \text{ mole Cl}}{1.86} = 1.00 \text{ mole Cl}$$

$$H: 4.30 \text{ g} \times \frac{1 \text{ mole}}{1.008 \text{ g}} = 4.27 \text{ mole H};\quad \frac{4.27 \text{ mole H}}{1.86} = 2.30 \text{ mole H}$$

Empirical Formula = $C_4H_7Cl_3$

3.46 $$\frac{0.1 \text{ mm}}{\text{bill}} \times \frac{1 \times 10^{-3} \text{ m}}{\text{mm}} \times \frac{6.022 \times 10^{23} \text{ bills}}{\text{mole}} \times 1 \text{ mole} = 6 \times 10^{19} \text{ meters}$$

3.47 a) NH_4Cl, ammonium chloride
b) XeF_4, xenon tetrafluoride
c) Fe_2O_3, iron(III) oxide
d) SO_2, sulfur dioxide
e) $KClO_4$, potassium perchlorate
f) $KClO_3$, potassium chlorate
g) $KClO_2$, potassium chlorite
h) $KClO$, potassium hypochlorite
i) KCl, potassium chloride
j) Na_2HPO_4, sodium hydrogen phosphate

3.48 a) $Na_4Al_3Si_3O_{12}Cl$

MM = (4 x 22.99) + (3 x 26.98) + (3 x 28.09) + (12 x 16.00) + (1 x 35.45)
= 484.62

% Na = ((4 x 22.99)/484.62) x 100 = 18.98%

% Al = 16.70% % Si = 17.39% % O = 39.62% % Cl = 7.315%

b) $CuAl_6(PO_4)_4(OH)_8 \cdot 4H_2O$

MM = (1 x 63.55) + (6 x 26.98) + (4 x 30.97) + (28 x 16.00) + (16 x 1.008)
= 813.44

% Cu = 7.813% % Al = 19.90% % P = 15.23% % O = 55.07% % H = 1.983%

c) $Mg_3Al_2(SiO_4)_3$

MM = (3 x 24.31) + (2 x 26.98) + (3 x 28.09) + (12 x 16.00) = 403.16

% Mg = 18.09% % Al = 13.38% % Si = 20.90% % O = 47.62%

d) $Ca_2Mg_5Si_8O_{22}F_2$

MM = (2 x 40.08) + (5 x 24.31) + (8 x 28.09) + (22 x 16.00) + (2 x 19.00)
= 816.43

% Ca = 9.818% % Mg = 14.89% % Si = 27.52% % O = 43.11%
% F = 4.654%

3.49 Verapamil $C_{27}H_{38}O_4N_2$

a) MM = (27 x 12.01) + (38 x 1.008) + (4 x 16.00) + (2 x 14.01) = 454.59

b) $120.0 \text{ mg} \times \frac{10^{-3}\text{ g}}{\text{mg}} \times \frac{1 \text{ mole}}{454.6 \text{ g}} = 2.640 \times 10^{-4}$ mole of $C_{27}H_{38}O_4N_2$

c) 2.640×10^{-4} mole of $C_{27}H_{38}O_4N_2 \times \frac{6.022 \times 10^{23} \text{ molecules}}{\text{mole}} \times \frac{2 \text{ atoms N}}{\text{molecule}}$

$= 3.180 \times 10^{20}$ atoms N

3.50 a) $CaBr_2$ b) $NaBrO_3$ c) $NaBrO_2$ d) $NaHCO_3$ e) Al_2O_3

f) PF_5 g) $(NH_4)_2SO_4$ h) $CoCl_2$

3.51 $$\% C = \left[(\text{mass of } CO_2) \times \left(\frac{\text{At. Wt. C}}{\text{MM } CO_2}\right) / \text{ mass of sample}\right] \times 100 = ?\%$$

$$= \left[(54.9140 - 54.4375)\left(\frac{12.01}{44.01}\right) / (2.8954 - 2.7534)\right] \times 100 = 91.57\%$$

$$\% H = \left[(47.9961 - 47.8845)\left(\frac{2 \times 1.008}{18.02}\right) / (2.8954 - 2.7534)\right] \times 100 = 8.79\%$$

$$C = 91.57 \text{ g} \times \frac{1 \text{ mole}}{12.01 \text{ g}} = 7.62 \text{ mole C}; \ \frac{7.62}{7.62} = 1.00; \ 1.00 \times 7 = 7.00 \text{ moles C}$$

$$H = 8.79 \text{ g} \times \frac{1 \text{ mole}}{1.008 \text{ g}} = 8.72 \text{ mole H}; \ \frac{8.72 \text{ mole}}{7.62} = 1.14; \ 1.14 \times 7 = 7.98 \text{ moles H}$$

Empirical Formula = C_7H_8

3.52 a) Some of the carbon is not oxidized to CO_2; the CO_2 amount is low; C is low; too little C and too much H in formula.

b) H_2O and $\therefore$ H too low; CO_2 and $\therefore$ C too high; too high C and too little H in formula.

c) CO_2 and $\therefore$ C too low; therefore, C too low in formula.

3.53 a) $NaNO_2$ and $NaNO_3$ b) K_2CO_3 and $KHCO_3$

c) FeO and Fe_2O_3 d) I_2 and I^-

3.54 a) (3 x 55.847) + (4 x 16.00) = 231.54

b) (2 x 12.01) + (6 x 1.008) + (1 x 16.00) = 46.07

c) (3 x 12.01) + (8 x 1.008) + (3 x 16.00) = 92.09

d) (2 x 26.98) + (3 x 32.07) + (12 x 16.00) = 342.17

e) (1 x 58.69) + (1 x 32.07) + (10 x 16.00) + (12 x 1.008) = 262.9

3.55 a) $\frac{25.0 \text{ g } Fe_3O_4}{231.54 \text{ g / mol}} = 0.108 \text{ mol } Fe_3O_4$ b) $\frac{25.0 \text{ g } C_2H_6O}{46.07 \text{ g / mol}} = 0.543 \text{ mol } C_2H_6O$

c) $\frac{25.0 \text{ g } C_3H_5(OH)_3}{92.09 \text{ g / mol}} = 0.271 \text{ mol } C_3H_5(OH)_3$

d) $\frac{25.0 \text{ g } Al_2(SO_4)_3}{342.17 \text{ g / mol}} = 0.0731 \text{ mol } Al_2(SO_4)_3$

e) $\frac{25.0 \text{ g } NiSO_4 \bullet 6H_2O}{262.9 \text{ g / mol}} = 0.0951 \text{ mol } NiSO_4 \bullet 6H_2O$

3.56 a) % Fe = [(3 x 55.85)/231.54] x 100 = 72.36%
% O = [(4 x 16.00)/231.54] x 100 = 27.64%

b) % C = [(2 x 12.01)/46.07] x 100 = 52.14%
% H = [(6 x 1.008)/46.07] x 100 = 13.13%
% O = [(16.00)/46.07] x 100 = 34.73%

c) % C = [(3 x 12.01)/92.09] x 100 = 39.12%
% H = [(8 x 1.008)/92.09] x 100 = 8.757%
% O = [(3 x 16.00)/92.09] x 100 = 52.12%

d) % Al = [(2 x 26.98)/342.17] x 100 = 15.77%
% S = [(3 x 32.07)/342.17] x 100 = 28.12%
% O = [(12 x 16.00)/342.17] x 100 = 56.11%

e) % Ni = [(58.69)/262.85] x 100 = 22.33%
% S = [(32.07)/262.85] x 100 = 12.20%
% O = [(10 x 16.00)/262.85] x 100 = 60.87%
% H = [(12 x 1.008)/262.85] x 100 = 4.602%

3.57 a) $\frac{16.0 \text{ g O}}{1.00 \text{ mol O}} \text{x} \frac{100.00 \text{ g } Fe_3O_4}{27.64 \text{ g O}} = 57.9 \text{ g } Fe_3O_4 \text{ / mol O}$

b) $\frac{16.0 \text{ g O}}{1.00 \text{ mol O}} \text{x} \frac{100.00 \text{ g } C_2H_6O}{34.73 \text{ g O}} = 46.1 \text{ g } C_2H_6O \text{ / mol O}$

c) $\frac{16.0 \text{ g O}}{1.00 \text{ mol O}} \text{x} \frac{100.00 \text{ g } C_3H_5(OH)_3}{52.12 \text{ g O}} = 30.7 \text{ g } C_3H_5(OH)_3 \text{ / mol O}$

d) $\frac{16.0 \text{ g O}}{1.00 \text{ mol O}} \text{x} \frac{100.00 \text{ g } Al_2(SO_4)_3}{56.12 \text{ g O}} = 28.5 \text{ g } Al_2(SO_4)_3 \text{ / mol O}$

e) $\frac{16.0 \text{ g O}}{1.00 \text{ mol O}} \text{x} \frac{100.00 \text{ g } NiSO_4 \bullet 6H_2O}{60.87 \text{ g O}} = 26.3 \text{ g } NiSO_4 \bullet 6H_2O \text{ / mol O}$

3.58 Abscisic Acid; $C_{15}H_{20}O_4$; Indole acetic acid; $C_{10}H_9NO_2$
Zeatin; $C_{10}H_{13}N_5O$

3.59 $1 \text{ L sea water} \times \frac{2.2 \times 10^{-9} \text{ mol}}{1 \text{ L}} \times \frac{6.022 \times 10^{23} \text{ molecules}}{1 \text{ mol}} = 1.3 \times 10^{15} \text{ molecules}$

$$1 \text{ kg Rb} \times \frac{10^3 \text{ g}}{1 \text{ kg}} \times \frac{1 \text{ mol Rb}}{85.47 \text{ g Rb}} \times \frac{1 \text{ L}}{2.2 \times 10^{-9} \text{ mol}} = 5.3 \times 10^9 \text{ L} \cong 5 \times 10^9 \text{ L}$$

3.60 Mg^{2+} = ○ , Cl^- = ●

3.61 a) IO_3^- b) IO_4^- c) PO_4^{3-} d) ClO_4^- e) SO_4^{2-} f) $CH_3CO_2^-$ g) HCO_3^- h) NO_3^-

3.62 a) MM of $CuSO_4 \cdot 5H_2O$ = (1 mol Cu)(63.55 g/mol) + (1 mol S)(32.07 g/mol) + (9 mol O)(16.00 g/mol) + (10 mol H)(1.008 g/mol) = 249.70 g/mol

b) $4.59 \text{ g } CuSO_4 \cdot 5H_2O \times \frac{1 \text{ mol}}{249.70 \text{ g}} \times \frac{1 \text{ mol S}}{1 \text{ mol}} \times \frac{6.022 \times 10^{23} \text{ atoms}}{1 \text{ mol}}$

$= 1.11 \times 10^{22}$ atoms S

c) $4.59 \text{ g } CuSO_4 \cdot 5H_2O \times \frac{1 \text{ mol}}{249.70 \text{ g}} \times \frac{9 \text{ mol O}}{1 \text{ mol } CuSO_4 \cdot 5H_2O} = 0.165 \text{ mol O}$

d) $4.59 \text{ g } CuSO_4 \cdot 5H_2O \times \frac{10 \times 1.008 \text{ g H}}{249.70 \text{ g } CuSO_4 \cdot 5H_2O} = 0.185 \text{ mol H}$

3.63 MM of C_3H_8 = (3 x 12.01) + (8 x 1.008) = 44.09
% C = [(3 x 12.01)/44.09] x 100 = 81.72%

3.64 MM $Na_2C_4H_3O_4SAu$ = (2 mol Na)(22.99 g/mol) + (4 mol C)(12.01 g/mol) + (3 mol H)(1.008 g/mol) + (4 mol O)(16.00 g/mol) + (1 mol S)(32.07 g/mol) + (1 mol Au)(196.97 g/mol) = 390.08 g/mol

a) $\frac{\frac{0.25 \text{ g}}{390.08 \text{ g/mol}}}{0.0100 \text{ L}} = 0.064\text{M (stock solution)}$

$\frac{(2.00 \text{ mL})(0.064 \text{ M})}{(10.00 \text{ mL})} = 0.013\text{M (dilute solution)}$

$0.40 \text{ mL injection} \times \frac{1 \text{ L}}{1000 \text{ mL}} \times \frac{0.013 \text{ mol}}{1 \text{ L}} \times \frac{390.08 \text{ g}}{1 \text{ mol}} = 0.0020 \text{ g}$

b) $5.0 \text{ mL} \times \frac{30.0 \text{ μg Au}}{10.0 \text{ mL}} \times \frac{1 \text{ g}}{10^6 \text{ μg}} \times \frac{6.022 \times 10^{23} \text{ atoms}}{1 \text{ mol Au}} = 4.6 \times 10^{16} \text{ atoms Au}$

3.65 NH_4NO_3; MM = (2 x 14.01) + (4 x 1.008) + (3 x 16.00) = 80.05 g/mol

$(NH_4)_2SO_4$; MM = (2 x 14.01) + (8 x 1.008) + (1 x 32.07) + (4 x 16.00) = 132.15 g/mol

$(NH_2)_2CO$; MM = (2 x 14.01) + (4 x 1.008) + 12.01 + 16.00) = 60.06 g/mol

$(NH_4)_2HPO_4$; MM = (2 x 14.01) + (9 x 1.008) + (30.97) + (4 x 16.00) = 132.06 g/mol

$$NH_4NO_3 \quad 1\ \text{kg N} \times \frac{10^3\ \text{g}}{\text{kg}} \times \frac{80.05\ \text{g}\ NH_4NO_3}{(2 \times 14.01)\ \text{g N}} \times \frac{1\ \text{kg}}{10^3\ \text{g}} = 2.857\ \text{kg}\ NH_4NO_3$$

$$(NH_4)_2SO_4 \quad \frac{132.15}{2 \times 14.01} = 4.716\ \text{kg}\ (NH_4)_2SO_4$$

$$(NH_2)_2CO \quad \frac{60.06}{2 \times 14.01} = 2.143\ \text{kg}\ (NH_2)_2CO$$

$$(NH_4)_2HPO_4 \quad \frac{132.06}{2 \times 14.01} = 4.713\ \text{kg}\ (NH_4)_2HPO_4$$

3.66 $$1.80 \times 10^2\ \text{g}\ H_2SO_4 \times \frac{(4 \times 16.00)\ \text{g O}}{98.09\ \text{g}\ H_2SO_4} = 1.17 \times 10^2\ \text{g Oxygen}$$

3.67 In addition to the atomic weights of carbon, oxygen and hydrogen and the weight of the sample burned, one would also need to know if any elements other than hydrogen, carbon and, possibly, oxygen are present and what the approximate molecular mass is before the molecular formula can be determined.

3.68 $$1.00\ \text{mton Cu} \times \frac{1000\ \text{kg}}{\text{mton}} \times \frac{10^3\ \text{g}}{\text{kg}} \times \frac{183.54\ \text{g}\ CuFeS_2\ \text{mole}^{-1}}{63.55\ \text{g Cu mole}^{-1}} \times \frac{1\ \text{kg}}{10^3\ \text{g}} = 2.89 \times 10^3\ \text{kg}\ CuFeS_2$$

3.69 $$\frac{8.3\ \text{mg}\ O_2}{\text{L}} \times \frac{10^{-3}\ \text{g}}{\text{mg}} \times \frac{1\ \text{mole}\ O_2}{32.00\ \text{g}} = 2.6 \times 10^{-4}\ \frac{\text{mole}\ O_2}{\text{L}}$$

2.6×10^{-4} M = molarity of oxygen as dissolved O_2

3.70 $$\%\ C = \left(\left(3.83\ \text{g}\ CO_2 \times \left[\frac{12.01\ \text{g/mol C}}{44.01\ \text{g/mol}\ CO_2}\right]\right) \Big/ 1.45\ \text{g}\right) \times 100 = 72.1\ \%\ C$$

$$\%\ H = \left(\left(1.56\ \text{g}\ H_2O \times \left[\frac{2(1.008)\ \text{g/mol H}}{18.02\ \text{g/mol}\ H_2O}\right]\right) \Big/ 1.45\ \text{g}\right) \times 100 = 12.0\ \%\ H$$

% O = 100.0 - 72.1 - 12.0 = 15.9%

C: 72.1 g x 1 mole/12.01 g = 6.00
H: 12.0 g x 1 mole/1.008 g = 11.9
O: 15.9 g x 1 mole/16.00 g = 0.994
$C_6H_{12}O$

3.71 MM of $C_{16}H_{18}N_2O_4S$ =
$(16 \times 12.01) + (18 \times 1.008) + (2 \times 14.01) + (4 \times 16.00) + 32.07 = 334.4$

$$50.0 \text{ mg} = \frac{10^{-3}\text{g}}{\text{mg}} \times \frac{1 \text{ mole}}{334.4 \text{ g}} = 1.50 \times 10^{-4} \text{ moles}$$

$$50.0 \text{ mg} = \frac{10^{-3}\text{g}}{\text{mg}} \times \frac{1 \text{ mole } C_{16}H_{18}N_2O_4S}{334.4 \text{ g}} \times \frac{6.022 \times 10^{23} \text{ molecules}}{\text{mole}}$$

$$\times \frac{16 \text{ C atoms}}{\text{molecule}} = 1.45 \times 10^{21} \text{ atoms C}$$

$$75.0 \text{ mg} = \frac{10^{-3}\text{g}}{\text{mg}} \times \frac{32.07 \text{ g S}}{334.4 \text{ g penicillin}} = 7.19 \times 10^{-3} \text{ g S or 7.19 mg S}$$

3.72 $$\frac{5.69 \times 10^{-22} \text{ g}}{\text{molecule}} \times \frac{6.022 \times 10^{23} \text{ molecules}}{\text{mole}} = 343 \text{ g / mole}$$

3.73 $$\% \text{ C} = \left(\left(9.1 \text{ g } CO_2O \times \left[\frac{12.01 \text{ g / mol C}}{44.01 \text{ g / mol } CO_2}\right]\right) \Big/ 5.00 \text{ g}\right) \times 100 = 49.8\% \text{ C}$$

$$\% \text{ H} = \left(\left(1.80 \text{ g } H_2O \times \left[\frac{2(1.008) \text{ g / mol H}}{18.02 \text{ g / mol } H_2O}\right]\right) \Big/ 5.00 \text{ g}\right) \times 100 = 4.03 \text{ \% H}$$

% Fe = 100.0 - 49.8 - 4.03 = 46.2

C: 49.8 g x 1 mole/12.01 g = 4.15 mole C; 4.15/0.8272 = 5.02 moles C
H: 4.03 g x 1 mole/1.008 g = 3.998 mole H; 3.998/0.8272 = 4.83 moles H
Fe: 46.2 g x 1 mole/55.85 g = 0.8272 mole Fe; 0.8272/0.8272 = 1.00 mole Fe

C_5H_5Fe

3.74 mol Ag^+ = (0.595 L)(1.75×10^{-2} mol / L) = 0.0104 mol Ag^+
0.0104 mol Ag^+ produces 0.0104 mol AgCl

$$0.0104 \text{ mol AgCl} \times \frac{(107.87 + 35.45) \text{ g AgCl}}{1 \text{ mol AgCl}} = 1.49 \text{ g AgCl}$$

3.75 $$0.138 \text{ g } CO_2 \times \frac{12.01 \text{ g C}}{44.01 \text{ g } CO_2} = 0.0377 \text{ g C}$$

$$0.0566 \text{ g } H_2O \times \frac{2 \times 1.008 \text{ g H}}{18.02 \text{ g } H_2O} = 0.00633 \text{ g H}$$

$$0.0238 \text{ g } NH_3 \times \frac{14.01 \text{ g N}}{17.03 \text{ g } NH_3} = 0.0196 \text{ g N}$$

$$1.004 \text{ g AgCl} \times \frac{35.45 \text{ g Cl}}{143.32 \text{ g AgCl}} = 0.2483 \text{ g Cl}$$

$$\% \text{ C} = \frac{0.0377 \text{ g C}}{0.150 \text{ g sample}} \times 100\% = 25.1\%$$

(continued)

(3.75 continued)

$$\% \text{ H} = \frac{0.00633 \text{ g H}}{0.150 \text{ g sample}} \text{x } 100\% = 4.22\%, \ \% \text{ N} = \frac{0.0196 \text{ g N}}{0.200 \text{ g sample}} \text{x } 100\% = 9.80\%$$

$$\% \text{ Cl} = \frac{0.2483 \text{ g Cl}}{0.500 \text{ g sample}} \text{x } 100\% = 49.7\%$$

% O = 100.0% - 25.1% C - 4.22% H - 9.80% N - 49.7% Cl = 11.2%

For 100 g sample;

$$\frac{25.1 \text{ g C}}{12.01 \text{ g / mol}} = 2.09 \text{ mol C} \qquad \frac{4.22 \text{ g H}}{1.008 \text{ g / mol}} = 4.19 \text{ mol H}$$

$$\frac{49.7 \text{ g Cl}}{35.45 \text{ g / mol}} = 1.40 \text{ mol Cl} \qquad \frac{11.2 \text{ g O}}{16.00 \text{ g / mol}} = 0.700 \text{ mol O}$$

$$\frac{9.80 \text{ g N}}{14.01 \text{ g / mol}} = 0.700 \text{ mol N}$$

$$\frac{2.09}{0.700} = 2.99 \qquad \frac{4.19}{0.700} = 5.99 \qquad \frac{0.700}{0.700} = 1.00$$

$$\frac{1.40}{0.700} = 2.00 \qquad \frac{0.700}{0.700} = 1.00 \qquad C_3H_6Cl_2NO$$

3.76 $5.75 \text{ g Pt x} \frac{1 \text{ mole Pt}}{195.08 \text{ g Pt}} \text{x} \frac{1 \text{ mole Li}}{1 \text{ mole Pt}} \text{x} \frac{6.94 \text{ g Li}}{\text{mole Li}} = 0.205 \text{ g}$

3.77 % Hg = (0.302 g Hg/0.350 g sample) x 100 = 86.3%

$$\% \text{ S} = \left(\left(0.0964 \text{ g x}\left[\frac{32.07 \text{ g / mol S}}{64.07 \text{ g / mol } SO_2}\right]\right)\Big/ 0.350 \text{ g}\right) \text{x } 100 = 13.8 \%$$

Hg: 86.3 g x (1 mole/200.6 g) = 0.430 mole; 0.430 mole/0.43 = 1.0
S: 13.8 g x (1 mole/32.07 g) = 0.430 mole; 0.430 mole/0.43 = 1.0
HgS

3.78 Sevin, $C_{12}H_{11}NO_2$; MM = 201.2 g/mole
Malathion, $C_{10}H_{19}O_6PS_2$; MM = 330.3 g/mole

a) $8.3 \text{ g Sevin x} \frac{1 \text{ mole Sevin}}{201.2 \text{ g}} \text{x} \frac{12 \text{ mole C}}{\text{mole Sevin}} = 0.50 \text{ mole}$

b) $6.5 \text{ g Malathion x} \frac{1 \text{ mole Malathion}}{330.3 \text{ g}} \text{x} \frac{2 \text{ moles S}}{\text{mole}} \text{x} \frac{6.022 \text{ x } 10^{23} \text{ atoms S}}{\text{mole}}$

$= 2.4 \text{ x } 10^{22}$ atoms S

(continued)

(3.78 continued)

c) $17.8 \text{ g Malathion} \times \frac{1 \text{ mole Malathion}}{330.3 \text{ g}} \times \frac{6 \text{ moles O}}{\text{mole}} \times \frac{16.00 \text{ g O}}{\text{mole O}} = 5.17 \text{ g O}$

d) $75 \text{ g} \times \frac{0.01 \text{ g}}{100 \text{ g}} \times \frac{1 \text{ mole}}{201.2 \text{ g}} = 3.7 \times 10^{-5} \text{ mole Sevin}$

$3.7 \times 10^{-5} \text{ mole} \times \frac{6.022 \times 10^{23} \text{ molecules}}{\text{mole}} = 2.2 \times 10^{19} \text{ molecules}$

e) $15 \text{ gal} \times \frac{1 \text{ mL}}{\text{gal}} \times \frac{1.00 \text{ g}}{\text{mL}} \times \frac{0.01 \text{ g}}{100 \text{ g}} \times \frac{1 \text{ mole}}{201.2 \text{ g}} = 7.5 \times 10^{-6} \text{ moles}$

3.79 a) $C_{44}H_{69}O_{12}N$

MM = (44 x 12.01) + (69 x 1.008) + (12 x 16.00) + 14.01 = 804.0 g/mol

b) $5.0 \text{ mg} \times \frac{10^{-3} \text{ g}}{\text{mg}} \times \frac{1 \text{ mole}}{804.0 \text{ g}} = 6.2 \times 10^{-6} \text{ moles FK - 506}$

c) $6.2 \times 10^{-6} \text{ moles FK - 506} \times \frac{12 \text{ mole O}}{\text{mole}} \times \frac{6.022 \times 10^{23} \text{ atoms O}}{\text{mole O}}$

$= 4.5 \times 10^{19} \text{ atoms O}$

3.80 $\% \text{ C} = \left(\left(8.13 \text{ mg} \times \left[\frac{12.01 \text{ g C}}{44.01 \text{ g } CO_2}\right]\right) \Big/ 5.00 \text{ mg}\right) \times 100 = 44.37\% \text{ C}$

$\% \text{ H} = \left(\left(2.76 \text{ mg} \times \left[\frac{2(1.008) \text{ g H}}{18.02 \text{ g } H_2O}\right]\right) \Big/ 5.00 \text{ mg}\right) \times 100 = 6.18\% \text{ H}$

$\% \text{ S} = \left(\left(3.95 \text{ mg} \times \left[\frac{32.07 \text{ g S}}{64.07 \text{ g } SO_2}\right]\right) \Big/ 5.00 \text{ mg}\right) \times 100 = 39.54\% \text{ S}$

% O = 100.00 - 44.37 - 6.18 - 39.54 = 9.91% O

C: $44.37 \text{ g} \times \frac{1 \text{ mole}}{12.01 \text{ g}} = 3.69 \text{ mole C}$; 3.69/0.619 = 5.96

H: $6.18 \text{ g} \times \frac{1 \text{ mole}}{1.008 \text{ g}} = 6.13 \text{ mole H}$; 6.13/0.619 = 9.90

S: $39.54 \text{ g} \times \frac{1 \text{ mole}}{32.07 \text{ g}} = 1.23 \text{ mole S}$; 1.23/0.619 = 1.99

O: $9.91 \text{ g} \times \frac{1 \text{ mole}}{16.00 \text{ g}} = 0.619 \text{ mole O}$; 0.619/0.619 = 1.00

Empirical Formula = $C_6H_{10}OS_2$

Mass of Emp. Form. = 162 Since this is very near the molar mass obtained in the other experiment, the Molecular Formula is $C_6H_{10}OS_2$.

3.81 $8.50 \times 10^4 \text{ L} \times \frac{10^3 \text{ mL}}{\text{L}} \times \frac{1.0 \text{ g}}{\text{mL}} \times \frac{1 \text{ mol}}{18.02 \text{ g}} = 4.7 \times 10^6 \text{ mol } H_2O$

3.82 $\% \text{ C} = \left(\left(139.9 \text{ mg} \times \left[\frac{12.01 \text{ g C}}{44.01 \text{ g / mol } CO_2}\right]\right) \Big/ 50.50 \text{ mg}\right) \times 100 = 75.60\% \text{ C}$

$\% \text{ H} = \left(\left(60.91 \text{ mg} \times \left[\frac{(2)(1.008 \text{ g H})}{18.02 \text{ g / mol } H_2O}\right]\right) \Big/ 50.50 \text{ mg}\right) \times 100 = 13.49\% \text{ H}$

% N = (8.35 mg/75.62 mg) x 100 = 11.04

% O = 100.00 - 75.60 - 13.49 - 11.04 = -0.13 (No oxygen present)

C: $75.60 \text{ g} \times \frac{1 \text{ mole}}{12.01 \text{ g}} = 6.295 \text{ mole C}; \quad 6.295/0.788 = 7.99$

H: $13.49 \text{ g} \times \frac{1 \text{ mole}}{1.008 \text{ g}} = 13.38 \text{ mole H}; \quad 13.38/0.788 = 16.98$

N: $11.04 \text{ g} \times \frac{1 \text{ mole}}{14.01 \text{ g}} = 0.788 \text{ mole N}; \quad 0.788/0.788 = 1.00$

Empirical formula = $C_8H_{17}N$ Emp. Form. mass = 127.2
Therefore, the molecular formula must be $C_8H_{17}N$ if its molar mass is to be below 200.

3.83 a) carbon dioxide b) potassium nitrate c) sodium chloride
d) sodium bicarbonate or sodium hydrogencarbonate e) sodium carbonate
f) sodium hydroxide g) calcium oxide h) magnesium hydroxide

3.84 Only these five line structures are possible.

or

or

or

(continued)

(3.84 continued)

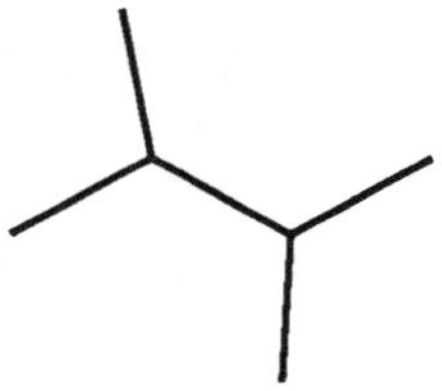

or

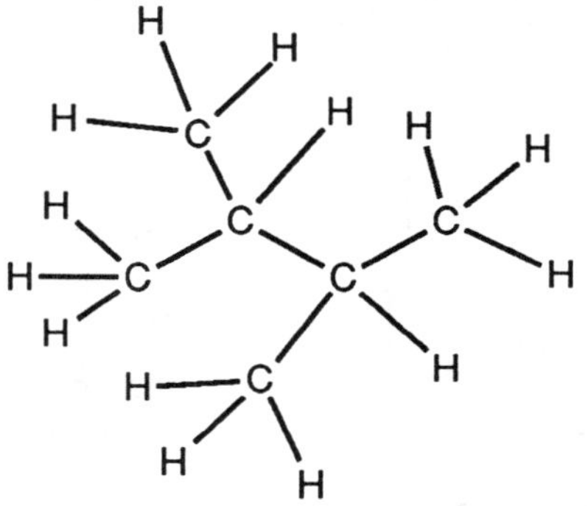

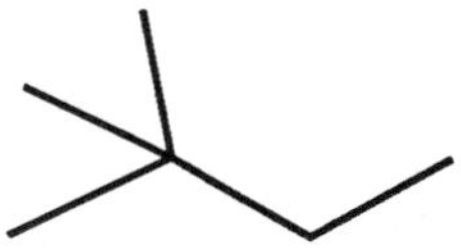

or

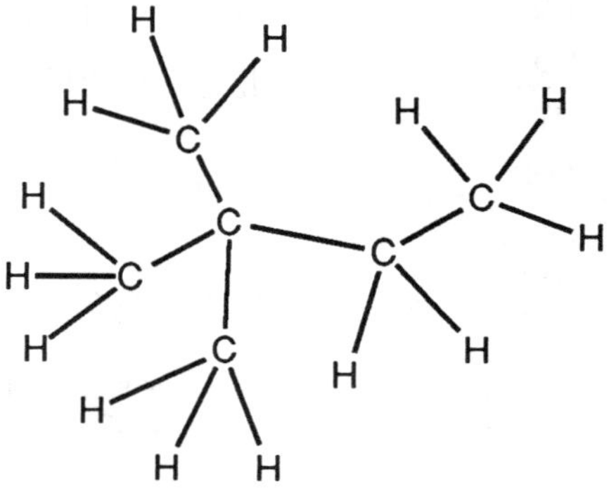

3.85 $\frac{100 \text{ g } B_{12}}{4.34 \text{ g Co}} \times \frac{58.93 \text{ g Co}}{1 \text{ mole Co}} \times \frac{1 \text{ mole Co}}{1 \text{ mole } B_{12}} = 1.36 \times 10^3 \text{ g } B_{12} / \text{mole } B_{12}$

3.86 C_7H_6O, MM = 106.05

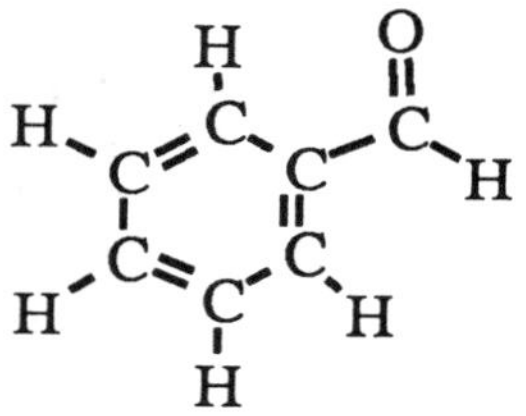

$C_7H_{14}O_2$, MM = 130.18

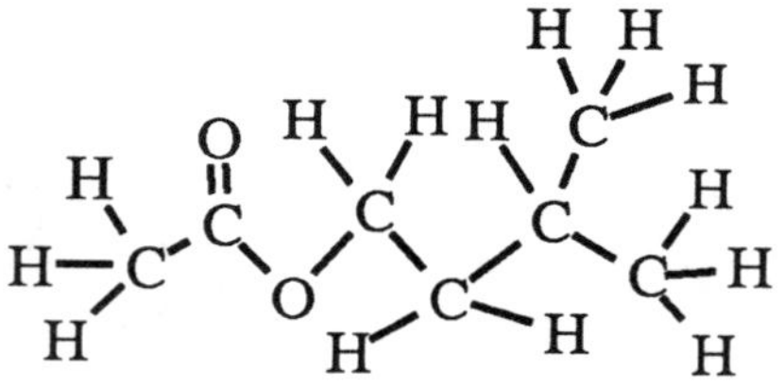

$C_{11}H_{16}O$, MM = 164.24

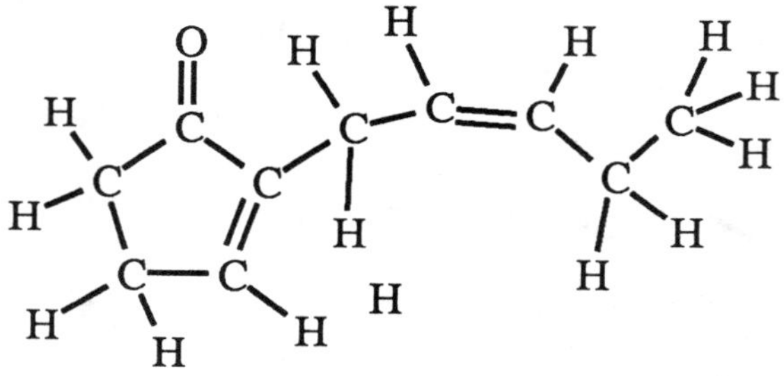

(continued)

(3.86 continued)

$C_{10}H_{16}$, MM = 136.23

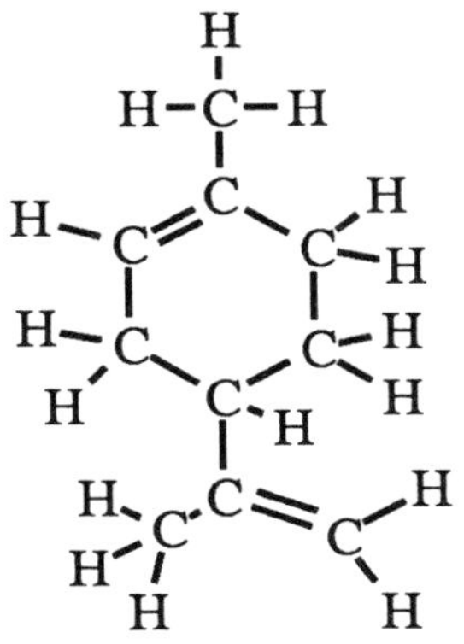

$C_8H_8O_3$, MM = 152.14

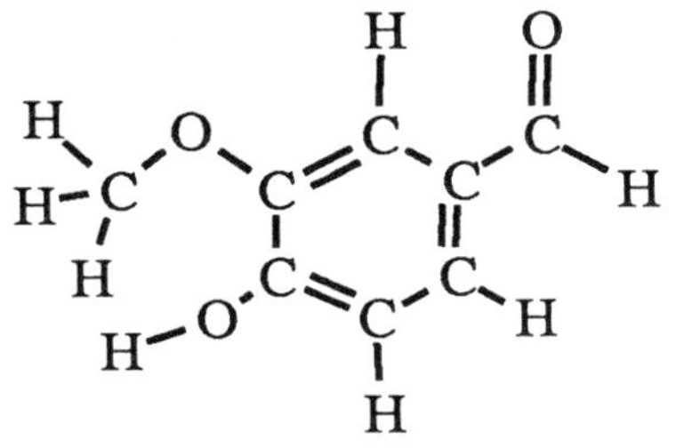

3.87 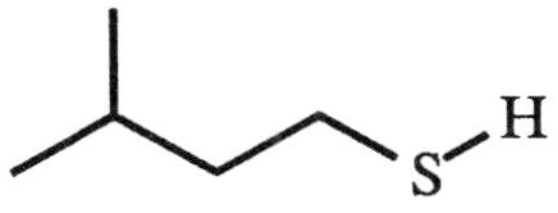 3-methylbutane thiol 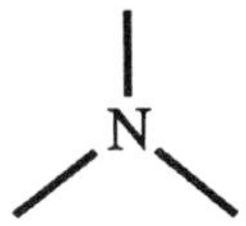 — trimethylamine  — cadaverine 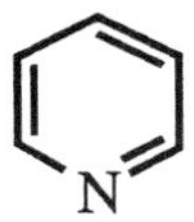 — pyridine

3.88 a) titanium(IV) oxide b) lead(II) sulfide c) aluminum oxide
d) calcium carbonate e) barium sulfate f) magnesium hydroxide
g) mercury(II) sulfide h) diantimony trisulfide

3.89 $$\frac{\$275}{2000 \text{ lb } SO_2} \times \frac{1 \text{ lb}}{454 \text{ g}} \times \frac{64.07 \text{ g } SO_2}{1 \text{ mole } SO_2} = \$0.0194 / \text{mol } SO_2 \approx 2¢$$

$$\$1.00 \times \frac{1 \text{ mol } SO_2}{\$0.0194} \times \frac{6.022 \times 10^{23} \text{ molecule}}{1 \text{ mole}} = 3.10 \times 10^{25} \text{ molecules of } SO_2 / \$1.00$$

3.90 a) ammonium and sulfate ions
b) carbon dioxide molecules (You will learn later that some of the CO_2 does react with the water.)
c) sodium and fluoride ions
d) potassium and carbonate ions
e) sodium and hydrogen sulfate ions (plus some hydrogen and sulfate ions)
f) chlorine molecules (You will learn later that chlorine will react with the water.)
g) sodium and dichromate ions
h) copper(II) and chloride ions
i) barium and hydroxide ions

3.91 $CH_3–CH_2–CH_2–CH_2–CH_2–CH_2–OH$
1-hexanol

$CH_3–CH_2–CH_2–CH_2–CH(OH)–CH_3$
2-hexanol

$CH_3–CH_2–CH_2–CH(OH)–CH_2–CH_3$
3-hexanol

Any others would require that some of the carbons not be in the largest (straight) chain.

3.92 1 part per billion is usually 1 g per 10^9 g, but for this problem it is implied to be 1 mole of contaminant per 10^9 moles.

$$\frac{55.5 \text{ mole } H_2O}{L} \times \frac{1 \text{ mole contaminant}}{10^9 \text{ mole } H_2O} = \frac{5.5 \times 10^{-8} \text{ mole contaminant}}{L}$$

$$\frac{5.5 \times 10^{-8} \text{ moles contaminant}}{L} \times \frac{6.022 \times 10^{23} \text{ molecules}}{\text{mole}} = \frac{3.34 \times 10^{16} \text{ molecules}}{L}$$

CHAPTER 4: CHEMICAL REACTIONS AND STOICHIOMETRY

4.1 Follow the methods and notations given in Section 4.1

a) $5H_2 + 2NO \rightarrow 2NH_3 + 2H_2O$ b) $2CO + 2NO \rightarrow N_2 + 2CO_2$

c) $2NH_3 + 2O_2 \rightarrow N_2O + 3H_2O$ d) $6NO + 4NH_3 \rightarrow 5N_2 + 6H_2O$

4.2 Follow the methods and notations given in Section 4.1

a) $NH_4NO_3 \rightarrow N_2O + 2H_2O$ b) $P_4O_{10} + 6H_2O \rightarrow 4H_3PO_4$

c) $2HIO_3 \rightarrow I_2O_5 + H_2O$ d) $2As + 5Cl_2 \rightarrow 2AsCl_5$

e) $Mg_3N_2 + 6H_2O \rightarrow 3Mg(OH)_2 + 2NH_3$

4.3 Follow the methods and notations given in Section 4.1

a) $N_2O_{5(g)} + H_2O_{(l)} \rightarrow 2HNO_{3(aq)}$ b) $2KClO_{3(s)} \rightarrow 2KCl_{(s)} + 3O_{2(g)}$

c) $2Fe_{(s)} + O_{2(g)} + 2H_2O_{(l)} \rightarrow 2Fe(OH)_{2(s)}$ d) $Au_2S_{3(s)} + 3H_{2(g)} \rightarrow 3H_2S_{(g)} + 2Au_{(s)}$

4.4 Follow the methods and notations given in Section 4.1

a) $N_2 + 3H_2 \rightarrow 2NH_3$ b) $2H_2 + CO \rightarrow CH_3OH$

c) $CaO + 3C \rightarrow CO + CaC_2$ d) $2C_2H_4 + O_2 + 4HCl \rightarrow 2C_2H_4Cl_2 + 2H_2O$

4.5 Follow the methods and notations given in Section 4.1

a) $4NH_3 + 5O_2 \rightarrow 4NO + 6H_2O$ b) $2NO + O_2 \rightarrow 2NO_2$

c) $3NO_2 + H_2O \rightarrow 2HNO_3 + NO$ d) $4NH_3 + 3O_2 \rightarrow 2N_2 + 6H_2O$

e) $4NH_3 + 6NO \rightarrow 5N_2 + 6H_2O$

4.6 For Problems 4.6, 4.7 & 4.8 the molecular pictures need to show reactants and product in the indicated proportions. For a discussion of notations used, see Chapter 2. In Problem 4.4a, molecular nitrogen ($N_{2(g)}$) reacts with hydrogen ($H_{2(g)}$) in the ratio of one molecule of nitrogen with three molecules of hydrogen to produce two molecules of ammonia ($NH_{3(g)}$). The simplest molecular picture would illustrate that smallest number of reactants and products. Draw a picture that shows one molecule of nitrogen and three molecules of hydrogen as reactants and two molecules of ammonia as the product.

4.7 For problems 4.6, 4.7, & 4.8 the molecular pictures need to show reactants and products in the indicated proportions. For a discussion of notations used, see chapter 2. In problem 4.5d, molecular ammonia ($NH_{3(g)}$) reacts with oxygen ($O_{2(g)}$) in the ratio of four molecules of ammonia with three molecules of oxygen to produce two molecules of nitrogen ($N_{2(g)}$) and six molecules of water ($H_2O_{(g)}$). The simplest molecular picture would illustrate that smallest number of reactants and products. Draw a picture showing that ratio of reactants and products.

4.8 For Problems 4.6, 4.7 & 4.8 the molecular pictures need to show reactants and product in the indicated proportions. For a discussion of notations used, see Chapter 2. In this problem solid magnesium ($Mg_{(s)}$) reacts with oxygen ($O_{2(g)}$) in the ratio of two atoms of magnesium with one molecule of oxygen to produce two formula representations of magnesium oxide ($MgO_{(s)}$). This problem specifies that the molecular picture show six magnesium atoms and four oxygen molecules. The products must then be shown in your molecular picture to be six $MgO_{(s)}$ and one excess $O_{2(g)}$.

4.9 In the following problems, obtain the number of moles of the first reactant by dividing 5.00 grams of that substance by its molecular mass. Then use the coefficients to convert to moles of the second reactant. Use the molecular mass of the second reactant to obtain the mass of that reactant.

a) $\dfrac{5.00 \text{ g } H_2}{2.016 \text{ g / mol}} \text{x} \dfrac{2 \text{ mol NO}}{5 \text{ mol } H_2} \text{x} \dfrac{30.01 \text{ g NO}}{\text{mol}} = 29.8 \text{ g NO}$

b) $\dfrac{5.00 \text{ g CO}}{28.01 \text{ g / mol}} \text{x} \dfrac{2 \text{ mol NO}}{2 \text{ mol CO}} \text{x} \dfrac{30.01 \text{ g NO}}{\text{mol}} = 5.36 \text{ g NO}$

c) $\dfrac{5.00 \text{ g } NH_3}{17.03 \text{ g / mol}} \text{x} \dfrac{2 \text{ mol } O_2}{2 \text{ mol } NH_3} \text{x} \dfrac{32.00 \text{ g } O_2}{\text{mol}} = 9.40 \text{ g } O_2$

d) $\dfrac{5.00 \text{ g NO}}{30.01 \text{ g / mol}} \text{x} \dfrac{4 \text{ mol } NH_3}{6 \text{ mol NO}} \text{x} \dfrac{17.03 \text{ g } NH_3}{\text{mol}} = 1.89 \text{ g } NH_3$

4.10 Use a procedure like that used in Problem 4.9, except first convert kg to grams.

a) $875 \text{ kg } H_2 \text{ x} \dfrac{10^3 \text{ g}}{\text{kg}} \text{x} \dfrac{1 \text{ mol } H_2}{2.016 \text{ g}} \text{x} \dfrac{1 \text{ mol } N_2}{3 \text{ mol } H_2} \text{x} \dfrac{28.02 \text{ g } N_2}{\text{mol}} \text{x} \dfrac{10^{-3} \text{ kg}}{\text{g}}$

$= 4.05 \text{ x } 10^3 \text{ kg } N_2$

b) $875 \text{ kg CO x} \dfrac{10^3 \text{ g}}{\text{kg}} \text{x} \dfrac{1 \text{ mol CO}}{28.01 \text{ g}} \text{x} \dfrac{2 \text{ mol } H_2}{1 \text{ mol CO}} \text{x} \dfrac{2.016 \text{ g } H_2}{\text{mol}} \text{x} \dfrac{10^{-3} \text{ kg}}{\text{g}}$

$= 1.26 \text{ x } 10^2 \text{ kg } H_2$

c) $875 \text{ kg C x} \dfrac{10^3 \text{ g}}{\text{kg}} \text{x} \dfrac{1 \text{ mol C}}{12.01 \text{ g}} \text{x} \dfrac{1 \text{ mol CaO}}{3 \text{ mol C}} \text{x} \dfrac{56.08 \text{ g CaO}}{\text{mol}} \text{x} \dfrac{10^{-3} \text{ kg}}{\text{g}}$

$= 1.36 \text{ x } 10^3 \text{ kg CaO}$

(continued)

(4.10 continued)

d) $875 \text{ kg } O_2 \times \frac{10^3 \text{ g}}{\text{kg}} \times \frac{1 \text{ mol } O_2}{32.00 \text{ g}} \times \frac{2 \text{ mol } C_2H_4}{1 \text{ mol } O_2} \times \frac{28.05 \text{ g } C_2H_4}{\text{mol}} \times \frac{10^{-3} \text{ kg}}{\text{g}}$

$= 1.53 \times 10^3 \text{ kg } C_2H_4$

4.11 Perform three calculations similar to those in Problems 4.9 and 4.10.

$CCl_4 + 2\ HF \rightarrow CCl_2F_2 + 2\ HCl$

$$175 \text{ kg } CCl_4 \times \frac{10^3 \text{ g}}{\text{kg } CCl_4} \times \frac{1 \text{ mol } CCl_4}{153.8 \text{ g}} \times \frac{2 \text{ mol HF}}{1 \text{ mol } CCl_4} \times \frac{20.01 \text{ g HF}}{\text{mol}} \times \frac{10^{-3} \text{ kg}}{\text{g}}$$

$= 45.5$ kg HF required

$$175 \text{ kg } CCl_4 \times \frac{10^3 \text{ g}}{\text{kg } CCl_4} \times \frac{1 \text{ mol } CCl_4}{153.8 \text{ g}} \times \frac{1 \text{ mol } CCl_2F_2}{1 \text{ mol } CCl_4} \times \frac{120.9 \text{ g } CCl_2F_2}{\text{mol}}$$

$$\times \frac{10^{-3} \text{ kg}}{\text{g}} = 138 \text{ kg } CCl_2F_2 \text{ obtained}$$

$$175 \text{ kg } CCl_4 \times \frac{10^3 \text{ g}}{\text{kg}} \times \frac{1 \text{ mol } CCl_4}{153.8 \text{ g}} \times \frac{2 \text{ mol HCl}}{1 \text{ mol } CCl_4} \times \frac{36.46 \text{ g HCl}}{\text{mol}}$$

$$\times \frac{10^{-3} \text{ kg}}{\text{g}} = 83.0 \text{ kg HCl obtained}$$

4.12 This problem is similar to the previous three problems.

$$150 \text{ g } I_2 \times \frac{1 \text{ mol } I_2}{253.8 \text{ g}} \times \frac{2 \text{ mol NaI}}{\text{mol } I_2} \times \frac{149.9 \text{ g NaI}}{\text{mol NaI}} = 177 \text{ g NaI}$$

4.13 Convert metric tons to grams, grams to moles of ammonium sulfate, then to moles of ammonia, and then convert the moles of ammonia to grams of ammonia, etc.

$$3.50 \text{ mton } (NH_4)_2SO_4 \times \frac{1000 \text{ kg}}{\text{mton}} \times \frac{10^3 \text{ g}}{\text{kg}} \times \frac{1 \text{ mol } (NH_4)_2SO_4}{132.2 \text{ g } (NH_4)_2SO_4}$$

$$\times \frac{2 \text{ mol } NH_3}{1 \text{ mol } (NH_4)_2SO_4} \times \frac{17.03 \text{ g } NH_3}{\text{mol}} \times \frac{10^{-3} \text{ kg}}{\text{g}} = 902 \text{ kg } NH_3$$

4.14 Follow procedure similar to that used to solve Problem 4.13.

$$1.00 \text{ kg } C_6H_{12}O_6 \times \frac{10^3 \text{ g}}{\text{kg}} \times \frac{1 \text{ mol } C_6H_{12}O_6}{180.2 \text{ g } C_6H_{12}O_6} \times \frac{2 \text{ mol } C_2H_5OH}{1 \text{ mol } C_6H_{12}O_6}$$

$$\times \frac{46.07 \text{ g } C_2H_5OH}{\text{mol } C_2H_5OH} = 511 \text{ g } C_2H_5OH$$

4.15 Convert mass to moles, moles to moles, and back to mass.

$C_2H_5OH + 3\ O_2 \rightarrow 2\ CO_2 + 3\ H_2O$

$$\frac{5.75\ g\ C_2H_5OH}{46.07\ g/mol} x \frac{2\ mol\ CO_2}{mol\ C_2H_5OH} x \frac{44.01\ g\ CO_2}{mol\ CO_2} = 11.0\ g\ CO_2$$

$$\frac{5.75\ g\ C_2H_5OH}{46.07\ g/mol} x \frac{3\ mol\ H_2O}{mol\ C_2H_5OH} x \frac{18.02\ g\ H_2O}{mol\ H_2O} = 6.75\ g\ H_2O$$

4.16 $4\ C_5H_5N + 25\ O_2 \rightarrow 10\ H_2O + 20\ CO_2 + 2\ N_2$

$$2.95\ mL\ C_5H_5N\ x \frac{0.982\ g}{mL} x \frac{1\ mol}{79.1\ g} x \frac{10\ mol\ H_2O}{4\ mol\ C_5H_5N} = 0.0916\ mol\ H_2O$$

4.17 Assuming the fluoroapatite is the limiting reagent: $Ca_5(PO_4)_3F \rightarrow 3\ H_3PO_4$

$$1.00\ kg\ Ca_3(PO_4)_3F\ x \frac{10^3\ g}{kg} x \frac{1\ mol\ Ca_3(PO_4)_3F}{504.3\ g} x \frac{3\ mol\ H_3PO_4}{1\ mol}$$

$$x \frac{97.99\ g\ H_3PO_4}{mol} = \text{theoretical yield of } H_3PO_4 = 583\ g$$

= percent yield = (400 / 583) x 100 = 68.6%

4.18

Reaction:	2 HgO	$\rightarrow$	2 Hg	+	O_2
Start (mol)	0.369		0		
Change	-0.369		+0.369		+0.185
Final	0g		74.0 g		5.92 g

Percent yield = (60/74) x 100 = 81%

4.19 $C_7H_8 \rightarrow (1) \rightarrow (2) \rightarrow (3) \rightarrow (4) \rightarrow (5) \rightarrow (6) \rightarrow (7) \rightarrow C_{12}H_{12}N_2O_3$

Finish 107.6 mol

Working backwards one step at a time: if 25 kg of $C_{12}H_{12}N_2O_3$ is 107.6 moles, then x moles or 107.9 divided by 90% of the intermediate in the last step must have been available. Repeating this backward approach yields: 250.1 mol $\rightarrow$ 225.1 $\rightarrow$ 202.6 $\rightarrow$ 182.3 $\rightarrow$ 164.1 $\rightarrow$ 147.7 $\rightarrow$ 132.9 $\rightarrow$ 119.6 mol

Check: 250.1 mol x $(.90)^8$ = 107.7

Answer: 250.1 mol C_7H_8 x (92.13 g/mol) x (1 kg/10^3 g) = 23 kg

4.20 a)

Reaction:	CCl_4	+	2 HF	$\rightarrow$	CCl_2F_2	+	2 HCl
MM	153.8		20.01		120.9		36.46
Start	1,138 mol						
Change	-1,138 mol				+1,138 mol		2,236 mol
Theoretical yield					137.6 kg		

Percent yield = (105 kg/137.6 kg) x 100 = 76.3%

b) If 175 kg of CCl_4 yielded 105 kg of CCl_2F_2, then (155/105) x 175 or 258 kg of CCl_4 will yield 155 kg of CCl_2F_2.
258 kg of CCl_4 will react with 67.4 kg of HF.

4.21

Reaction:	$C_6H_{12}O_6$	→	2 C_2H_5OH
MM	180.2		46.07
Th. Yield			4.58 kg
Finish			99.4 mol
Start	49.7 mole		
Start	8.92 kg		

4.22

	roll +	2 x 1/4 lb patties +	cheese +	1/4 tomato +	15 g lettuce →	cheeseburger
Start	144	40 lb	130	40	1000 g	0
Change	-66.7	-33.3	-66.7	-16.7	-1000 g	66.7
or	-66	-33	-66	-16.5	-990 g	66
Final	78	7 lb	64 slices	23.5 lb	10 g	66

The lettuce runs out first. Only 66 complete cheeseburgers can be made.

4.23

	#8 [1/2"]	#8 [1"]	#10 [1/2"]	#10 [11/2"]	#12 [2"]
# per lb	38	21	31	15	10
# per assort	10	8	8	6	4
2 assort's	20	16	16	12	8

One could prepare 2 “handyman assortments” per pound of each type of screw.

4.24

a)

Reaction:	5 H_2 +	2 NO	→	2 NH_3 +	2 H_2O
Start	3.72 mol	0.250 mol		0 mol	0 mol
	(Limiting reagent = NO)				
Change	-0.625 mol	-0.250 mol		+0.250 mol	0.250 mol
Final	6.23 g	0 g		4.26 g	4.51 g

b)

Reaction:	2 CO +	2 NO	→	N_2 +	2 CO_2
Start	0.268 mol	0.250 mol		0 mol	0 mol
	(Limiting reagent = NO)				
Change	-0.250 mol	-0.250 mol		+0.125 mol	0.250 mol
Final	0.50 g	0 g		3.50 g	11.00 g

c)

Reaction:	2 NH_3 +	2 O_2	→	N_2O +	3 H_2O
Start	0.440 mol	0.234 mol		0 mol	0 mol
	(Limiting reagent = O_2)				
Change	-0.234 mol	-0.234 mol		+0.117 mol	0.351 mol
Final	3.51 g	0 g		5.15 g	6.33 g

d)

Reaction:	6 NO +	4 NH_3	→	5 N_2 +	6 H_2O
Start	0.250 mol	0.440 mol		0 mol	0 mol
	(Limiting reagent = NO)				
Change	-0.250 mol	-0.167 mol		+0.208mol	0.250 mol
Final	0 g	4.65 g		5.83 g	4.51 g

4.25

a)

Reaction:	N_2 +	3 H_2	→	2 NH_3
Start (10^3)	26.4 mol	367		0
Change	-26.4	-26.4 x 3		+26.4 x 2
Finish (10^3)	0	287.8 mol		52.8 mol
Mass	0	580 kg		899 kg

(continued)

(4.25 continued)

b)

Reaction:	$2\,H_2$ +	CO →	CH_3OH
Start (10^3)	367	26.4	0
Change	-52.8	-26.4	+26.4
Finish (10^3)	314 mol	0	26.42 mol
Mass	633 kg	0	847 kg

c)

Reaction:	CaO +	3 C →	CO +	CaC_2
Start (10^3)	13.2 mol	61.6 mol	0	0
Change	-13.2	-39.6 mol	+13.2 mol	+13.2
Finish (10^3)	0	22.0 mol	13.2 mol	13.2
Mass	0	264 kg	370 kg	846 kg

d)

Reaction:	$2\,C_2H_4$ +	O_2 +	4 HCl →	$2\,C_2H_4Cl_2$ +	$2\,H_2O$
Start (10^3)	26.4 mol	23.1 mol	20.3 mol	0	0
Change	-10.15	-5.08	-20.3	+10.15	+10.15
Finish (10^3)	16.25 mol	18.02 mol	0	10.15 mol	10.15
Mass	456 kg	577 kg	0	1.01×10^3 kg	184 kg

4.26

Reaction:	SiO_2 +	2 C →	SiC +	CO_2
Start	0.832 mol	4.16 mol	0	0
Change	-0.832	-1.66 mol	+0.832 mol	+0.832
Finish	0	2.50 mol	0.832 mol	0.832
Mass		30.0 g	33.4 g	36.6 g

4.27

Reaction:	P_4 +	$5\,O_2$ →	P_4O_{10}
Start	0.0303 mol	0.205 mol	0
Change	-0.0303 mol	-0.151 mol	+0.0303 mol
Finish	0	0.054 mol	0.0303 mol
Mass	0	1.7 g	8.60 g

4.28

Reaction:	$2\,C_3H_6$ +	$2\,NH_3$ +	$3\,O_2$ →	$2\,C_3H_3N$ +	$6\,H_2O$
Start (10^3)	35.6 mol	39.9	60.0	0	0
Change	-35.6 mol	-35.6 mol	-53.5 mol	+35.6 mol	107 mol
Finish (10^3)	0	4.3 mol	6.5 mol	35.6 mol	107 mol
Mass	0	73 kg	208 kg	1.89×10^3 kg	1.93×10^3 kg

4.29

Reaction:	SiO_2 +	2 C +	$2\,Cl_2$ →	$SiCl_4$ +	2 CO
Start	1.25 mol	6.24 mol	1.06	0	0
Change	-0.53 mol	-1.06 mol	-1.06 mol	+0.53 mol	+1.06 mol
Finish	0.72	5.18 mol	0 mol	0.53 mol	1.06 mol
Mass			0	90.0 g	29.7 g

Yield: 90.0 g of $SiCl_4$ x 95.7% = 86.1 g of $SiCl_4$

4.30

Reaction:	$B_{10}H_{18}$	+	$12\ O_2$	→	$5\ B_2O_3$	+	$9\ H_2O$
Start	0.792 mol		8.59 mol		0		0
Change	-0.716 mol		-8.59 mol		3.58 mol		6.44
Finish	0.076 mol		0 mol		3.58 mol		6.44
	<u>9.6 g</u>		0		<u>249 g</u>		<u>116 g</u>

4.31 Follow the solubility guidelines given in Section 4.5.

a) $H_2O_{(l)}$, $NH_4^+{}_{(aq)}$ and $Cl^-{}_{(aq)}$
b) $H_2O_{(l)}$, $Fe^{2+}{}_{(aq)}$ and $ClO_4^-{}_{(aq)}$
c) $H_2O_{(l)}$, $Na^+{}_{(aq)}$ and $SO_4^{2-}{}_{(aq)}$
d) $H_2O_{(l)}$, $K^+{}_{(aq)}$ and $Br^-{}_{(aq)}$

4.32 Follow the solubility guidelines given in Section 4.5.

a) $K^+{}_{(aq)}$, $HPO_4^{2-}{}_{(aq)}$ and $H_2O_{(l)}$
b) $CH_3CO_2H_{(aq)}$ and $H_2O_{(l)}$
c) $Na^+{}_{(aq)}$, $ClO^-{}_{(aq)}$ and $H_2O_{(l)}$
d) $NH_{3(aq)}$ and $H_2O_{(l)}$
e) $Mg^{2+}{}_{(aq)}$, $Cl^-{}_{(aq)}$ and $H_2O_{(l)}$

4.33 a) $AgNO_3$ will form a precipitate when mixed with NH_4Cl, Na_2SO_4 or KBr solutions. The precipitates will be AgCl, Ag_2SO_4 and AgBr, respectively.
b) Na_2CO_3 would be expected to form a precipitate when mixed with $Fe(ClO_4)_2$. The precipitate would be $FeCO_3$.
c) $Ba(OH)_2$ would be expected to form precipitates when mixed with $Fe(ClO_4)_2$ and Na_2SO_4 solutions. The precipitates would be $Fe(OH)_2$ and $BaSO_4$ respectively.

4.34 a) $2\ Ag^+{}_{(aq)} + SO_4^{2-}{}_{(aq)} \rightarrow Ag_2SO_{4(s)}$
b) $2\ Fe^{3+}{}_{(aq)} + 3\ C_2O_4^{2-}{}_{(aq)} \rightarrow Fe_2(C_2O_4)_{3(s)}$
c) $Pb^{2+}{}_{(aq)} + 2\ Br^-{}_{(aq)} \rightarrow PbBr_{2(s)}$

4.35 a) $Al^{3+}{}_{(aq)} + 3\ OH^-{}_{(aq)} \rightarrow Al(OH)_{3(s)}$
b) $3Mg^{2+}{}_{(aq)} + 2\ PO_4^{3-}{}_{(aq)} \rightarrow Mg_3(PO_4)_{2(s)}$
c) $Ba^{2+}{}_{(aq)} + SO_4^{2-}{}_{(aq)} \rightarrow BaSO_{4(s)}$

4.36

Reaction:	$2\ Ag^+$	+	CO_3^{2-}	→	$Ag_2CO_{3(s)}$
Start	2.75×10^{-3} mol		3.33×10^{-3} mol		0
Change	-2.75×10^{-3} mol		-1.38×10^{-3} mol		$+1.38 \times 10^{-3}$
Finish	0		1.95×10^{-3} mol		1.38×10^{-3}
Mass					0.381 g

0.381 g of Ag_2CO_3 will form. Remaining in solution will be K^+, NO_3^- and the excess CO_3^{2-}.

4.37

Reaction:	Pb^{2+}	+	$2\ Cl^-$	→	$PbCl_{2(s)}$
Start	5.625×10^{-2} mol		0.1069 mol		0
Change	-5.344×10^{-2} mol		-0.1069 mol		$+5.344 \times 10^{-2}$
Finish	2.81×10^{-3} mol		0		5.344×10^{-2}
Mass					14.9 g

14.9 g of $PbCl_2$ will form. Remaining in solution will be ammonium ions, nitrate ions and excess lead(II) ions.

4.38 For this problem the molecular pictures need to show reactants and products in the indicated proportions. See Chapter 2 for a discussion of notations used. In this problem lead nitrate will exist as lead(II) ions and nitrate ions in the ratio of 1 to 2 and ammonium chloride will exist as ammonium ions and chloride ions in the ratio of 1 to 1 in the reactants. There should be slightly more than half as many of the lead ions as there are chloride ions. For the products show solid lead chloride and soluble ammonium and nitrate ions plus the slight excess of soluble lead ions. The same number of each type of atom must be shown in the products and unused reactants at the end of the reaction as was present in the reactants at the beginning of the reaction.

4.39 $2\,H_3O^+ + 2\,NO_3^- + Ba^{2+} + 2\,OH^- \rightarrow 4\,H_2O + Ba^{2+} + 2\,NO_3^-$

Net equation: $H_3O^+ + OH^- \rightarrow 2\,H_2O$

5×10^{-4} mole + 5×10^{-4} mol →

For the molecular pictures show reactants and products in the indicated proportions. See Chapter 2 for a discussion of notations used. Hydrogen (or hydronium) ions will react with hydroxide ions in the ratio of 1 to 1. In the reaction described in this problem the hydrogen ion and hydroxide ion are being mixed in equal numbers of moles; therefore, they both will be all reacted in going from the reactants to the products side of your molecular picture. For your molecular picture show an equal number of hydrogen (hydronium), nitrate, and hydroxide ions and half that number of barium ions in the reacting solution. For the products show soluble barium and nitrate ions and the water molecules formed. The same number of each type of atom must be shown in the products as was present in the reactants.

4.40 See Section 4.6 for a discussion of strong acids, weak acids, etc.

a) H_2CO_3 weak acid
b) CH_4 none of these
c) LiOH strong base
d) NH_3 weak base
e) C_2H_5OH none of these

4.41 Total Ionic Reaction: $Ba^{2+} + 2\,OH^- + 2\,H_3O^+ + 2\,Cl^- \rightarrow 4\,H_2O + Ba^{2+} + 2\,Cl^-$

Net Ionic:	H_3O^+	+ OH^-	$\rightarrow$ $2\,H_2O$
Start	5.00×10^{-3} mol	6.00×10^{-3} mol	
Change	-5.00×10^{-3} mol	-5.00×10^{-3} mol	
Finish	0	1.00×10^{-3} mol	

Final Volume 2.50×10^2 mL

Molarities: $H_3O^+ = 0$

$$OH^- = \frac{1.00 \times 10^{-3}\ \text{mol}}{0.250\ \text{L}} = 4.00 \times 10^{-3}\ \text{M}$$

$$Ba^{2+} = \frac{(1.50 \times 10^2\ \text{mL})(2.00 \times 10^{-2}\ \text{M})}{(2.50 \times 10^2\ \text{mL})} = 1.20 \times 10^{-2}\ \text{M}$$

$$Cl^- = \frac{(1.00 \times 10^2\ \text{mL})(5.00 \times 10^{-2}\ \text{M})}{(2.50 \times 10^2\ \text{mL})} = 2.00 \times 10^{-2}\ \text{M}$$

Note: Ba^{2+} and Cl^- are spectator ions and change concentration only by dilution.

4.42 Total Ionic Reaction: $CH_3CO_2H + Na^+ + OH^- \rightarrow CH_3CO_2^- + Na^+ + H_2O$

Net Ionic:	CH_3CO_2H	+	OH^-	$\rightarrow$	$CH_3CO_2^-$	+	H_2O
Start	3.225×10^{-2} mol		3.825×10^{-2} mol				
Change	-3.225×10^{-2} mol		-3.225×10^{-2} mol		$+ 3.225 \times 10^{-2}$ mol		
Finish	0		0.600×10^{-2} mol		3.225×10^{-2} mol		

Final Volume: 215 mL; assuming no volume change.

Molarities: $$CH_3CO_2^- = \frac{3.225 \times 10^{-2} \text{ mol}}{0.215 \text{ L}} = 0.150 \text{ M}$$

$$OH^- = \frac{0.600 \times 10^{-2} \text{ mol}}{0.215 \text{ L}} = 0.028 \text{ M}$$

$$Na^+ = \frac{3.825 \times 10^{-2} \text{ mol}}{0.215 \text{ L}} = 0.178 \text{ M}$$

Note: Na^+ is a spectator ion and does not change concentration.

4.43 $$\text{mol KHP} = \frac{0.7996 \text{ g}}{204.2 \text{ g / mol}} = 3.916 \times 10^{-3} \text{ mol}$$

$\therefore$ mol of NaOH added in titration = 3.916×10^{-3} mol

volume of NaOH added = 43.75 mL - 0.15 mL = 43.60 mL (or 43.60×10^{-3} L)

$$\text{Molarity of NaOH solution} = \frac{3.916 \times 10^{-3} \text{ mol}}{43.60 \times 10^{-3} \text{ L}} = 0.08982 \text{ M}$$

4.44 See Section 4.7

a) no reaction b) no reaction

c) $Cu_{(s)} + 2\,Ag^+_{(aq)} \rightarrow 2\,Ag_{(s)} + Cu^{2+}_{(aq)}$

d) $2\,K_{(s)} + 2\,H_2O_{(l)} \rightarrow H_{2(g)} + 2\,K^+_{(aq)} + 2\,OH^-_{(aq)}$

4.45 a) $2\,Al_{(s)} + 6\,H_3O^+_{(aq)} \rightarrow 3\,H_{2(g)} + 2\,Al^{3+}_{(aq)} + 6\,H_2O_{(l)}$

b) $3\,Zn_{(s)} + 2\,Au^{3+}_{(aq)} \rightarrow 2\,Au_{(s)} + 3\,Zn^{2+}_{(aq)}$

c) no reaction

d) $2\,Na_{(s)} + 2\,H_2O_{(l)} \rightarrow 2\,Na^+_{(aq)} + H_{2(g)} + 2\,OH^-_{(aq)}$

4.46 (From Problem 4.44)

a) no reaction b) no reaction

c) oxid.: $Cu_{(s)} \rightarrow Cu^{2+}_{(aq)} + 2\,e^-$ red.: $Ag^+_{(aq)} + e^- \rightarrow Ag_{(s)}$

d) oxid.: $K_{(s)} \rightarrow K^+_{(aq)} + e^-$ red.: $2\,H_2O_{(l)} + 2\,e^- \rightarrow H_{2(g)} + 2\,OH^-_{(aq)}$

(From Problem 4.45)

a) oxid.: $Al_{(s)} \rightarrow Al^{3+}_{(aq)} + 3\,e^-$ red.: $2\,H_3O^+_{(aq)} + 2\,e^- \rightarrow H_{2(g)} + 2\,H_2O$

b) oxid.: $Zn_{(s)} \rightarrow Zn^{2+}_{(aq)} + 2\,e^-$ red.: $Au^{3+}_{(aq)} + 3\,e^- \rightarrow Au_{(s)}$

c) no reaction

d) oxid.: $Na_{(s)} \rightarrow Na^+_{(aq)} + e^-$ red.: $2\,H_2O_{(l)} + 2\,e^- \rightarrow H_{2(g)} + 2\,OH^-_{(aq)}$

4.47 a) $2\,Sr_{(s)} + O_{2(g)} \rightarrow 2\,SrO_{(s)}$ b) $4\,Cr_{(s)} + 3\,O_{2(g)} \rightarrow 2\,Cr_2O_{3(s)}$

c) $Sn_{(s)} + O_{2(g)} \rightarrow SnO_{2(s)}$

4.48 $2\,Al + 6\,H^+ \rightarrow 2\,Al^{3+} + 3\,H_2$

$$\frac{0.355\text{ g Al}}{26.98\text{ g/mol}} \times \frac{3\text{ mol }H_2}{2\text{ mol Al}} \times \frac{2.01\text{ g }H_2}{\text{mol }H_2} = 0.0397\text{ g }H_2$$

or

$$\frac{6.00\text{ mole HCl}}{\text{L}} \times 8.00 \times 10^{-3}\text{ L} \times \frac{1\text{ mol }H_2}{2\text{ mol HCl}} \times \frac{2.01\text{ g }H_2}{\text{mol }H_2} = 0.0482\text{ g }H_2$$

Al is the limiting reagent and 0.0397g is the maximum amount of H_2 possible.

4.49 Counting atoms and molecules: $16\,X + 6\,Y \rightarrow 4\,X + 6\,YX_2$

Canceling unreacted atoms: $12\,X + 6\,Y \rightarrow 6\,YX_2$

Reducing to smallest set of whole-number coefficients:

$2\,X + Y \rightarrow YX_2$

Answer (d) is the best answer.

4.50 Counting atoms and molecules: $6\,X + 4\,Y_2 \rightarrow 2\,Y + 6\,XY$

Reducing to smallest set of whole-number coefficients:

$3\,X + 2\,Y_2 \rightarrow Y + 3\,XY$

Answer (c) is the best answer by default.

4.51

Reaction:	$Ca_3(PO_4)_2$	+	$2\,H_2SO_4$	$\rightarrow$	$Ca(H_2PO_4)_2$	+	$2\,CaSO_4$
Start	0.822 mol		1.539 mol		0		0
Change	-0.770 mol		-1.539 mol		+0.770 mol		+1.539 mol
Finish	0.052 mol		0		+0.770 mol		+1.539 mol

Grams of $Ca(H_2PO_4)_2$ = 0.770 mol x 234 g/mol = 180 g

Grams of $CaSO_4$ = 210 g

Grams of superphosphate = 210 + 180 = 390 g

4.52 a) $HCO_3^-{}_{(aq)} + Na^+{}_{(aq)} \rightarrow NaHCO_{3(s)}$ b) Spectator ions are: NH_4^+ and Cl^-

c)

Reaction:	NH_4HCO_3	+	NaCl	$\rightarrow$	$NaHCO_3$	+	NH_4Cl
Start	7.50×10^2 mol		30×10^2 mol		0		0
Change	-7.50×10^2 mol		-7.50×10^2 mol		$+7.50 \times 10^2$ mol		$+7.50 \times 10^2$ mol
Finish	0		22.5×10^2 mol		7.50×10^2 mol		7.50×10^2 mol

Theoretical yield of $NaHCO_3$ = 7.50×10^2 mol x 84.0 g/mol = 63.0 kg

Percent yield = (61.7 kg/63.0 kg) x 100 = 97.9%

4.53 $C_2H_5OH + 3\,O_2 \rightarrow 2\,CO_2 + 3\,H_2O$

a) $$\frac{4.6\text{ g }C_2H_5OH}{46.07\text{ g/mol}} \times \frac{3\text{ mole }H_2O}{1\text{ mol }C_2H_5OH} = 0.30\text{ mol }H_2O$$

b) 0.30 mole H_2O x 6.022×10^{23} molecules/mole = 1.8×10^{23} molecules H_2O

c) 0.30 mol H_2O x 18.02 g/mol = 5.4 g H_2O

4.54 $P_4O_{10} + 6\,H_2O \rightarrow 4\,H_3PO_4$

$$\frac{1.00 \times 10^2 \text{ g } P_4O_{10}}{283.9 \text{ g / mol}} \times \frac{4 \text{ mol } H_3PO_4}{\text{mol } P_4O_{10}} \times \frac{98.0 \text{ g } H_3PO_4}{\text{mol } H_3PO_4} = 1.38 \times 10^2 \text{ g}$$

4.55 a) The reaction is: $N_2 + 3\,H_2 \rightarrow 2\,NH_3$

Therefore, if the number of molecules shown is used there will be unreacted N_2 and H_2. The reaction becomes $6\,N_2 + 16\,H_2 \rightarrow 10\,NH_3 + N_2 + H_2$

Your drawing should show 10 ammonia molecules, one nitrogen molecule, and one hydrogen.

b) Hydrogen (H_2) is the limiting reagent.

c)	N_2	+	$3\,H_2 \rightarrow$	$2\,NH_3$
Start:	6 mol		16 mol	
Change:	-5.333 mol		-16 mol	+10.667 mol
Finish:	0.667 mol		0	10.667 mol

Mass of NH_3 produced = (10.667 mol)(17.034 g/mol) = 181.7 g NH_3

Mass of excess N_2 = (0.667 mol)(28.02 g/mol) = 18.7 g N_2

4.56 $Xe + 2F_2 \rightarrow XeF_4$

$$\frac{5.00 \text{ g Xe}}{131.3 \text{ g / mol}} \times \frac{1 \text{ mol } XeF_4}{1 \text{ mol Xe}} \times \frac{207.3 \text{ g } XeF_4}{\text{mol}} = 7.89 \text{ g } XeF_4$$

Percent yield = (4.00/7.89) x 100 = 50.7%

4.57 Reaction:	$B_{10}H_{18}$	+	$12\,O_2$	$\rightarrow$
Start	3.96×10^5 mol		2.03×10^6 mol	
Change	-1.69×10^5 mol		-2.03×10^6	
Finish	2.27×10^5 mol		0 mol	

O_2 will empty first. There will be 2.27×10^5 mol of $B_{10}H_{18}$ (or 2.87×10^5 kg) remaining after all the oxygen has been reacted.

4.58 Reaction: $B_{10}H_{18} + 12\,O_2 \rightarrow$

(Let x equal the grams of $B_{10}H_{18}$, then 12.0×10^4 - x will be the grams of O_2.)

Start $\quad \dfrac{x \text{ g}}{126 \text{ g / mol}} \quad \dfrac{(12.0 \times 10^4) - x \text{ g}}{32}$

$$\frac{x \text{ g}}{126 \text{ g / mol}} \times 12 = \frac{(12.0 \times 10^4) - x \text{ g}}{32}$$

Answer: 2.97×10^4 kg $B_{10}H_{18}$ and 9.04×10^4 kg O_2

4.59 1 mton earth x 95% eff x 15% ilmenite x 1,000 kg/mton = 142.5 kg ilmenite

Reaction: $FeTiO_3 \rightarrow Ti$

Start 142.5 kg

$$\frac{142.5 \text{ kg } FeTiO_3}{151.7 \text{ g / mol}} \times \frac{10^3 \text{ g}}{\text{kg}} \times \frac{1 \text{ mol Ti}}{1 \text{ mol } FeTiO_3} \times \frac{47.88 \text{ g Ti}}{\text{mol Ti}} \times \frac{10^{-3} \text{ kg}}{\text{g}} = 45.0 \text{ kg Ti}$$

4.60 $\frac{750 \text{ kg } NH_3}{17.0 \text{ g / mol}} \times \frac{10^3 \text{ g}}{\text{kg}} \times \frac{4 \text{ mol NO}}{4 \text{ mol } NH_3} \times 0.945 \text{ eff} = 4.17 \times 10^4 \text{ mol NO}$

4.17×10^4 mol NO x (2 mol NO_2/2 mol NO) x 0.945 eff = 3.94×10^4 mol NO_2

3.94×10^4 mol NO_2 x (2 mol HNO_3/3 mol NO_2) x 0.945 eff

x 63.0 g HNO_3/mole x 10^{-3} kg/g = 1.56×10^3 kg HNO_3

4.61 $CH_{4(g)} + 2\ H_2O_{(g)} \rightarrow CO_{2(g)} + 4\ H_{2(g)}$

4.62

Reaction:	8 Al	+	3 Fe_3O_4	→	9 Fe	+	4 Al_2O_3
Start	7.41 mol		3.02 mol		0		0
Change	-7.41 mol		-2.78 mol		+8.34 mol		+3.71 mol
Finish	0		0.24 mol		8.34 mol		3.71 mol
Masses at finish	0		56 g		466g		378 g

4.63

Reaction:	As_2O_3	+	3 C	→	2 As	+	3 CO
Start	0.250 mol		0.600 mol				
Change	-0.050 mol		-0.600		+0.400 mol		+0.600 mol
Finish	0.050 mol		0		0.400 mol		0.600 mol
Masses at finish	9.9 g		0 g		30.0 g		16.8 g

4.64

Reaction:	N_2	+	3 H_2	→	2 NH_3
Start (10^3)	3.0 mol		11.9 mol		
Change(10^3)	-3.0 mol		-9.0 mol		+6.0 mol
Th. yield					1.02×10^2 kg

Percent yield = (68/102) x 100 = 67%

(Repeat the calculation at 67% yield.)

Reaction:	N_2	+	3 H_2	→	2 NH_3
Start (10^3)	3.0 mol		11.9 mol		0
Finish (10^3)	?		?		4.0 mol
Change (10^3)	-2.0 mol		-6.0 mol		+4.0 mol
Finish (10^3)	1.0 mol		5.9 mol		4.0 mol
Masses at finish	28 kg		12 kg		68 kg

4.65 $Ca(NO_3)_{2(aq)} + (NH_4)_2SO_{4(aq)} \rightarrow CaSO_{4(s)} + 2\ NH_4NO_3$

precipitate = $CaSO_4$

net ionic equation = $Ca^{2+}{}_{(aq)} + SO_4{}^{2-}{}_{(aq)} \rightarrow CaSO_{4(s)}$

(continued)

(4.65 continued)

Reaction:	$Ca(NO_3)_2$	+	$(NH_4)_2SO_4$	$\rightarrow$	$CaSO_4$	+	$2\ NH_4NO_3$
Start	0.150 mol		0.225 mol		0		0
Change	-0.150 mol		-0.150 mol		+0.150 mol		+0.300 mol
Finish	0		0.075 mol		0.150 mol		0.300 mol

These are in 175 mL of solution.

$$NH_4^+ \text{ as a spectator ion: } \frac{(75.0\ \text{mL})(3.00\ \text{mol / L})\left(\dfrac{2NH_4^+}{1\ \text{mol}\ (NH_4)SO_3}\right)}{175\ \text{mL}} = 2.57\ \text{M}$$

SO_4^{2-} = 0.075 mol/0.175 L = 0.43 M

$$NO_3^- \text{ as spectator ion: } \frac{(100.0\ \text{mL})(1.50\ \text{M})2\ \text{mol}\ NO_3^-/1\ \text{mol}\ Ca(NO_3)_2}{175\ \text{mL}} = 1.71\ \text{M}$$

$CaSO_4$ = 0.150 mol x 136.14 g/mol = 20.4 g

4.66 a) $Hg^{2+}{}_{(aq)} + S^{2-}{}_{(aq)} \rightarrow HgS_{(s)}$

b) $Ba^{2+}{}_{(aq)} + SO_4^{2-}{}_{(aq)} \rightarrow BaSO_{4(s)}$

c) $OH^-{}_{(aq)} + CH_3CO_2H_{(aq)} \rightarrow H_2O_{(l)} + CH_3CO_2^-{}_{(aq)}$

d) $3\ OH^-{}_{(aq)} + Fe^{3+}{}_{(aq)} \rightarrow Fe(OH)_{3(s)}$

e) $Ba^{2+}{}_{(aq)} + 2\ BrO_3^-{}_{(aq)} \rightarrow Ba(BrO_3)_{2(s)}$

4.67

Reaction:	CH_4O	+	C_4H_8	$\rightarrow$	$C_5H_{12}O$
Sold (10^9)					3.4×10^2 g
Th. yield (10^9)					3.96×10^2 g
Moles (10^9)					4.49 mol
Reagents (10^9)	4.49 mol	+	4.49 mol	$\rightarrow$	4.49 mol
a) Mass			2.52×10^{11} g		

b) $$\frac{2.52 \times 10^{11}\ \text{g}\ C_4H_8}{56.1\ \text{g / mol}} \times \frac{1\ \text{mol}\ C_5H_{12}O}{1\ \text{mol}\ C_4H_8} \times 0.93\ \text{eff} \times \frac{88.1\ \text{g}}{\text{mol}}$$

$$\times \frac{1\ \text{lb}}{454\ \text{g}} \times \frac{10^{-6}\ \text{million pounds}}{\text{lb}} = 811\ \text{million pounds}$$

(Note: could also use a ratio to solve this problem.)

c) $$(811 - 750)\ \text{million lb} \times \frac{10^6\ \text{pounds}}{\text{million lb}} \times \$0.10 / \text{lb} \times \frac{10^{-6}\ \text{million dollars}}{\text{dollar}}$$

= 6.1 million dollars

Note: This assumes that MTBE was sold at cost (no profit) the first year.

4.68 $C_4H_{10}O_2 + C_3H_6 \rightarrow C_3H_6O + C_4H_{10}O$

a) $$\frac{75\ kg\ C_4H_{10}O_2}{90.1\ g/mol} x \frac{10^3\ g}{kg} x \frac{1\ mol\ C_3H_6O}{1\ mol\ C_4H_{10}O_2} x \frac{58.1\ g}{mole} x \frac{10^{-3}\ kg}{g} = 48\ kg\ C_3H_6O$$

b) $$\frac{75\ kg\ C_4H_{10}O_2}{90.1\ g/mol} x \frac{10^3\ g}{kg} x \frac{1\ mol\ C_3H_6}{1\ mol\ C_4H_{10}O_2} x \frac{42.1\ g}{mole} x \frac{10^{-3}\ kg}{g} = 35\ kg\ C_3H_6$$

4.69 a) $Na^+_{(aq)}$, $Cl^-_{(aq)}$, $H_3O^+_{(aq)}$ and $NO_3^-{}_{(aq)}$, no reaction

b) $Ca^{2+}_{(aq)}$, $Cl^-_{(aq)}$, $Na^+_{(aq)}$ and $SO_4^{2-}{}_{(aq)}$

$Ca^{2+}_{(aq)} + SO_4^{2-}{}_{(aq)} \rightarrow CaSO_{4(s)}$

c) $K^+_{(aq)}$, $OH^-_{(aq)}$, $H_3O^+_{(aq)}$ and $Cl^-_{(aq)}$

$H_3O^+_{(aq)} + OH^-_{(aq)} \rightarrow 2\ H_2O_{(l)}$

d) $NH_{3(aq)}$, $H_3O^+_{(aq)}$ and $Cl^-_{(aq)}$

$NH_{3(aq)} + H_3O^+_{(aq)} \rightarrow NH_4^+{}_{(aq)} + H_2O_{(l)}$

4.70 Following the notations given in Chapter 2, the answer to this problem would involve a picture that shows widely separated calcium ions and nitrate ions in the ratio of 1 to 2.

4.71 a) $Fe^{3+}_{(aq)} + PO_4^{3-}{}_{(aq)} \rightarrow FePO_{4(s)}$

$Fe^{3+} + 3\ Cl^- + 3\ Na^+ + PO_4^{3-} \rightarrow FePO_{4(s)} + 3\ Na^+ + 3\ Cl^-$

$$2.50\ kg\ FePO_4\ x \frac{1000\ g}{1\ kg} x \frac{1\ mol\ FePO_4}{150.77\ g\ FePO_4} = 16.6\ mol\ FePO_4$$

$$16.6\ mol\ FePO_4\ x \frac{1\ mol\ FeCl_3}{1\ mol\ FePO_4} x \frac{162.2\ g\ FeCl_3}{1\ mol\ FeCl_3} = 2.69 \times 10^3\ g\ FeCl_3$$

$= 2.69\ kg\ FeCl_3$

$$16.6\ mol\ FePO_4\ x \frac{1\ mol\ Na_3PO_4}{1\ mol\ FePO_4} x \frac{163.94\ g\ Na_3PO_4}{1\ mol\ Na_3PO_4} = 2.72 \times 10^3\ g\ Na_3PO_4$$

$= 2.72\ kg\ Na_3PO_4$

b) $Zn^{2+}_{(aq)} + 2\ OH^-_{(aq)} \rightarrow Zn(OH)_{2(s)}$

$Zn^{2+} + 2\ NO_3^- + 2\ Na^+ + 2\ OH^- \rightarrow Zn(OH)_{2(s)} + 2\ Na^+ + 2\ NO_3^-$

$$2.50\ kg\ Zn(OH)_2\ x \frac{1000\ g}{1\ kg} x \frac{1\ mol\ Zn(OH)_2}{99.41\ g\ Zn(OH)_2} = 25.1\ mol\ Zn(OH)_2$$

$$25.1\ mol\ Zn(OH)_2\ x \frac{1\ mol\ Zn(NO_3)_2}{1\ mol\ Zn(OH)_2} x \frac{189.41\ g\ Zn(NO_3)_2}{1\ mol\ Zn(NO_3)_2}$$

$= 4.75 \times 10^3\ g\ Zn(NO_3)_2 = 4.75\ kg\ Zn(NO_3)_2$

$$25.1\ mol\ Zn(OH)_2\ x \frac{2\ mol\ NaOH}{1\ mol\ Zn(OH)_2} x \frac{40.00\ g\ NaOH}{1\ mol\ NaOH} = 2.01 \times 10^3\ g\ NaOH$$

$= 2.01\ kg\ NaOH$

(continued)

(4.71 continued)

c) $Ni^{2+}{}_{(aq)} + CO_3{}^{2-}{}_{(aq)} \rightarrow NiCO_{3(s)}$

$Ni^{2+}{}_{(aq)} + 2\,Cl^-{}_{(aq)} + 2\,Na^+{}_{(aq)} + CO_3{}^{2-}{}_{(aq)} \rightarrow NiCO_{3(s)} + 2\,Na^+{}_{(aq)} + 2\,Cl^-{}_{(aq)}$

$$2.50 \text{ kg } NiCO_3 \times \frac{1000 \text{ g}}{1 \text{ kg}} \times \frac{1 \text{ mol } NiCO_3}{118.70 \text{ g } NiCO_3} = 21.1 \text{ mol } NiCO_3$$

$$21.1 \text{ mol } NiCO_3 \times \frac{1 \text{ mol } NiCl_2}{1 \text{ mol } NiCO_3} \times \frac{129.59 \text{ g } NiCl_2}{1 \text{ mol } NiCl_2} = 2.73 \times 10^3 \text{ g } NiCl_2$$

$$= 2.73 \text{ kg } NiCl_2$$

$$21.1 \text{ mol } NiCO_3 \times \frac{1 \text{ mol } Na_2CO_3}{1 \text{ mol } NiCO_3} \times \frac{105.99 \text{ g } Na_2CO_3}{1 \text{ mol } Na_2CO_3} = 2.24 \times 10^3 \text{ g } Na_2CO_3$$

$$= 2.24 \text{ kg } Na_2CO_3$$

4.72 From procedure C described in the problem:

$$\frac{185.9 \text{ mg} \times 10^{-3} \text{ g/mg KHP}}{204.2 \text{ g/mol}} \times \frac{1 \text{ mol KOH}}{1 \text{ mol KHP}} \times \frac{1}{0.02567 \text{ L}} = 0.03546 \text{ M KOH}$$

From procedure D:

$$\frac{0.03546 \text{ mol KOH}}{\text{L}} \times 0.03402 \text{ L} \times \frac{1 \text{ mol HCl}}{1 \text{ mol KOH}} \times \frac{1}{0.05000 \text{ L}} = 0.02413 \text{ M HCl}$$

From procedure E: $2HCl + Ca(OH)_2 \rightarrow 2H_2O + CaCl_2$

$$\frac{0.02413 \text{ mol HCl}}{\text{L}} \times 0.02928 \text{ L} \times \frac{1 \text{ mol } Ca(OH)_2}{2 \text{ mol HCl}} \times \frac{1}{0.02500 \text{ L}}$$

$$= 0.01413 \text{ M } Ca(OH)_2$$

$$\frac{0.01413 \text{ mole } Ca(OH)_2}{\text{L}} \times 1 \text{ L} \times \frac{74.10 \text{ g}}{\text{mole}} = 1.05 \text{ g } Ca(OH)_2 \text{ in the 1 liter of solution}$$

4.73

$$\frac{1.632 \text{ mol NaOH}}{1 \text{ L}} \times 0.05000 \text{ L} \times \frac{1}{1.000 \text{ L}} \times 0.04000 \text{ L}$$

$$\times \frac{1 \text{ mol } H_2C_2O_4 \cdot 2H_2O}{2 \text{ mol NaOH}} \times \frac{126.1 \text{ g } H_2C_2O_4 \cdot 2H_2O}{\text{mole}} = 0.2058 \text{ g } H_2C_2O_4 \cdot 2H_2O$$

% purity = (0.2058 g/0.2500) x 100 = 82.3%

4.74 $Mg_{(s)} + 2\,HCl_{(aq)} \rightarrow MgCl_{2(aq)} + H_{2(g)}$

a) $$\frac{1.215 \text{ g Mg}}{24.31 \text{ g/mol}} \times \frac{1 \text{ mol } H_2}{1 \text{ mol Mg}} \times \frac{2.016 \text{ g } H_2}{\text{mol } H_2} = 0.101 \text{ g } H_2$$

b)

Reaction:	$Mg_{(s)}$	+	$2HCl_{(aq)}$	$\rightarrow$	$MgCl_{2(aq)}$	+	$H_{2(g)}$
Start	0.04998 mol		0.400 mol		0		0
Change	-0.04998 mol		-0.09996		+0.04998 mol		
Final	0		0.3000		0.04998		

(All in 100 mL) Concentrations: $H_3O^+ = 3.000$ M, $Mg^{2+} = 0.500$ M

$Cl^- = 3.000 \text{ M} + 0.9996 \text{ M} = 4.00 \text{ M}$

4.75 a) $Mg_{(s)} + 2\,HCl_{(aq)} \rightarrow MgCl_{2(aq)} + H_{2(g)}$

Net ionic eq.: $Mg_{(s)} + 2\,H_3O^+_{(aq)} \rightarrow Mg^{2+}_{(aq)} + H_{2(g)} + 2\,H_2O_{(l)}$

b) $KOH_{(aq)} + HCl_{(aq)} \rightarrow H_2O_{(l)} + KCl_{(aq)}$

Net ionic eq.: $OH^-_{(aq)} + H_3O^+_{(aq)} \rightarrow 2\,H_2O_{(l)}$

c) $BaCl_{2(aq)} + HCl_{(aq)} \rightarrow$ no reaction

4.76 a) $2\,H_2S_{(g)} + 3\,O_{2(g)} \rightarrow 2\,SO_{2(g)} + 2\,H_2O_{(g)}$

b) $SO_{2(g)} + 2\,H_2S_{(g)} \rightarrow 3\,S_{(s)} + 2\,H_2O_{(g)}$

4.77 $$\frac{1.25\text{ kg S}}{32.07\text{ g / mol}} \times \frac{\text{g}}{10^3\text{kg}} \times \frac{1\text{ mol }SO_2}{3\text{ mol S}} \times \frac{6\text{ mol }H_2S}{2\text{ mol }SO_2} \times \frac{34.09\text{ g }H_2S}{\text{mol}}$$

$$\times \frac{\text{kg}}{10^3\text{ g}} = 1.33\text{ kg }H_2S$$

4.78 a) $2\,H_2SO_4 + Ca_3(PO_4)_2 \rightarrow 2\,CaSO_4 + Ca(H_2PO_4)_2$

b) Note that this question refers to the reaction:

$3\,H_2SO_{4(aq)} + Ca_3(PO_4)_{2(s)} \rightarrow 3\,CaSO_{4(s)} + 2\,H_3PO_{4(aq)}$

$[3 \times MM(CaSO_4)] + [2 \times MM(H_3PO_4)] = 604.4$ g (mass of product combined)

$$\frac{50.0\text{ kg product}}{604.4\text{ g / min. mol product}} \times \frac{10^3\text{ g}}{\text{kg}} \times \frac{3\text{ mol }H_2SO_4}{\text{min. mol product}} \times \frac{98.09\text{ g }H_2SO_4}{\text{mol}}$$

$$\times \frac{10^{-3}\text{ kg}}{\text{g}} = 24.3\text{ kg }H_2SO_4$$

$$\frac{50.0\text{ kg product}}{604.4\text{ g / min. mol product}} \times \frac{10^3\text{ g}}{\text{kg}} \times \frac{1\text{ mol }Ca_3(PO_4)_2}{\text{min. mol product}} \times \frac{310.2\text{ g }Ca_3(PO_4)_2}{\text{mol}}$$

$$\times \frac{10^{-3}\text{ kg}}{\text{g}} = 25.7\text{ kg }Ca_3(PO_4)_2$$

To produce 50.0 kg of the $CaSO_4$ - $Ca(H_2PO_4)_2$ mixture:

$[2 \times MM(CaSO_4)] + [MM\ Ca(H_2PO_4)_2] = 506.4$ g

$$\frac{50.0\text{ kg product}}{506.4\text{ g / min. mol product}} \times \frac{10^3\text{ g}}{\text{kg}} \times \frac{2\text{ mol }H_2SO_4}{\text{min. mol product}} \times \frac{98.09\text{ g }H_2SO_4}{\text{mol}}$$

$$\times \frac{10^{-3}\text{ kg}}{\text{g}} = 19.4\text{ kg }H_2SO_4$$

$$\frac{50.0\text{ kg product}}{506.4\text{ g / min. mol product}} \times \frac{10^3\text{ g}}{\text{kg}} \times \frac{1\text{ mol }Ca_3(PO_4)_2}{\text{min. mol product}} \times \frac{310.2\text{ g }Ca_3(PO_4)_2}{\text{mol}}$$

$$\times \frac{10^{-3}\text{ kg}}{\text{g}} = 30.6\text{ kg }Ca_3(PO_4)_2$$

(continued)

(4.78 continued)

c) For: $3\ H_2SO_{4(aq)} + Ca_3(PO_4)_{2(s)} \rightarrow 3\ CaSO_{4(s)} + 2\ H_3PO_{4(aq)}$

$$\frac{50.0 \text{ kg product}}{604.4 \text{ g / min. mol product}} \times \frac{10^3 \text{ g}}{\text{kg}} \times \frac{2 \text{ mol } PO_4^{3-}}{\text{min. mol product}} = 165 \text{ mol } PO_4^{3-}$$

For: $2\ H_2SO_4 + Ca_3(PO_4)_2 \rightarrow 2\ CaSO_4 + Ca(H_2PO_4)_2$

$$\frac{50.0 \text{ kg product}}{506.4 \text{ g / min. mol product}} \times \frac{10^3 \text{ g}}{\text{kg}} \times \frac{2 \text{ mol } PO_4^{3-}}{\text{min. mol product}} = 197 \text{ mol } PO_4^{3-}$$

4.79 a) $H_3PO_{4(aq)} + 3\ KOH_{(aq)} \rightarrow 3\ H_2O_{(l)} + K_3PO_{4(aq)}$

$H_3PO_{4(aq)} + 3\ OH^-_{(aq)} \rightarrow 3\ H_2O_{(l)} + PO_4^{3-}{}_{(aq)}$ acid/base

b) $2\ Sr_{(s)} + O_{2(g)} \rightarrow 2\ SrO_{(s)}$ redox

c) $2\ C_4H_8O_{(l)} + 11\ O_{2(g)} \rightarrow 8\ CO_{2(g)} + 8\ H_2O_{(g)}$ redox

d) $Mg_{(s)} + 2\ HBr_{(aq)} \rightarrow MgBr_{2(aq)} + H_{2(g)}$

$Mg_{(s)} + 2\ H_3O^+_{(aq)} \rightarrow Mg^{2+}_{(aq)} + H_{2(g)} + 2\ H_2O_{(l)}$ redox

e) $Pb(NO_3)_{2(aq)} + (NH_4)_2S_{(aq)} \rightarrow PbS_{(s)} + 2\ NH_4NO_{3(aq)}$

$Pb^{2+}_{(aq)} + S^-_{(aq)} \rightarrow PbS_{(s)}$ precipitation

f) $Ag_{(s)} + HCl_{(aq)} \rightarrow$ "nr"

g) $Ni_{(s)} + 2HCl_{(aq)} \rightarrow NiCl_{2(aq)} + H_{2(g)}$

$Ni_{(s)} + 2\ H_3O^+_{(aq)} \rightarrow Ni^{2+}_{(aq)} + H_{2(g)} + 2\ H_2O_{(l)}$ redox

h) $AgNO_{3(aq)} + KCH_3CO_{2(aq)} \rightarrow$ "nr"

4.80 $2\ C_2H_4 + 2\ HCl + O_2 \rightarrow 2\ C_2H_3Cl + 2\ H_2O$

$$\frac{8 \times 10^9 \text{ lb } C_2H_3Cl}{62.49 \text{ g / mol}} \times \frac{454 \text{ g}}{\text{lb}} \times \frac{2 \text{ mol } C_2H_4}{2 \text{ mol } C_2H_3Cl} \times \frac{28.05 \text{ g}}{\text{mol}} \times \frac{1 \text{ lb}}{454 \text{ g}}$$

$= 3.6 \times 10^9$ lb C_2H_4 = 3.6 billion pounds ethylene

$$\frac{8 \times 10^9 \text{ lb } C_2H_3Cl}{62.49 \text{ g / mol}} \times \frac{454 \text{ g}}{\text{lb}} \times \frac{2 \text{ mol HCl}}{2 \text{ mol } C_2H_3Cl} \times \frac{36.46 \text{ g}}{\text{mol}} \times \frac{1 \text{ lb}}{454 \text{ g}}$$

$= 4.7 \times 10^9$ lb HCl = 4.7 billion pounds HCl

4.81 $Fe + 2\ HCl \rightarrow Fe^{2+} + 2\ Cl^- + H_2$

$$\frac{5.8 \text{ g Fe}}{55.85 \text{ g / mol}} \times \frac{2 \text{ mol HCl}}{1 \text{ mol Fe}} \times \frac{1 \text{ L soln.}}{1.5 \text{ mol HCl}} = 0.14 \text{ L or } 140 \text{ mL}$$

4.82 $4\,FeS_2 + 11\,O_2 \rightarrow 2\,Fe_2O_3 + 8\,SO_2$

$$\frac{175 \text{ mton } FeS_2}{120.0 \text{ g / mol}} \times \frac{10^6 \text{ g}}{\text{mton}} \times \frac{2 \text{ mol } Fe_2O_3}{4 \text{ mol } FeS_2} \times \frac{159.7 \text{ g } Fe_2O_3}{\text{mol}} \times \frac{1 \text{ mton}}{10^6 \text{ g}}$$

$$= 116 \text{ metric tons } Fe_2O_3$$

$$\frac{175 \text{ mton } FeS_2}{120.0 \text{ g / mol}} \times \frac{10^6 \text{ g}}{\text{mton}} \times \frac{8 \text{ mol } SO_2}{4 \text{ mol } FeS_2} \times \frac{64.1 \text{ g } SO_2}{\text{mol}} \times \frac{1 \text{ mton}}{10^6 \text{ g}}$$

$$= 187 \text{ metric tons } SO_2$$

4.83 $Al(OH)_3 + 3HCl \rightarrow AlCl_3 + 3H_2O$

$$\frac{0.175 \text{ mol HCl}}{\text{L}} \times 0.155 \text{ L} \times \frac{1 \text{ mol } Al(OH)_3}{3 \text{ mol HCl}} \times \frac{78.0 \text{ g } Al(OH)_3}{\text{mol}}$$

$$= 0.704 \text{ g } Al(OH)_3$$

4.84

Reaction:	$3\,HCO_3^-$ +	$H_3C_6H_5O_7$ →	$3\,CO_2$
Start	0.02281 mol	0.005206 mol	
Change	-0.01562 mol	-0.005206 mol	+0.01562 mol

0.01562 mol CO_2 x 44.0 g/mol = 0.687 g CO_2

4.85

$$\frac{375 \text{ kg } C_6H_{12}}{84.14 \text{ g / mol}} \times \frac{10^3 \text{ g}}{\text{kg}} \times \frac{2 \text{ mol } C_6H_{10}O_4}{2 \text{ mol } C_6H_{12}} \times \frac{146.16 \text{ g}}{\text{mol}}$$

$$\times \frac{10^{-3} \text{ kg}}{\text{g}} = 651 \text{ kg adipic acid}$$

4.86 $Na_2Cr_2O_7{\cdot}2H_2O + 2\,H_2SO_4 + H_2O \rightarrow 2\,HCrO_4^- + 2\,Na^+ + 2HSO_4^- + 2\,H_3O^+$

$$\frac{18 \text{ mol } H_2SO_4}{\text{L}} \times 0.80 \text{ L} \times \frac{1 \text{ mol } H_3O^+}{1 \text{ mol } H_2SO_4} \times \frac{1}{1.260 \text{ L}} = 11.4 \text{ M } H_3O^+$$

11.4 M H_3O^+ = 11.4 M HSO_4^-

$$\frac{92 \text{ g } Na_2Cr_2O_7 \bullet 2H_2O}{298.0 \text{ g / mol}} \times \frac{2 \text{ mol } HCrO_4^-}{1 \text{ mol } Na_2Cr_2O_7 \bullet 2H_2O} \times \frac{1}{1.260 \text{ L}} = 0.490 \text{ M}$$

0.490 M $HCrO_4^-$ = 0.490 M Na^+ Note: $HCrO_4^-$ may be reported as 0.245 M $Cr_2O_7^{2-}$ or as 0.490 M H_2CrO_4. If reported as H_2CrO_4, one would also need to adjust the concentration for H_3O^+ (10.9 M).

4.87

Reaction:	$CaCO_3$ +	2 HCl →	$CaCl_2$ +	H_2O +	$CO_{2(g)}$
Start	0.050 mol	0.050 mol	0		0
Change	-0.025 mol	-0.050 mol	+0.025 mol		+0.025
Finish	0.0250 mol	0	0.025 mol		0.025

$CaCO_3$: 2.5 g remain unchanged
CO_2: 1.1 g of gaseous CO_2 escaped
Solution: Ca^{2+} = 0.050 M, Cl^- = 0.10 M

4.88 See Appendix I.

Bases: ammonia - fertilizers
calcium oxide - cement
sodium hydroxide - various uses

Acids: sulfuric acid - fertilizers, detergents, drugs, explosives and other uses.
phosphoric acid - fertilizers, detergents
nitric acid - fertilizers, explosives, nitrogenous chemicals
hydrochloric acid - pickling metals, chemicals preparation
perchloric acid -
hydrobromic acid -
hydriodic acid -
acetic acid - chemicals, fibers
terephthalic acid - polymers
adipic acid - fibers

4.89 1. Incomplete reaction 2. Non-quantitative isolation of the product
3. Impure starting materials 4. The presence of a completing reaction

4.90 $ZnSO_{4(aq)} + BaS_{(aq)} \rightarrow BaSO_{4(s)} + ZnS_{(s)}$

$$\frac{1000\ \text{g product}}{331\ \text{g / mol product}} \text{x} \frac{1\ \text{mol } ZnSO_4}{1\ \text{mol product}} \text{x} \frac{161.4\ \text{g } ZnSO_4}{\text{mol}} = 488\ \text{g } ZnSO_4$$

$$\frac{1000\ \text{g product}}{331\ \text{g / mol product}} \text{x} \frac{1\ \text{mol BaS}}{1\ \text{mol product}} \text{x} \frac{169.4\ \text{g BaS}}{\text{mol}} = 512\ \text{g BaS}$$

Mix a solution that contains 488 g of $ZnSO_4$ with a solution that contains 512 g BaS.

4.91 a) $CaO_{(s)} + H_2O_{(l)} \rightarrow Ca^{2+}{}_{(aq)} + 2\ OH^-{}_{(aq)}$

b) $3\ HPO_4^{2-} + 5\ Ca^{2+} + 4\ OH^- \rightarrow Ca_5(PO_4)_3OH + 3\ H_2O$

c) 1.00×10^4 L x (0.0156 mol HPO_4^-/L) x (5 mol CaO/3 mol HPO_4^{2-})
x (56.1 g CaO/mol) x (1 kg/10^3 g) = 14.6 kg lime

4.92 $CH_4 + 2\ O_2 \rightarrow CO_2 + 2\ H_2O$

$C_2H_6 + 7/2\ O_2 \rightarrow 2\ CO_2 + 3\ H_2O$

$C_3H_8 + 5\ O_2 \rightarrow 3\ CO_2 + 4\ H_2O$

$$CH_4\text{:}\ \frac{750\ \text{g x } 74\%\ CH_4}{16.0\ \text{g / mol}} \text{x} \frac{1\ \text{mol } CO_2}{1\ \text{mol } CH_4} = 34.7\ \text{mol } CO_2$$

$$C_2H_6\text{:}\ \frac{750\ \text{g x } 18\%\ C_2H_6}{30.0\ \text{g / mol}} \text{x} \frac{2\ \text{mol } CO_2}{1\ \text{mol } C_2H_6} = 9.0\ \text{mol } CO_2$$

$$C_3H_8\text{:}\ \frac{750\ \text{g x } 8\%\ C_3H_8}{44.0\ \text{g / mol}} \text{x} \frac{3\ \text{mol } CO_2}{1\ \text{mol } C_3H_8} = 4.1\ \text{mol } CO_2$$

Total = 34.7 + 9.0 + 4.1 mol CO_2 = 47.8 mol CO_2

4.93 $Zn + 2\ HCl \rightarrow ZnCl_2 + H_{2(g)}$

$$\frac{21.3\ mg\ H_2}{2.02\ g/mol} \times \frac{10^{-3}\ g}{mg} \times \frac{1\ mol\ Zn}{1\ mol\ HCl} \times \frac{65.39\ g\ Zn}{mol} = 0.690\ g\ Zn$$

% Zn = (0.690/5.73) x 100 = 12.0%   % Cu = [(5.73 - 0.690)/5.73] x 100 = 88.0%

4.94 [0.156 g AgCl/(143.4 g AgCl/mol)] x (1 mol Ag/1 mol AgCl) x (107.9 g Ag/mol Ag) = 0.1174 g Ag

% Ag = (0.1174/0.135) x 100 = 87.0% silver

% Cu = 100.0 - 87.0 = 13.0% copper

4.95 a) $3\ Na_2CO_{3(aq)} + 2\ Fe(NO_3)_{3(aq)} \rightarrow 6\ NaNO_{3(aq)} + Fe_2(CO_3)_{3(s)}$

b) $HClO_{4(aq)} + KOH_{(aq)} \rightarrow KClO_{4(aq)} + H_2O_{(l)}$

c) $NaCl + Ba(OH)_2 \rightarrow$ "nr"

4.96 a) $H_3O^+_{(aq)}$ (or $H^+_{(aq)}$) and $NO_3^-{}_{(aq)}$

b) $NH_4^+{}_{(aq)}$ and $SO_4^{2-}{}_{(aq)}$

c) $K^+_{(aq)}$ and $HCO_3^{2-}{}_{(aq)}$

d) $CO_{2(aq)}$ and $H_2CO_{3(aq)}$

4.97 Your drawing must be based upon the drawing given in the problem. The initial solution has 7 HSO_4^- and 7 H_3O^+. The new drawings that show what the solution looks like after the reaction is complete will contain:

a) (after the addition of 6 OH^-) 7 HSO_4^- + 12 H_2O + H_3O^+

$7\ HSO_4^- + 7\ H_3O^+ + 6\ OH^- \rightarrow 7\ HSO_4^- + H_3O^+ + 12\ H_2O$

b) (after the addition of 12 OH^-) 2 HSO_4^- + 5 SO_4^{2-} + 19 H_2O.

$7\ HSO_4^- + 7\ H_3O^+ + 12\ OH^- \rightarrow 2\ HSO_4^- + 5\ SO_4^{2-} + 19\ H_2O$

c) (after the addition of 18 OH^-) 7 SO_4^{2-} + 4 OH^- + 21 H_2O.

$7\ HSO_4^- + 7\ H_3O^+ + 18\ OH^- \rightarrow 7\ SO_4^{2-} + 4\ OH^- + 21\ H_2O$

4.98
$$\frac{1.0\ kg\ CO_2/day}{astronaut} \times 6\ days \times 5\ astronauts \times \frac{1\ mol\ CO_2}{44.0\ g} \times \frac{10^3\ g}{kg} \times \frac{2\ mol\ LiOH}{1\ mol\ CO_2}$$

$$\times \frac{23.9\ g\ LiOH}{mol\ LiOH} \times \frac{10^{-3}\ kg}{g} = 32.6\ kg\ LiOH$$

4.99 $1.0 \times 6 \times 5 \times 1/44.0 \times 10^3 \times 2 \times 56.1 \times 10^{-3} = 76$ kg KOH

4.100 $$N_2H_4 + 2\,H_2O_2 \rightarrow N_2 + 4\,H_2O$$

MM = 32.05   34.02

volumes = 22.3 mL   2 x 33.7 mL

67.4 mL/22.3 = 3   3 volumes H_2O_2 for every 1 volume N_2H_4

4.101 $$\frac{5.96 \times 10^{-3}\text{ mol NaOH}}{1\text{ L}} \times 0.00570\text{ L} \times \frac{1\text{ mol }H_2SO_4}{2\text{ mol NaOH}} \times \frac{1\text{ mol }SO_2}{1\text{ mol }H_2SO_4}$$

$$\times \frac{64.1\text{ g }SO_2}{\text{mol }SO_2} = 1.09 \times 10^{-3}\text{ g }SO_2$$

4.102 $4\,C_3H_5N_3O_9 \rightarrow 6\,N_2 + 12\,CO_2 + 10\,H_2O + O_2$

4.103 a) $2\,C_4H_{10} + 13\,O_2 \rightarrow 8\,CO_2 + 10\,H_2O$

b) $2\,C_6H_6 + 15\,O_2 \rightarrow 12\,CO_2 + 6\,H_2O$

c) $C_2H_6O + 3\,O_2 \rightarrow 2\,CO_2 + 3\,H_2O$

d) $C_5H_{12} + 8\,O_2 \rightarrow 5\,CO_2 + 6\,H_2O$

e) $2C_6H_{12}O + 17\,O_2 \rightarrow 12\,CO_2 + 12\,H_2O$

4.104 a) $$\frac{1.50\text{ g }C_4H_{10}}{58.1\text{ g / mol}} \times \frac{8\text{ mol }CO_2}{2\text{ mol }C_4H_{10}} \times \frac{44.0\text{ g }CO_2}{\text{mol}} = 4.54\text{ g }CO_2$$

$$\frac{1.50\text{ g }C_4H_{10}}{58.1\text{ g / mol}} \times \frac{10\text{ mol }H_2O}{2\text{ mol }C_4H_{10}} \times \frac{18.0\text{ g }H_2O}{\text{mol}} = 2.32\text{ g }H_2O$$

b) $$\frac{1.50\text{ g }C_6H_6}{78.1\text{ g / mol}} \times \frac{12\text{ mol }CO_2}{2\text{ mol }C_6H_6} \times \frac{44.0\text{ g }CO_2}{\text{mol}} = 5.07\text{ g }CO_2$$

$$\frac{1.50\text{ g }C_6H_6}{78.1\text{ g / mol}} \times \frac{6\text{ mol }H_2O}{2\text{ mol }C_6H_6} \times \frac{18.0\text{ g }H_2O}{\text{mol}} = 1.04\text{ g }H_2O$$

c) $$\frac{1.50\text{ g }C_2H_6O}{46.1\text{ g / mol}} \times \frac{2\text{ mol }CO_2}{1\text{ mol }C_2H_6O} \times \frac{44.0\text{ g }CO_2}{\text{mol}} = 2.86\text{ g }CO_2$$

$$\frac{1.50\text{ g }C_2H_6O}{46.1\text{ g / mol}} \times \frac{3\text{ mol }H_2O}{1\text{ mol }C_2H_6O} \times \frac{18.0\text{ g }H_2O}{\text{mol}} = 1.76\text{ g }H_2O$$

d) $$\frac{1.50\text{ g }C_5H_{12}}{72.1\text{ g / mol}} \times \frac{5\text{ mol }CO_2}{1\text{ mol }C_5H_{12}} \times \frac{44.0\text{ g }CO_2}{\text{mol}} = 4.58\text{ g }CO_2$$

$$\frac{1.50\text{ g }C_5H_{12}}{72.1\text{ g / mol}} \times \frac{6\text{ mol }H_2O}{1\text{ mol }C_5H_{12}} \times \frac{18.0\text{ g }H_2O}{\text{mol}} = 2.25\text{ g }H_2O$$

(continued)

(4.104 continued)

e) $\frac{1.50 \text{ g } C_6H_{12}O}{100.2 \text{ g / mol}} \text{x} \frac{12 \text{ mol } CO_2}{2 \text{ mol } C_6H_{12}O} \text{x} \frac{44.0 \text{ g } CO_2}{\text{mol}} = 3.95 \text{ g } CO_2$

$\frac{1.50 \text{ g } C_6H_{12}O}{100.2 \text{ g / mol}} \text{x} \frac{12 \text{ mol } H_2O}{2 \text{ mol } C_6H_{12}O} \text{x} \frac{18.0 \text{ g } H_2O}{\text{mol}} = 1.62 \text{ g } H_2O$

4.105 $C_9H_4O_3 + 11O_2 \rightarrow 9\ CO_2 + 7\ H_2O$

$\frac{3.00 \text{ g } C_9H_{14}O_3}{170.2 \text{ g / mol}} \text{x} \frac{11 \text{ mol } O_2}{1 \text{ mol } C_9H_{14}O_3} \text{x} \frac{32.0 \text{ g } O_2}{\text{mol}} = 6.20 \text{ g } O_2 \text{ (consumed)}$

$\frac{3.00 \text{ g } C_9H_{14}O_3}{170.2 \text{ g / mol}} \text{x} \frac{7 \text{ mol } H_2O}{1 \text{ mol } C_9H_{14}O_3} \text{x} \frac{18.02 \text{ g } H_2O}{\text{mol}} = 2.22 \text{ g } H_2O \text{ (produced)}$

CHAPTER 5: THE BEHAVIOR OF GASES

5.1

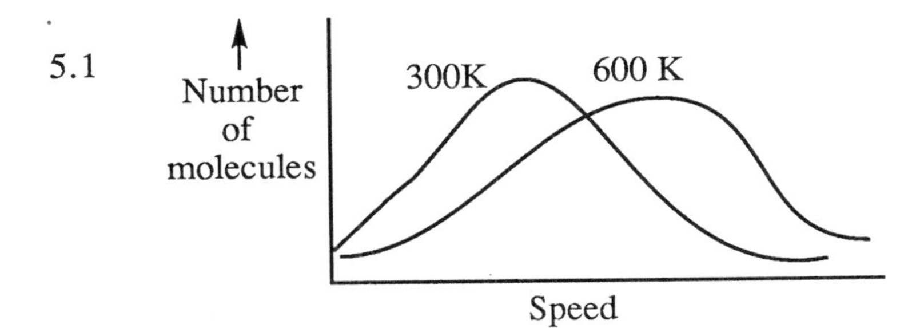

5.2

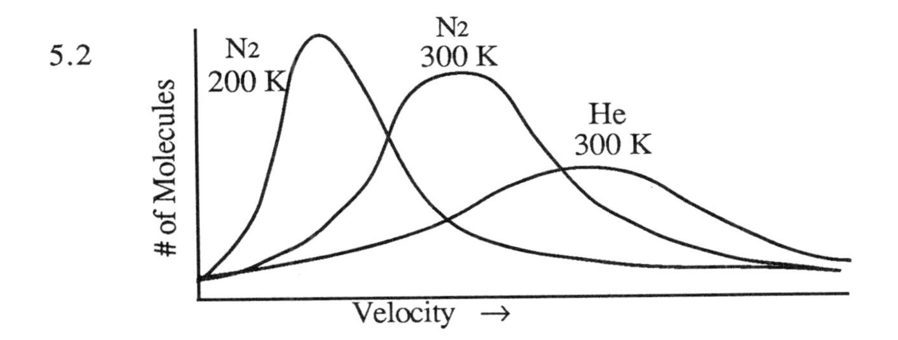

5.3

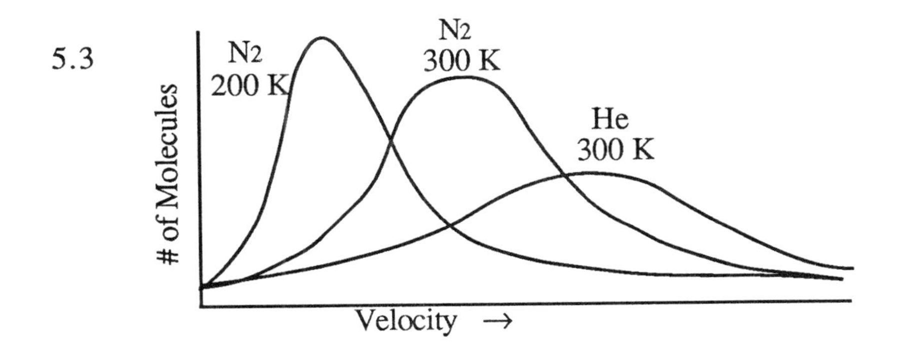

5.4 From Figure 5.4 one would conclude that the probable kinetic energy at 300 K is about 5×10^{-21} J/molecule and at 900 K it is about 12×10^{-21} J/molecule

a) For He at 900 K $\quad 12 \times 10^{-21}\ \text{J/molecule} = 1/2\left(\dfrac{4.00 \times 10^{-3}\ \text{kg}}{6.02 \times 10^{23}\ \text{molecules}}\right)v^2$

$$v^2 = 3.6 \times 10^6\ \text{m}^2/\text{s}^2$$

$$v = 1.9 \times 10^3\ \text{m/s}$$

b) For O_2 at 300 K $\quad 5 \times 10^{-21}\ \text{J/molecule} = 1/2\left(\dfrac{32.00 \times 10^{-3}\ \text{kg}}{6.02 \times 10^{23}\ \text{molecules}}\right)v^2$

$$v^2 = 1.88 \times 10^5\ \text{m}^2/\text{s}^2$$

$$v = 4.3 \times 10^2\ \text{m/s}$$

c) For SF_6 at 900 K $\quad 12 \times 10^{-21}\ \text{J/molecule} = 1/2\left(\dfrac{146 \times 10^{-3}\ \text{kg}}{6.02 \times 10^{23}\ \text{molecules}}\right)v^2$

$$v^2 = 9.9 \times 10^4\ \text{m}^2/\text{s}^2$$

$$v = 3.1 \times 10^2\ \text{m/s}$$

5.5 $\text{k.E.}_{molar} = 3/2\ RT$

a) For He at 900 K, k.E._{molar} = (3/2) (8.314 J/mol K) (900 K) = 1.12×10^4 J/mol
b) For O_2 at 300 K, k.E._{molar} = (3/2) (8.314 J/mol K) (300 K) = 3.74×10^3 J/mol
c) For SF_6 at 900 K, k.E._{molar} = (3/2) (8.314 J/mol K) (900 K) = 1.12×10^4 J/mol

5.6 $$\frac{\text{rate } U^{235}F_6}{\text{rate } U^{238}F_6} = \left(\frac{\text{MM } U^{238}F_6}{\text{MM } U^{235}F_6}\right)^{1/2} = \left(\frac{352.0}{349.0}\right)^{1/2} = 1.004$$

5.7 a) They have the same force (F) since F is directly proportional to temperature (T). $\dfrac{F}{A} = P\alpha\dfrac{n}{V}T$. Pressure (P) is force per area (F/A) and is proportional to (n/V) times T. As T varies so will both the k.E. and the F since k.E. is also directly proportional to T.

b) The methane molecule will have the higher force at 700 K. F α v and v α T The molecule at the higher temperature will have the higher v and ∴ greater F.

c) The F_2 will have the greater force because F α mv, the mass of F_2 is greater than that of H_2.

5.8 a) At high pressure molecules are close to each other and are a high proportion of the space.
b) At a low temperature the speed of the molecules is slow enough that the forces between molecules is a factor.

5.9 a) The pressure inside the container is 1 atm. Otherwise the piston will move. The collisions must be sufficient to keep the piston in position.
b) If the temperature in kelvins is doubled, the molecules will collide twice as many times per unit of time forcing the piston out until twice the volume is obtained and the number of collisions per unit time would then equal the original number and the pressure would be 1 atm again.
The redrawn picture would need to have the piston raised to a point that would yield twice the volume shown in the original figure in the textbook.

5.10 a) The pressure increases because the number of molecular collisions increases for the smaller volume.
b) The pressure decreases because the fewer number of molecules collide with the piston less times.
c) The pressure increases because there are more molecular collisions since the molecules are moving faster at the higher temperature and the force is greater because of the higher speed.

5.11 The level of the column of mercury would slowly drop until it reached the level of the mercury in the dish.

5.12 $18.4 \text{ cm } H_2O \times \dfrac{1.00 \text{ g / mL } H_2O}{13.59 \text{ g / mL Hg}} = 1.35 \text{ cm Hg}$

764.4 torr - 1.4 torr = 763.0 torr

763.0 torr x 1 atm/760 torr = 1.004 atm

$$1.004 \text{ atm} \times \frac{1.01325 \times 10^5 \text{ Pa}}{1 \text{ atm}} = 1.017 \times 10^5 \text{ Pa} \times \frac{1 \text{ kPa}}{10^3 \text{ Pa}} = 1.017 \times 10^2 \text{ kPa}$$

5.13 See Section 5.3 for conversion factors and a discussion of SI units.

a) $455 \text{ torr} \times \dfrac{1.01325 \times 10^5 \text{ Pa}}{760 \text{ torr}} = 6.07 \times 10^4 \text{ Pa or } 6.07 \times 10^4 \text{ N / m}^2$

b) $2.45 \text{ atm} \times \dfrac{1.01325 \times 10^5 \text{ Pa}}{1 \text{ atm}} = 2.48 \times 10^5 \text{ Pa or } 2.48 \times 10^5 \text{ N / m}^2$

c) $0.46 \text{ torr} \times \dfrac{1 \text{ atm}}{760 \text{ torr}} \times \dfrac{1.01325 \times 10^5 \text{ Pa}}{1 \text{ atm}} = 61 \text{ Pa or } 61 \text{ N / m}^2$

d) $1.33 \times 10^{-3} \text{ atm} \times \dfrac{1.01325 \times 10^5 \text{ Pa}}{1 \text{ atm}} = 135 \text{ Pa or } 135 \text{ N / m}^2$

5.14 See Section 5.3 for conversion values.

a) $1.00 \text{ Pa} \times \dfrac{1 \text{ atm}}{1.01325 \times 10^5 \text{ Pa}} \times \dfrac{760 \text{ torr}}{1 \text{ atm}} = 7.50 \times 10^{-3} \text{ torr}$

b) $125.6 \text{ kPa} \times \dfrac{1000 \text{ Pa}}{1 \text{ kPa}} \times \dfrac{1 \text{ atm}}{1.01325 \times 10^5 \text{ Pa}} \times \dfrac{760 \text{ torr}}{1 \text{ atm}} = 942.1 \text{ torr}$

c) 75.0 atm x 760 torr/1 atm = 5.70×10^4 torr

d) 4.55×10^{-10} atm x 760 torr/1 atm = 3.46×10^{-7} torr

5.15 $PV = nRT$

$$n = \frac{PV}{RT} = \frac{(5.00 \text{ atm})(20.0 \text{ L})}{(0.08206 \text{ L atm mol}^{-1} \text{ K}^{-1})(298 \text{ K})} = 4.09 \text{ mol}$$

$$V = \frac{nRT}{P} = \frac{(4.09 \text{ mol})(0.08206 \text{ L atm mol}^{-1} \text{ K}^{-1})(298 \text{ K})}{(1 \text{ atm})} = 100 \text{ L}$$

5.16 $PV = nRT$

a) $n = PV/RT$ b) $V = nRT/P$ c) $n/V = P/RT$

5.17 a) $\frac{P_i}{T_i} = \frac{P_f}{T_f}$ or $\frac{T_i}{P_i} = \frac{T_f}{P_f}$ b) $\frac{n_i}{V_i} = \frac{n_f}{V_f}$ or $\frac{V_i}{n_i} = \frac{V_f}{n_f}$

c) $P_iV_i = P_fV_f$

5.18 See Problem 5.17 for the following equation: $P_f = (P_i/T_i)\ T_f$

Given: $P_i = 1.075$ atm, $T_i = 273 + 25 = 298$ K, $T_f = 273 - 10 = 263$ K

Solving: $P_f = \frac{(1.075 \text{ atm})(263 \text{ K})}{298 \text{ K}} = 0.949 \text{ atm}$

5.19 $V_i/T_i = V_f/T_f$ $V_f = (V_i/T_i)\ T_f = (V_i \times T_f) / T_i$

Given: $V_i = 0.255$ L, $T_i = 273 + 25 = 298$ K, $T_f = 273 - 10 = 263$ K

$$V_f = \frac{(0.255 \text{ L})(263 \text{ K})}{298 \text{ K}} = 0.225 \text{ L}$$

5.20 At the lower temperature the molecules are moving more slowly. Therefore, the pressure is less because the molecules strike the walls of the container less often and with less force.

5.21 Assume that the temperature of the ice bath is 0°C or 273 K.

V and n are constant, $\therefore P_i/T_i = P_f/T_f$ and $T_f = P_f(T_i/P_i)$.

$T_f = \frac{(745 \text{ torr})(273 \text{ K})}{(345 \text{ torr})} = 590 \text{ K}$ $T_f = 590 \text{ K} - 273 \text{ K} = 317°\text{C}$

5.22 The equation would not be used if: b) T is in °C and d) n is changing.

5.23 $PV = \frac{m}{MM}RT \therefore MM = \frac{mRT}{PV}$

$$MM = \frac{(2.55 \text{ g})(0.08206 \text{ L atm mol}^{-1} \text{ K}^{-1})(298 \text{ K})}{\left(\frac{262 \text{ torr}}{760 \text{ torr / atm}}\right)(1.50 \text{ L})} = 121 \text{ g / mol}$$

121 g/mol is the molecular mass of CCl_2F_2

5.24 $n_{Ar} = \dfrac{1.25 \text{ g}}{39.95 \text{ g / mol}} = 3.13 \times 10^{-2}$ moles

$n_{CO} = \dfrac{1.25 \text{ g}}{28.01 \text{ g / mol}} = 4.46 \times 10^{-2}$ moles

$n_{CH_4} = \dfrac{1.25 \text{ g}}{16.04 \text{ g / mol}} = 7.79 \times 10^{-2}$ moles

$$P_{Ar} = \frac{(3.13 \times 10^{-2} \text{ mol})(8.206 \times 10^{-2} \text{ L atm mol}^{-1} \text{ K}^{-1})(648 \text{ K})}{4.00 \text{ L}} = 0.416 \text{ atm}$$

$$P_{CO} = \frac{(4.46 \times 10^{-2} \text{ mol})(8.206 \times 10^{-2} \text{ L atm mol}^{-1} \text{ K}^{-1})(648 \text{ K})}{4.00 \text{ L}} = 0.593 \text{ atm}$$

$$P_{CH_4} = \frac{(7.79 \times 10^{-2} \text{ mol})(8.206 \times 10^{-2} \text{ L atm mol}^{-1} \text{ K}^{-1})(648 \text{ K})}{4.00 \text{ L}} = 1.04 \text{ atm}$$

$P_{total} = 0.416 \text{ atm} + 0.593 \text{ atm} + 1.04 \text{ atm} = 2.05 \text{ atm}$

Or one can also calculate the total pressure in the following way:

$P_{total} =$

$$\frac{(4.46 \times 10^{-2} + 3.13 \times 10^{-2} + 7.79 \times 10^{-2})(8.206 \times 10^{-2} \text{ L atm mol}^{-1} \text{ K}^{-1})(648 \text{ K})}{4.00 \text{ L}}$$

$= 2.05$ atm

$$X_{Ar} = \frac{3.13 \times 10^{-2} \text{ mol}}{0.1538 \text{ mol}} = 0.204 \text{ or } \frac{0.416 \text{ atm}}{2.05 \text{ atm}} = 0.203$$

$$X_{CO} = \frac{4.46 \times 10^{-2} \text{ mol}}{0.1538 \text{ mol}} = 0.290 \text{ or } \frac{0.593 \text{ atm}}{2.05 \text{ atm}} = 0.289$$

$$X_{CH_4} = \frac{7.79 \times 10^{-2} \text{ mol}}{0.1538 \text{ mol}} = 0.507 \text{ or } \frac{1.04 \text{ atm}}{2.05 \text{ atm}} = 0.507$$

5.25 $P_{NO_2} = X_{NO_2} P_{total} = \dfrac{0.78}{10^6}(758.4 \text{ torr}) = 5.9 \times 10^{-4}$ torr

5.26 Sample I has 4 A, 3 B and 3 C; II has 3 A, 6 B and 7 C; III has 3 A, 5 B and 18 C.

	I	II	III
P_{total}	10(RT/V)	16(RT/V)	26(RT/V)
X_A	4/10 = 2/5	3/16	3/26
X_B	3/10 = 0.30	6/16 = 3/8 = 0.375	5/26 = 0.19

a) Sample I: $P_A = X_A P_{total} = \left(\dfrac{2}{5}\right)10\left(\dfrac{RT}{V}\right) = 4\left(\dfrac{RT}{V}\right)$

Sample II: $P_A = \left(\dfrac{3}{16}\right)16\left(\dfrac{RT}{V}\right) = 3\left(\dfrac{RT}{V}\right)$

Sample III: $P_A = \left(\dfrac{3}{26}\right)26\left(\dfrac{RT}{V}\right) = 3\left(\dfrac{RT}{V}\right)$

Sample I has the highest partial pressure of gas A.

(continued)

(5.26 continued)

b) Sample II has the highest mole fraction of gas B.

c) In Sample III the concentration of gas A in ppm is:

$$\left(\frac{3}{26}\right)x\frac{10^6}{10^6} = \frac{0.115 \text{ x } 10^6}{10^6} = 0.115 \text{ x } 10^6 \text{ ppm}$$

5.27 For standard conditions P_{total} = 760 torr; $P_{gas} = X_{gas}P_{total}$.

P_{N_2} = (0.7808)(760 torr) = 593.4 torr P_{O_2} = (0.2095)(760 torr) = 159.2 torr

P_{Ar} = (9.34 x 10^{-5})(760 torr) = 7.10 x 10^{-2}torr

P_{CO_2} = (3.25 x 10^{-4})(760 torr) = 0.247 torr

5.28 Assuming standard pressure and ppm in terms of moles:

$X_{hydrocarbons}$ = 220/10^6 X_{CO} = 1.2/100

P_{total} = 760 torr = 1 atmosphere

$P_{hydrocarbons}$ = (220/10^6)760 torr = 0.167 torr or (220/10^6)1 atm = 2.20 x 10^{-4} atm

P_{CO} =(1.2/100)(760 torr) = 9.1 torr or (1.2/100)1 atm = 0.012 atm

5.29 $$n_{CH_4} = \frac{1.57 \text{ g}}{16.04 \text{ g / mol}} = 9.79 \text{ x } 10^{-2} \text{ mol}$$

$$n_{C_2H_6} = \frac{0.41 \text{ g}}{30.068 \text{ g / mol}} = 1.36 \text{ x } 10^{-2} \text{ mol}$$

$$n_{C_3H_8} = \frac{0.020 \text{ g}}{44.094 \text{ g / mol}} = 4.54 \text{ x } 10^{-4} \text{ mol}$$

X_{CH_4} = 0.874 $X_{C_2H_6}$ = 0.121 $X_{C_3H_8}$ = 4.06 x 10^{-3}

P_{CH_4} = (0.874)(2.35 atm) = 2.05 atm $P_{C_2H_6}$ = (0.121)(2.35 atm) = 0.284 atm

$P_{C_3H_8}$ = (4.06 x 10^{-3})(2.35 atm) = 9.5 x 10^{-3} atm

5.30 $$n_{KClO_3} = \frac{1.57 \text{ g } KClO_3}{122.55 \text{ g / mol}} = 1.28 \text{ x } 10^{-2} \text{ mol } KClO_3$$

$$n_{O_2} = (1.28 \text{ x } 10^{-2} \text{ mol } KClO_3)\left(\frac{3 \text{ mol } O_2}{2 \text{ mol } KClO_3}\right) = 1.92 \text{ x } 10^{-2} \text{ mol } O_2$$

$$V = \frac{nRT}{P} = \frac{(1.92 \text{ x } 10^{-2} \text{ mol})(0.08206 \text{ L atm / mol K})(293 \text{ K})}{\left(\frac{765.1 \text{ torr}}{760 \text{ torr / atm}}\right)} = 0.459 \text{ L}$$

5.31 $C_6H_{12}O_{6(s)} + 6\ O_{2(g)} \rightarrow 6\ CO_{2(g)} + 6\ H_2O_{(l)}$

$MM_{glucose} = 180.16 \qquad n_{glucose} = \dfrac{4.65\ \text{g glucose}}{180.16\ \text{g / mol}} = 2.581 \times 10^{-2}\ \text{mol glucose}$

$$n_{CO_2} = (2.581 \times 10^{-2}\ \text{mol glucose})\left(\frac{6\ \text{mol } CO_2}{1\ \text{mol glucose}}\right) = 0.1549\ \text{mol } CO_2$$

$$V = \frac{(0.1549\ \text{mol } CO_2)(0.08206\ \text{L atm mol}^{-1}\ \text{K}^{-1})(310\ \text{K})}{(1.00\ \text{atm})} = 3.94\ \text{L}$$

5.32 $P_{CO} = 1.00$ atm $\qquad P_{O_2} = 3.56\ \text{atm} - 1.00\ \text{atm} = 2.56\ \text{atm}$

$$n_{CO} = \frac{PV}{RT} = \frac{(1.00\ \text{atm})(50.0\ \text{L})}{(0.08206\ \text{L atm mol}^{-1}\ \text{K}^{-1})(298\ \text{K})} = 2.04\ \text{mol CO}$$

$$n_{O_2} = \frac{PV}{RT} = \frac{(2.56\ \text{atm})(50.0\ \text{L})}{(0.08206\ \text{L atm mol}^{-1}\ \text{K}^{-1})(298\ \text{K})} = 5.23\ \text{mol } O_2$$

Reaction:	2 CO	+	O_2	$\rightarrow$	2 CO_2
Initial amounts:	2.04 mol		5.23 mol		0
Change in amounts:	-2.04 mol		-1.02 mol		+2.04 mol
Final amounts:	0 mol		4.21 mol		2.04 mol

$P_{CO} = 0$

$$P_{O_2} = \frac{n_{O_2}RT}{V} = \frac{(4.21\ \text{mol})(0.08206\ \text{L atm mol}^{-1}\ \text{K}^{-1})(298\ \text{K})}{50.0\ \text{L}} = 2.06\ \text{atm}$$

$$P_{CO_2} = \frac{(2.04\ \text{mol})(0.08206\ \text{L atm mol}^{-1}\ \text{K}^{-1})(298\ \text{K})}{50.0\ \text{L}} = 0.998\ \text{atm}$$

5.33 $V = 3.00 \times 10^3\ \text{mL} = 3.00\ \text{L}$

$$n_{Cl_2} = \frac{PV}{RT} = \frac{\left(\dfrac{1.25 \times 10^3\ \text{torr}}{760\ \text{torr / atm}}\right)(3.00\ \text{L})}{(0.08206\ \text{L atm mol}^{-1}\ \text{K}^{-1})(300\ \text{K})} = 0.200\ \text{mol } Cl_2$$

$$n_{Na} = \frac{6.90\ \text{g}}{22.99\ \text{g / mol}} = 0.300\ \text{mol}$$

Reaction:	2 $Na_{(s)}$	+	$Cl_{2(g)}$	$\rightarrow$	2 $NaCl_{(s)}$
Initial amounts:	0.300 mol		0.200 mol		0 mol
Change in amounts:	-0.300 mol		-0.150 mol		+0.300 mol
Final amounts:	0 mol		0.050 mol		0.300 mol

Because Cl_2 is the only gas,

$$P_{total} = P_{Cl_2} = \frac{nRT}{V}$$

$$= \frac{(0.050\ \text{mol})\ (0.08206\ \text{L atm mol}^{-1}\text{K}^{-1})(320\text{K})}{(3.00\ \text{L})} = 0.438\ \text{atm}$$

5.34 Because the mole ratio is 1:1 and P_{total} = 1 atm; $P_{O_2} = P_{C_2H_4}$ = 0.5 atm

$$n_{C_2H_4} = n_{O_2} = \frac{PV}{RT} = \frac{(0.5\text{ atm})(5.00 \times 10^4\text{ L})}{(0.08206\text{ L atm mol}^{-1}\text{ K}^{-1})(553\text{ K})} = 550.9\text{ mol}$$

(It is assumed that P_{total} = 1 atm, exactly).

Reaction:	$2\,C_2H_{4(g)}$	+	$O_{2(g)}$	→	$2\,C_2H_4O_{(g)}$
Initial amounts:	550.9 mol		550.9 mol		0 mol
Change in amounts:	-550.9 mol		-275.5 mol		+550.9 mol
Final amounts:	0 mol		275.4 mol		550.9 mol

$$(550.9\text{ mol})\left(\frac{44.05\text{ g }C_2H_4O}{1\text{ mol}}\right)\left(\frac{1\text{ kg}}{10^3\text{ g}}\right) = 24.3\text{ kg }C_2H_4O_6$$

5.35 Repeat Problem 5.34, then: (24.3 kg $C_2H_4O_6$)(0.65) = 16 kg

5.36 Given at 0% relative humidity, $P_{N2} = X_{N2}P_{total}$, P_{N2} = 0.7808 atm, P_{H2O} = 0 atm, and P_{total} = 1.00 atm

At 30°C, 100% relative humidity, $P_{H_2O} = \frac{31.824\text{ torr}}{760\text{ torr / atm}} = 0.04187\text{ atm}$

At 30°C, 100% relative humidity, $P_{dry\ air}$ = 1.00 atm - P_{H2O} = 1.00 atm - 0.04 atm = 0.96 atm

P_{N2} (at 100%) = X_{N2} (at 0%) $P_{dry\ air}$ = (0.7808)(0.96 atm) = 0.75 atm

Change = 0.7808 - 0.75 = 0.03 atm decrease

5.37 VP at 25°C = 23.765 torr

$$\text{relative humidity} = 78\% = \frac{P_{H_2O}}{VP_{H_2O}} \times 100 = \frac{P_{H_2O}}{23.756} \times 100$$

$$P_{H_2O} = \frac{(78)(23.756)}{100} = 18.5\text{ torr}$$

This VP falls between 20°C and 25°C ($VP_{20°C}$ = 17.535)

$$\left(\frac{18.5 - 17.535}{23.756 - 17.535}\right)(5°C) + 20°C = 20.8°C$$

5.38 $X_{Kr} = 1.14 \times 10^{-6}$ $P_{Kr} = (X_{Kr})(1.0\text{ atm}) = (1.14 \times 10^{-6})(1.0\text{ atm}) = 1.1 \times 10^{-6}$ atm

1.0 km^3 = $(10^3\text{ m})^3$ = 10^9m^3 = $10^9(10^2\text{ cm})^3$ = $(10^{15}\text{ cm}^3)(1\text{ L}/10^3\text{cm}^3)$ = 1.0×10^{12} L

$$n_{Kr} = \frac{PV}{RT} = \frac{(1.1 \times 10^{-6}\text{ atm})(1.0 \times 10^{12}\text{ L})}{(0.08206\text{ L atm mol}^{-1}\text{ K}^{-1})(298\text{ K})} = 4.5 \times 10^4\text{ mol} \quad \text{(Assuming 25°C)}$$

$m_{Kr} = (n_{Kr})(MM) = (4.5 \times 10^4\text{ mol})(83.80\text{ g/mol})(1\text{ kg}/1000\text{ g}) = 3.8 \times 10^3$ kg Kr

5.39 At 35°C, VP_{H2O} = 42.175 torr; at 40°C, VP_{H2O} = 55.324 torr.
At 37°C, VP_{H2O} = (2/5)(55.324 - 42.175) torr + 42.175 torr = 47.435 torr

$$\frac{47.435\text{ torr}}{55.324\text{ torr}} \times 100\% = 85.74\% \approx 86\%$$

5.40 $(1.00 \text{ mton coal})\left(\frac{10^3 \text{ kg}}{1 \text{ mton}}\right)\left(\frac{10^3 \text{ g}}{1 \text{ kg}}\right)\left(\frac{4.55 \text{ g S}}{100 \text{ g coal}}\right) = 4.55 \times 10^4 \text{ g S}$

$S + O_2 \rightarrow SO_2$

$$n_{SO_2} = n_S = \frac{4.55 \times 10^4 \text{ g S}}{32.07 \text{ g S / mol S}} = 1.419 \times 10^3 \text{ mol S} = 1.419 \times 10^3 \text{ mol } SO_2$$

$$V_{SO_2} = \frac{nRT}{P} = \frac{(1.419 \times 10^3 \text{ mol})(0.08206 \text{ L atm mol}^{-1} \text{ K}^{-1})(323 \text{ K})}{1 \text{ atm}}$$

$$= 3.76 \times 10^4 \text{ L } SO_2$$

$CaO + SO_2 \rightarrow CaSO_3$

$n_{SO_2} = n_{CaO} = 1.419 \times 10^3$ mol $\qquad MM_{CaO} = 56.08$ g/mol

$m_{CaO} = (1.419 \times 10^3 \text{ mol})(56.08 \text{ g/mol}) = 7.96 \times 10^4 \text{ g CaO}$

$m_{CaSO_3} = (1.419 \times 10^3 \text{ mol})(120.144 \text{ g/mol}) = 1.70 \times 10^5 \text{ g } CaSO_3$

5.41 a) Particles must not suffer collisions while traveling their paths.
b) The materials will be used under the vacuum of space and, therefore, should be tested under such conditions.
c) Materials at high vacuum will not deposit on the metal surfaces but will, in fact, be released into the atmosphere.

5.42 $$\frac{n}{V} = \left(\frac{10^{10} \text{ molecules}}{m^3}\right)\left(\frac{1 \text{ mol}}{6.022 \times 10^{23} \text{ molecules}}\right)\left(\frac{m^3}{(10^2 \text{ cm})^3}\right)\left(\frac{10^3 \text{ cm}^3}{1 \text{ L}}\right)$$

$$= 1.66 \times 10^{-17} \frac{\text{mol}}{\text{L}}$$

$$P = \frac{nRT}{V} = \left(1.66 \times 10^{-17} \frac{\text{mol}}{\text{L}}\right)\left(0.08206 \text{ L atm mol}^{-1} \text{ K}^{-1}\right)(25 \text{ K})$$

$$= 3.4 \times 10^{-17} \text{ atm}$$

5.43 $$n_{Kr} = \frac{PV}{RT} = \frac{(10.0 \text{ atm})(0.600 \text{ L})}{(0.08206 \text{ L atm mol}^{-1} \text{ K}^{-1})(1273 \text{ K})} = 5.74 \times 10^{-2} \text{ mol}$$

$m_{Kr} = n_{Kr}MM_{Kr} = (5.74 \times 10^{-2} \text{ mol})(83.80 \text{ g/mol}) = 4.81 \text{ g Kr}$

atoms of Kr $= (5.74 \times 10^{-2} \text{ mol})(6.022 \times 10^{23} \text{ atoms/mol}) = 3.46 \times 10^{22}$ atoms

5.44 $$P = \frac{nRT}{V} = \frac{\left(\frac{96.0 \text{ g } O_2}{32.0 \text{ g / mol}}\right)(0.08206 \text{ L atm mol}^{-1} \text{ K}^{-1})(300 \text{ K})}{3.00 \text{ L}} = 24.63 \text{ atm}$$

$$(24.63 \text{ atm})\left(\frac{760 \text{ torr}}{1 \text{ atm}}\right) = 1.87 \times 10^4 \text{ torr}$$

5.45 Test whether 16 or 32 g of oxygen satisfied the ideal gas equation.

5.46 $V_{tire} = (2.00 \times 10^2\ in^3)\left(\frac{2.54\ cm}{in}\right)^3\left(\frac{1\ L}{10^3\ cm^3}\right) = 3.277\ L$

$V_{gas\ at\ 1\ atm} = (0.35\ ft^3)\left(\frac{12\ in}{1\ ft}\right)^3\left(\frac{2.54\ cm}{1\ in}\right)^3\left(\frac{1\ L}{10^3\ cm^3}\right) = 9.911\ L$

$P_{air} = (1\ atm)(9.911\ L/3.277\ L) = 3.02\ atm$

5.47 $X_{C_3H_6} = \frac{1}{1+4} = \frac{1}{5} = 0.200 \qquad X_{O_2} = \frac{4}{1+4} = \frac{4}{5} = 0.800$

(Assuming mole ratio to at least 3 sig. fig.)

$P_{C_3H_6} = (0.200)(1.00\ atm) = 0.200\ atm \qquad P_{O_2} = (0.800)(1.00\ atm) = 0.800\ atm$

$n_{C_3H_6} = \frac{PV}{RT} = \frac{(0.200\ atm)(2.00\ L)}{(0.08206\ L\ atm\ mol^{-1}\ K^{-1})(296\ K)} = 1.647 \times 10^{-2}\ mol$

$n_{O_2} = \frac{(0.800\ atm)(2.00\ L)}{(0.08206\ L\ atm\ mol^{-1}\ K^{-1})(296\ K)} = 6.587 \times 10^{-2}\ mol$

$m_{C_3H_6} = (n_{C_3H_6})(MM_{C_3H_6}) = 1.647 \times 10^{-2}\ mol)(42.078\ g/mol) = 0.693\ g\ C_3H_6$

$m_{O_2} = (n_{O_2})(MM_{O_2}) = (6.587 \times 10^{-2}\ mol)(32.00\ g/mol) = 2.11\ g\ O_2$

5.48 a) $n_{N_2} = \frac{PV}{RT} = \frac{(2.0\ atm)(10\ L)}{(0.08206\ L\ atm\ mol^{-1}\ K^{-1})(300\ K)} = 0.812\ mol\ N_2$

$n_{O_2} = \frac{PV}{RT} = \frac{(3.0\ atm)(1.0\ L)}{(0.08206\ L\ atm\ mol^{-1}\ K^{-1})(300\ K)} = 0.122\ mol\ O_2$

b) $P_{N_2} = \frac{nRT}{V} = \frac{(0.812\ mol)(0.08206\ L\ atm\ mol^{-1}\ K^{-1})(300\ K)}{11\ L} = 1.8\ atm$

$P_{O_2} = \frac{(0.12\ mol)(0.08206\ L\ atm\ mol^{-1}\ K^{-1})(300\ K)}{11\ L} = 0.27\ atm$

or

$P_{N2} = (2.0\ atm)(10\ L/11L) = 1.8\ atm$

$P_{O2} = (3.0\ atm)(1.0\ L/11L) = 0.27\ atm$

c) 1/11 of the O_2 molecules will be in smaller chamber after mixing.

5.49 0°C = 273 K 22°C = 295 K

$(0.963\ L)\left(\frac{273\ K}{295\ K}\right)\left(\frac{0.969\ atm}{1\ atm}\right) = 0.864\ L$

5.50 $P_{air} = (1.0\text{ atm})(298\text{ K}/90\text{ K}) = 3.31\text{ atm}$

$m_{He} = (0.147\text{ g/mL})(100\text{ mL}) = 14.7\text{ g He}$

$$n_{He} = \frac{14.7\text{ g He}}{4.00\text{ g / mol}} = 3.675\text{ mol He}$$

$$P_{He} = \frac{(3.675\text{ mol})(0.08206\text{ L atm mol}^{-1}\text{ K}^{-1})(298\text{ K})}{2.00\text{ L}} = 44.93\text{ atm}$$

$P_{total} = P_{He} + P_{air} = 44.93\text{ atm} + 3.31\text{ atm} = 48.2\text{ atm}$

5.51 The decrease in pressure is due to the consumption of O_2 by the mouse.

$\therefore$ 760 torr - 720 torr = 40 torr of O_2 consumed.

$$n_{O_2\text{ cons.}} = \frac{PV}{RT} = \frac{\left(\frac{40\text{ torr}}{760\text{ torr / atm}}\right)(2.05\text{ L})}{(0.08206\text{ L atm mol}^{-1}\text{ K}^{-1})(300\text{ K})} = 4.38 \times 10^{-3}\text{ mol}$$

$m_{O_2} = n_{O_2}MM_{O_2} = (4.38 \times 10^{-3}\text{ mol})(32.0\text{ g }O_2) = 0.14\text{ g }O_2$

5.52 $P_{air} = (1.00\text{ atm})(500\text{ K}/300\text{ K}) = 1.67\text{ atm}$

$$n_{CuSO_4\bullet 5H_2O} = \left(\frac{2.50\text{ g}}{249.69\text{ g / mol}}\right) = 1.00 \times 10^{-2}\text{ mol}$$

$CuSO_4\bullet 5H_2O \rightarrow CuSO_4 + 5H_2O$ $\quad n_{H_2O} = 5 \times 1.00 \times 10^{-2}\text{ mol} = 5.00 \times 10^{-2}\text{ mol}$

$$P_{H_2O} = \frac{(5.00 \times 10^{-2}\text{ mol})(0.0821\text{ L atm mol}^{-1}\text{ K}^{-1})(500\text{ K})}{4.00\text{ L}} = 0.513\text{ atm}$$

$P_{total} = P_{H_2O} + P_{air} = 0.513\text{ atm} + 1.67\text{ atm} = 2.18\text{ atm}$

5.53
$$n_{H_2} = \frac{PV}{RT} = \frac{(20.0\text{ atm})(100\text{ L})}{(0.0821\text{ L atm mol}^{-1}\text{ K}^{-1})(600\text{ K})} = 40.6\text{ mol }H_2$$

$$n_{CO} = \frac{(10.0\text{ atm})(100\text{ L})}{(0.0821\text{ L atm mol}^{-1}\text{ K}^{-1})(600\text{ K})} = 20.3\text{ mol CO}$$

Reaction:	$3\ H_2$	+	CO	$\rightarrow$	CH_4	+	H_2O
Initial amounts:	40.6 mol		20.3 mol		0		0
Change in amounts:	-40.6 mol		-13.5 mol		13.5 mol		13.5 mol
Final amounts:	0 mol		6.8 mol		13.5 mol		13.5 mol

$m_{CH_4} = (n_{CH_4})(MM_{CH_4}) = (13.5\text{ mol})(16.042\text{ g/mol}) = 216.6\text{ g }CH_4$ = theor. yield

$$\text{percent yield} = \frac{150\text{ g }CH_4}{216.6\text{ g }CH_4} \times 100\% = 69.3\%$$

5.54 VP_{H_2O} at 30°C = 31.824 torr (from Table 5-4)

Pressure after drying = 756 torr - 31.824 torr = 724 torr

$$P_{H_2O} = X_{H_2O}P_{total} \qquad X_{H_2O} = \frac{P_{H_2O}}{P_{total}} = \frac{31.824 \text{ torr}}{756 \text{ torr}} = 0.0421$$

$$\frac{n}{V} = \frac{P_{H_2O}}{RT} = \frac{\left(\dfrac{31.824 \text{ torr}}{760 \text{ torr / atm}}\right)}{(0.0821 \text{ L atm mol}^{-1} \text{ K}^{-1})(303 \text{ K})} = \frac{1.68 \times 10^{-3} \text{ mol } H_2O}{\text{L}}$$

$$\frac{m}{V} = \left(\frac{n}{V}\right)(MM_{H_2O}) = \frac{1.68 \times 10^{-3} \text{ mol } H_2O}{\text{L}} \text{x } (18.016 \text{ g / mol}) = 0.0303 \text{ g / L}$$

5.55 (The MnO_2 is a catalyst in this reaction.)

$$\text{a) } n_{O_2} = \frac{PV}{RT} = \frac{\left(\dfrac{759.2 \text{ torr}}{760 \text{ torr / atm}}\right)(0.02296 \text{ L})}{(0.08206 \text{ L atm mol}^{-1} \text{ K}^{-1})(298 \text{ K})} = 9.379 \times 10^{-4} \text{ mol } O_2$$

b) $2KClO_3 \rightarrow 2KCl + 3O_2$

$$9.379 \times 10^{-4} \text{ mol } O_2 \text{ x} \frac{2 \text{ mol } KClO_3}{3 \text{ mol } O_2} = 6.253 \times 10^{-4} \text{ mol } KClO_3$$

c) $(6.253 \times 10^{-4} \text{ mol } KClO_3)(122.55 \text{ g/mol}) = 0.07663 \text{ g } KClO_3$

$$\text{mass \%} = \frac{0.07663 \text{ g } KClO_3}{0.1054 \text{ g mixture}} \text{x } 100\% = 72.70\%$$

5.56 VP_{H_2O} at 25°C = 23.756 torr

$$\text{Rel.Hum.} = \frac{P_{H_2O}}{VP_{H_2O}} \text{x } 100 \qquad P_{H_2O} = \frac{(50)(23.756 \text{ torr})}{100} = 11.878 \text{ torr}$$

$$P_{H_2O} = 12 \text{ torr} = \left(\frac{12 \text{ torr}}{760 \text{ torr / atm}}\right) = 1.6 \times 10^{-2} \text{ atm} = 0.016 \text{ atm}$$

$$P_{dry\ air} = 765 \text{ tor} - 12 \text{ torr} = 753 \text{ torr} = \left(\frac{753 \text{ torr}}{760 \text{ torr / atm}}\right) = 0.991 \text{ atm}$$

$P_{N_2} = X_{N_2}\ P_{dry\ air} = (0.7808)(0.991 \text{ atm}) = 0.774 \text{ atm}$

$P_{O_2} = X_{O_2}\ P_{dry\ air} = (0.2095)(0.991 \text{ atm}) = 0.208 \text{ atm}$

P_{H_2O}= 0.016 atm (from the calculation above)

$P_{Ar} = (9.34 \times 10^{-3})(0.991 \text{ atm}) = 0.00926 \text{ atm}$

$P_{CO_2} = (3.25 \times 10^{-4})(0.991 \text{ atm}) = 0.000322 \text{ atm}$

$P_{Ne} = (1.82 \times 10^{-5})(0.991 \text{ atm}) = 0.0000180 \text{ atm}$

$P_{He} = (5.24 \times 10^{-6})(0.991 \text{ atm}) = 0.00000510 \text{ atm}$

$P_{CH_4} = (1.4 \times 10^{-6})(0.991 \text{ atm}) = 0.0000013 \text{ atm}$

5.57 a) for dry air

$m_{N_2} = (28.02 \text{ g/mol})(0.7808 \text{ mol}) = 21.88 \text{ g } N_2$

$m_{O_2} = (32.00 \text{ g/mol})(0.2095 \text{ mol}) = 6.704 \text{ g } O_2$

$m_{Ar} = (39.948 \text{ g/mol})(9.34 \times 10^{-3} \text{ mol}) = 0.373 \text{ g Ar}$

$m_{CO_2} = (44.01 \text{ g/mol})(3.25 \times 10^{-4} \text{ mol}) = 0.0143 \text{ g } CO_2$

Total mass = 28.97 g

$$V = \frac{nRT}{P} = \frac{(1)(0.0821 \text{ L atm mol}^{-1} \text{ K}^{-1})(300 \text{ K})}{1 \text{ atm}} = 24.6 \text{ L}$$

density of dry air = 28.97 g/24.6 L = 1.18 g/L

b) VP_{H_2O} at 300 K = (31.824 - 23.756) torr (2/5) + 23.756 torr = 26.983 torr

or (26.983 torr) ÷ (760 torr/atm) = 0.0355 atm

$P_{\text{dry air}}$ = (760 torr - 26.983 torr) = 733 torr = (733 torr)/(760 torr/atm) = 0.964 atm

$$n_{\text{dry air}} = \frac{PV}{RT} = \frac{(0.964 \text{ atm})(24.6 \text{ L})}{(0.0821 \text{ L atm mol}^{-1} \text{ K}^{-1})(300 \text{ K})} = 0.963 \text{ mol}$$

$$n_{H_2O} = \frac{(0.0355 \text{ atm})(24.6 \text{ L})}{(0.0821 \text{ L atm mol}^{-1} \text{ K}^{-1})(300 \text{ K})} = 0.0355 \text{ mol}$$

mass of 0.963 mol of dry air ≈ (0.963 mol)(28.97 g/mol) = 27.90 g

mass of 0.0355 mol of H_2O = (0.0355 mol)(18.016 g/mol) = 0.64 g

mass of wet air = 27.90 g + 0.64 g = 28.54 g

density of wet air = 28.54 g/24.6 L = 1.16 g/L

5.58 $$(100 \text{ L } O_2)(1.14 \text{ g / mL})\left(\frac{1000 \text{ mL}}{1 \text{ L}}\right)\left(\frac{1 \text{ mol}}{32.00 \text{ g}}\right) = 3.56 \times 10^3 \text{ mol } O_2$$

$$V_{O_2} = \frac{nRT}{P} = \frac{(3.56 \times 10^3 \text{ mol})(0.08206 \text{ L atm mol}^{-1} \text{ K}^{-1})(298 \text{ K})}{(750 / 760) \text{ atm}} = 8.82 \times 10^4 \text{ L}$$

$V_{Air} = 8.82 \times 10^4 / 0.2095 = 4.21 \times 10^5$ L

5.59 $$\text{Relative humidty} = \frac{P_{H_2O}}{VP_{H_2O}} \times 100 \qquad P_{H_2O} = \frac{(\text{rel. hum.})(VP_{H_2O})}{100}$$

a) $$P_{H_2O} = \frac{(80)(42.175 \text{ torr})}{100} = 33.74 \text{ torr}$$

$$\left(\frac{33.74 - 31.824}{42.175 - 31.824}\right)(5) + 30°\text{C} = 30.9°\text{C} \approx 31°\text{C}$$

b) $$P_{H_2O} = \frac{(50)(12.788 \text{ torr})}{100} = 6.394 \text{ torr}$$

$$\left(\frac{6.394 - 4.579}{6.543 - 4.579}\right)(5) + 0°\text{C} = 4.62°\text{C} \approx 4.6°\text{C}$$

c) $$P_{H_2O} = \frac{(30)(23.756 \text{ torr})}{100} = 7.1268 \text{ torr}$$

$$\left(\frac{7.1268 - 6.543}{9.209 - 6.543}\right)(5) + 5°\text{C} = 6.09°\text{C} \approx 6.1°\text{C}$$

5.60 Assuming ppm in terms of moles/moles $\therefore$ 0.50 ppm $= \dfrac{0.50 \text{ mol } O_3}{10^6 \text{ mol air}} = X_{O_3}$

$$P_{O_3} = X_{O_3} P_{total} = \left(\frac{0.50}{10^6}\right)(762 \text{ torr}) = 3.81 \times 10^{-4} \text{ torr or } 5.013 \times 10^{-7} \text{ atm}$$

$$\frac{n_{O_3}}{V} = \frac{P}{RT} = \frac{(5.013 \times 10^{-7} \text{ atm})}{(0.0821 \text{ L atm mol}^{-1} \text{ K}^{-1})(301 \text{ K})} = 2.03 \times 10^{-8} \frac{\text{mol}}{\text{L}}$$

$$\left(2.03 \times 10^{-8} \frac{\text{mol}}{\text{L}}\right)\left(\frac{6.022 \times 10^{23} \text{ molecules}}{1 \text{ mol}}\right)\left(\frac{1 \text{ L}}{1000 \text{ cm}^3}\right)$$

$$= 1.22 \times 10^{13} \text{ molecules / cm}^3$$

5.61 $$n_{unk} = \frac{PV}{RT} = \frac{\left(\dfrac{435 \text{ torr}}{760 \text{ torr / atm}}\right)(0.150 \text{ L})}{(0.0821 \text{ L atm mol}^{-1} \text{ K}^{-1})(423 \text{ K})} = 2.472 \times 10^{-3} \text{ mol}$$

$$MM_{unkn} = \frac{m}{n} = \frac{0.250 \text{ g}}{2.472 \times 10^{-3} \text{ mol}} = 101 \text{ g / mol}$$

H: 14.94 14.94/1.008 = 14.82; 14.82 ÷ 0.9879 = 15.00
C: 71.22 71.22/12.01 = 5.93; 5.93 ÷ 0.9879 = 6.00
N: 13.84 13.84/14.01 = 0.9879; 0.9879 ÷ 0.9879 = 1.00 $C_6H_{15}N$

$MM_{C_6H_{15}N}$ = 101 g/mol $\therefore$ molecular formula = $C_6H_{15}N$

5.62 6,000 x 10^6 mtons C =
(6,000 x 10^6 mtons C)(1000 kg/mton)(44.01 kg CO_2 /12.01 kg C)
= 2.199 x 10^{13} kg CO_2 or (2.199 x 10^{13} kg)(10^3g/1 kg) = 2.199 x 10^{16} g CO_2

$$n_{CO_2} = \frac{2.199 \times 10^{16} \text{ g } CO_2}{44.01 \text{ g / mol}} = 5.00 \times 10^{14} \text{ mol } CO_2$$

5.63 $He_A : He_B : He_C$ = 6 : 12 : 9 Assume each chamber contains only He.

a) B has the highest pressure because it contains the greatest amount of gas (He).

b) The pressure of A would be 0.5 atm (1/2 of 1.0 atm) because it contains half as much He (6 : 12).

c) The new pressure would be (1.0 atm)(27/6) = 4.5 atm because the amount of He is increased 4.5 (27/6) times.

d) If the valves were opened, the 27 portions of He (6 + 12 + 9) would divide equally between the chambers, 27/3 = 9. The contents and, therefore, the pressure would be 3/4 (9/12) of the original pressure. The pressure would be (0.50 atm) x (9/12) or 0.38 atm.

5.64 $$n_{N_2} = \frac{PV}{RT} = \frac{(150 \text{ atm})(9.50 \text{ L})}{(0.08206 \text{ L atm mol}^{-1} \text{ K}^{-1})(298 \text{ K})} = 58.27 \text{ mol } N_2$$

m_{N_2} = (58.27 mol N_2)(28.02 g/mol) = 1.63 x 10^3 g N_2

5.65 $n_{air} = \frac{PV}{RT} = \frac{(170\text{ atm})(12.5\text{ L})}{(0.0821\text{ L atm mol}^{-1}\text{ K}^{-1})(300\text{ K})} = 86.28\text{ mol air}$

$n_{O_2} = (X_{O_2})(n_{air}) = (0.20)(86.28\text{ mol}) = 17.3\text{ mol }O_2$

$m_{O_2} = (32.00\text{ g/mol})(17.3\text{ mol }O_2) = 554\text{ g }O_2$

$(554\text{ g }O_2) \div (14.0\text{ g }O_2/\text{min}) = 39.6\text{ min}$

39.6 min - 6.0 min = 33.6 min of diving

5.66 $v_{avg} = \left(\frac{3\,RT}{MM}\right)^{1/2}$ Equation 5-2 $MM_{O_3} = 48.00\text{ g/mol} = 48.00 \times 10^{-3}\text{ kg/mol}$

$J = kg{\bullet}m^2/s^2$

$$v_{avg} = \left[\frac{3(8.314\text{ J / mol K})(248\text{ K})}{48.00 \times 10^{-3}\text{ kg / mol}}\right]^{1/2} = 359\text{ m / s}$$

5.67 $n_{Ar} = \frac{1.00\text{ g}}{39.95\text{ g / mol}} = 0.0250\text{ mol}$

$n_{Ne} = \frac{0.050\text{ g}}{20.18\text{ g / mol}} = 0.0025\text{ mol}$

$X_{Ar} = \frac{0.0250\text{ mol}}{0.0250\text{ mol} + 0.0025\text{ mol}} = 0.9091$

$X_{Ne} = \frac{0.0025\text{ mol}}{0.0250\text{ mol} + 0.0025\text{ mol}} = 0.0909$

$P_{total} = \frac{nRT}{V}$

$$= \frac{(0.0250\text{ mol} + 0.0025\text{ mol})(0.08206\text{ L atm mol}^{-1}\text{ K}^{-1})(275\text{ K})}{5.00\text{ L}} = 0.124\text{ atm}$$

$P_{Ar} = (X_{Ar})(P_{total}) = (0.9091)(0.124\text{ atm}) = 0.113\text{ atm}$

$P_{Ne} = (X_{Ne})(P_{total}) = (0.0909)(0.124\text{ atm}) = 0.0113\text{ atm}$

5.68

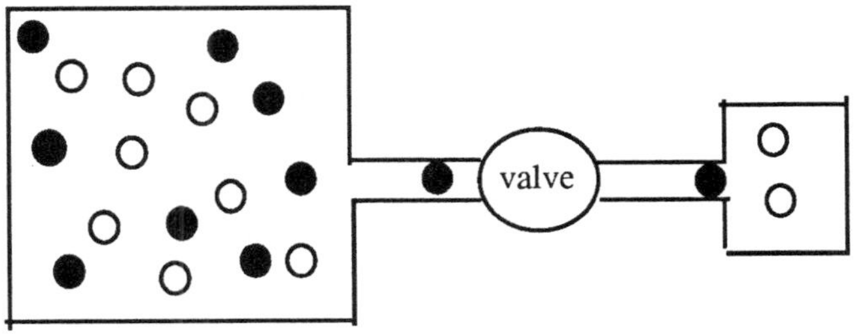

"After equilibrium is reached."

a) The 20 particles would be evenly spaced throughout the total volume of 60 L. That would require ≈ 3 1/3 molecules in small tank and ≈ 16 2/3 molecules in larger tank. Assuming no change:

(continued)

(5.68 continued)

b) $P_{total} = \frac{nRT}{V} = \frac{(20)(0.0821 \text{ L atm mol}^{-1}\text{ K}^{-1})(273 \text{ K})}{60 \text{ L}}$

$= 7.47$ atm (7.5 atm)

c) $X_O = 1/2$ $X_\bullet = 1/2$

$$P_{total} = \frac{nRT}{V} = \frac{(20)(0.0821 \text{ L atm mol}^{-1}\text{ K}^{-1})(273 \text{ K})}{10 \text{ L}} = 44.8 \text{ atm}$$

$P_O = (1/2)(44.8 \text{ atm}) = 22$ atm $P_\bullet = (1/2)(44.8 \text{ atm}) = 22$ atm

5.69 The sulfur burns when the coal is burned: $S + O_2 \rightarrow SO_2$

The SO_2 then either dissolves in rain: $SO_2 + H_2O \rightarrow H_2SO_3$ (sulfurous acid)

or is oxidized by O_2 and uv light: $2SO_2 + O_2 \rightarrow 2SO_3$

The SO_3 then dissolves in rain: $SO_3 + H_2O \rightarrow H_2SO_4$ (sulfuric acid)

5.70 Equation 5-1; $\text{k.E.}_{ave.} = \frac{3RT}{2N_A}$; solving for $RT = \left(\frac{2}{3}\right)(\text{k.E.}_{ave.})(N_A)$

$$P = \frac{nRT}{V} = \frac{n}{V}\left(\frac{2}{3}\right)(\text{k.E.}_{ave.})(N_A)$$

$$= \frac{\left(\frac{1.55 \text{ g Ar}}{39.95 \text{ g Ar / mol}}\right)}{5.00 \text{ L}}\left(\frac{2}{3}\right)\left(\frac{1.02 \times 10^{-22} \text{ kg m}^2/\text{s}^2}{1 \text{ atom}}\right)\left(\frac{6.022 \times 10^{23} \text{ atoms}}{1 \text{ mol}}\right)$$

$$= 0.3178 \frac{\text{kg m}^2/\text{s}^2}{\text{L}} = \left(0.3178 \frac{\text{kg m}^2/\text{s}^2}{\text{L}}\right)\left(\frac{1 \text{ L}}{10^{-3} \text{m}^3}\right)$$

$= 0.3178 \times 10^3 \text{ kg/m}\cdot\text{s}^2$ $\{L = 1000 \text{ cm}^3 = (10^3)(10^{-2} \text{ m})^3 = 10^{-3} \text{ m}^3\}$

$= 0.3178 \times 10^3$ Pa $\{1 \text{ Pa} = 1 \text{ kg/m}\cdot\text{s}^2\}$; therefore $\{1 \text{ atm} = 1.013 \times 10^5 \text{ Pa}\}$

$$= (0.3178 \times 10^3 \text{ Pa})\left(\frac{1 \text{ atm}}{1.013 \times 10^5 \text{ Pa}}\right) = 3.14 \times 10^{-3} \text{ atm}$$

$\text{k.E.}_{ave.} = 1/2m\, v^2_{ave.} = 1.02 \times 10^{-22}$ J/atm $= 1.02 \times 10^{-22} \text{ kg}\cdot\text{m}^2/\text{s}^2\cdot\text{atom}$

$$\frac{1}{2}\left(\frac{39.95 \times 10^{-3} \text{ kg Ar / mol}}{6.022 \times 10^{23} \text{ atoms Ar / mol}}\right)v^2_{ave.} = 1.02 \times 10^{-22} \text{ kg m}^2/\text{s}^2 \bullet \text{atom}$$

$$v^2_{ave.} = \frac{2(1.02 \times 10^{-22} \text{ kg m}^2/\text{s}^2 \bullet \text{atom})}{\left(\frac{39.95 \times 10^{-3} \text{ kg Ar / mol}}{6.022 \times 10^{23} \text{ atoms Ar / mol}}\right)} = \frac{2.04 \times 10^{-22} \text{ kg m}^2/\text{s}^2 \bullet \text{atom}}{6.634 \times 10^{-26} \text{ kg / atom}}$$

$$v^2_{ave.} = 3.075 \times 10^3 \text{ m}^2/\text{s}^2$$

$$v_{ave.} = \sqrt{3.075 \times 10^3 \text{ m}^2/\text{s}^2} = 55.5 \text{ m / s}$$

5.71 a) ~400 m/s

b) $$\text{k.E.} = 1/2mv^2 = \left(\frac{1}{2}\right)\left(\frac{17.03 \times 10^{-3}\ \text{kg/mol}}{6.022 \times 10^{23}\ \text{molecules/mol}}\right)(400\ \text{m/s})^2$$

$$= 2.26 \times 10^{-21}\ \text{kg m}^2/\text{s}^2\ \text{molecule}$$

$$= 2.26 \times 10^{-21}\ \text{J/molecule}$$

5.72 $$(363\ \text{mg } CO_2)\left(\frac{12.01\ \text{mg C}}{44.01\ \text{mg } CO_2}\right) = 99.06\ \text{mg C}$$

$$\frac{99.06\ \text{mg C}}{12.01\ \text{g/mol}} \times \frac{10^{-3}\ \text{g}}{\text{mg}} = 8.25 \times 10^{-3}\ \text{mol C};\ \frac{8.25 \times 10^{-3}}{1.17 \times 10^{-3}} = 7\ \text{C}$$

$$(63.7\ \text{mg } H_2O)\left(\frac{2.016\ \text{mg H}}{18.016\ \text{mg } H_2O}\right) = 7.13\ \text{mg H}$$

$$\frac{7.13\ \text{mg H}}{1.008\ \text{g/mol}} \times \frac{10^{-3}\ \text{g}}{\text{mg}} = 7.07 \times 10^{-3}\ \text{mol H};\ \frac{7.07 \times 10^{-3}}{1.17 \times 10^{-3}} = 6\ \text{H}$$

mg O = 125 mg benzaldehyde - 99.06 mg C - 7.13 mg H = 18.81 mg O

$$\frac{18.81\ \text{mg O}}{16.0\ \text{g/mol}} \times \frac{10^{-3}\ \text{g}}{\text{mg}} = 1.17 \times 10^{-3}\ \text{mol O};\ \frac{1.17 \times 10^{-3}}{1.17 \times 10^{-3}} = 1\ \text{O}$$

$$n = \frac{PV}{RT} = \frac{\left(\frac{274\ \text{torr}}{760\ \text{torr/atm}}\right)(0.100\ \text{L})}{(0.0821\ \text{L atm mol}^{-1}\ \text{K}^{-1})(423\ \text{K})} = 1.038 \times 10^{-3}\ \text{mol}$$

$MM = (110 \times 10^{-3}\ \text{g})/1.038 \times 10^{-3}\ \text{mol} = 106\ \text{g/mol}$

MM of C_7H_6O is 106; therefore the formula of benzaldehyde is C_7H_6O.

5.73 $$n = \frac{PV}{RT} = \frac{m}{MM} \qquad m = \frac{MM\ PV}{RT} \qquad m_{He} + 300 \times 10^3\ \text{g} = m_{air}$$

$MM_{He} = 4.0026$ g/mol; $MM_{air} = 28.97$ g/mol (See Prob. 5.57)

$$\frac{MM_{He}\ PV}{RT} + 300 \times 10^3\ \text{g} = \frac{MM_{air}\ PV}{RT}$$

$$\frac{(4.0026\ \text{g/mol})(1.0\ \text{atm})V}{(0.0821\ \text{L atm mol}^{-1}\ \text{K}^{-1})(298\ \text{K})} + 300 \times 10^{-3}\ \text{g}$$

$$= \frac{(28.97\ \text{g/mol})(1.0\ \text{atm})V}{(0.0821\ \text{L atm mol}^{-1}\ \text{K}^{-1})(298\ \text{K})}$$

(continued)

(5.73 continued)

$$300 \times 10^3 \text{ g} = (28.97 - 4.0026 \text{ g/mol})\left(\frac{(1.0 \text{ atm})}{(0.0821 \text{ L atm mol}^{-1} \text{ K}^{-1})(298 \text{ K})}\right)V$$

$$300 \times 10^3 \text{ g} = (24.97 \text{g/mol})(0.04087 \frac{\text{mol}}{\text{L}})V$$

$$V = \frac{300 \times 10^3 \text{ g}}{(24.97 \text{ g/mol})(0.04087 \text{ mol/L})} = 2.9 \times 10^5 \text{ L}$$

5.74 Assuming the balloon is to have the same volume, the density of the hot air must equal the density of He at 1.0 atm and 25°C.

$$\text{density} = \frac{\text{mass}}{V} = \frac{(n_{He})(MM_{He})}{V} = \left(\frac{P}{RT}\right)(MM_{He})$$

$$= \frac{1.0 \text{ atm}}{(0.0821 \text{ L atm mol}^{-1} \text{ K}^{-1})(298 \text{ K})}(4.0026 \text{ g/mol})) = 0.1636 \text{ g/L}$$

$$\text{density of hot air} = \frac{0.1636 \text{ g}}{\text{L}} = \frac{n_{air} MM_{air}}{V} = \left(\frac{P}{RT}\right)(MM_{air})$$

(See Problem 5.57 for the value of MM_{air})

$$\frac{0.1636 \text{ g}}{\text{L}} = \left(\frac{P}{RT}\right)(MM_{air}) = \left[\frac{1.0 \text{ atm}}{(0.0821 \text{ L atm mol}^{-1} \text{ K}^{-1})(T)}\right](28.97 \text{ g/mol})$$

$$T = \left[\frac{1.0 \text{ atm}}{(0.0821 \text{ L atm mol}^{-1} \text{ K}^{-1})}\right](28.97 \text{ g/mol})/(0.1636 \text{ g/L})$$

$= 2157 \text{ K} \approx 2.2 \times 10^3 \text{ K} = 1884°\text{C} \approx 1.9 \times 10^3 °\text{C}$

5.75 a) $MM_{TNT} = 227.14$ g/mol $\qquad n_{TNT} = \dfrac{1.0 \times 10^3 \text{ g}}{227.14 \text{ g/mol}} = 4.40 \text{ mol}$

$$2C_7H_5(NO_2)_{3(s)} \rightarrow 12CO_{(g)} + 2C_{(s)} + 5H_{2(g)} + 3N_{2(g)}$$

4.40 mol $\qquad$ 26.4 mol $\quad$ 4.4 mol $\quad$ 11.0 mol $\quad$ 6.60 mol

moles of gas = 26.4 + 11.0 mol + 6.60 mol = 44.0 mol

b) $V = \dfrac{nRT}{P} = \dfrac{(44.0 \text{ mol})(0.0821 \text{ L atm mol}^{-1} \text{ K}^{-1})(298 \text{ K})}{1.0 \text{ atm}} = 1076 \text{ L} = 1.1 \times 10^3 \text{ L}$

c) $X_{CO} = \dfrac{26.4 \text{ mol}}{44.0 \text{ mol}} = 0.60 \qquad X_{H_2} = \dfrac{11.0 \text{ mol}}{44.0 \text{ mol}} = 0.25$

$X_{N_2} = \dfrac{6.60 \text{ mol}}{44.0 \text{ mol}} = 0.15$

$P_{CO} = (0.60)(1.0 \text{ atm}) = 0.60 \text{ atm} \qquad P_{H_2} = (0.25)(1.0 \text{ atm}) = 0.25 \text{ atm}$

$P_{N_2} = (0.15)(1.0 \text{ atm}) = 0.15 \text{ atm}$

5.76 $X_{CO_2} = 351.5/10^6 = 3.515 \times 10^{-4}$ Assume P = 1 atm

$P_{CO_2} = X_{CO_2} \times P_{total} = 3.515 \times 10^{-4} \times 1 \text{ atm} = 3.515 \times 10^{-4} \text{ atm}$

$$n_{CO_2} = \frac{PV}{RT} = \frac{(3.515 \times 10^{-4} \text{ atm})(1.0 \text{ L})}{(0.0821 \text{ L atm mol}^{-1} \text{ K}^{-1})(223 \text{ K})} = 1.92 \times 10^{-5} \text{ mol}$$

CO_2 molecules = $(1.92 \times 10^{-5} \text{ mol})(6.022 \times 10^{23} \text{ molecules/mol})$
= 1.2×10^{19} molecules

5.77 $$P_{SF_6} = \left(\frac{1}{10^9}\right)(1 \text{ atm}) = 10^{-9} \text{ atm}$$

$$n = \frac{PV}{RT} = \frac{(10^{-9} \text{ atm})(10^{-3} \text{ L})}{(0.0821 \text{ L atm mol}^{-1} \text{ K}^{-1})(294 \text{ K})} = 4.14 \times 10^{-14} \text{ mol}$$

molecules of SF_6 = $(4.14 \times 10^{-14} \text{ mol})(6.022 \times 10^{23} \text{ molecules/mol})$

= 2.49×10^{10} molecules

5.78 $$n_{Ni(CO)_x} = \frac{PV}{RT} = \frac{\left(\frac{552 \text{ torr}}{760 \text{ torr / atm}}\right)(0.100 \text{ L})}{(0.0821 \text{ L atm mol}^{-1} \text{ K}^{-1})(303 \text{ K})} = 2.920 \times 10^{-3} \text{ mol}$$

$$MM_{Ni(CO)_x} = \frac{0.500 \text{ g}}{2.920 \times 10^{-3} \text{ mol}} = 171 \text{ g / mol}$$

171 g = 58.70 g + (x)(28.01 g); (28.01 g) x = 112.3 g; x = 4; Answer: $Ni(CO)_4$

5.79 a)

P

V

V = nRT / P

The product of P x V is equal to the constant nRT.

b)

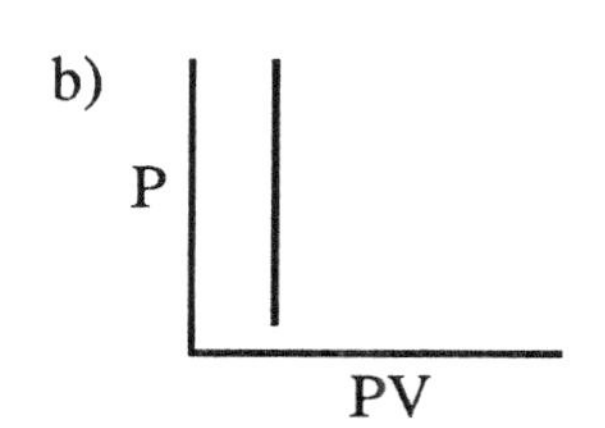

PV = nRT

PV = constant

c)

P

1 / V

P x (1/nRT) = 1/V

5.80 Use the ideal gas law in the form: $n = PV/RT$. By measuring the P, V and T of a weighed sample of gas, the number of moles of gas can be determined. Then the molar mass equals the weight of sample divided by the number of moles ($MM = m/n$).

5.81 The figure in the textbook shows:

$$\underset{4\ \text{mol}}{N_2} + \underset{12\ \text{mol}}{3H_2} \rightarrow \underset{8\ \text{mol}}{2NH_3}$$

Your drawing of the system after reaction must show 8 representations of NH_3.

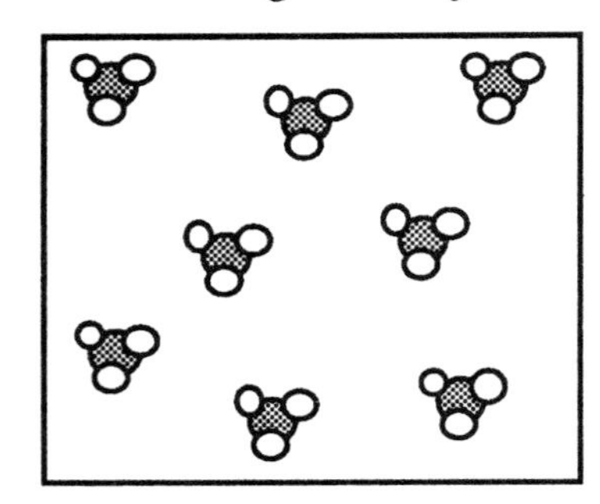

5.82 a) $V = (3.00\ \text{L})\left(\dfrac{500\ \text{torr}}{750\ \text{torr}}\right) = 2.00\ \text{L}$

b) $P = (500\ \text{torr})\left(\dfrac{3.00\ \text{L}}{2.00\ \text{L}}\right) = 750\ \text{torr}$

c) $P = (500\ \text{torr})\left(\dfrac{323\ \text{K}}{273\ \text{K}}\right) = 592\ \text{torr}$

d) $P = (500\ \text{torr})\left(\dfrac{3.00\ \text{L}}{1.50\ \text{L}}\right)\left(\dfrac{223\ \text{K}}{273\ \text{K}}\right) = 817\ \text{torr}$

e) $P = (500\ \text{torr})\left(\dfrac{1/2n_{gas}}{n_{gas}}\right) = 250\ \text{torr}$

5.83 $n_{CO_2} = \dfrac{15.00\ \text{g}}{44.01\ \text{g/mol}} = 0.3408\ \text{mol}$

$$P = \frac{nRT}{V} = \frac{(0.3408\ \text{mol})(0.0821\ \text{L atm mol}^{-1}\ \text{K}^{-1})(273\ \text{K})}{(0.750\ \text{L})} = 10.2\ \text{atm}$$

5.84 $\dfrac{n}{V} = \dfrac{P}{RT}$ $\qquad$ density $= \dfrac{m}{V} = \dfrac{nMM}{V} = \dfrac{P\,MM}{RT}$

$$\text{density of } H_2 = \frac{P\,MM_{H_2}}{RT} = \frac{\left(\dfrac{380\ \text{torr}}{760\ \text{torr/atm}}\right)(2.016\ \text{g/mol})}{(0.0821\ \text{L atm mol}^{-1}\ \text{K}^{-1})(250\ \text{K})} = 0.0491\ \text{g/L}$$

5.85 See Problem 5.84 for the formula that can be used in this problem.

$$\text{Density of } SF_6 = \frac{P\,MM_{SF_6}}{RT} = \frac{\left(\dfrac{755 \text{ torr}}{760 \text{ torr / atm}}\right)(146.05 \text{ g / mol})}{(0.0821 \text{ L atm mol}^{-1} \text{ K}^{-1})(300 \text{ K})} = 5.89 \text{ g / L}$$

5.86 a) At constant T and V, P is (directly) proportional to the number of moles of gas, n.

b) True

c) When gas is added to a chamber at fixed V and T, the pressure increases as n.

d) At fixed n and P, V is (directly) proportional to T.

e) True

5.87 in auto: $N_2 + O_2 \rightarrow 2\ NO$

in atmosphere: $2\ NO + O_2 \rightarrow 2\ NO_2$

5.88 H_2 gas at 0°C and 2 atm $\qquad$ O_2 gas at 25°C and 1 atm

equal volume bulbs

a) $n = \dfrac{PV}{RT} = \dfrac{P}{T}\left(\dfrac{V}{R}\right)$

$$n_{H_2} = \frac{2 \text{ atm}}{273 \text{ K}}\left(\frac{V}{R}\right) = 7.3 \text{ x } 10^{-3}\frac{\text{atm}}{\text{K}}\left(\frac{V}{R}\right)$$

$$n_{O_2} = \frac{1 \text{ atm}}{298 \text{ K}}\left(\frac{V}{R}\right) = 3.36 \text{ x } 10^{-3}\frac{\text{atm}}{\text{K}}\left(\frac{V}{R}\right)$$

molecules = nN_A, $\therefore$ more molecules of H_2 than O_2.

b) $m_{H_2} = \left(n_{H_2}\right)\left(MM_{H_2}\right) = 7.3 \text{ x } 10^{-3}\dfrac{\text{atm}}{\text{K}}\left(\dfrac{V}{R}\right)(2.016 \text{ g / mol}) = 0.0147\dfrac{\text{atm} \bullet \text{g}}{\text{K} \bullet \text{mol}}\left(\dfrac{V}{R}\right)$

$m_{O_2} = \left(n_{O_2}\right)\left(MM_{O_2}\right) = 3.36 \text{ x } 10^{-3}\dfrac{\text{atm}}{\text{K}}\left(\dfrac{V}{R}\right)(32.00 \text{ g / mol}) = 0.1075\dfrac{\text{atm} \bullet \text{g}}{\text{K} \bullet \text{mol}}\left(\dfrac{V}{R}\right)$

More mass of O_2 than H_2

c) Higher average kinetic energy of O_2 molecules because at higher temperature (25°C vs 0°C)

$$\text{k.E.}_{\text{average}} = \frac{3RT}{2N_A} \quad \text{Equation 5.1}$$

(continued)

(5.88 continued)

O_2 at 25°C, $\text{k.E.}_{average} = \dfrac{(3)(8.314\ \text{J / mol K})(298\ \text{K})}{(2)(6.022 \times 10^{23}\ \text{molecules / mol})}$

$= 6.17 \times 10^{-21}$ J / molecule

H_2 at 0°C, $\text{k.E.}_{average} = \dfrac{(3)(8.314\ \text{J / mol K})(273\ \text{K})}{(2)(6.022 \times 10^{23}\ \text{molecules / mol})}$

$= 5.65 \times 10^{-21}$ J / molecule

d) $\text{k.E.}_{average} = \frac{1}{2} m v^2_{average}$ $\quad$ $J = kg \cdot m^2/s^2$

$$v_{average} = \sqrt{\frac{2\ \text{k.E.}_{average}}{m}}$$

$$v_{average,\ O_2} = \sqrt{\frac{(2)(6.17 \times 10^{-21}\ \text{kg} \cdot \text{m}^2 / \text{s}^2 \cdot \text{molecule})}{(32.00 \times 10^{-3}\ \text{kg} / 6.022 \times 10^{23}\ \text{molecule})}} = 482\ \text{m / s}$$

$$v_{average,\ H_2} = \sqrt{\frac{(2)(5.65 \times 10^{-21}\ \text{kg} \cdot \text{m}^2 / \text{s}^2 \cdot \text{molecule})}{(2.016 \times 10^{-3}\ \text{kg} / 6.022 \times 10^{23}\ \text{molecule})}} = 1837\ \text{m / s}$$

H_2 has a higher average molecular velocity.

CHAPTER 6: ATOMS AND LIGHT

6.1 Knowing the grams per volume (density) and the grams per mole (atomic mass), one can calculate the volume per mole of atoms (molar volume, in this case where they are all treated as single atom units). From molar volume one can conclude which element has the largest atoms since each molar volume will contain the same number of atoms.

Molar Volume = MM ÷ density

Molar Volume of Al = 26.98 g/mol ÷ 2.70 g/cm^3 = 9.99 cm^3/mol

Molar Volume of Hg = 200.6 g/mol ÷ 13.55 g/cm^3 = 14.80 cm^3/mol

Molar Volume of Pb = 207.2 g/mol ÷ 11.34 g/cm^3 = 18.27 cm^3/mol

Each mole has the same number of atoms, ∴ Pb has the largest atoms;

Al has the smallest atoms.

6.2 From the molar volumes obtained in Problem 6.1 one can calculate atomic volumes, then one can calculate atomic radii, and then the thickness of 6.5 x 10^6 atoms.

Vol. of an Al atom = (9.99 cm^3/mol) ÷ (6.022 x 10^{23} atoms/mol)
= 1.659 x 10^{-23} cm^3/atom

Vol. of a Hg atom = (14.80 cm^3/mol) ÷ (6.022 x 10^{23} atoms/mol)
= 2.458 x 10^{-23} cm^3/atom

Vol. of a Pb atom = (18.27 cm^3/mol) ÷ (6.022 x 10^{23} atoms/mol)
= 3.034 x 10^{-23} cm^3/atom

$$\text{Radius of atom} = r = \left(\frac{3}{4\pi}V\right)^{\frac{1}{3}} = \sqrt[3]{(0.2387)\ V}$$

$$r_{Al} = \sqrt[3]{(0.2387)(1.659 \times 10^{-23}\ cm^3)} = 1.58 \times 10^{-8}\ cm$$

$$r_{Hg} = \sqrt[3]{(0.2387)(2.458 \times 10^{-23}\ cm^3)} = 1.80 \times 10^{-8}\ cm$$

$$r_{Pb} = \sqrt[3]{(0.2387)(3.034 \times 10^{-23}\ cm^3)} = 1.93 \times 10^{-8}\ cm$$

Thickness of film = (2r)(6.5 x 10^6)
Thickness of Al film = (2)(1.58 x 10^{-8} cm)(6.5 x 10^6) = 0.21 cm
Thickness of Hg film = (2)(1.80 x 10^{-8} cm)(6.5 x 10^6) = 0.23 cm
Thickness of Pb film = (2)(1.93 x 10^{-8} cm)(6.5 x 10^6) = 0.25 cm

6.3 Everything is made of atoms and everything has mass and volume.

6.4 $V = 4/3\ \pi\ r^3$ $\quad V_{atom} = \left(\frac{4}{3}\pi r^3\right)(10^{-10}\ m)^3 = 4 \times 10^{-30}\ m^3$

$$V_{nucleus} = \left(\frac{4}{3}\pi r^3\right)(10^{-15}\ m)^3 = 4 \times 10^{-45}\ m^3$$

6.5

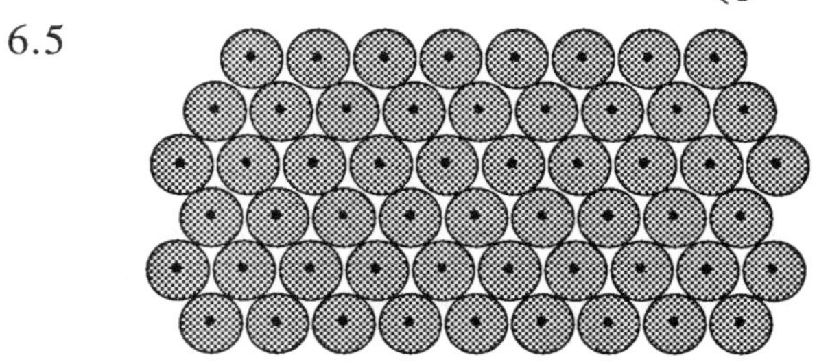

This picture is only 2-dimensional. The thinnest layer would be only one atom thick.
The nuclei would also be smaller and exactly in the center of each atom.

6.6 Given that: $E = h\nu = \frac{hc}{\lambda}$, $h = 6.626 \times 10^{-34}$ J•s, $c = 2.998 \times 10^8$ m•s^{-1}, and 1 nm = 10^{-9} m.

a) $E = \frac{(6.626 \times 10^{-34}\ J \bullet s)(2.998 \times 10^8\ m \bullet s^{-1})}{(490 \times 10^{-9}\ m)} = 4.05 \times 10^{-19}\ J$

b) $E = \frac{(6.626 \times 10^{-34}\ J \bullet s)(2.998 \times 10^8\ m \bullet s^{-1})}{(665 \times 10^{-9}\ m)} = 2.99 \times 10^{-19}\ J$

c) $E = \frac{(6.626 \times 10^{-34}\ J \bullet s)(2.998 \times 10^8\ m \bullet s^{-1})}{(25.5 \times 10^{-9}\ m)} = 7.79 \times 10^{-18}\ J$

d) $E = \frac{(6.626 \times 10^{-34}\ J \bullet s)(2.998 \times 10^8\ m \bullet s^{-1})}{(1250 \times 10^{-9}\ m)} = 1.59 \times 10^{-19}\ J$

6.7 $E_{mole\ of\ photons} = E_{photon} \times N_A$

a) $E_{mole\ of\ photons} = (4.05 \times 10^{-19}\ J)(6.022 \times 10^{23}\ mol^{-1})(1\ kJ / 10^3\ J)$

$$= 2.44 \times 10^2\ kJ / mol$$

b) $E_{mole\ of\ photons} = (2.99 \times 10^{-19}\ J)(6.022 \times 10^{23}\ mol^{-1})(1\ kJ / 10^3\ J)$

$$= 1.80 \times 10^2\ kJ / mol$$

c) $E_{mole\ of\ photons} = (7.79 \times 10^{-18}\ J)(6.022 \times 10^{23}\ mol^{-1})(1\ kJ / 10^3\ J)$

$$= 4.69 \times 10^3\ kJ / mol$$

d) $E_{mole\ of\ photons} = (1.59 \times 10^{-19}\ J)(6.022 \times 10^{23}\ mol^{-1})(1\ kJ / 10^3\ J)$

$$= 95.7\ kJ / mol$$

6.8 Given: $E = \frac{hc}{\lambda}$ or $\lambda = \frac{hc}{E}$:

$$\lambda = \frac{(6.626 \times 10^{-34}\ J \bullet s)(2.998 \times 10^{8}\ m \bullet s^{-1})}{(2.00 \times 10^{-15}\ J)} = 9.93 \times 10^{-11}\ m \text{ or } 99.3\ pm$$

6.9 Given: E_{photon} = 745 kJ/mole. From the value of E_{photon} per photon one can calculate its wavelength.

$$E_{photon} = \frac{(745\ kJ / mol)(10^{3}\ J / kJ)}{(6.022 \times 10^{23}\ mol^{-1})} = 1.237 \times 10^{-18}\ J$$

$$\lambda = \frac{hc}{E} = \frac{(6.626 \times 10^{-34}\ J \bullet s)(2.998 \times 10^{8}\ m \bullet s^{-1})}{(1.237 \times 10^{-18}\ J)}$$

$$= 1.61 \times 10^{-7}\ m \text{ or } 161\ nm$$

6.10 Infrared light does not have enough energy (high enough frequency) to free the electrons from the surface of the metal (i.e., the energy does not exceed the binding energy of the electrons).

6.11 a) Given: $c = \nu\lambda$ and $\lambda = \frac{c}{\nu}$

$$\lambda = \frac{(2.998 \times 10^{8}\ m \bullet s^{-1})}{(1.30 \times 10^{15}\ s^{-1})} = 2.31 \times 10^{-7}\ m \text{ or } 231\ nm$$

b) $h\nu = h\nu_o + k.E._{electron}$

$$h\nu_o = h\nu - k.E._{electron} = (6.626 \times 10^{-34}\ J \bullet s)(1.30 \times 10^{15}\ s^{-1}) - 5.2 \times 10^{-19}\ J$$

$$= 8.6 \times 10^{-19}\ J - 5.2 \times 10^{-19}\ J = 3.4 \times 10^{-19}\ J$$

c) $3.4 \times 10^{-19}\ J = \frac{hc}{\lambda}$ $\qquad \lambda = \frac{hc}{3.4 \times 10^{-19}\ J}$

$$\lambda = \frac{(6.626 \times 10^{-34}\ J \bullet s)(2.998 \times 10^{8}\ m \bullet s^{-1})}{(3.4 \times 10^{-19}\ J)} = 5.8 \times 10^{-7}\ m \text{ or } 580\ nm$$

6.12 Given: $E = \frac{hc}{\lambda}$ and $\lambda = \frac{hc}{E}$

$$\lambda = \frac{(6.626 \times 10^{-34}\ J \bullet s)(2.998 \times 10^{8}\ m \bullet s^{-1})}{(7.21 \times 10^{-19}\ J)} = 2.76 \times 10^{-7}\ m \text{ or } 276\ nm$$

6.13 For this answer the drawing should look like Figure 6-8 except the answer needs to have a binding energy and a kinetic energy of cesium and a separate indication for the binding energy and kinetic energy of chromium.

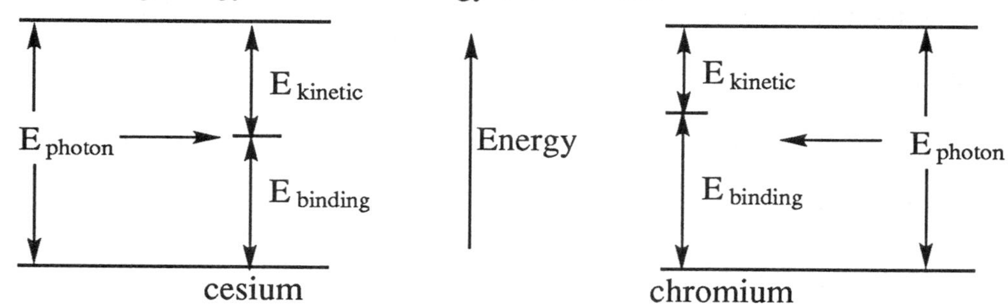

6.14 $$\nu = \frac{c}{\lambda} = \frac{(2.998 \times 10^{8}\ \mathrm{m \bullet s^{-1}})}{(606\ \mathrm{nm})(10^{-9}\ \mathrm{m/nm})} = 4.95 \times 10^{14}\ \mathrm{s^{-1}}$$

$$E = h\nu$$

$$= (6.626 \times 10^{-34}\ \mathrm{J \bullet s})(4.95 \times 10^{14}\ \mathrm{s^{-1}})(6.022 \times 10^{23}\ \mathrm{mol^{-1}})(1\ \mathrm{kJ}/10^{3}\ \mathrm{J})$$

$$= 198\ \mathrm{kJ/mol}$$

6.15 To elevate an electron to the state from which it can emit 404 nm radiation involves absorption of 254 nm radiation then 436 nm radiation.

$$E = \frac{hc}{\lambda}N_A \qquad \Delta E = \Delta E_1 + \Delta E_2$$

$$\Delta E = hcN_A\left(\frac{1}{\lambda_1} + \frac{1}{\lambda_2}\right) = (6.626 \times 10^{-34}\ \mathrm{J \bullet s})(2.998 \times 10^{8}\ \mathrm{m \bullet s^{-1}}) \times$$

$$(6.022 \times 10^{23}\ \mathrm{mol^{-1}})\left(\frac{1\ \mathrm{kJ}}{10^{3}\ \mathrm{J}}\right)\left(\frac{1}{\lambda_1} + \frac{1}{\lambda_2}\right) = 1.196 \times 10^{-4}\ \frac{\mathrm{kJ \bullet m}}{\mathrm{mol}}\left(\frac{1}{\lambda_1} + \frac{1}{\lambda_2}\right)$$

$$\Delta E = 1.196 \times 10^{-4}\ \frac{\mathrm{kJ \bullet m}}{\mathrm{mol}}\left(\frac{1}{436\ \mathrm{nm}(10^{-9}\mathrm{m/nm})} + \frac{1}{254\ \mathrm{nm}(10^{-9}\mathrm{m/nm})}\right)$$

$$= 745\ \mathrm{kJ/mol}$$

6.16 The energy needed to reach the lowest excited state of the mercury atom would be the difference between the energy needed to reach the state that emits the 404 nm light (See Prob. 6.15) and the energy of the 404 nm transition.
For the 404 nm transition the energy is:

$$E = \frac{hc}{\lambda} \frac{(6.626 \times 10^{-34}\mathrm{J \bullet s})(2.998 \times 10^{8}\ \mathrm{m \bullet s^{-1}})}{(404\ \mathrm{nm})(10^{-9}\ \mathrm{m/nm})} = 4.917 \times 10^{-19}\ \mathrm{J}$$

For the 745 nm transition the energy is (See Problem 6.15):

$$E = \left[(745\ \mathrm{kJ/mol})(10^{3}\ \mathrm{J/kJ})\right] \div (6.022 \times 10^{23}\ \mathrm{atom/mol}) = 1.237 \times 10^{-18}\ \mathrm{J}$$

$$E_{746\ nm} - E_{404\ nm} = 7.453 \times 10^{-19}\ \mathrm{J};\ \lambda = 2.665 \times 10^{-7}\ \mathrm{m} = 266.5\ \mathrm{nm} = 2665\mathrm{Å}$$

6.17 The energy needed to reach the $n = 8$ state is the difference between the energy for $n = 1$ and the energy for $n = 8$. A similar calculation is needed for the $n = 1$ to the $n = 9$ transition. Equation 6-5 $\left(E_n = -\frac{2.18 \times 10^{-18}\ J}{n^2}\right)$ is used to calculate these energies.

For $n = 8$: $\Delta E = E_8 - E_1 = \left(-\frac{2.18 \times 10^{-18}\ J}{8^2}\right) - \left(-\frac{2.18 \times 10^{-18}\ J}{1^2}\right)$

$= (2.18 \times 10^{-18}\ J)(1/1^2 - 1/8^2) = (2.18 \times 10^{-18}\ J)(0.984375) = 2.146 \times 10^{-18}\ J$

$$\lambda = \frac{hc}{E} = \frac{(6.626 \times 10^{-34}\ J \bullet s)(2.998 \times 10^8\ m \bullet s^{-1})}{(2.146 \times 10^{-18}\ J)} = 9.26 \times 10^{-8}\ m$$

$$= 92.6 \times 10^{-9}\ m = 92.6\ nm = 0.926 \times 10^{-7}\ m$$

$$\nu = \frac{E}{h} = \frac{2.146 \times 10^{-18}\ J}{6.626 \times 10^{-34}\ J \bullet s} = 3.239 \times 10^{15}\ s^{-1}$$ ultraviolet photons (See Table 6-1)

For $n = 9$: $\Delta E = E_9 - E_1 = (2.18 \times 10^{-18}\ J)(1/1^2 - 1/9^2)$

$= (2.18 \times 10^{-18}\ J)(0.987654) = 2.153 \times 10^{-18}\ J$

$$\lambda = \frac{hc}{E} = \frac{(6.626 \times 10^{-34}\ J \bullet s)(2.998 \times 10^8\ m \bullet s^{-1})}{(2.153 \times 10^{-18}\ J)} = 9.23 \times 10^{-8}\ m$$

$$= 92.3 \times 10^{-9}\ m = 92.3\ nm = 0.923 \times 10^{-7}\ m$$

$$\nu = \frac{E}{h} = \frac{2.153 \times 10^{-18}\ J}{6.626 \times 10^{-34}\ J \bullet s} = 3.25 \times 10^{15}\ s^{-1}$$ ultraviolet photons (See Table 6-1)

6.18 a) Recall that: $E = \left(\frac{hc}{\lambda}\right)N_A$

For 589.6 nm

$$E = \frac{(6.626 \times 10^{-34}\ J \bullet s)(2.998 \times 10^8\ m \bullet s^{-1})}{(589.6\ nm)(10^{-9}\ m / nm)}(6.022 \times 10^{23}\ mol^{-1})$$

$$= 2.029 \times 10^5\ J / mol = 202.9\ kJ / mol$$

For 590.0 nm

$$E = \frac{(6.626 \times 10^{-34}\ J \bullet s)(2.998 \times 10^8\ m \bullet s^{-1})}{(590.0\ nm)(10^{-9}\ m / nm)}(6.022 \times 10^{23}\ mol^{-1})$$

$$= 2.028 \times 10^5\ J / mol = 202.8\ kJ / mol$$

(continued)

(6.18 continued)

b) See Figure 6-15 for a similar drawing:

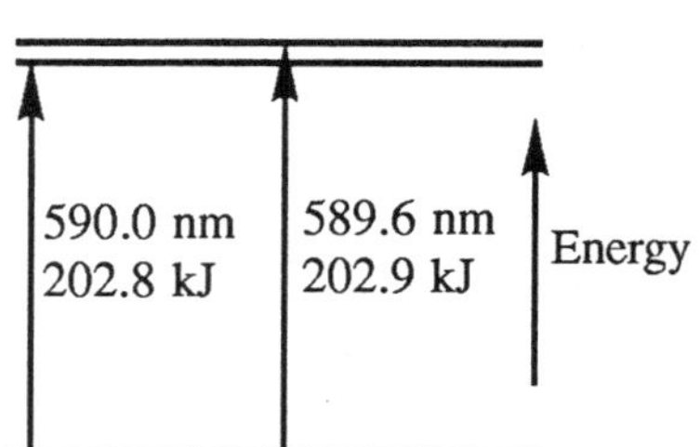

c) The second photon's energy: $\Delta E = 486 \text{ kJ/mol} - 203 \text{ kJ/mol} = 283 \text{ kJ/mol}$

$$\Delta E = \frac{(283 \text{ kJ / mol})(10^3 \text{ J / kJ})}{(6.022 \times 10^{23} \text{ mol}^{-1})} = 4.699 \times 10^{-19} \text{ J}$$

$$\lambda = \frac{hc}{E} = \frac{(6.626 \times 10^{-34} \text{ J} \bullet \text{s})(2.998 \times 10^8 \text{ m} \bullet \text{s}^{-1})}{(4.699 \times 10^{-19} \text{ J})} = 4.23 \times 10^{-7} \text{ m}$$

$$= 423 \times 10^{-9} \text{ m} = 423 \text{ nm}$$

6.19

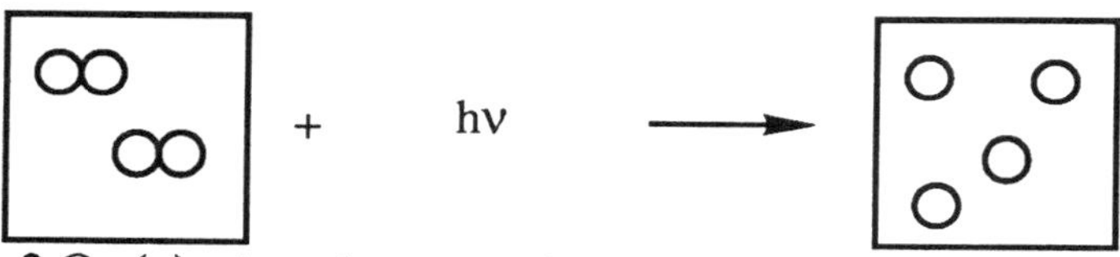

2 N_2 (g) plus short wavelength light yields 2 N_2^+ and 2 e^-

+ hν →

2 O_2 (g) plus short wavelength light yields 4 O

6.20 To break oxygen molecules into atoms:

$$E = \frac{(496 \text{ kJ / mol})(10^3 \text{ J / kJ})}{(6.022 \times 10^{23} \text{ atoms / mol})} = 8.236 \times 10^{-19} \text{ J / atom}$$

$$\lambda = \frac{hc}{E} = \frac{(6.626 \times 10^{-34} \text{ J} \bullet \text{s})(2.998 \times 10^8 \text{ m} \bullet \text{s}^{-1})}{(8.236 \times 10^{-19} \text{ J})} = 2.41 \times 10^{-7} \text{ m}$$

$$= 241 \times 10^{-9} \text{ m} = 241 \text{ nm (ultraviolet)}$$

To break nitrogen molecules into atoms:

$$E = \frac{(945 \text{ kJ / mol})(10^3 \text{ J / kJ})}{(6.022 \times 10^{23} \text{ atoms / mol})} = 1.569 \times 10^{-18} \text{ J / atom}$$

$$\lambda = \frac{(6.626 \times 10^{-34} \text{ J} \bullet \text{s})(2.998 \times 10^8 \text{ m} \bullet \text{s}^{-1})}{(1.569 \times 10^{-18} \text{ J})} = 1.27 \times 10^{-7} \text{ m}$$

$$= 127 \times 10^{-9} \text{ m} = 127 \text{ nm (ultraviolet)}$$

6.21

Spectral region	Gas in atmosphere that absorbs
a) <200 nm	O_2, N_2
b) 200 - 240 nm	O_3, O_2
c) 240 - 310 nm	O_3
d) 310 - 700 nm	none
e) 700 - 2000 nm	CO_2, H_2O

6.22

Spectral region	Region where absorbed
a) <200 nm	thermosphere
b) 200 - 240 nm	thermosphere
c) 240 - 310 nm	stratosphere
d) 310 - 700 nm	none
e) 700 - 2000 nm	troposphere

6.23 -30°C = 243 K This is not sufficient information to determine the altitude. Possibliities are: ~7 km, ~38 km, ~68 km, ~107 km.

6.24 There is now enough information. The altitude is approximately 38 km.

6.25 Mass of one e^- = 9.109×10^{-31} kg

Mass of one mole of e^- = $(9.109 \times 10^{-31}\ kg/e^-)(6.022 \times 10^{23}\ e^-/mol)$
$= 5.485 \times 10^{-7}$ kg/mol

6.26 Charge of one e^- = 1.602×10^{-19} C

Charge of one mole of e^- = $(1.602 \times 10^{-19}\ C/e^-)(6.022 \times 10^{23}\ e^-/mol)$
$= 9.647 \times 10^4$ C/mol

6.27 From Table 6-2: $\lambda = h/mv$ $\quad$ $J\bullet s = (kg\ m^2/s^2)\bullet s = kg\ m^2/s$

$$\lambda = \frac{(6.626 \times 10^{-34}\ kg\ m^2/s)}{(9.109 \times 10^{-31}\ kg)(4.8 \times 10^5\ m/s)} = 1.5 \times 10^{-9}\ m = 1.5\ nm$$

6.28 An atom with an odd number of electrons must have magnetism. An atom with an even number of electrons might not have magnetism.
a) Be might have no net magnetism (has 4 electrons).
b) K must have magnetism (19 electrons).
c) I must have magnetism (53 electrons).
d) Kr might have no net magnetism (has 36 electrons).

6.29 **H**: Hamiltonian, expression for the total energy
E: Total energy of the system and has only certain definite values (quantized)
ψ: The wave function has wave properties and is generally a function of all spatial variables (x, y and z). Its square, ψ^2, describes the distribution of an electron in space.

6.30 0.80 nm = 8.0×10^{-10} m

a) lowest energy, $n = 1$

$$E_1 = \frac{1^2 h^2}{8\ ma^2} = \frac{(1)(6.626 \times 10^{-34}\ kg\ m^2/s)^2}{(8)(9.109 \times 10^{-31}\ kg)(8.0 \times 10^{-10}\ m)^2}$$

$$= 9.4 \times 10^{-20}\ kg\ m^2/s^2 = 9.4 \times 10^{-20}\ J$$

b) $$\lambda = \frac{h}{mv} \qquad v = \left(\frac{2E}{m}\right)^{\frac{1}{2}} \qquad \lambda = \frac{h}{m}\left(\frac{m}{2E}\right)^{\frac{1}{2}} = \left(\frac{h^2}{m\ 2E}\right)^{\frac{1}{2}}$$

$$\lambda = \left[\frac{(6.626 \times 10^{-34}\ kg\ m^2/s)^2}{(9.109 \times 10^{-31}\ kg)(2)(9.4 \times 10^{-20}\ kg\ m^2/s^2)}\right]^{\frac{1}{2}} = 1.6 \times 10^{-9}\ m = 1.6\ nm$$

c) Energy of next higher energy state:

$$E_2 = \frac{2^2 h^2}{8\ ma^2} = \frac{(4)(6.626 \times 10^{-34}\ kg\ m^2/s)^2}{(8)(9.109 \times 10^{-31}\ kg)(8.0 \times 10^{-10}\ m)^2}$$

$$= 3.8 \times 10^{-19}\ kg\ m^2/s^2 = 3.8 \times 10^{-19}\ J$$

$$\Delta E = E_2 - E_1 = 3.8 \times 10^{-19}\ J - 9.4 \times 10^{-20}\ J = 2.9 \times 10^{-19}\ J$$

$$\lambda = \frac{hc}{E}$$

$$= \frac{(6.626 \times 10^{-34}\ J \bullet s)(2.998 \times 10^{8}\ m \bullet s^{-1})}{(2.9 \times 10^{-19}\ J)} = 6.8 \times 10^{-7} m = 680\ nm$$

6.31 The emission transitions can occur in steps back to the ground state while the absorptions occur from the ground state to the excited state in one step.

6.32 Photon of wavelength 500 nm

a) $$v = \frac{c}{\lambda} = \frac{2.998 \times 10^{8}\ m \bullet s^{-1}}{(500\ nm)(10^{-9}\ m/nm)} = 6.00 \times 10^{14}\ s^{-1}$$

b) $E = hv = (6.626 \times 10^{-34}\ J \bullet s)(6.00 \times 10^{14}\ s^{-1}) = 3.98 \times 10^{-19}\ J$

c) $$\frac{(93 \times 10^{6}\ miles)(1.609\ km/mi)(10^{3}\ m/km)}{(2.998 \times 10^{8}\ m/s)} = 499\ s = 8.3\ min$$

6.33 $E = hc / \lambda$

For 488 nm: $E = \dfrac{(6.626 \times 10^{-34}\ J \cdot s)(2.998 \times 10^{8}\ m/s)}{(488\ nm)(10^{-9}\ m/nm)} = 4.07 \times 10^{-19}\ J$

For 514 nm: $E = \dfrac{(6.626 \times 10^{-34}\ J \cdot s)(2.998 \times 10^{8}\ m/s)}{(514\ nm)(10^{-9}\ m/nm)} = 3.86 \times 10^{-19}\ J$

a) 488 nm   514 nm

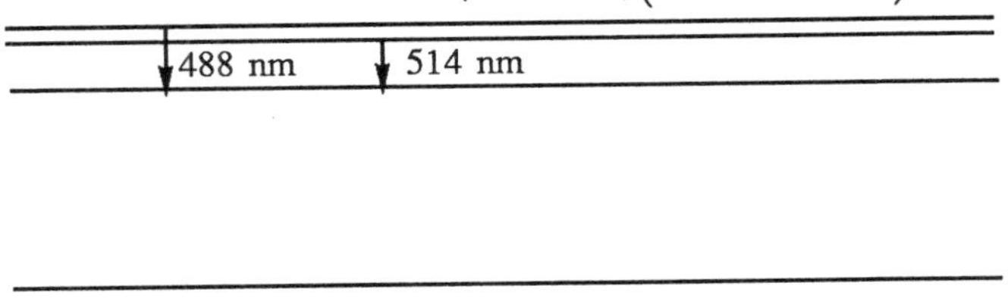

b) After the above emission, the Ar^{+} ion is in an energy level 2.76×10^{-18} J above the ground state. The radiation emitted on returning to the ground state:

$$\nu = E/h = \frac{(2.76 \times 10^{-18}\ J)}{(6.626 \times 10^{-34}\ J \cdot s)} = 4.17 \times 10^{15}\ s^{-1}$$

$$\lambda = \frac{c}{\nu} = \frac{2.998 \times 10^{8}\ m/s}{4.17 \times 10^{15}\ s^{-1}} = 7.19 \times 10^{-8}\ m = 71.9 \times 10^{-9}\ m = 71.9\ nm$$

6.34 E of a photon of wavelength 510 nm =

$$\frac{hc}{\lambda} = \frac{(6.626 \times 10^{-34}\ J \cdot s)(2.998 \times 10^{8}\ m/s)}{(510\ nm)(10^{-9}\ m/nm)} = 3.90 \times 10^{-19}\ J/photon$$

$$\frac{2.35\ 10^{-18}\ J}{3.90 \times 10^{-19}\ J/photon} = 6.03\ photons \approx 6\ photons$$

6.35 E from wavelength =

$$E = \frac{hc}{\lambda} = \frac{(6.626 \times 10^{-34}\ J \cdot s)(2.998 \times 10^{8}\ m/s)}{(\lambda(nm))(10^{-9}\ m/nm)} = \frac{1.9865 \times 10^{-16}\ J}{\lambda(nm)}$$

422.7 nm: $E = 4.700 \times 10^{-19}$ J

272.2 nm: $E = 7.298 \times 10^{-19}$ J

239.9 nm: $E = 8.281 \times 10^{-19}$ J

671.8 nm: $E = 2.957 \times 10^{-19}$ J

504.2 nm: $E = 3.940 \times 10^{-19}$ J

(continued)

(6.35 continued)

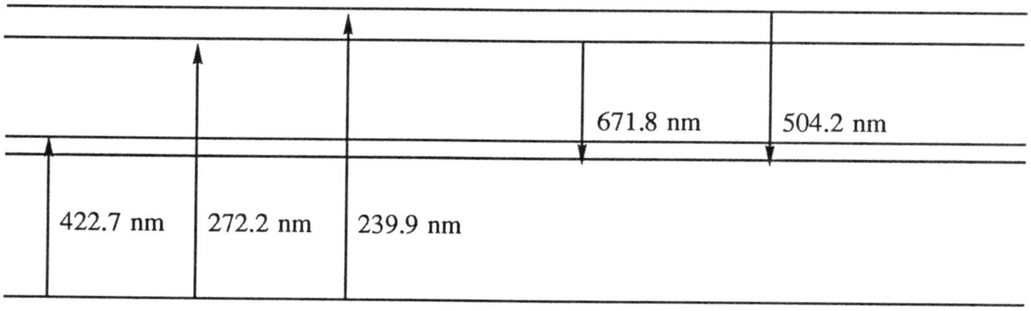

422.7 nm | 272.2 nm | 239.9 nm | 671.8 nm | 504.2 nm

a) $E_{239.9\ nm} - E_{504.2\ nm} = 8.281 \times 10^{-19}\ J - 3.940 \times 10^{-19}\ J = 4.341 \times 10^{-19}\ J$

$$\lambda = \frac{hc}{E} = \frac{(6.626 \times 10^{-34}\ J \cdot s)(2.998 \times 10^{8}\ m/s)}{(4.341 \times 10^{-19}\ J)} = 4.576 \times 10^{-7}\ m$$

$$= 457.6 \times 10^{-9}\ m = 457.6\ nm$$

b) $E_{272.2\ nm} - E_{422.7\ nm} = 7.298 \times 10^{-19}\ J - 4.700 \times 10^{-19}\ J = 2.598 \times 10^{-19}\ J$

$$\lambda = \frac{hc}{E} = \frac{(6.626 \times 10^{-34}\ J \cdot s)(2.998 \times 10^{8}\ m/s)}{(2.598 \times 10^{-19}\ J)} = 7.646 \times 10^{-7}\ m$$

$$= 764.6 \times 10^{-9}\ m = 764.6\ nm$$

c) The 272.2 nm absorption followed by the 671.8 nm emission and the 239.9 nm absorption followed by the 504.2 nm emission both lead to an energy level of 4.34×10^{-19} J above the lowest state.

6.36 Given: $\frac{(364\ kJ/mol)(10^{3}\ J/kJ)}{(6.022 \times 10^{23}\ HBr/mol)} = 6.045 \times 10^{-19}\ J/HBr$

$$\lambda = \frac{hc}{E} = \frac{(6.626 \times 10^{-34}\ J \cdot s)(2.998 \times 10^{8}\ m \cdot s^{-1})}{(6.045 \times 10^{-19}\ J)} = 3.29 \times 10^{-7}\ m$$

$$= 329 \times 10^{-9}\ m = 329\ nm$$

6.37 Equation 6-5: $E_n = -\frac{2.18 \times 10^{-18}\ J}{n^2}$

a) for n = 4: $E_4 = -\frac{2.18 \times 10^{-18}\ J}{4^2} = -1.36 \times 10^{-19}\ J$

b) for n = 2: $E_2 = -\frac{2.18 \times 10^{-18}\ J}{2^2} = -5.45 \times 10^{-19}\ J$

$E_{emitted\ photon} = E_2 - E_4 = (-5.45 \times 10^{-19}\ J) - (-1.36 \times 10^{-19}\ J) = -4.09 \times 10^{-19}\ J$

(continued)

(6.37 continued)

$$\lambda = \frac{hc}{E} = \frac{(6.626 \times 10^{-34}\ J \bullet s)(2.998 \times 10^{8}\ m/s)}{(4.09 \times 10^{-19}\ J)} = 4.86 \times 10^{-7}\ m$$

$$= 486 \times 10^{-9}\ m = 486\ nm$$

c) $E_{\text{emitted photon}} = E_{\text{threshold}} + E_{\text{of ejected electron}}$

$E_{\text{of ejected electron}} = E_{\text{emitted photon}} - E_{\text{threshold}} = 4.09 \times 10^{-19}\ J - 3.2 \times 10^{-19}\ J$

$= 0.9 \times 10^{-19}\ J = 9 \times 10^{-20}\ J$

d) $$\lambda = \frac{h}{mv} = \frac{h}{m}\left(\frac{m}{2E}\right)^{\frac{1}{2}} = \left(\frac{h^2}{2mE}\right)^{\frac{1}{2}} = \left[\frac{(6.626 \times 10^{-34}\ J \bullet s)^2}{(2)(9.109 \times 10^{-31}\ kg))(9 \times 10^{-20}\ J)}\right]^{\frac{1}{2}}$$

$$= \left[2.678 \times 10^{-18}\ \frac{kg\ m^2/s^2 \bullet s^2}{kg}\right]^{\frac{1}{2}} = \sqrt{2.7 \times 10^{-18}\ m^2}$$

$$= 1.6 \times 10^{-9}\ m = 1.6\ nm$$

6.38 For 486 nm: $E = \frac{hc}{\lambda} = \frac{(6.626 \times 10^{-34}\ J \bullet s)(2.998 \times 10^{8}\ m/s)}{(486\ nm)(10^{-9}\ m/nm)} = 4.087 \times 10^{-19}\ J$

Equation 6-5: $E_n = -\frac{2.18 \times 10^{-18}\ J}{n^2}$ and $E = E_{n_f} - E_{n_i}$

$$-4.087 \times 10^{-19}\ J = \left(-\frac{2.18 \times 10^{-18}\ J}{n_f^2}\right) - \left(-\frac{2.18 \times 10^{-18}\ J}{n_i^2}\right)$$

$$= +2.18 \times 10^{-18}\ J\left(\frac{1}{n_i^2} - \frac{1}{n_f^2}\right)$$

$$\left(\frac{1}{n_i^2} - \frac{1}{n_f^2}\right) = \frac{-4.087 \times 10^{-19}\ J}{2.18 \times 10^{-18}\ J} = -0.1875$$

n_i	n_f	$\left(\frac{1}{n_i^2} - \frac{1}{n_f^2}\right)$
2	1	-0.75
3	1	-0.8889
3	2	-0.1389
4	2	-0.1875

Answer: $n_f = 2$ and $n_i = 4$

6.39 $E_{\text{photon at 337.1 nm}} = \dfrac{hc}{\lambda}$

$$= \frac{(6.626 \times 10^{-34}\ \text{J} \bullet \text{s})(2.998 \times 10^{8}\ \text{m/s})}{(337.1\ \text{nm})(10^{-9}\ \text{m/nm})} = 5.893 \times 10^{-19}\ \text{J}$$

$$\frac{E_{laser}}{E_{photon}} = \text{number of photons} = \frac{10\ \text{mJ}}{5.893 \times 10^{-19}\ \text{J/photons}}$$

$$= \frac{10 \times 10^{-3}\ \text{J}}{5.893 \times 10^{-19}\ \text{J/photons}} = 1.7 \times 10^{16}\ \text{photons}$$

6.40 $E = hc/\lambda$

For 1014 nm: $E = \dfrac{(6.626 \times 10^{-34}\ \text{J} \bullet \text{s})(2.998 \times 10^{8}\ \text{m/s})}{(1014\ \text{nm})(10^{-9}\ \text{m/nm})} = 1.959 \times 10^{-19}\ \text{J}$

For 469 nm: $E = \dfrac{(6.626 \times 10^{-34}\ \text{J} \bullet \text{s})(2.998 \times 10^{8}\ \text{m/s})}{(469\ \text{nm})(10^{-9}\ \text{m/nm})} = 4.236 \times 10^{-19}\ \text{J}$

For 26 nm: $E = \dfrac{(6.626 \times 10^{-34}\ \text{J} \bullet \text{s})(2.998 \times 10^{8}\ \text{m/s})}{(26\ \text{nm})(10^{-9}\ \text{m/nm})} = 7.64 \times 10^{-18}\ \text{J}$

Excitation Energy

For the state that emits x-ray of 26 nm: 7.64×10^{-18} J

For the state that emits green photon of 469 nm:

$$7.64 \times 10^{-18}\ \text{J} + 4.236 \times 10^{-19}\ \text{J} = 8.06 \times 10^{-18}\ \text{J}$$

For the state that emits infrared photon of 1014 nm (fifth energy level):

$$7.64 \times 10^{-18}\ \text{J} + 4.236 \times 10^{-19}\ \text{J} + 1.959 \times 10^{-19}\ \text{J} = 8.26 \times 10^{-18}\ \text{J}$$

6.41 Adapting equations for electrons from Table 6-2:

a) $v = \dfrac{h}{m\lambda} = \dfrac{(6.626 \times 10^{-34}\ \text{kg m}^2/\text{s})}{(1.6749 \times 10^{-27}\ \text{kg})(75\ \text{pm})(10^{-12}\ \text{m/pm})} = 5.3 \times 10^{3}\ \text{m/s}$

b) $\lambda = \dfrac{h}{mv} = \dfrac{(6.626 \times 10^{-34}\ \text{kg m}^2/\text{s})}{(1.6749 \times 10^{-27}\ \text{kg})\left(\dfrac{1.25}{100}\right)(2.998 \times 10^{8}\ \text{m/s})} = 1.06 \times 10^{-13}\ \text{m}$

$$= 0.106 \times 10^{-12}\ \text{m} = 0.106\ \text{pm}$$

6.42 For energy per photon:

$$E = \frac{hc}{\lambda} = \frac{(6.626 \times 10^{-34} \text{ J} \bullet \text{s})(2.998 \times 10^{8} \text{ m/s})}{(634 \text{ nm})(10^{-9} \text{ m/nm})}$$

$$= 3.133 \times 10^{-19} \text{ J/photon}$$

Energy per minute: $E = \left(\frac{1 \text{ mJ}}{\text{s}}\right)\left(\frac{60 \text{ s}}{1 \text{ minute}}\right) = 60 \text{ mJ/minute}$

Number of photons $= \left(\frac{60 \text{ mJ}}{\text{minute}}\right)\left(\frac{10^{-3} \text{ J}}{\text{mJ}}\right) \div (3.133 \times 10^{-19} \text{ J/photon})$

$$= 1.91 \times 10^{17} \text{ photons}$$

6.43 The H_2O in clouds absorbs the heat being radiated from the earth and keeps it from being lost.

6.44 Coins and bills (of money)
Laboratory glassware (such as beakers or flasks)
Gauges of wire, sizes of nails and screws

6.45 $E_{\text{per electron}} = (216.4 \text{ kJ/mol})(10^{3} \text{ J/kJ}) \div (6.022 \times 10^{23} \text{ electrons/mol})$

$$= 3.593 \times 10^{-19} \text{ J/electron}$$

$$\lambda = \frac{hc}{E} = \frac{(6.626 \times 10^{-34} \text{ J} \bullet \text{s})(2.998 \times 10^{8} \text{ m/s})}{(3.593 \times 10^{-19} \text{ J})} = 5.529 \times 10^{-7} \text{ m}$$

$$= 552.9 \times 10^{-9} \text{ m} = 552.9 \text{ nm}$$

6.46 Energy of uv light of 250 nm wavelength = hc/λ

$$= \frac{(6.626 \times 10^{-34} \text{ J} \bullet \text{s})(2.998 \times 10^{8} \text{ m/s})}{(250 \text{ nm})(10^{-9} \text{ m/nm})} = 7.946 \times 10^{-19} \text{ J}$$

In Problem 6.45, the threshold energy for potassium was determined to be: 3.593×10^{-19} J/electron, $\therefore$ the kinetic energy of the ejected electron = 7.946×10^{-19} J - 3.593×10^{-19} J or 4.353×10^{-19} J.

The kinetic energy: $E = \frac{mv^2}{2}, \quad \therefore \; v = \left(\frac{2E}{m_e}\right)^{\frac{1}{2}}$

$$v = \sqrt{\frac{(2)(4.353 \times 10^{-19} \text{ kg m}^2/\text{s}^2)}{(9.109 \times 10^{-31} \text{ kg})}} = 9.78 \times 10^{5} \text{ m/s or } 978 \text{ km/s}$$

6.47 As we move upward through the Earth's atmosphere, the pressure continually decreases because there are less and less molecules present.

The temperature of the troposphere decreases as we go upward because the portion near the earth is heated by the earth and has enough molecules present to hold that heat. As we go higher and the troposphere "thins", there are fewer molecules to hold heat.

The stratosphere becomes warmer as we move upward because O_3 strongly absorbs ultraviolet light. Although the concentration of O_3 in the stratosphere decreases on moving upward, it absorbs uv light so efficiently that the uv light is almost completely absorbed by the O_3 in the upper part of the stratosphere and little uv light reaches the more concentrated O_3 in the lower part.

The mesosphere's temperature decreases on moving upward. Sunlight is not absorbed in the mesosphere and the temperature decreases as the density of molecules decreases on moving upward.

The thermosphere's temperature increases on moving upward. Molecules of N_2 and O_2 absorb high frequency solar light. As the solar light passes through the thermosphere, its high energy photons are progressively removed by this absorption. Thus, the intensity of high-energy light decreases and less energy is deposited on moving downward in the thermosphere.

6.48 $E = hc/\lambda$

$$E_{323\text{ nm}} = \frac{(6.626 \times 10^{-34}\ \text{J}\bullet\text{s})(2.998 \times 10^{8}\ \text{m/s})}{(323\ \text{nm})(10^{-9}\ \text{m/nm})} = 6.15 \times 10^{-19}\ \text{J}$$

$$E_{812.7\text{ nm}} = \frac{(6.626 \times 10^{-34}\ \text{J}\bullet\text{s})(2.998 \times 10^{8}\ \text{m/s})}{(812.7\ \text{nm})(10^{-9}\ \text{m/nm})} = 2.444 \times 10^{-19}\ \text{J}$$

$$E_{670.8\text{ nm}} = \frac{(6.626 \times 10^{-34}\ \text{J}\bullet\text{s})(2.998 \times 10^{8}\ \text{m/s})}{(670.8\ \text{nm})(10^{-9}\ \text{m/nm})} = 2.961 \times 10^{-19}\ \text{J}$$

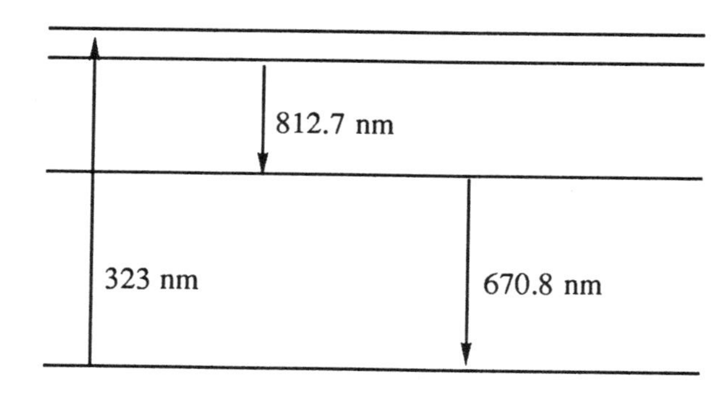

$$E_{\text{lost in collisions}} = 6.15 \times 10^{-19}\ \text{J} - 2.444 \times 10^{-19}\ \text{J} - 2.961 \times 10^{-19}\ \text{J} = 7.45 \times 10^{-20}\ \text{J}$$

$$\text{Fraction of energy lost in collisions} = \frac{7.45 \times 10^{-20}\ \text{J}}{6.15 \times 10^{-19}\ \text{J}} = 0.12$$

6.49 a) The binding (threshold) energy of the electrons =

$$6.00 \times 10^{-19}\ J - 2.70 \times 10^{-19}\ J = 3.30 \times 10^{-19}\ J$$

b) $E = hc/\lambda \qquad \lambda = hc/E$

$$\lambda = \frac{(6.626 \times 10^{-34}\ J \bullet s)(2.998 \times 10^{8}\ m/s)}{(6.00 \times 10^{-19}\ J)} = 3.31 \times 10^{-7}\ m$$

$$= 331 \times 10^{-9}\ m = 331\ nm$$

c) Given: $\lambda_e = \frac{h}{mv}$, $v = \left(\frac{2E}{m}\right)^{\frac{1}{2}}$, and $\lambda_e = \left(\frac{h}{m}\right)\left(\frac{m}{2E}\right)^{\frac{1}{2}}$

$$\lambda_e = \frac{(6.626 \times 10^{-34}\ kg\ m^2/s)}{(9.109 \times 10^{-31}\ kg)}\left[\frac{9.109 \times 10^{-31}\ kg}{(2)(2.70 \times 10^{-19}\ kg \bullet m^2/s^2)}\right]^{\frac{1}{2}}$$

$$= 9.45 \times 10^{-10}\ m = 945 \times 10^{-12}\ m = 945\ pm$$

6.50 a) The binding energy of the electrons: $E = h\nu$

$$= (6.626 \times 10^{-34}\ J \bullet s)(7.5 \times 10^{14}\ s^{-1}) = 5.0 \times 10^{-19}\ J$$

b) $$E_{366\ nm\ photon} = \frac{hc}{\lambda} = \frac{(6.626 \times 10^{-34}\ J \bullet s)(2.998 \times 10^{8}\ m/s)}{(366\ nm)(10^{-9}\ m/nm)}$$

$$= 5.43 \times 10^{-19}\ J$$

$$k.E. = E_{photon} - E_{binding} = 5.43 \times 10^{-19}\ J - 5.00 \times 10^{-19}\ J = 0.43 \times 10^{-19}\ J$$

$$= 4.3 \times 10^{-20}\ J$$

c) $$\lambda = \frac{h}{mv} = \left(\frac{h^2}{2mE}\right)^{\frac{1}{2}} = \left[\frac{(6.626 \times 10^{-34}\ J \bullet s)^2}{(2)(9.109 \times 10^{-31}\ kg)(4.3 \times 10^{-20}\ J)}\right]^{\frac{1}{2}}$$

$$= \left[5.604x10^{-18}\ \frac{J\ kg\ m^2 s^2/s^2}{kg\ J}\right]^{\frac{1}{2}}$$

$$= 2.367 \times 10^{-9}\ m = 2.4 \times 10^{-9}\ m = 2.4\ nm$$

$k.E. = 1/2\ m_e v^2$;

$$v = \sqrt{2\ k.E./m_e} = [(2)(4.3 \times 10^{-20}\ J)/9.109 \times 10^{-31}\ kg]^{1/2}$$

$$= \left[9.44x10^{10}\ \frac{kg\ m^2/s^2}{kg}\right]^{1/2} = 3.1 \times 10^{5}\ m/s$$

6.51 5% velocity of light = (0.05)(3.0×10^8 m/s) = 1.5×10^7 m/s

$m_e = 9.109 \times 10^{-31}$ kg

$m_p = 1.673 \times 10^{-27}$ kg

$\lambda = h/mv$

$$\lambda_e = \left(\frac{(6.626 \times 10^{-34}\ \text{kg} \cdot \text{m}^2 \cdot \text{s}/\text{s}^2)}{(9.109 \times 10^{-31}\ \text{kg})(1.5 \times 10^7\ \text{m}/\text{s})} \right) = 4.8 \times 10^{-11}\ \text{m}$$

$$= 48 \times 10^{-12}\ \text{m} = 48\ \text{pm}$$

$$\lambda_p = \left(\frac{(6.626 \times 10^{-34}\ \text{kg} \cdot \text{m}^2 \cdot \text{s}/\text{s}^2)}{(1.673 \times 10^{-27}\ \text{kg})(1.5 \times 10^7\ \text{m}/\text{s})} \right) = 2.6 \times 10^{-14}\ \text{m}$$

$$= 26 \times 10^{-15}\ \text{m}\ (= 26\ \text{fm})$$

6.52 $$E_{\text{molecule}} = \frac{(242.7\ \text{kJ}/\text{mol})(10^3\ \text{J}/\text{kJ})}{(6.022 \times 10^{23}\ \text{molecules}/\text{mole})} = 4.030 \times 10^{-19}\ \text{J}/\text{molecule}$$

$$\lambda = \frac{hc}{E} = \frac{(6.626 \times 10^{-34}\ \text{J} \cdot \text{s})(2.998 \times 10^8\ \text{m}/\text{s})}{(4.030 \times 10^{-19}\ \text{J})} = 4.929 \times 10^{-7}\ \text{m}$$

$$= 492.9 \times 10^{-9}\ \text{m} = 492.9\ \text{nm}$$

Yes, Cl_2 molecules are likely to be fragmented in the troposphere because sunlight of this wavelength will penetrate through to the troposphere.

6.53 From Equation 6-5: $E_n = 2.18 \times 10^{-18}\text{J} \left(\frac{1}{n_f^2} - \frac{1}{n_i^2} \right)$

$$\lambda = \frac{hc}{E} = \frac{(6.626 \times 10^{-34}\ \text{J} \cdot \text{s})(2.998 \times 10^8\ \text{m}/\text{s})}{E\ (\text{in J})}$$

$$\lambda = (1.9865 \times 10^{-25}/\ E)$$

For the Paschen Series $n_{\text{final}} = 3$ and n_{initial} is equal to integers 4 or greater.

$n = 4$: $E_n = 2.18 \times 10^{-18}\text{J} \left(\frac{1}{3^2} - \frac{1}{4^2} \right) = 1.06 \times 10^{-19}$ J

$\lambda = 1.9 \times 10^{-6}$ m = 1.9 μm

$n = 5$: $E_n = 2.18 \times 10^{-18}\text{J} \left(\frac{1}{3^2} - \frac{1}{5^2} \right) = 1.55 \times 10^{-19}$ J

$\lambda = 1.3 \times 10^{-6}$ m = 1.3 μm

(continued)

(6.53 continued)

$$n = 6: \quad E_n = 2.18 \times 10^{-18} J \left(\frac{1}{3^2} - \frac{1}{6^2} \right) = 1.81 \times 10^{-19} \text{ J}$$

$$\lambda = 1.1 \times 10^{-6} \text{ m} = 1.1 \ \mu\text{m}$$

$$n = 7: \quad E_n = 2.18 \times 10^{-18} J \left(\frac{1}{3^2} - \frac{1}{7^2} \right) = 1.98 \times 10^{-19} \text{ J}$$

$$\lambda = 1.0 \times 10^{-6} \text{ m} = 1.0 \ \mu\text{m}$$

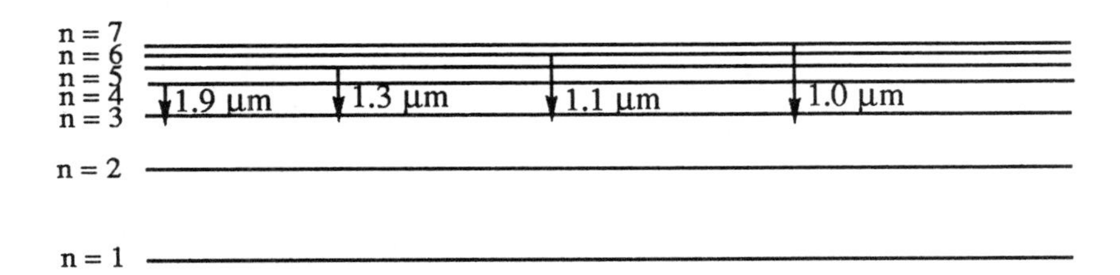

6.54 No. According to Table 6-1 the range of frequencies for visible light is 10^{13} to 10^{14} sec^{-1} (i.e., 0.1×10^{14} to 1×10^{14} sec^{-1}). The threshold frequency of 8.95×10^{14} sec^{-1} falls outside this range.

6.55 27.3 MHz = 27.3×10^6 Hz = 27.3×10^6 s^{-1}

$$\lambda = \frac{c}{\nu} = \frac{(2.998 \times 10^8 \text{ m/s})}{(27.3 \times 10^6 \text{ s}^{-1})} = 11.0 \text{ m}$$

$$E = h\nu = (6.626 \times 10^{-34} \text{ J}\cdot\text{s})(27.3 \times 10^6 \text{ s}^{-1}) = 1.81 \times 10^{-26} \text{ J}$$

6.56
$$\text{minimum k.E.}_{\text{electrons}} = \frac{hc}{\lambda} = \frac{(6.626 \times 10^{-34} \text{ J} \bullet \text{s})(2.998 \times 10^8 \text{ m/s})}{(253.7 \text{ nm})(10^{-9} \text{ m/nm})}$$

$$= 7.83 \times 10^{-19} \text{ J}$$

$$v = \left(\frac{2E}{m} \right)^{\frac{1}{2}} = \left[\frac{(2)(7.83 \times 10^{-19} \text{ kg m}^2/\text{s}^2)}{9.109 \times 10^{-31} \text{kg}} \right]^{\frac{1}{2}} = 1.31 \times 10^6 \text{ m/s}$$

6.57 frequencies: $\nu = \frac{c}{\lambda} = \frac{2.998 \times 10^{8} \text{ m/s}}{\lambda}$

$$\nu_{487 \text{ nm}} = \frac{2.998 \times 10^{8} \text{ m/s}}{(487 \text{ nm})(10^{-9} \text{ m/nm})} = 6.156 \times 10^{14} \text{ s}^{-1} = 6.16 \times 10^{14} \text{ s}^{-1}$$

$\nu_{514 \text{ nm}} = 5.833 \times 10^{14} \text{ s}^{-1} = 5.83 \times 10^{14} \text{ s}^{-1}$

$\nu_{543 \text{ nm}} = 5.521 \times 10^{14} \text{ s}^{-1} = 5.52 \times 10^{14} \text{ s}^{-1}$

$\nu_{553 \text{ nm}} = 5.421 \times 10^{14} \text{ s}^{-1} = 5.42 \times 10^{14} \text{ s}^{-1}$

$\nu_{578 \text{ nm}} = 5.187 \times 10^{14} \text{ s}^{-1} = 5.19 \times 10^{14} \text{ s}^{-1}$

Energies per mole:

$$E = h\nu N_A = (6.626 \times 10^{-34} \text{ J}\bullet\text{s})\nu(6.022 \times 10^{23} \text{ mol}^{-1})(\text{kJ}/10^{3} \text{ J})$$

$$= 3.99 \times 10^{-13} \text{ kJ}\bullet\text{s/mol} \times \nu$$

$E_{487 \text{ nm}} = (3.99 \times 10^{-13} \text{ kJ}\bullet\text{s/mol})(6.156 \times 10^{14} \text{ s}^{-1}) = 246 \text{ kJ/mol}$

$E_{514 \text{ nm}} = (3.99 \times 10^{-13} \text{ kJ}\bullet\text{s/mol})(5.833 \times 10^{14} \text{ s}^{-1}) = 233 \text{ kJ/mol}$

$E_{543 \text{ nm}} = (3.99 \times 10^{-13} \text{ kJ}\bullet\text{s/mol})(5.521 \times 10^{14} \text{ s}^{-1}) = 220 \text{ kJ/mol}$

$E_{553 \text{ nm}} = (3.99 \times 10^{-13} \text{ kJ}\bullet\text{s/mol})(5.421 \times 10^{14} \text{ s}^{-1}) = 216 \text{ kJ/mol}$

$E_{578 \text{ nm}} = (3.99 \times 10^{-13} \text{ kJ}\bullet\text{s/mol})(5.187 \times 10^{14} \text{ s}^{-1}) = 207 \text{ kJ/mol}$

CHAPTER 7: ATOMIC STRUCTURE AND PERIODICITY

7.1 For 6p; $n = 6, l = 1$

n	l	m_l	m_s
6	1	1	+1/2
6	1	1	-1/2
6	1	0	+1/2
6	1	0	-1/2
6	1	-1	+1/2
6	1	-1	-1/2

7.2 For 4f; $n = 4, l = 3$

n	l	m_l	m_s
4	3	3	+1/2
4	3	3	-1/2
4	3	2	+1/2
4	3	2	-1/2
4	3	1	+1/2
4	3	1	-1/2
4	3	0	+1/2

and

n	l	m_l	m_s
4	3	0	-1/2
4	3	-1	+1/2
4	3	-1	-1/2
4	3	-2	+1/2
4	3	-2	-1/2
4	3	-3	+1/2
4	3	-3	-1/2

7.3 For $n = 3$

$l = 0$	$m_l = 0$	$m_s = -1/2, +1/2$
$l = 1$	$m_l = -1, 0, +1$	$m_s = -1/2, +1/2$
$l = 2$	$m_l = -2, -1, 0, +1, +2$	$m_s = -1/2, +1/2$

7.4 For $m_l = -2$

$n = 3, 4, 5, \ldots, \infty$

$l = 2, 3, 4, \ldots, (n-1)$

$m_s = -1/2, +1/2$

7.5 a) nonexistent, m_s can be only -1/2 or +1/2

b) describes actual orbital

c) nonexistent, largest possible value of l is $(n-1)$, 2 in this case

d) describes actual orbital

7.6 a) nonexistent, l has only positive values

b) describes actual orbital

c) nonexistent, m_l cannot be larger than l

d) nonexistent, m_s can be only -1/2, +1/2

7.7 a) See Figure 7-8 (a) for a similar drawing.

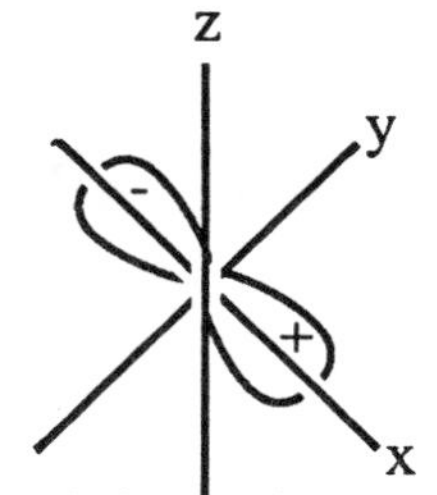

A 2p orbital (could also have been the $2p_y$ or $2p_z$ orbital)

b) See Figure 7-5 (d) for a similar drawing.

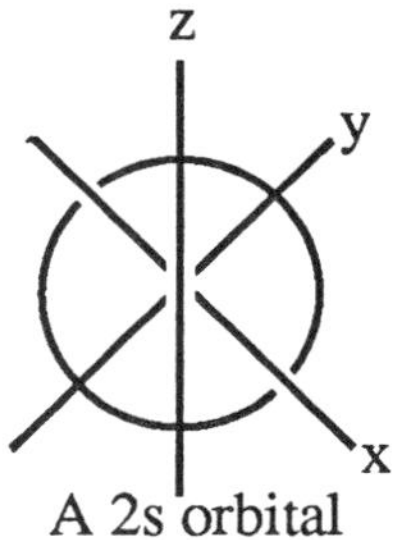

A 2s orbital

c) See Figure 7-5 (d) for a similar drawing except 3s will be larger than 2s.

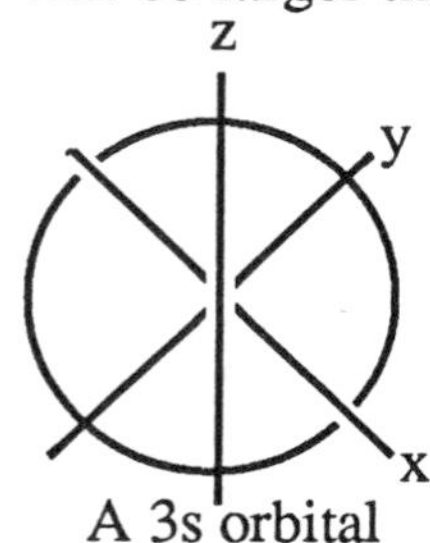

A 3s orbital

d) See Figure 7-9 for a similar drawing.

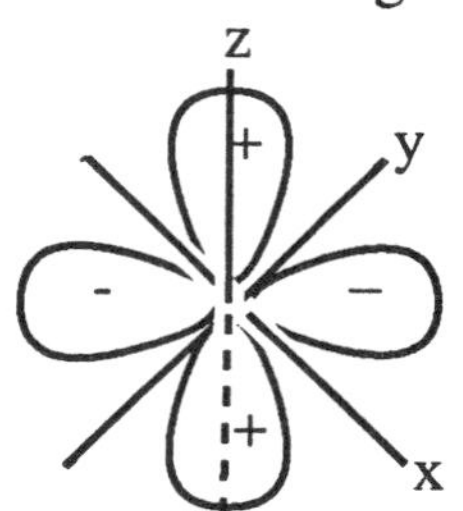

A $3d_{xy}$ orbital (could also have been one of the other d orbitals.)

7.8

	n	l	m_l	m_s
3s	3	0	0	-1/2 or +1/2
a) $3p_x$	3	1	-1	-1/2 or +1/2
b) $3d_{x^2-y^2}$	3	2	1	-1/2 or +1/2
c) $3d_{z^2}$	3	2	2	-1/2 or +1/2

7.9 1s Figure C; 2s Figure B; 2p Figure D; 3p Figure A

7.10 The electron contour surface does not represent all details of electron density inside the surface. It encloses about 90% of all the electron density. It uses 2 dimensions to represent 3 dimensions.

7.11

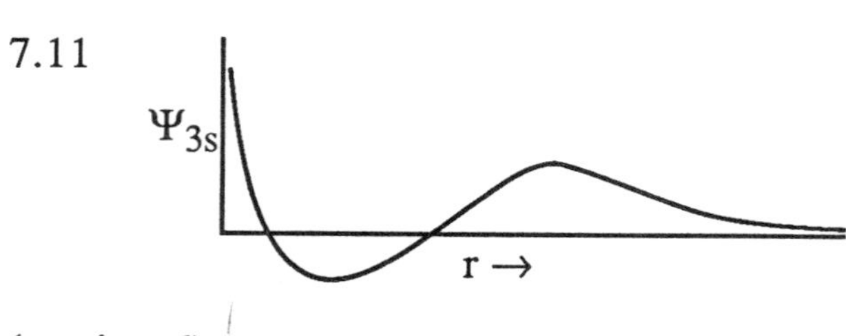

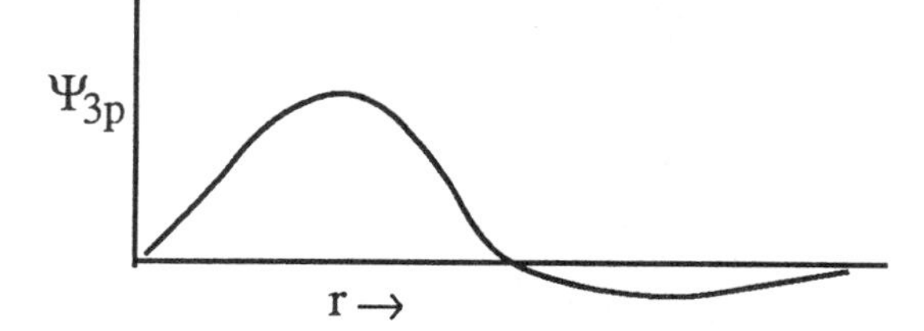

(continued)

(7.11 continued)

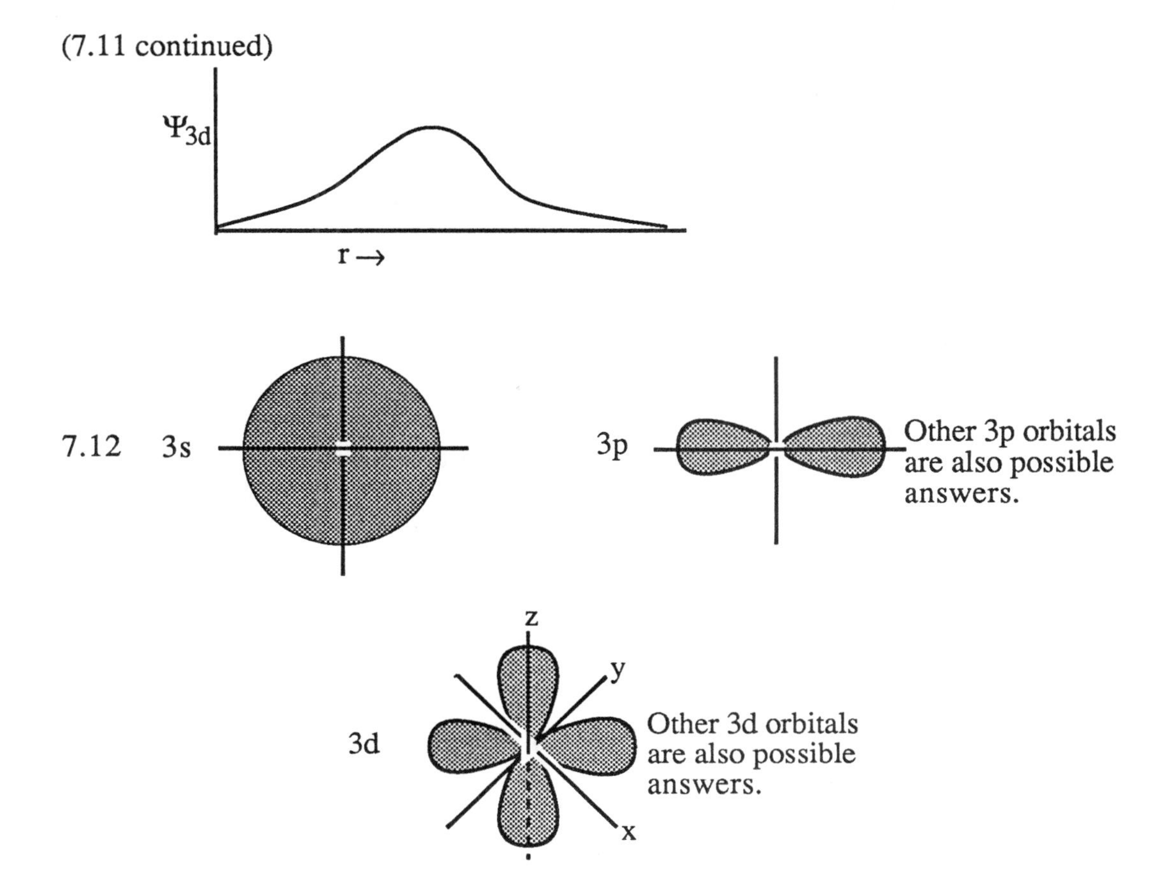

7.13 a) He 1s is more stable than the 2s orbital because orbital stability decreases with increasing *n* value.
b) Kr 5p is more stable than Kr 6p because orbital stability decreases with increasing *n* value
c) He^+ 2s is more stable because the electron in the 2s orbital of a He atom has a 1s electron screening the electron in the 2s orbital. The Z_{eff} is approximately +1 in the He atom and +2 in the He^+.
d) Ar 4s is more stable because the higher the *l* quantum number, the more that orbital is screened by electrons in smaller more stable orbitals.

7.14 a) For hydrogen (a one electron atom) energy depends on *n* and Z which are the same for the 2s and 2p orbitals of hydrogen.
b) In a helium atom the 1s electron more effectively screens the 2p orbital than it does the 2s orbital.
c) The three 2p orbitals of the helium atom have identical energy because the p orbitals are perpendicular to one another in distinctly different regions of space which results in little mutual screening, but they each receive the same screening from electrons in smaller orbitals.
d) The 2p orbital of the He atom has nearly the same energy as the 2p orbital in H because they have the same value of *n* (= 2), *l*(=1), and Z_{eff} (= +1). (See Table 7-2 and Figure 7-14.)

7.15 In a multi-electron atom (such as helium) each electron affects the properties of all other electrons. These electron-electron interactions make the orbital energies different and unique for each element. The lower the l for an orbital, the more stable the orbital because of the difference in screening by the inner electron. The hydrogen atom has no inner electron, no screening, and the energy of orbitals depends only on the size of the orbitals (therefore, on the value of n).

7.16 a) The ionization energy for the 2p electron in the He^+ ion (with no screening 1s electron) is nearly four times the ionization energy for the 2p electron in the He atom ($Z = +2$, $Z_{eff} = +1$) and also the 2p electron in the H atom ($Z = +1$ and $Z_{eff} = +1$).
b) The ionization energies of the 2p electrons of the H atom and the He^+ ion increase by a factor of four as Z increases from 1 to 2 (Z^2 increases from 1 to 4).
c) The ionization energies of the 1s and 2p electrons of the He atom and the 2p electron of the H atom show that the 2p electron is much less stable than the 1s electron of the He atom and about matches the ionization energy of the H atom 2p electron that indicates they both have $Z_{eff} \approx +1$.

7.17 For Be: $Z = 4$ $1s^22s^2$

n	l	m_l	m_s
1	0	0	+1/2
1	0	0	-1/2
2	0	0	+1/2
2	0	0	-1/2

For O: $Z = 8$ $1s^22s^22p^4$

n	l	m_l	m_s
1	0	0	+1/2
1	0	0	-1/2
2	0	0	+1/2
2	0	0	-1/2
2	1	+1	+1/2
2	1	+1	-1/2
2	1	0	+1/2
2	1	-1	+1/2

For P: $Z = 15$ $1s^22s^22p^63s^23p^3$

n	l	m_l	m_s
1	0	0	+1/2
1	0	0	-1/2
2	0	0	+1/2
2	0	0	-1/2
2	1	+1	+1/2
2	1	+1	-1/2
2	1	0	+1/2
2	1	0	-1/2
2	1	-1	+1/2
2	1	-1	-1/2
3	0	0	+1/2
3	0	0	-1/2
3	1	+1	+1/2
3	1	0	+1/2
3	1	-1	+1/2

For Ne: $Z = 10$ $1s^22s^22p^6$

n	l	m_l	m_s
1	0	0	+1/2
1	0	0	-1/2
2	0	0	+1/2
2	0	0	-1/2
2	1	+1	+1/2
2	1	+1	-1/2
2	1	0	+1/2
2	1	0	-1/2
2	1	-1	+1/2
2	1	-1	-1/2

7.18 No other set for Be

For O:

n	l	m_l	m_s
2	1	+1	+1/2
2	1	+1	-1/2
2	1	0	-1/2
2	1	-1	-1/2
2	1	+1	+1/2 (or -1/2)
2	1	0	+1/2
2	1	0	-1/2
2	1	-1	+1/2 (or -1/2)
	or		
2	1	1	+1/2 (or -1/2)
2	1	0	+1/2 (or -1/2)
2	1	-1	+1/2
2	1	-1	-1/2

No other set for Ne

For P:

n	l	m_l	m_s
3	1	+1	-1/2
3	1	0	-1/2
3	1	-1	-1/2

7.19 Z = 4 for Be $1s^22s^2$

a) $1s^3\ 2s^1$ Pauli-forbidden

b) $1s^1\ 2s^3$ Pauli-forbidden

c) $1s^1\ 2p^3$ excited state

d) $1s^2\ 2s^12p^1$ excited state

e) $1s^2\ 2s^2$ ground state

f) $1s^2\ 1p^2$ nonexistent orbital

g) $1s^2\ 2s^12d^1$ nonexistent orbital

7.20 Z = 9 for F $1s^2\ 2s^22p^5$

a) $1s^2\ 2s^22p^4$ Only 8 electrons, fluorine has 9 electrons.

b) $1s^2\ 2s^12p^6$ The 2s orbital is not filled with 2 electrons.

c) $1s^3\ 2s^22p^4$ The 1s orbital can only contain 2 electrons.

d) $1s^2\ 2s^21p^5$ 1p is a nonexistent orbital.

7.21 Mo [Kr] $5s^14d^5$ Tc [Kr] $5s^24d^5$

5s ↑ 4d ↑ ↑ ↑ ↑ ↑ 5s ↑↓ 4d ↑ ↑ ↑ ↑ ↑

7.22 a) Nb [Kr] $5s^14d^4$, Mo [Kr] $5s^14d^5$, Ru [Kr] $5s^14d^7$, Rh [Kr] $5s^14d^8$, Pd [Kr] $5s^04d^{10}$, Ag [Kr] $5s^14d^{10}$ all indicate near-degeneracy of 5s and 4d orbitals.

b) La [Kr] $6s^25d^14f^0$ and Ce [Kr] $6s^25d^14f^1$ both indicate near-degeneracy of 6s, 5d and 4f orbitals. Ce and Pr, Eu and Gd, Gd and Tb also indicate near-degeneracy of 5d and 4f orbitals. Ir [Kr] $6s^25d^74f^{14}$ and Pt [Kr] $6s^15d^94f^{14}$ indicate near-degeneracy of 6s and 5d orbitals.

c) Ni [Ar]$4s^23d^8$, Pd [Kr] $5s^04d^{10}$ and Pt [Xe] $6s^15d^9$ are in the same column but have different valence configurations.

7.23 O is paramagnetic and P is paramagnetic

2s ↑↓ 2p ↑↓ ↑ ↑
[He]

3s ↑↓ 3p ↑ ↑ ↑
[Ne]

7.24 F has the ground state of ($1s^22s^22p^7$). It has two excited states in which no electron has n >2, they are: $1s^22s^12p^8$ and $1s^12s^22p^8$.

7.25 N has the ground state of ($1s^22s^22p^3$). It has 7 excited states in which no electron has n >2; they are: $1s^22s^12p^4$, $1s^22s^02p^5$, $1s^12s^22p^4$, $1s^12s^12p^5$, $1s^12s^02p^6$, $1s^02s^12p^6$ and $1s^02s^22p^5$

7.26 Your drawing would include the following features: (a) all elements have $n = 1$ (The only ones in which that is the energy quantum number used are H and He.), (b) the elements for which the 5f orbitals are filling are elements with atomic numbers 91 through 102, (c) elements with s^2p^4 configurations are O, S, Se, Te and Po, and (d) the elements with filled s orbitals are He plus those in column II (Be, Mg, Ca, Sr, Ba and Ra).

7.27 Your drawing would include the following features: (a) the elements that are one electron short of having a filled set of p orbitals are all the elements of column VII (F, Cl, Br, I and At), (b) the elements for which $n = 3$ is filling are: Na, Mg, Al, Si, P, S, Cl, Ar, Sc, Ti, V, Cr, Mn, Fe, Co, Ni, Cu and Zn, (c) the elements with half filled d orbitals are: C r, Mn, Mo, Tc, Re and Uns, and (d) the first element that contains a 5s electron is Rb.

7.28 Z = 119 $8s^1$

7.29 Using Figure 7-20, one should conclude that since Z = 111 the configuration might be: $1s^22s^22p^63s^23p^64s^23d^{10}4p^65s^24d^{10}5p^66s^24f^{14}5d^{10}6p^67s^25f^{14}6d^9$. One might also have predicted it to be $7s^15f^{14}6d^{10}$ since the three elements above it have the s^1d^{10} configuration.

7.30 F<Cl<S<P

7.31 Ar>Cl>K>Cs

7.32 This element probably occupies the second column (II). By comparing the second ionization energy to the first, we see an increase as would be expected from removing the second electron from a positive ion. The much larger increase of the third ionization compared to the second suggests the third electron is a core electron. The negative electron affinity suggests an element on the left side of the periodic chart (a metal) which does not form a negative ion which is more stable than the neutral atom.

7.33 A negative electron affinity means that the neutral atom is more stable than the anion.

Be — $2s$ ↑↓ ; $2p$ — — —

Forming the anion Be^- requires adding an electron to a p orbital which is of higher energy than the s orbital (screening less for p orbitals).

N — $2s$ ↑↓ ; $2p$ ↑ ↑ ↑

Forming the anion N^- requires pairing two electrons in one of the p orbitals. This requires energy; therefore, the atom is more stable than the anion.

Mg — $3s$ ↑↓ ; $3p$ — — —

Forming the anion Mg^- requires adding an electron to a p orbital which is of higher energy than the s orbital (screening less for p orbitals).

Ar — $4s$ — ; $3s$ ↑↓ ; $3p$ ↑↓ ↑↓ ↑↓

Forming the anion Ar^- requires adding an electron to an s orbital with *n* one greater than the valence orbitals; therefore, the energy of the electron in the ion is greater than the energy of the electrons in the atom.

Zn — $4p$ — — — ; $3d$ ↑↓ ↑↓ ↑↓ ↑↓ ↑↓ ; $4s$ ↑↓

Forming the anion Zn^- requires adding an electron to a 4p orbital which is of higher energy than the 4s or 3d orbitals; therefore, the anion is more unstable than the neutral atom.

7.34 a) Ionization of boron involves removing the only 2p electron leaving a relatively stable $2s^2$ ion while beryllium involves removing a 2s electron from the relatively stable $2s^2$ atom.
b) Sulfur ion, S^+, is formed by removing an electron from $3p^4$ to form the more stable $3p^3$ (containing 3 unpaired electrons with the same spin orientation which minimizes the electron-electron repulsion). Phosphorus starts in the relatively stable $3p^3$ configuration and removing an electron results in the $3p^2$ configuration.

7.35 $Y^{3+} < Sr^{2+} < Rb^+ < Kr < Br^- < Se^{2-} < As^{3-}$

7.36 Elements that you might expect to form anions with -1 charge are: H^-, F^-, Cl^-, Br^-, I^- and (At^-)
Elements that you might expect to form anions with -2 charge are: O^{2-}, S^{2-}, Se^{2-} and Te^{2-}

7.37 $K_{(g)} + 1/2I_{2(g)} \rightarrow K^+I^-_{(g)}$

$$E_{coulomb} = \frac{(1.389 \times 10^5 \text{ kJ pm mol}^{-1})(q^+)(q^-)}{d}$$

$$= \frac{(1.389 \times 10^5 \text{ kJ pm mol}^{-1})(+1)(-1)}{(133 + 220) \text{ pm}} = -393 \text{ kJ / mol}$$

(continued)

(7.37 continued)

$K_{(g)} \rightarrow K^+_{(g)} + e^-$ $\Delta E = IE = 418.8$ kJ/mol

$I_{(g)} + e^- \rightarrow I^-_{(g)}$ $\Delta E = -EA = -295.3$ kJ/mol

$\Delta E = IE + (-EA) + E_{coulomb} = 418.8$ kJ/mol $- 295.3$ kJ/mol $- 393$ kJ/mol $= -270$ kJ/mol

7.38 $$E_{coulomb} = \frac{(1.389 \times 10^5 \text{ kJ pm mol}^{-1})(+2)(-2)}{(133 + 220) \text{ pm}} = -1574 \text{ kJ / mol}$$

$\Delta E = IE_1 + IE_2 - EA_1 - EA_2 + E_{coulomb}$

$\Delta E = 418.8$ kJ/mol $+ 3051$ kJ/mol -295.3 kJ/mol $+ 500$ kJ/mol $- 1574$ kJ/mol

$= +2.10 \times 10^3$ kJ/mol

7.39 From the list given, the stable cations and anions would be: Ca^{2+}, Cu^+ or Cu^{2+}, Cs^+, Cl^- and Cr^{3+} or Cr^{2+}.

7.40 From the list given, the stable cations and anions would be: Ba^{2+}, Bi^{3+} and Br^-.

7.41 As n increases, the electrons in the p orbitals are held more and more loosely and, therefore, can be easily removed. This is especially true if these p orbitals contain 1 or 2 electrons.

7.42 Other metals that might be expected to show metalloid-like properties are: Bi, Al, Ga, and Sn. These elements might have metalloid properties since their positions on the Periodic Table are in that region that separates the metals from the nonmetals.

7.43 From column VI the nonmetals are O, S and Se; the only metalloid is Te; Po is a metal. See Problem 7.42.

7.44 Of the elements Ca, C, Cu, Cs, Cl and Cr; Cl and C are nonmetals; none are metalloids; and four are metals (Ca, Cu, Cs and Cr).

7.45 There are 6 sets of quantum numbers for any set of np electrons. Therefore, there would also be 6 possible sets of quantum numbers for any 4p electron.

7.46 $Fe^{3+} \equiv 1s^2 2s^2 2p^6 3s^2 3p^6 4s^0 3d^5$

n	l	m_l	m_s	
3	2	-2	-1/2	
3	2	-1	-1/2	
3	2	0	-1/2	(or m_s may all be +1/2)
3	2	+1	-1/2	
3	2	+2	-1/2	

7.47 For 4d

n	l	m_l	m_s
4	2	-2	±1/2
4	2	-1	±1/2
4	2	0	±1/2
4	2	+1	±1/2
4	2	+2	±1/2

7.48 In order of increasing ionization energy: Na < O < N < Ne < Na^+

7.49 After 7p the next two orbitals to fill would be 8s and 5g.

7.50 The number of valence electrons for each of the following is: H = 1, Cs = 1, Ca = 2, Ge = 4, Br = 7 and Xe = 0 (or 8).

7.51 From the following electron configurations one can see that S^+ will have the most unpaired electrons of these three since it has one electron in each of the 3p orbitals. S has 2 unpaired electrons (and 2 paired), while S^- has 1 unpaired electron (and 4 paired electrons).

S	$1s^22s^22p^63s^23p^4$	S^+	$1s^22s^22p^63s^23p^3$
S^-	$1s^22s^22p^63s^23p^5$		

7.52 Your sketch should look like the 3s and 3p orbital plots in Figure 7-7 plus have an added plot for the 2p orbital. The maximum Ψ^2 value for the 2p orbital will occur at a smaller value of r. Therefore, the plot shows that the 2p orbital screens the 3s and 3p orbitals.

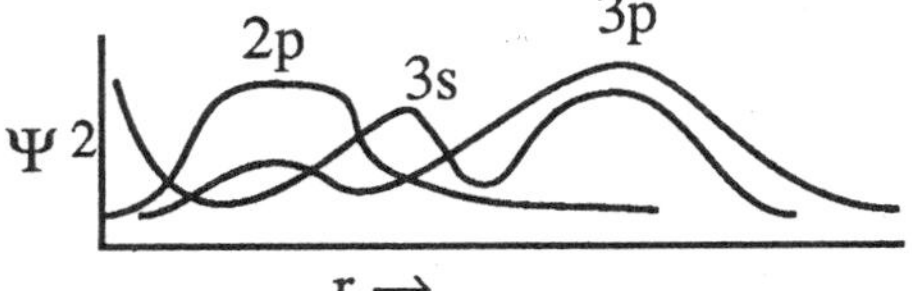

7.53 Rivers and most lakes have some dissolved salts in their waters. These waters eventually flow into the oceans. Some of the oceans' waters evaporate leaving behind the salts. This evaporated water forms precipitation which supplies water for the rivers and lakes. This water dissolves more salts as it flows to the oceans. This process continues, thus increasing the concentrations of the dissolved salts in the ocean. This same process can make a lake 'salty' if the lake has no outflow.

7.54 Rb would have a larger radius than Br because it has an electron in an orbital with a larger principal quantum number, *n* (*n* = 5 vs. *n* = 4).

Rb ≡ $1s^22s^22p^63s^23p^64s^23d^{10}4p^65s^1$

Br ≡ $1s^22s^22p^63s^23p^64s^23d^{10}4p^5$

The electron configurations for Rb^+ and Br^- are the same:
$1s^22s^22p^63s^23p^64s^23d^{10}4p^6$
The Z for Rb (Z = 37) is greater than the Z for Br (Z = 35). The 2 extra protons in Rb exert more attraction for the electrons and, therefore, Rb^+ is smaller than Br^-.

7.55

Quantum Number	Associated Property	Restrictions
n	energy of electron	positive integers
l	shape of atomic orbitals	positive integers less than n
m_l	direction orientation of orbitals	all integers from $-l$ to $+l$
m_s	spin orientation of electrons	either +1/2 or -1/2

7.56 $Z = 9$ (F) $1s^22s^22p^5$ or [He] $2s^22p^5$

$Z = 20$ (Ca) $1s^22s^22p^63s^23p^64s^2$ or [Ar] $4s^2$

$Z = 26$ (Fe) $1s^22s^22p^63s^23p^64s^23d^6$ or [Ar] $4s^23d^6$

$Z = 33$ (As) $1s^22s^22p^63s^23p^64s^23d^{10}4p^3$ or [Ar] $4s^23d^{10}4p^3$

7.57 F^{2+} $1s^22s^22p^3$ Ca^{2+} $1s^22s^22p^63s^23p^6$

Fe^{2+} $1s^22s^22p^63s^23p^63d^6$ As^{2+} $1s^22s^22p^63s^23p^64s^23d^{10}4p^1$

7.58 For the $3d_{xy}$ orbital a graph of Ψ^2 vs. r should show very little if any electron density along the (a) z axis or (b) x axis. The maximum amount of electron density for this orbital will be about a line between the x and the y axis.

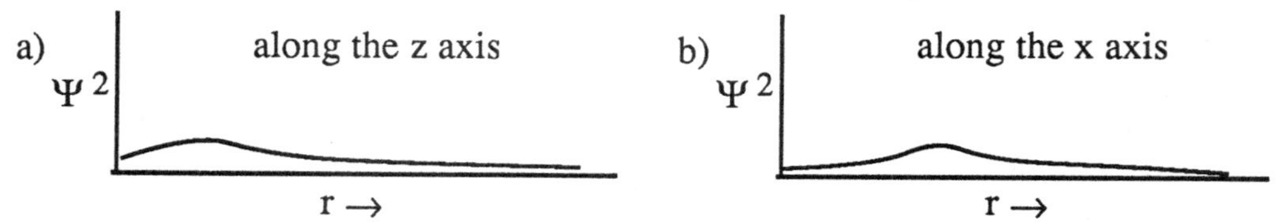

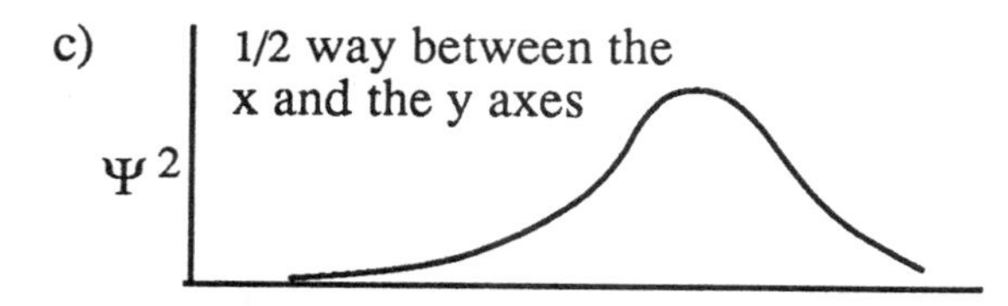

7.59 Below are drawings of the contour diagrams for the 1s orbital and the $2p_z$ orbital drawn to about the same scale.

7.60 Cu^+ 4s — 3d ↓↑ ↓↑ ↓↑ ↓↑ ↓↑

Mn^{2+} 4s — 3d ↑ ↑ ↑ ↑ ↑

Au^{3+} 6s — 5d ↓↑ ↓↑ ↓↑ ↑ ↑

7.61 The electron configuration for P is: $1s^22s^22p^63s^23p^3$. Its total spin = 3(1/2) = 3/2 or 3p $\uparrow$ $\uparrow$ $\uparrow$.

The electron configuration for Br^- is: $1s^22s^22p^63s^23p^64s^23d^{10}4p^6$. Its total spin = 0 or 4p $\downarrow\uparrow$ $\downarrow\uparrow$ $\downarrow\uparrow$.

The electron configuration for Cu^+ is: $1s^22s^22p^63s^23p^64s^03d^{10}$. Its total spin = 0 or 3d $\downarrow\uparrow$ $\downarrow\uparrow$ $\downarrow\uparrow$ $\downarrow\uparrow$ $\downarrow\uparrow$. (See Problem 7.60 above)

The electron configuration for Gd is: [Xe] $6s^24f^75d^1$. Its total spin = 8(1/2) = 4 or 4f $\uparrow$ $\uparrow$ $\uparrow$ $\uparrow$ $\uparrow$ $\uparrow$ $\uparrow$ 5d $\uparrow$ — — — —.

The electron configuration for Sr is: [Kr] $5s^2$ Its total spin = 0 or 5s $\downarrow\uparrow$.

7.62 Ar^+ > Ne > Ar > Cl

Ar^+ is the largest because ionization is removing an electron from a charged ion. Ne and Ar both have filled valence orbitals but it is easier to remove an electron from Ar because it has larger valence orbitals (larger value of n) and, therefore, holds the electron more loosely. Cl has the smallest ionization because it has less protons than the Ar has and larger valence orbitals than the Ne.

7.63 Ca > Br > Cl^- > K^+ Valence electrons of Ca and Br occupy orbitals with n = 4 which means their radii are greater than the radii for Cl^- and K^+ with occupied orbitals with n = 3. Br is smaller than Ca because Br has more protons (Z = 35) than Ca (Z = 20) which attract the electrons more strongly and make the radius smaller. Cl^- and K^+ have the same electron configuration ($1s^22s^22p^63s^23p^6$). Again, the extra protons of K^+ (Z = 19) vs. Cl^- (Z = 17) pull the electrons closer to the nucleus making K^+ smaller.

7.64 Ne < K^+ < Ar < Na < K Ne is the smallest with electrons in orbitals with n = 2. Ar, K^+ and Na have electrons in orbitals with n = 3. K is the largest of this group because its valence electron is in the 4s orbital. Ar and K^+ have the same electron configuration ($1s^22s^22p^63s^23p^6$) but the extra proton of K attracts the electrons more making K^+ smaller than Ar. The Na atom also has an electron with n = 3 and fewer protons than Ar and K^+ and is, therefore, larger than Ar and K^+.

7.65 Francium should most closely resemble Cesium. It is an alkali metal. It should form a plus one ion (Fr^+). It should be a soft, silvery to slightly golden, corrosive metal with relatively low melting and boiling points. It should be a good electrical and thermal conductor and show good photoelectric properties. Its first ionization energy should be small but the second ionization energy should be very large. Its compounds should virtually all be ionic and exhibit the +1 oxidation state. Unlike the other alkali metals, it is radioactive.

7.66 The metals with s^2p^1 configurations are aluminum (Al), gallium (Ga), indium (In), and thallium (Tl).

7.67 Four electrons is the maximum number of valence p electrons in the metal polonium (Po).

7.68 Li has a $1s^22s^1$ configuration and Li^{2+} has a $1s^1$ configuration. The Li^{2+} cation would have the lower energy 2s orbital because there is less screening by the one electron in the 1s orbital than by the 2 electrons in the 1s orbital of the Li atom.

7.69 Make a plot of Figure 7.7 superimposed on Figure 7.6. Your plot should be similar to the answer to Section Exercise 7.3.1. Using your plot (drawing), describe the reduced screening arising from the electron density lobes of the 3s and 3p orbitals that lie close to the nucleus.

7.70 The noble gases have high ionization energies so it is unlikely they will form cations. The formation of negative ions (-EA) is also endothermic so anions are also formed with difficulty.

7.71

5d electron	$l = 2$	$m_l = +2, +1, 0, -1, -2$
4f electron	$l = 3$	$m_l = +3, +2, +1, 0, -1, -2, -3$

7.72

	ground state	excited state	orbital
O^{2+}	$1s^22s^22p^2$	$1s^22s^22p^13s^1$	
O^{2-}	$1s^22s^22p^6$	$1s^22s^22p^53s^1$	
Br^-	$1s^22s^22p^63s^23p^64s^23d^{10}4p^6$	$1s^22s^22p^63s^23p^64s^23d^{10}4p^55s^1$	
Ca^+	$1s^22s^22p^63s^23p^64s^1$	$1s^22s^22p^63s^23p^64s^03d^1$	
Sb^{3+}	$1s^22s^22p^63s^23p^64s^23d^{10}4p^65s^24d^{10}5p^0$	(excited state)$1s^22s^22p^63s^23p^64s^23d^{10}4p^65s^14d^{10}5p^1$ or $5s^24d^95p^1$	

7.73

Ce^{3+} [Xe] $4f^1$	Ce [Xe] $4f^15d^16s^2$
La^{2+} [Xe] $5d^1$	La [Xe] $5d^16s^2$
Ba^+ [Xe] $6s^1$	Ba [Xe] $6s^2$

Ground state configurations are different. Orbitals are most effectively screened by electrons occupying orbitals with smaller values of *n*. Therefore, 5d and 4f electrons are lower energy than 6s electrons.

7.74 a) Two electrons is the maximum number of electrons that can be placed in any orbital.

b)

n	l	m_l	m_s
2	1	+1	
2	1	0	
2	1	-1	

c) B<C<A

d) There are 4 other d orbitals with the same energy.

e) $1s^22s^2$ Be beryllium

f) $1s^22s^22p^1$ B boron or $1s^22s^22p^2$ C carbon

g) O^{2-} or F^- (N^{3-} is not a common ion)

7.75 Removing an electron from Be ($1s^22s^2$, Z = 4) is more difficult than removing an electron from Li ($1s^22s^1$, Z = 3) because the Z_{eff} is greater for Be. It is more difficult to remove an electron from $Li^+(1s^2)$ than from Be^+ ($1s^22s^1$) because for Li^+ the electron is being removed from the much more stable 1s orbital (n = 1).

7.76 a) The rank in order of stability of the orbitals shown are: B > A > C.

b) A set of quantum numbers for orbital A is

n	l	m_l	m_s
3	0	0	+1/2
3	0	0	-1/2

c) The atomic number of an element that has two electrons in orbital A but no electrons in orbital C is Z = 12 (Mg).

d) Orbital C is a d_{z^2} orbital. An electron configuration that contains only one electron in this orbital is: $1s^22s^22p^63s^23p^64s^03d^5$. That is the electron configuration for Fe^{3+} or Mn^{2+}.

e) There are at least 9 orbitals that will have the same principal quantum number as orbital C because orbital C must have a principal quantum number of 3 or larger since it is a d orbital. The maximum is one 3s orbital, three 3p orbitals, and five 3d orbitals.

f) Orbital B will become smaller if one of its electrons is lost. If it contains only 1 electron, there will be no electron-electron repulsion and no screening.

7.77 $E = \frac{hc}{\lambda}$ $E_{molar} = \left(\frac{hc}{\lambda}\right)N_A$

$$E = \frac{(6.626 \times 10^{-34}\ J\ s)(2.998 \times 10^{8}\ m/s)}{(589\ nm)\left(10^{-9}\ m/nm\right)}(6.022 \times 10^{23}\ mol^{-1})$$

$$= 203.1\ kJ/mol$$

Ionization energy of excited atom = 496.5 kJ/mol - 203.1 kJ/mol = 293.4 kJ/mol

7.78

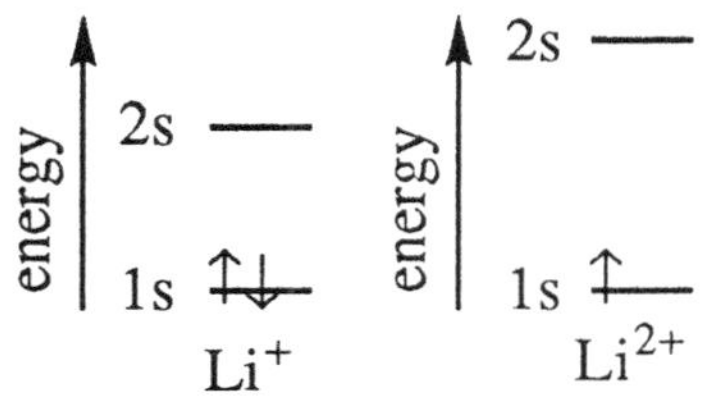

energy ↑: 2s —; 1s ↑↓ — Li^+

energy ↑: 2s —; 1s ↑ — Li^{2+}

The answer to this question needs to show energy diagrams that also include the 2p, 3s, 3p, and 3d orbitals as requested in the question. In the energy diagram for Li^{2+} it should look like that for H, except for greater energy spacing. The energy diagram for Li^+ should look like the generic energy level diagram for a multi-electron atom.

7.79 a) There would be 9 g orbitals when l = 4. This can be calculated by: number of orbitals = $(2l + 1) = 2(4) + 1 = 9$

b) The possible values of m_l for the g orbitals are: +4, +3, +2, +1, 0, -1, -2, -3, -4.

c) The lowest principal quantum number for which g orbitals could exist is n = 5.

(continued)

(7.79 continued)

d) The 8s orbital is the orbital closest to the lowest energy g orbital and may turn out to be nearly degenerate with that lowest g orbital.

e) For the requested drawing extend Figure 7-17 by adding the following orbitals as one goes to higher and higher energy levels: one 5s orbital, three 5p orbitals, one 6s orbital, seven 4f's, five 5d's, three 6p's, one 7s, seven 5f's, five 6d's, three 7p's, one 8s, and nine 5g's.
The atomic number of the first element which could include a g electron would be 121 if 5g fills after 8s. If 1 (or 2) electrons enter the 6f and/or the 7d before the 5g fills (analogous to lanthanides and actinides), the atomic number could be 124.

7.80 The element is probably a halogen and is located in column VII of the Periodic Table. The electron affinity shows that the formation of X^- is quite exothermic suggesting a stable X^- ion such as halogens form. The trend in ionization energies supports this choice. As we go from [ ] $np^5 \rightarrow$ [ ] $np^4 \rightarrow$ [ ] $np^3 \rightarrow$ [ ] np^2, the IE increases as we remove an electron from an atom, from a positive ion and from a doubly positive ion. There is no large increase as if we were removing an electron from a noble gas configuration although this is a slightly larger increase from $np^3 \rightarrow np^2$ vs. $np^4 \rightarrow np^3$ as expected.

7.81

Nonexistent wave function:	Reason for nonexistence:
4g	For g, $l = 4$, but $n = 4$ and largest value of $l = n - 1 = 3$
2d	For d, $l = 2$, but $n = 2$ and largest value of $l = n - 1 = 1$

7.82 Below are the correct ground-state electronic configurations:

C	$1s^22s^22p^2$	Cr	$1s^22s^22p^63s^23p^64s^13d^5$
Sb	[Kr] $5s^24d^{10}5p^3$	Br	$1s^22s^22p^63s^23p^64s^23d^{10}4p^5$
Xe	[Kr] $5s^24d^{10}5p^6$		

7.83 a)

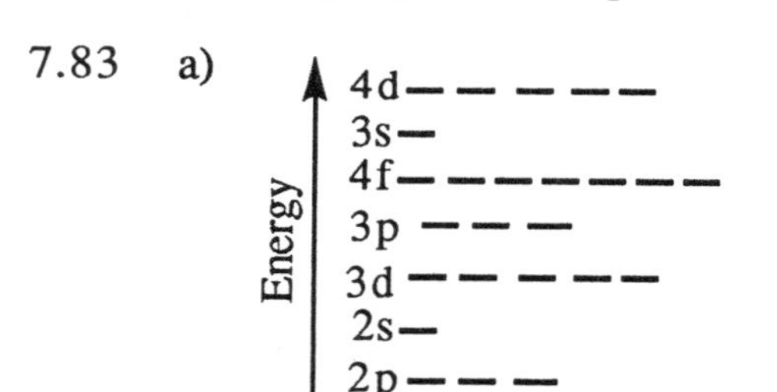

b) In this system the ground-state configuration for an atom containing 27 electrons would be: $1s^22p^62s^23d^{10}3p^64f^1$.

c)

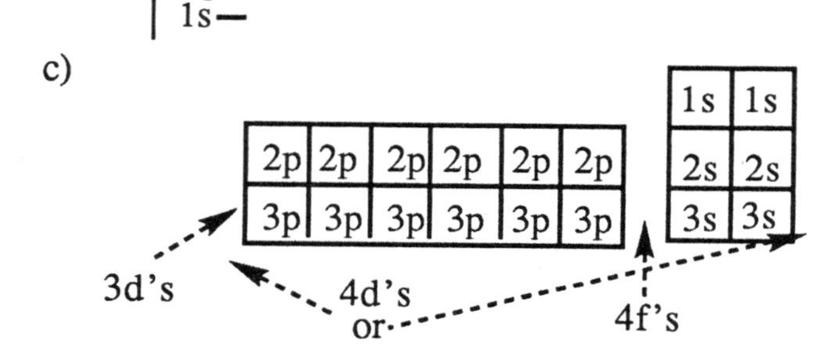

7.84 a) The electronic configuration of element 19 is: $1s^3 2s^3 2p^9 3s^3 3p^1$.

b) The electronic configuration of element 28 would be: $1s^3 2s^3 2p^9 3s^3 3p^9 4s^1$. Assuming that paired electrons means an orbital contains 2 electrons with +1/2 and -1/2 spins or 3 electrons with +1/2, -1/2 and 0 spins, element 28 would contain 1 "unpaired" electron.

c) The electronic configuration of element 30 would be: $1s^3 2s^3 2p^9 3s^3 3p^9 4s^3$
The electronic configuration of element 30's most highly charged stable cation would be: $1s^3 2s^3 2p^9 3s^3 3p^9$. Therefore, the highest positive charge of a cation of element 30 would be $3+$ when it lost 3 electrons from the 4s orbital.

7.85 a) Element 18 ($n = 3$) should have a larger ionization energy than element 30 ($n = 4$) because the electron is held more loosely as n increases and the orbitals become larger.

b) Element 15 and cation 17^{2+} have the same electronic configuration:
$1s^3 2s^3 2p^9$
This electronic configuration is analogous to a noble gas electronic configuration. Since 17^{2+} has 2 more protons than 15, it would be expected that the radius of 17^{2+} would be smaller than 15. Comparison of the measurement of the radius of an ion with the radius of the atom of a noble gas presents problems because the measurement of an ion is made on a salt containing that ion while it is not possible to measure the radius of the noble gas atom under similar conditions.

c) The electronic configuration of element 47 is: $1s^3 2s^3 2p^9 3s^3 3p^9 4s^3 3d^{15} 4p^2$.
The electronic configuration of element 48 is: $1s^3 2s^3 2p^9 3s^3 3p^9 4s^3 3d^{15} 4p^3$.
The electron affinity should be larger (more exothermic) for element 47 than for element 48 because adding an electron to 47 results in 3 unpaired electrons in the 4p orbitals while adding an electron to 48 requires pairing of 2 electrons in a 4 p orbital.

7.86 a) The electronic configuration of element 38 is: $1s^3 2s^3 2p^9 3s^3 3p^9 4s^3 3d^8$.
The electronic configuration of the ion 38^{3+} is: $1s^3 2s^3 2p^9 3s^3 3p^9 3d^8$.

b) There is no indication that the d_{xy} orbital should appear any different in that universe. It can accommodate three electrons with different spins (+1/2, 0 and -1/2) but it need not have a shape that is different. Therefore, make your drawing look like the d_{xy} orbital found in Figure 7-9.

c) The ground-state configuration of element 3 would be: $1s^3$.
The ground-state configuration of element 14 would be: $1s^3 2s^3 2p^8$.

7.87 a) Elements with 3, 15 and 27 electrons would be the first three noble gases.

b) Element 4:

n	l	m_l	m_s
1	0	0	+1/2
1	0	0	-1/2
1	0	0	0
2	0	0	+1/2

Element 14:

n	l	m_l	m_s
1	0	0	+1/2
1	0	0	-1/2
1	0	0	0
2	0	0	+1/2
2	0	0	-1/2
2	0	0	0
2	1	-1	+1/2
2	1	0	+1/2
2	1	+1	+1/2
2	1	-1	-1/2
2	1	0	-1/2
2	1	+1	-1/2
2	1	-1	0
2	1	0	0

Element 16:

n	l	m_l	m_s
1	0	0	+1/2
1	0	0	-1/2
1	0	0	0
2	0	0	+1/2
2	0	0	-1/2
2	0	0	0
2	1	-1	+1/2
2	1	0	+1/2
2	1	+1	+1/2
2	1	-1	-1/2
2	1	0	-1/2
2	1	+1	-1/2
2	1	-1	0
2	1	0	0
2	1	+1	0
3	0	0	+1/2

Element 18:

n	l	m_l	m_s
1	0	0	+1/2
1	0	0	-1/2
1	0	0	0
2	0	0	+1/2
2	0	0	-1/2
2	0	0	0
2	1	-1	+1/2
2	1	0	+1/2
2	1	+1	+1/2
2	1	-1	-1/2
2	1	0	-1/2
2	1	+1	-1/2
2	1	-1	0
2	1	0	0
2	1	+1	0
3	0	0	+1/2
3	0	0	-1/2
3	0	0	0

c) The electrons of Morspin would probably more effectively screen a 4s electron because there are 3/2 as many electrons (in presumably the same space) as in our universe.

CHAPTER 8: FUNDAMENTALS OF CHEMICAL BONDING

8.1 The electron configuration of Be is: $1s^2 2s^2$. The 2s electrons will be involved in bond formation.

8.2 The electron configurations of H and Li are: $1s^1$ and $1s^2 2s^1$. The bond is formed by the overlap of H 1s orbital with the Li 2s orbital.

Li + H → LiH

8.3 The electron configuration of Na is: $1s^2 2s^2 2p^6 3s^1$. The bond is formed by the overlap of the 3s orbitals of the 2 Na atoms.

Na + Na → Na_2

8.4

Atom	Valence Electrons	Orbitals That Participate in Bonding
C	$2s^2\ 2p^2$	2p orbitals
S	$3s^2\ 3p^4$	3p orbitals
Hg	$6s^2$	6s orbitals (and 6p orbitals)
Xe	$5s^2\ 5p^6$	Noble gas which does not commonly form bonds. When they do form bonds, 5s, 5p and 5d are used.

8.5 The electron configuration of Br is: $1s^2 2s^2 2p^6 3s^2 3p^6 4s^2 3d^{10} 4p^5$. The bond is formed by the overlap of two 4p orbitals (one from each atom) pointing along the bond axis (line joining the nuclei).

8.6 The normal periodic trend for electronegativities for elements of the p block are to increase from left to right across a row and to decrease down a column. The only exception to the trend for crossing a row is Pb (1.9) and Bi (1.9) with no change in electronegativities. The exceptions to the trend for descending a column are: Al (1.5) and Ga (1.6), Ga (1.6) and In (1.7), In (1.7) and Tl (1.8), and Sn (1.8) and Pb (1.9). In addition there is no change in electronegativities for the pairs: Si (1.8) and Ge (1.8), Ge (1.8) and Sn (1.8), and Sb (1.9) and Bi (1.8).

8.7 From the list given one would conclude, based upon the electronegativity differences of the two atoms involved, that F_2 is non-polar, HF and NaH are polar, and NaF and CaO are ionic.

8.8 For each of the pairs, the element attracting the greatest electron density is: a) O b) O c) C d) Cl

8.9 From smallest electronegativity difference to largest electronegativity difference: Si—H < C—H < N—H < O—H < F—H

8.10 The following bonds are arranged in order of increasing polarity: Cl—Cl < Se—H < C—Br < Al—Cl < H—F

8.11 The following molecules are arranged in order of increasing bond polarity: $PH_3 < H_2S < NH_3 < H_2O$

8.12 From the list given one should conclude that KBr and NH_4Br are ionic and that HBr, CBr_4, and Br_2 are covalent.

8.13

Compound	Valence Electrons
HBr	H : 1 Br : 7 total = 8
KBr	K : 1 Br : 7 K^+ : 1 - 1 = 0 Br^- : 7 + 1 = 8
NH_4Br	NH_4^+ : N : 5 H : 1 total: 5 + 4(1) -1 = 8 Br : 7 Br^-: 7 + 1 = 8
CBr_4	C : 4 Br : 7 total: 4 + 4(7) = 32
Br_2	Br : 7 total: (2)(7) = 14

8.14 From the list given one should conclude that Na_2SO_3 and NH_4PCl_6 are ionic and that SO_3, $AlCl_3$, PCl_3, and PCl_5 are covalent.

8.15

Compound	Valence Electrons
Na_2SO_4	Na:1 each Na^+:1 - 1 = 0 SO_4^{2-}: S:6 O:6 total SO_4^{2-}: 6 + 4(6) + 2 = 32
SO_3	S : 6 O : 6 total: 6 + 3(6) = 24
$AlCl_3$	Al : 3 Cl : 7 total: 3 + 3(7) = 24
PCl_3	P : 5 Cl : 7 total: 5 + 3(7) = 26
PCl_5	P : 5 Cl : 7 total: 5 + 5(7) = 40
NH_4PCl_6	NH_4^+ : N : 5 H:1 total 5 + 4(1) - 1 = 8 PCl_6^-: P : 5 Cl : 7 total: 5 + 6(7) + 1 = 48

8.16 NH_4NO_3 and KNO_2 are ionic. NH_3, HNO_3 and NO_2 are covalent.

8.17

Compound	Valence Electrons
NH_3	N : 5 H : 1 total: 5 + 3(1) = 8
NH_4NO_3	NH_4^+: N : 5 H : 1 total: 5 + 4(1) - 1 = 8 NO_3^-: N : 5 O : 6 total: 5 + 3(6) + 1 = 24
HNO_3	H : 1 N : 5 O : 6 total: 1 + 5 + 3(6) = 24
KNO_2	K^+: K : 1 total: 1 - 1 = 0 NO_2^-: N : 5 O : 6 total = 5 + 2(6) + 1 = 18
NO_2	N : 5 O : 6 total: 5 + 2(6) = 17

8.18

Compound	Cation	Anion
$NH_4(HSO_4)$	ammonium cation, NH_4^+	hydrogen sulfate anion, HSO_4^-
$NaClO_3$	sodium cation, Na^+ (not polyatomic)	chlorate anion, ClO_3^-
$LiNO_3$	(not polyatomic)	nitrate anion, NO_3^-
$(NH_4)_2CO_3$	ammonium cation, NH_4^+	carbonate anion, CO_3^{2-}
KPF_6	(not polyatomic)	hexafluorophosphate anion, PF_6^-
$KMnO_4$	(not polyatomic)	permanganate anion, MnO_4^-

8.19

Compound	Polyatomic Ions Present	
LiOH	hydroxide anion, OH^-	
KH_2PO_4	dihydrogenphosphate anion, $H_2PO_4^-$	
$NaBF_4$	fluoroborate anion (or tetrafluoroborate anion), BF_4^-	
$LiIO_4$	periodate anion, IO_4^-	
NaCN	cyanide anion, CN^-	
$(NH_4)_2Cr_2O_7$	ammonium cation, NH_4^+	dichromate anion, $Cr_2O_7^{2-}$

8.20

Compound	Valence Electrons		
N_2O	N : 5	O : 6	total: 2(5) + 6 = 16
O_3	O : 6		total 3(6) = 18
CCl_4	C : 4	Cl : 7	total: 4 + 4(7) = 32
H_2S	H : 1	S : 6	total: 2(1) + 6 = 8
PH_3	P : 5	H : 1	total: 5 + 3(1) = 8

8.21

Compound	Valence Electrons	
$NH_4(HSO_4)$	NH_4^+: N : 5 H : 1	total: 5 + 4(1) - 1 = 8
	HSO_4^-: H : 1 S : 6 O : 6	total: 1 + 6 + 4(6) + 1 = 32
$NaClO_3$	ClO_3^-: Cl : 7 O : 6	total 7 + 3(6) + 1 = 26
$LiNO_3$	NO_3^-: N : 5 O : 6	total: 5 + 3(6) + 1 = 24
$(NH_4)_2CO_3$	NH_4^+: N : 5 H : 1	total: 5 + 4(1) - 1 = 8
	CO_3^{2-}: C : 4 O : 6	total 4 + 3(6) + 2 = 24
KPF_6	PF_6^-: P : 5 F : 7	total: 5 + 6(7) + 1 = 48
$KMnO_4$	MnO_4^-: Mn : 7 O : 6	total: 7 + 4(6) + 1 = 32

8.22

Compound	Ions	Bonding Frameworks
$NH_4(HSO_4)$	NH_4^+ & HSO_4^-	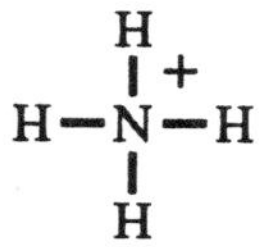 H (+) H—N—H H and O (-) O—S—O—H O 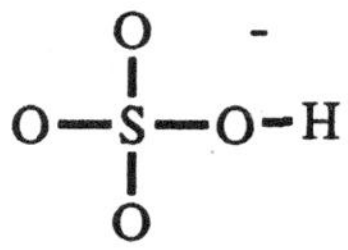
$NaClO_3$	Na^+ & ClO_3^-	Na+ and 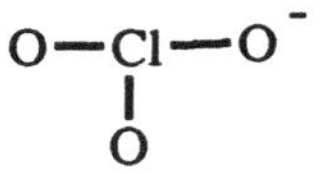 O—Cl—O (-) O
$LiNO_3$	Li^+ & NO_3^-	Li^+ and 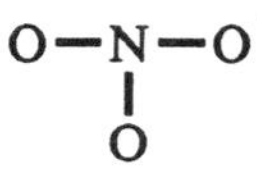 O—N—O (-) O
$(NH_4)_2CO_3$	NH_4^+ & CO_3^{2-}	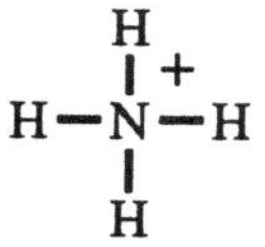 H (+) H—N—H H and O—C—O (2-) O 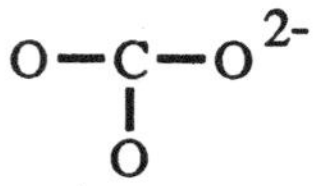
KPF_6	K^+ & PF_6^-	K^+ and 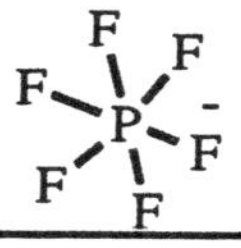 F F F—P—F (-) F F
$KMnO_4$	K^+ & MnO_4^-	K^+ and 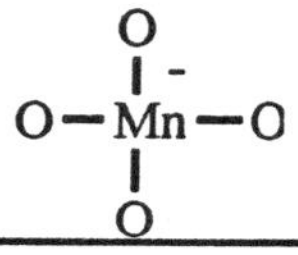 O (-) O—Mn—O O

8.23

Compound	Bonding Framework
N_2O	N—N—O
O_3	O—O—O
CCl_4	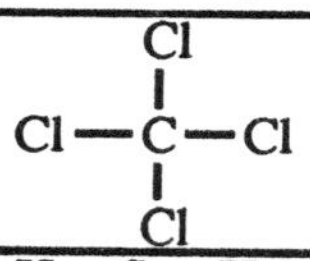 Cl Cl—C—Cl Cl
H_2S	H—S—H
PH_3	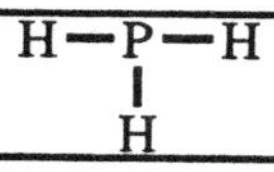 H—P—H H

8.24

Compounds	Provisional Lewis Structures and Formal Charges	
$NH_4(HSO_4)$ $40\ e^- - 4(2) - 5(2) - 3(6) - 4 = 0$	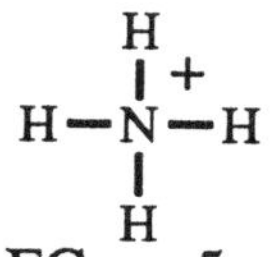 H (+) H—N—H H $FC_N = 5 - 4 = +1$ $FC_H = 1 - 1 = 0$	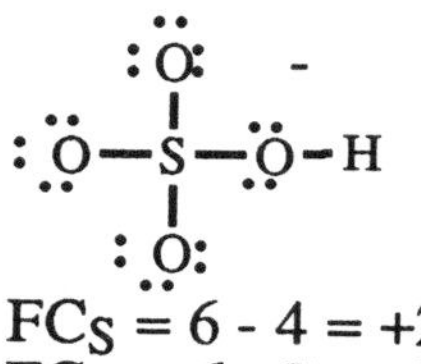 :Ö: (-) :Ö—S—Ö—H :Ö: $FC_S = 6 - 4 = +2$ $FC_O = 6 - 7 = -1$ $FC_{O—H} = 6 - 6 = 0$

(continued)

(8.24 continued)

$NaClO_3$ $26\ e^- - 3(2) - 3(6) - 2 = 0$	Na^+	$[:\ddot{O}-\ddot{Cl}-\ddot{O}:$ with $:\ddot{O}:$ below Cl$]^-$ $FC_{Cl} = 7 - 5 = +2$ $FC_O = 6 - 7 = -1$
$LiNO_3$ $24\ e^- - 3(2) - 3(6) = 0$	Li^+	$[:\ddot{O}-N-\ddot{O}:$ with $:\ddot{O}:$ below N$]^-$ $FC_N = 5 - 3 = +2$ $FC_O = 6 - 7 = -1$
$(NH_4)_2CO_3$ $40\ e^- - 8(2) - 3(2) - 3(6) = 0$	$[NH_4]^+$ (N bonded to four H) $FC_N = 5 - 4 = +1$ $FC_H = 1 - 1 = 0$	$[:\ddot{O}-C-\ddot{O}:$ with $:\ddot{O}:$ below C$]^{2-}$ $FC_C = 4 - 3 = +1$ $FC_O = 6 - 7 = -1$
KPF_6 $48\ e^- - 6(2) - 6(6) = 0$	K^+	$[PF_6]^-$ (P bonded to six $:\ddot{F}:$) $FC_P = 5 - 6 = -1$ $FC_F = 7 - 7 = 0$
$KMnO_4$ $32\ e^- - 4(2) - 4(6) = 0$	K^+	$[MnO_4]^-$ (Mn bonded to four $:\ddot{O}:$) $FC_{Mn} = 7 - 4 = +3$ $FC_O = 6 - 7 = -1$

8.25 Compound Provisional Lewis Structures and Formal Charges

N_2O $\quad :\ddot{N}-N-\ddot{O}:$

$16\ e^- - 2(2) - 2(6) = 0 \quad FC_{N-} = 5 - 7 = -2$

$FC_{-N-} = 5 - 2 = +3 \quad FC_O = 6 - 7 = -1$

O_3 $\quad :\ddot{O}-\ddot{O}-\ddot{O}: \quad 18\ e^- - 2(2) - 2(6) - 2 = 0$

$FC_{O-} = 6 - 7 = -1 \quad FC_{-O-} = 6 - 4 = +2$

CCl_4 $\quad$ C bonded to four $:\ddot{Cl}:$ $\quad 32\ e^- - 4(2) - 4(6) = 0$

$FC_C = 4 - 4 = 0 \quad FC_{Cl} = 7 - 7 = 0$

(continued)

(8.25 continued)

H_2S

```
  ..
H—S—H
  ..
```

$8\ e^- - 2(2) - 4 = 0$

$FC_S = 6 - 6 = 0$ $FC_H = 1 - 1 = 0$

PH_3

```
  ..
H—P—H
  |
  H
```

$8\ e^- - 3(2) - 2 = 0$

$FC_P = 5 - 5 = 0$ $FC_H = 1 - 1 = 0$

8.26 (See Problem 8.24)

Compound Adjusted Lewis Structures

$NH_4(HSO_4)$

```
   H                 ..
   | +              :O:   -
H—N—H            ..  |  ..
   |            :O—S—O—H
   H             ..  |  ..
                    :O:
                     ..

     :O:  -              :O:  -              ..
      ||                  ||                 :O:  -
 ..   |   ..              |   ..          ..  |   ..
:O—S—O—H  ↔  :O=S—O—H ↔  O=S—O—H
 ..   ||  ..              |   ..          ..  ||  ..
     :O:                 :O:                 :O:
                          ..
```

$FC_{O=} = 6 - 6 = 0$ $FC_{O—} = 6 - 7 = -1$ $FC_{—O—} = 6 - 6 = 0$
$FC_S = 6 - 6 = 0$

$NaClO_3$

```
       ..  ..  ..   -
Na+   :O—Cl—O:
       ..  |   ..
          :O:
           ..

 ..  ..  ..   -        ..  ..   -         ..  ..  ..   -
:O=Cl=O:  ↔  :O=Cl—O:  ↔  :O—Cl=O:
     |               ..  ||  ..          ..  ||
    :O:                 :O:                 :O:
     ..
```

$FC_{Cl} = 7 - 7 = 0$ $FC_{O=} = 6 - 6 = 0$ $FC_{O—} = 6 - 7 = -1$

$LiNO_3$

```
      ..     ..   -
Li+  :O—N—O:
      ..  |  ..
         :O:
          ..

..       ..   -      ..      ..   -      ..       ..   -
 O=N—O:   ↔   :O—N—O:   ↔   :O—N=O:
..   |   ..          ..  ||  ..          ..   |
    :O:                 :O:                  :O:
     ..                                       ..
```

$FC_N = 5 - 4 = +1$ $FC_{O=} = 6 - 6 = 0$ $FC_{O—} = 6 - 7 = -1$
Note: Although formation of a second N=O bond would further minimize the formal charges, this is not possible because N only has 4 orbitals (1s and 3-2p's) available.

(continued)

(8.26 continued)

$(NH_4)_2CO_3$

$$\left[H-\underset{H}{\overset{H}{N}}-H\right]^{+} \quad \left[:\ddot{O}-\underset{:\ddot{O}:}{C}-\ddot{O}:\right]^{2-}$$

$$\left[:\ddot{O}=\underset{:\ddot{O}:}{C}-\ddot{O}:\right]^{2-} \leftrightarrow \left[:\ddot{O}-\underset{:O:}{\overset{}{C}}-\ddot{O}:\right]^{2-} \leftrightarrow \left[:\ddot{O}-\underset{:\ddot{O}:}{C}=\ddot{O}:\right]^{2-}$$

$FC_C = 4 - 4 = 0$ $FC_{O=} = 6 - 6 = 0$ $FC_{O—} = 6 - 7 = -1$

KPF_6 K^+ $[PF_6]^-$ no change

$KMnO_4$ K^+

$$\left[:\ddot{O}-\underset{:\ddot{O}:}{\overset{:\ddot{O}:}{Mn}}-\ddot{O}:\right]^{-}$$

$$\left[:\ddot{O}=\underset{:O}{\overset{:\ddot{O}}{Mn}}-\ddot{O}:\right]^{-} \leftrightarrow \left[:\ddot{O}=\underset{:O:}{\overset{:\ddot{O}:}{Mn}}=\ddot{O}:\right]^{-} \leftrightarrow \left[:\ddot{O}-\underset{:O:}{\overset{:\ddot{O}}{Mn}}=\ddot{O}:\right]^{-} \leftrightarrow \left[:\ddot{O}=\underset{:\ddot{O}:}{\overset{:\ddot{O}}{Mn}}=\ddot{O}:\right]^{-}$$

$FC_{Mn} = 7 - 7 = 0$ $FC_{O=} = 6 - 6 = 0$ $FC_{O—} = 6 - 7 = -1$

8.27 (See Problem 8.25)

Compound	Adjusted Lewis Structures
N_2O	$:\ddot{N}-N-\ddot{O}:$

$$:\ddot{N}=N=\ddot{O}: \longleftrightarrow :N\equiv N-\ddot{O}:$$

$FC_{N=} = 5 - 6 = -1$ $FC_{N\equiv} = 5 - 5 = 0$
$FC_{=N=} = 5 - 4 = +1$ $FC_{\equiv N—} = 5 - 4 = +1$
$FC_{O=} = 6 - 6 = 0$ $FC_{—O} = 6 - 7 = -1$

As also described in Problem 8.26, N is limited to 4 available orbitals and, therefore, the formal charge cannot be minimized further.

O_3

$$:\ddot{O}-\ddot{O}-\ddot{O}: \qquad \ddot{O}=\ddot{O}-\ddot{O}: \leftrightarrow :\ddot{O}-\ddot{O}=\ddot{O}:$$

$FC_{O=} = 6 - 6 = 0$ $FC_{=O—} = 6 - 5 = +1$ $FC_{O—} = 6 - 7 = -1$

Note: O also is limited to 4 available orbitals and the formal charges cannot be minimized further.

(continued)

(8.27 continued)

CCl_4 — no change

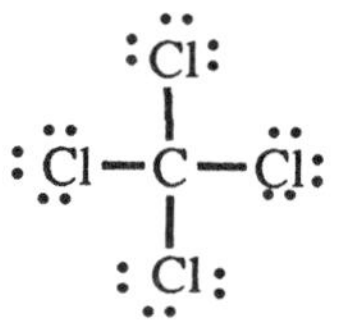

```
      ..
     :Cl:
  ..  |   ..
 :Cl—C—Cl:
  ..  |   ..
     :Cl:
      ..
```

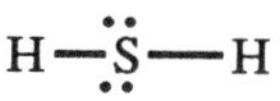

H_2S — H—S̤̈—H no change

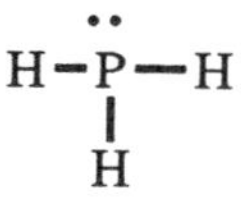

PH_3 — no change

```
    ..
 H—P—H
    |
    H
```

8.28 The steps followed in the solution below are like those steps followed in Sample Problem 8-10 in the text.

For H_3PO_4:

1. not ionic, but covalent
2. e⁻ H : 1 P : 5 O : 6 total $e^- = 3(1) + 5 + 4(6) = 32$
3. 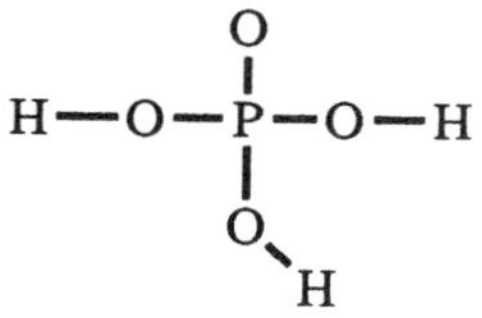 H—O—P(—O)(—O—H)—O—H $e^- = 32 - 7(2) = 18\ e^-$
4. 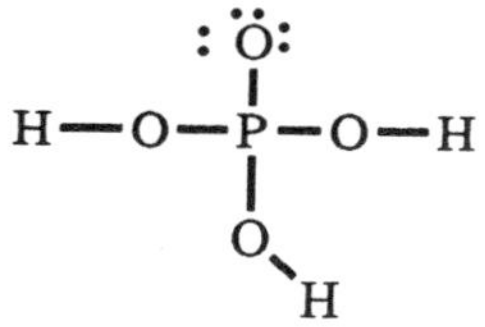 H—O—P(—:Ö:)(—O—H)—O—H $18\ e^- - 6 = 12\ e^-$
5. 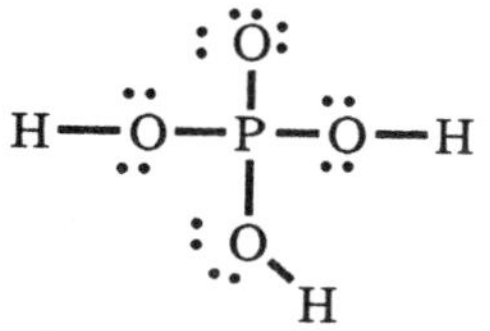 H—Ö—P(—:Ö:)(—:Ö—H)—Ö—H $12\ e^- - 3(4) = 0$
6. $FC_H = 1 - 1 = 0$ $FC_P = 5 - 4 = +1$ $FC_{-O-} = 6 - 6 = 0$ $FC_{O-} = 6 - 7 = -1$
7. 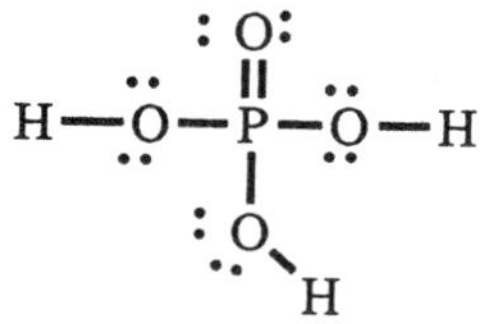 H—Ö—P(=Ö:)(—:Ö—H)—Ö—H $FC_P = 5 - 5 = 0$ $FC_{O=} = 6 - 6 = 0$

For $HClO_3$:

1. not ionic, but covalent
2. e⁻ H : 1 Cl : 7 O : 6 total $e^- = 1 + 7 + 3(6) = 26$

(continued)

(8.28 continued)

3. O—Cl—O—H $e^- = 26 - 4(2) = 18\ e^-$
 (with O bonded below Cl)

4. :Ö—Cl—O—H $18\ e^- - 6(2) = 6\ e^-$
 (with :Ö: bonded below Cl)

5. :Ö—C̈l—Ö—H $6\ e^- - 4 - 2 = 0$
 (with :Ö: bonded below Cl)

6. $FC_H = 1 - 1 = 0$ $FC_{Cl} = 7 - 5 = +2$ $FC_{O—} = 6 - 7 = -1$ $FC_{—O—} = 6 - 6 = 0$

7. :Ö=C̈l—Ö—H $FC_{Cl} = 7 - 7 = 0$ $FC_{O=} = 6 - 6 = 0$ $FC_{—O—} = 6 - 6 = 0$
 (with :Ö: double bonded below Cl)

For H_2SO_3:

1. not ionic, but covalent
2. e^- H : 1 S : 6 O : 6 total $e^- = 2(1) + 6 + 3(6) = 26$
3. H—O—S—O—H $e^- = 26 - 5(2) = 16\ e^-$
 (with O bonded below S)
4. H—O—S—O—H $16\ e^- - 6 = 10\ e^-$
 (with :Ö: bonded below S)
5. H—Ö—S̈—Ö—H $10\ e^- - 2(4) - 2 = 0$
 (with :Ö: bonded below S)
6. $FC_H = 1 - 1 = 0$ $FC_S = 6 - 5 = +1$ $FC_{O—} = 6 - 7 = -1$ $FC_{—O—} = 6 - 6 = 0$
7. H—Ö—S̈—Ö—H $FC_S = 6 - 6 = 0$ $FC_{O=} = 6 - 6 = 0$ $FC_{—O—} = 6 - 6 = 0$
 (with :Ö: double bonded below S)

8.29 The geometry of CH_2Cl_2 is a tetrahedron with the C atom at the center of the tetrahedron and the H and Cl atoms each at a corner of the tetrahedron. This is not a "perfect" tetrahedron because the C—H bond and C—Cl bonds are not exactly the same in length or polarity. The C—H bonds result from the overlap of an sp^3 orbital on the C and the 1s orbital of the H. The C—Cl bonds result from the overlap of an sp^3 orbital on the C and a 3p orbital of the Cl.

8.30 The geometry of C_2H_5Br consists of two overlapping tetrahedrons. One carbon is at the center of a tetrahedron and is bonded to 3 hydrogen atoms and the other carbon which are each at a corner of the tetrahedron. The second carbon is at the center of a second tetrahedron and is bonded to the first carbon, the bromine atom and 2 hydrogen atoms which are each at a corner of the second tetrahedron.
The C—H bonds are formed by overlap of a C sp^3 orbital with a H 1s orbital.
The C—Br bond is formed by overlap of a C sp^3 orbital with the Br 4p orbital.
The C—C bond is formed by the overlap of an sp^3 orbital from each C.

8.31 The steps followed in the solution below are like those steps followed in Sample Problem 8-10 in the text.

For C_6H_{12}

1. covalent

2. e^- H : 1 C : 6 total $e^- = 6(4) + 12(1) = 36\ e^-$

```
        H   H
     H  |   | H
      \ C---C /
 3.  H /     \ H
   H--C       C--H     36e- - 18(2) = 0
       \     / \
    H---C---C   H
        |   | \
        H   H  H
```

4. and 5. These steps do not change the structure given in step 3.

6. $FC_C = 4 - 4 = 0$ $FC_H = 1 - 1 = 0$

7. Based upon the FC's in step 6, there is no need to change the structure worked out in steps 3, 4, and 5.

"Ball-and-Stick" Model. See Figures 8-14 through 8-17 for examples of ball-and-stick models

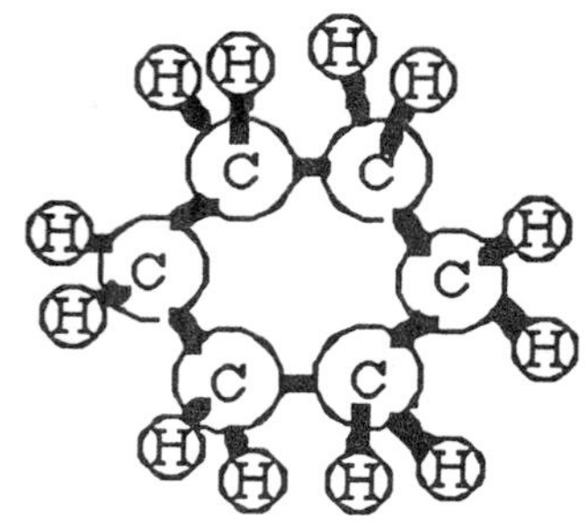

8.32 All possible structural isomers of hexane are:

```
 H       H H H  H               H H  H H  H                H H H H
H-C   C   C   C   H             H-C   C    C-H             H-C   C
 H  C   C   C   C-H           H-C   C   C                 H         C  H
   H  H H  H  H  H              H H   H H H  H            H>C-C       C-H
                                                          H  H  C    H  H
                                                               H H H

   H H                           H H      H  H H
 H  C  H  H                     H-C    H   C-H
H    C   C-H                  H   C-C-C
H-C-C   C  H                  H>C    H   C-H
H    H H H H                  H  H        H  H
```

8.33

Formulas	Use
C_8H_{18}	automobile fuel
C_3H_8	cooking gas
$C_{30}H_{62}$	asphalt
$C_{18}H_{38}$	lubricant
$C_{15}H_{32}$	jet fuel

8.34 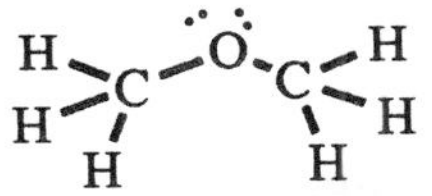 Both the oxygen and carbons have sp^3 hybrid orbitals which overlap to form the C—O bonds and hold the lone pairs on the oxygen. Three of these sp^3 orbitals on the carbons overlap with the 1s orbitals of the H atoms to form the C—H bonds. The orbital geometry is tetrahedral around the O atom and C atoms. The molecular shape is tetrahedral around the carbon atoms and bent for the molecule with the O atom at the center.

The ball-and-stick model requested in this problem should reflect the similarities of water and dimethyl ether by showing, as shown in the structural formulas below, that dimetyl ether's structure can be compared to water with the hydrogens replaced with methyl groups. See Figures 8-14 through 8-17 in the textbook for examples of how to draw ball-and-stick models.

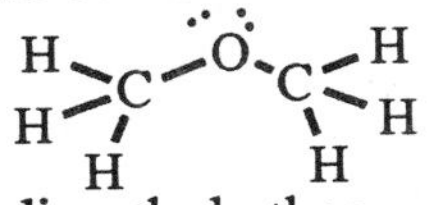

dimethyl ether

water

8.35 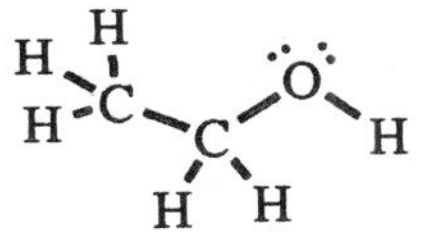 The hybridization of the two C atoms and the O atom are sp^3. For the end C, 3 sp^3 orbitals overlap with the 1s orbitals of H atoms to form C—H bonds and the fourth sp^3 orbital overlaps with an sp^3 orbital of the other C atom to form a C—C bond. The other C atom uses 2 of its sp^3 orbitals to overlap with the 1s orbitals of H atoms to form C—H bonds and the last sp^3 orbital overlaps with an sp^3 from the O atom to form the C—O bond. Two of the other 3 sp^3 orbitals of the O atom hold lone pairs of electrons and the last sp^3 orbital overlaps the 1s orbital of H to form the O—H bond. The geometry of the orbitals is tetrahedral around the 2 C atoms and the O atom. The molecular geometry is tetrahedral around the C atoms but bent around the O atom.

The ball-and-stick model requested in this problem should reflect the similarities of methane and ethyl alcohol by showing, as shown in the structural formulas below, that ethyl alcohol's structure can be compared to methane with two hydrogens replaced: one with a methyl group, the other with a O–H group. See Figures 8-14 through 8-17 in the textbook for examples of how to draw ball-and-stick models.

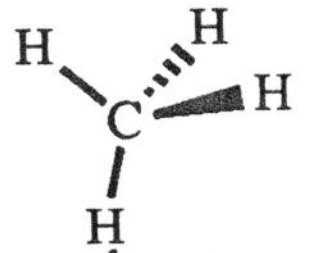

methane 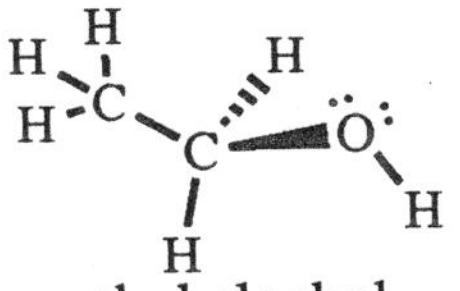 → ethyl alcohol

8.36 $[\ ]^{6-}$

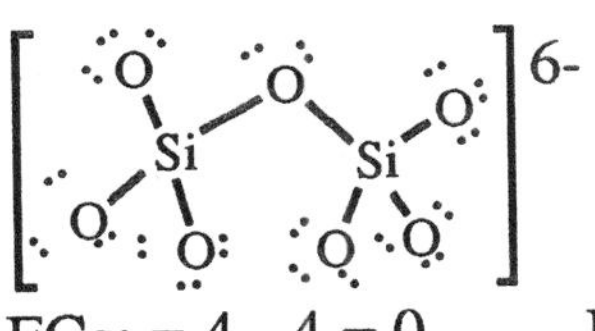

total $e^- = 2(4) + 7(6) + 6 = 56\ e^-$

$56\ e^- - 8(2 \text{ per bond}) = 40\ e^-$

$40\ e^- - 6(6 \text{ per outer O}) = 4\ e^-$

$4\ e^- - 1(4 \text{ per inner O}) = 0$

$FC_{Si} = 4 - 4 = 0$ $\quad FC_{-O-} = 6 - 6 = 0$ $\quad FC_{O-} = 6 - 7 = -1$

(continued)

(8.36 continued)

The ball-and-stick model requested in this problem should reflect features included in the structural formula shown below. See Figures 8-14 through 8-17 in the textbook for examples of how to draw ball-and-stick models.

$[Si_2O_7]^{6-}$

8.37 The hydridizations are sp^3 for both the Si atom and the C atoms (using 2s and 2p's for the C atoms and 3s and 3p's for the Si atom). The orbital and molecular geometries about the Si and C atoms are all tetrahedral.

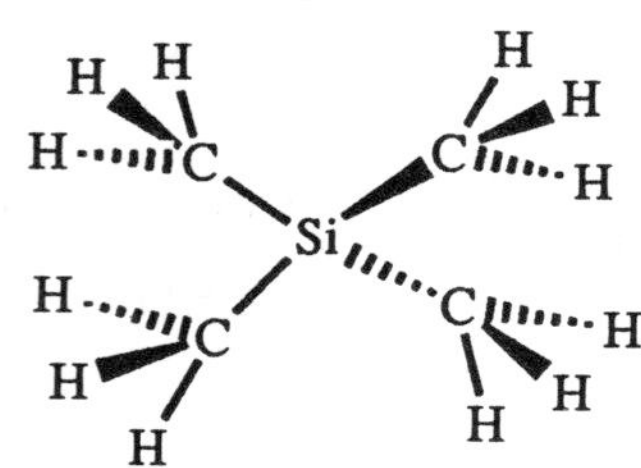

The ball-and-stick model requested in this problem should reflect features included in the structural formula shown below. See Figures 8-14 through 8-17 in the textbook for examples of how to draw ball-and-stick models.

8.38 Reaction: $2PCl_5 \rightarrow [PCl_4]^+ + [PCl_6]^-$

Structures: 2 PCl_5 → $[PCl_4]^+$ + $[PCl_6]^-$

	$2PCl_5$	$[PCl_4]^+$	$[PCl_6]^-$
	$e^- = 5 + 5(7) = 40$	$e^- = 5 + 4(7) - 1 = 32$	$e^- = 5 + 6(7) + 1 = 48$
Steric No.:	5	4	6
Orbital Arrangement:	trigonal bipyramid	tetrahedron	octahedron
Molecular Geometry:	trigonal bipyramid	tetrahedron	octahedron
Hybridization:	sp^3d	sp^3	sp^3d^2

8.39 $XeOF_4$ $e^- = 8 + 6 + 4(7) = 42$

$FC_{Xe} = 8 - 8 = 0$ $FC_F = 7 - 7 = 0$ $FC_O = 6 - 6 = 0$
Steric No. = 6
Orbital Arrangement: octahedron
Molecular Geometry: square pyramid
Hybridization: sp^3d^2 (for Xe); no hybridization for other atoms

(Second bond for Xe=O uses overlap of 2p on O atom and 4d on the Xe atom.)

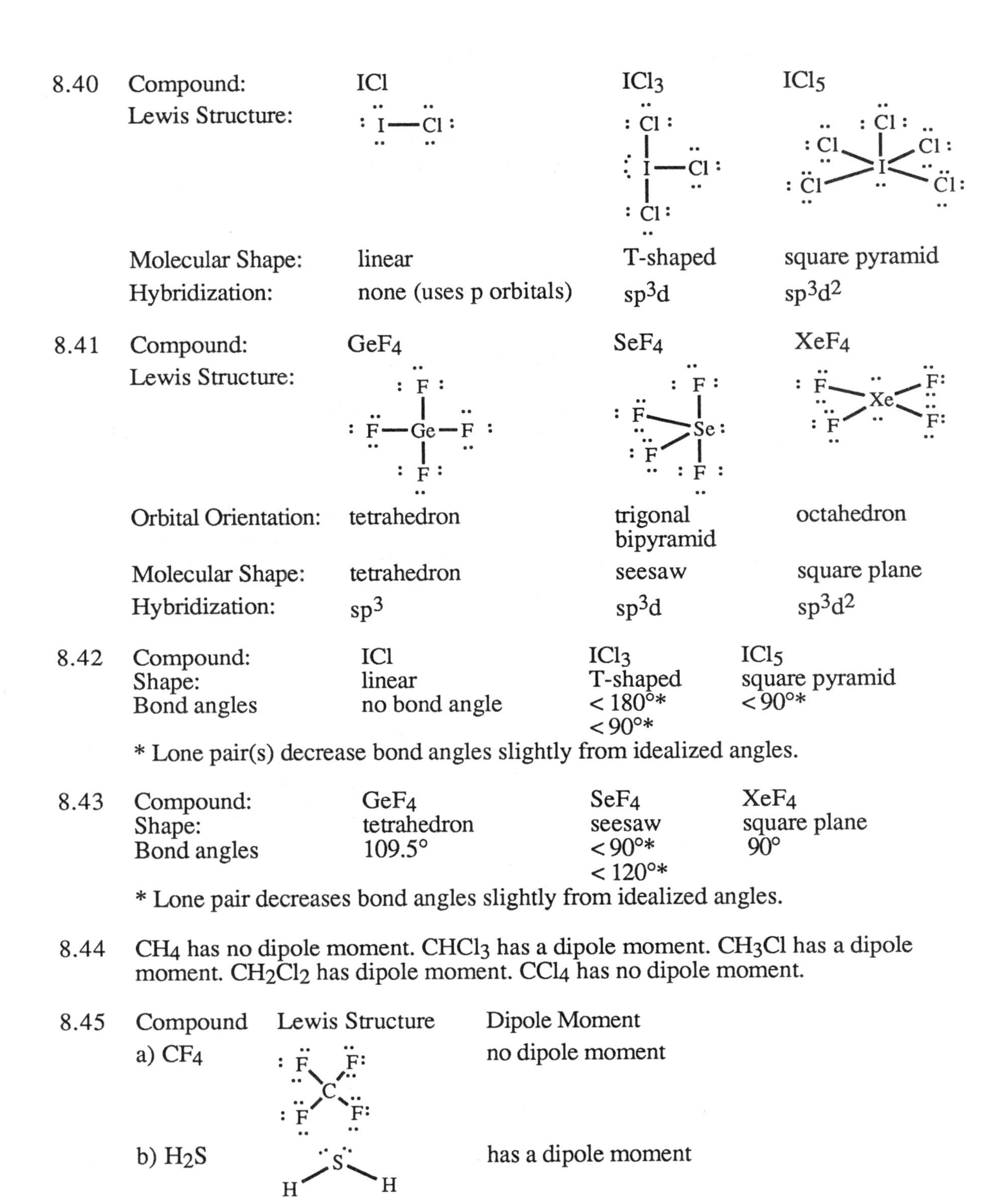

8.40

Compound:	ICl	ICl_3	ICl_5
Lewis Structure:			
Molecular Shape:	linear	T-shaped	square pyramid
Hybridization:	none (uses p orbitals)	sp^3d	sp^3d^2

8.41

Compound:	GeF_4	SeF_4	XeF_4
Lewis Structure:			
Orbital Orientation:	tetrahedron	trigonal bipyramid	octahedron
Molecular Shape:	tetrahedron	seesaw	square plane
Hybridization:	sp^3	sp^3d	sp^3d^2

8.42

Compound:	ICl	ICl_3	ICl_5
Shape:	linear	T-shaped	square pyramid
Bond angles	no bond angle	< 180°* < 90°*	< 90°*

* Lone pair(s) decrease bond angles slightly from idealized angles.

8.43

Compound:	GeF_4	SeF_4	XeF_4
Shape:	tetrahedron	seesaw	square plane
Bond angles	109.5°	< 90°* < 120°*	90°

* Lone pair decreases bond angles slightly from idealized angles.

8.44 CH_4 has no dipole moment. $CHCl_3$ has a dipole moment. CH_3Cl has a dipole moment. CH_2Cl_2 has dipole moment. CCl_4 has no dipole moment.

8.45

Compound	Lewis Structure	Dipole Moment
a) CF_4		no dipole moment
b) H_2S		has a dipole moment

(continued)

(8.45 continued)

c) XeF_2

: F :
|
: Xe :
|
: F :

no dipole moment

d) NF_3

N
: F — | — F :
: F :

has a dipole moment

8.46 The compound is probably IF_7. The center atom needs to be as large as possible with large s, p and d orbitals available for bonding; therefore, X is probably I with $n = 5$. The outer atoms (Y) should be small and highly electronegative in order to separate the atoms and bonding electrons of Y as much as possible in the pentagon bipyramidal geometry (at bond angles of 72°). F is our best candidate for Y.

8.47

Compound	Lewis Structures	3-D Ball-and-Stick
a) Cl_2O	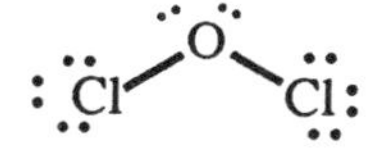	For the three-dimensional ball-and-stick structures one needs to draw 3-D drawings like those in Figures 8-14 through 8-17 in the textbook.
b) C_6H_6	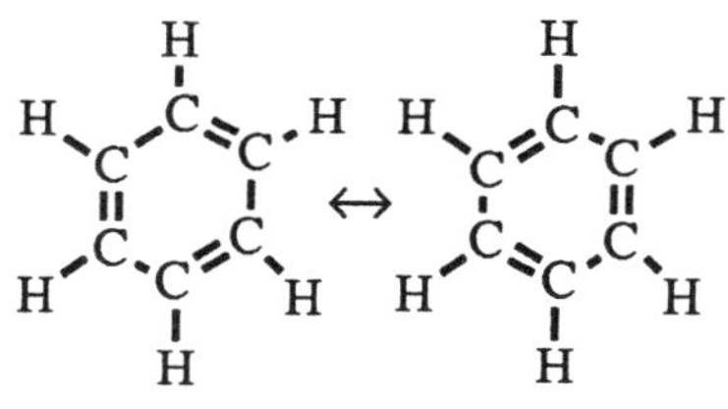	
c) C_2H_4O	H, H — C — C — H, H (ring with O)	

8.48 H, H — O — H (+) + (−) O — H → H, H — O — H — O — H → H, H — O + H — O — H

H_3O^+ + OH^- → $[H_4O_2]$ → H_2O + H_2O

8.49

: O :
||
: Cl — P — Cl :
|
: Cl :

total $e^- = 5 + 6 + 3(7) = 32$
$FC_P = 5 - 5 = 0$
$FC_O = 6 - 6 = 0$
$FC_{Cl} = 7 - 7 = 0$

$H-N_A-N_B-N_C$

total $e^- = 1 + 3(5) = 16$
$16 - 3(2) - 6 - 2(2) = 0$
$FC_{N_C} = 5 - 7 = -2$ $FC_{N_{A,B}} = 5 - 4 = +1$

$H-N_A=N_B-N_C$

$FC_{N_{A,B}} = 5 - 5 = 0$ $FC_{N_C} = 5 - 5 = 0$

(continued)

(8.49 continued)

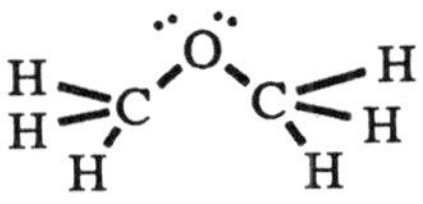

total e^- = 6(1) + 6 + 2(4) = 20 e^-
$FC_O = 6 - 6 = 0$ $\quad FC_C = 4 - 4 = 0$
$FC_H = 1 - 1 = 0$

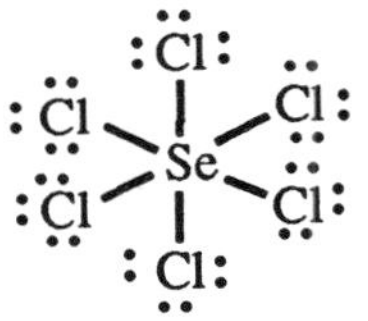

total e^- = 6(7) + 6 = 48 e^-
$FC_{Se} = 6 - 6 = 0$
$FC_{Cl} = 7 - 7 = 0$

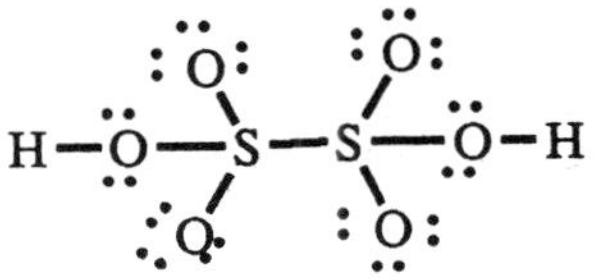

total e^- = 2(1) + 6(6) + 2(6) = 50 e^-
$50e^- - 9(2) - 4(6) - 2(4) = 0$
$FC_S = 6 - 4 = +2$ $\quad FC_{-O-} = 6 - 6 = 0$
$FC_{O-} = 6 - 7 = -1$

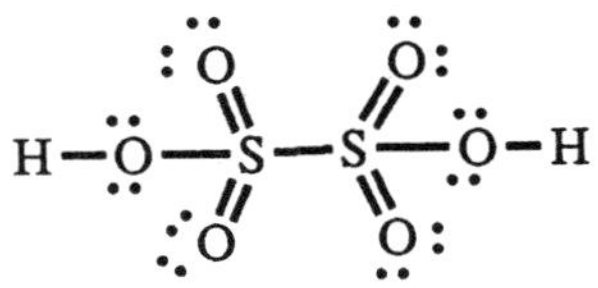

$FC_S = 6 - 6 = 0$
$FC_{O=} = 6 - 6 = 0$

8.50 PCl_3 + $N(CH_3)_3$ → $Cl_3PN(CH_3)_3$

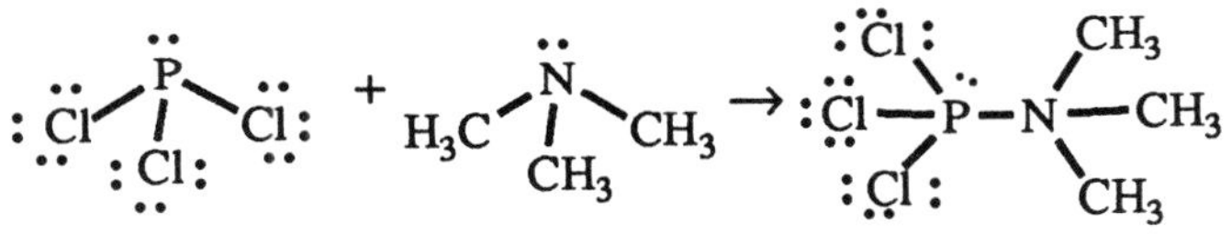

$FC_P = 5 - 6 = -1$
$FC_N = 5 - 4 = +1$

P is an acceptor.

PCl_3 + BBr_3 → Cl_3PBBr_3

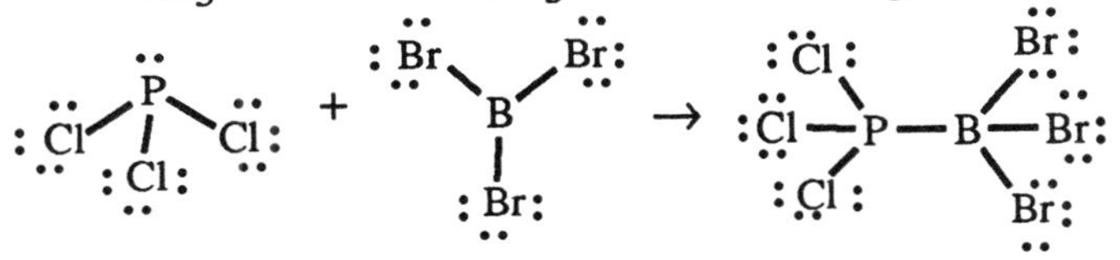

$FC_P = 5 - 4 = +1$
$FC_B = 3 - 4 = -1$

P is a donor.
Note: N and B are row two elements and are, therefore, limited to 4 bonds.

8.51 P may form hybrid orbitals from 3s, 3p and 3d atomic orbitals. N can form hybrid orbitals only from 2s and 2p atomic orbitals. Therefore, N is limited to 4 hybrid orbitals and to a total of 4 bonds and/or lone pairs. NF_5 would require 5 hybrid orbitals and; therefore, would not be expected to exist. PF_5 is possible because P can form more than 4 hybrid orbitals.

8.52 In the figures provided, molecules a and d are equivalent and molecules b and c are equivalent.

8.53 There are only 2 possible structural isomers of AX_3Y_3.

8.54 total e^- = 3(3) + 3(5) + 6(1) = 30 e^-

The B atoms have sp^2 hybridization and the N atoms have sp^3 hybridization. The N forms 2 bonds by overlap with B sp^2 orbitals, one bond by overlap with a H 1s orbital and the fourth sp^3 orbital holds a lone pair of electrons. The B forms 2 bonds by overlap of two of its sp^2 orbitals with the sp^3 orbitals of 2 N atoms and forms its third (and last) bond by overlap of the third sp^2 orbital with the 1s orbital of H.

8.55 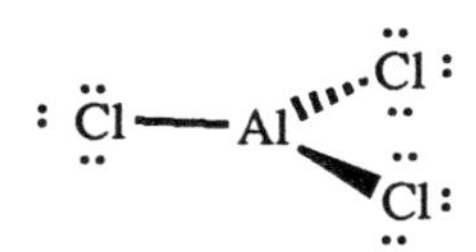

Bonds are formed by overlap of each of the sp^2 hybrid orbitals of the Al with a 3p orbital of Cl. The molecule is planar and the bond angles are 120°.

8.56 The C has sp^2 hybrid orbitals, 2 of which each overlap with a 1s orbital of H to form the C—H bonds. The third sp^2 orbital contains the lone pair of electrons. The fragment is bent. The H—C—H bond angle is probably slightly less than 120°.

8.57

	Molecular Geometry	Formula	Lewis Structure	
a)	bent	TeF_2	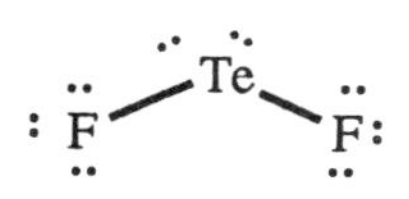	total available e^- from atoms = 6 + 2(7) = 20 e^- e^- needed = 2(2) + 2(6) + 4 = 20 e^-
b)	T-shape	TeF_3^-	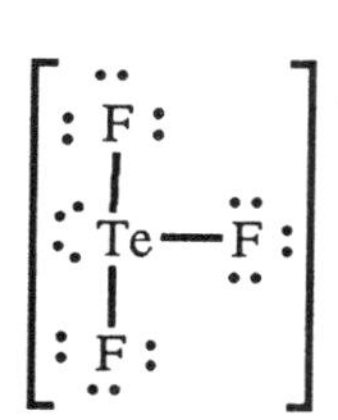	total available e^- = 6 + 3(7) + 1 = 28 e^- e^- needed = 3(2) + 3(6) + 4 = 28 e^-
c)	square pyramid	TeF_5^-	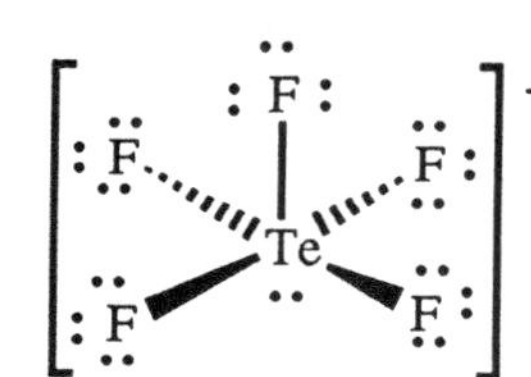	total available e^- = 6 + 5(7) + 1 = 42e^- e^- needed = 5(2) + 5(6) + 2 = 42 e^-

(continued)

(8.57 continued)

d) trigonal bipyramid TeF_5^+

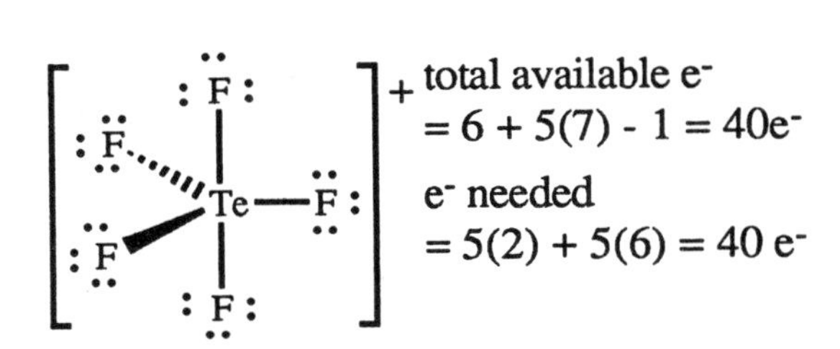

total available e$^-$
= 6 + 5(7) - 1 = 40e$^-$

e$^-$ needed
= 5(2) + 5(6) = 40 e$^-$

e) octahedron TeF_6

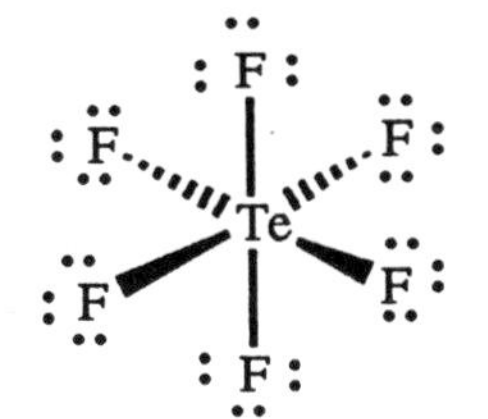

total available e$^-$ from atoms = 6 + 6(7) = 48 e$^-$

e$^-$ needed
= 6(2) + 6(6) = 48 e$^-$

f) seesaw TeF_4

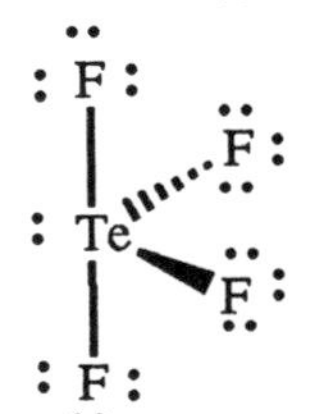

total available e$^-$ from atoms = 6 + 4(7) = 34 e$^-$

e$^-$ needed
= 4(2) + 4(6) + 2 = 34 e$^-$

8.58 Listed in order of increasing bond angles: $XeF_4 < H_2O < CH_4 < BeH_2$
$90° < (<109.5°) < 109.5° < 180°$

8.59 Examples of sulfur-containing compounds with various steric numbers:

Steric Number	Example	Lewis Structure
2	none	—
3	SO_2	
4	SO_4^{2-}	
5	SF_4	
6	SF_6	

8.60 Lewis Structure

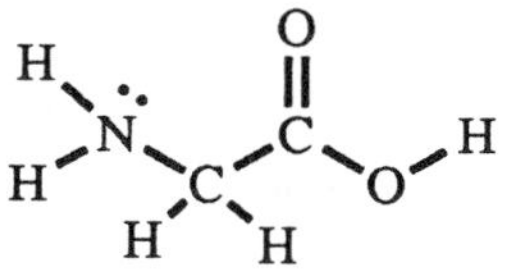

Bonds and Bond Angles

H—O—C, H—N—H, H—N—C < 109.5°
C—C—H, H—C—H, N—C—C = 109.5°
O=C—O, C—C—O, C—C=O = 120°

8.61 Compound Lewis Structures

a) NO_2

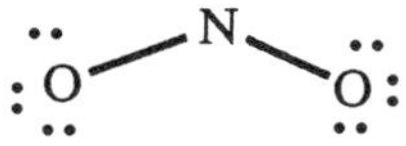

total $e^- = 5 + 2(6) = 17\ e^-$

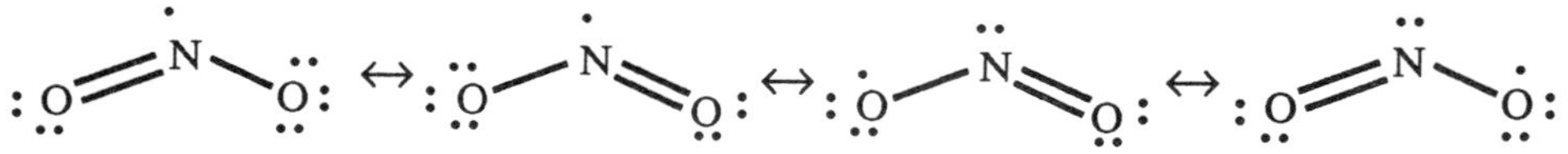

b) CH_3NCO

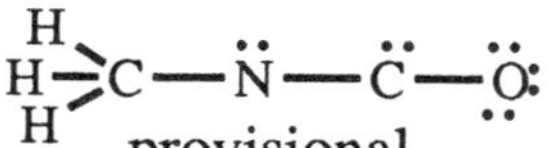

provisional

total $e^- = 3(1) + 2(4) + 5 + 6 = 22\ e^-$
$22\ e^- - 6(2) - 6 - 2(2) = 0$
$FC_{CH_3} = 4 - 4 = 0$ $FC_N = 5 - 4 = +1$
$FC_{\text{—C—}} = 4 - 4 = 0$ $FC_O = 6 - 7 = -1$

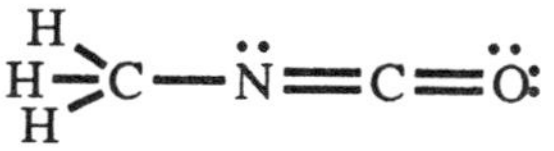

$FC_N = 5 - 5 = 0$ $FC_{=C=} = 4 - 4 = 0$
$FC_O = 6 - 6 = 0$

c) ClO_2

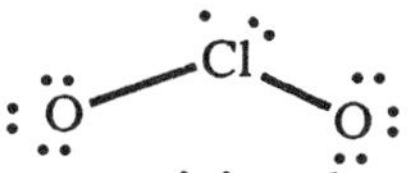

total $e^- = 7 + 2(6) = 19\ e^-$
$19\ e^- - 2(2) - 2(6) - 2 - 1 = 0$

provisional

$FC_{Cl} = 7 - 5 = +2$ $FC_O = 6 - 7 = -1$

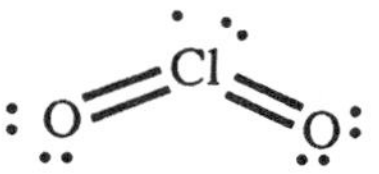

$FC_{Cl} = 7 - 7 = 0$ $FC_O = 6 - 6 = 0$

d) N_2F_4

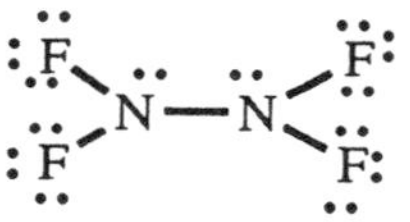

$FC_F = 7 - 7 = 0$ $FC_N = 5 - 5 = 0$

8.62

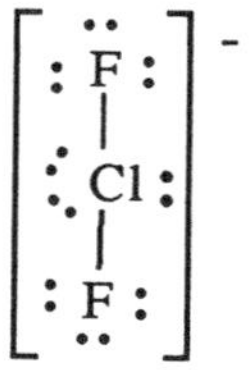

total $e^- = 7 + 2(7) + 1 = 22\ e^-$
$22\ e^- - 2(2) - 2(6) - 3(2) = 0$

Linear ion (orbitals are trigonal bipyramid)

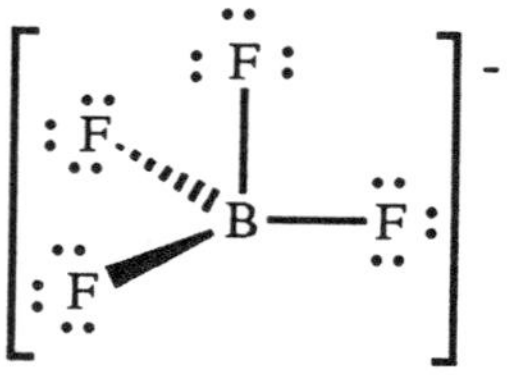

total $e^- = 3 + 4(7) + 1 = 32\ e^-$
$32\ e^- - 4(2) - 4(6) - 3(2) = 0$

Tetrahedral molecular geometry

(continued)

(8.62 continued)

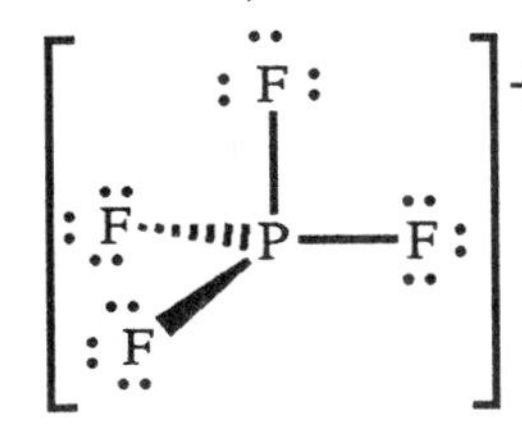

total $e^- = 5 + 4(7) - 1 = 32\ e^-$

$32\ e^- - 4(2) - 4(6) = 0$

Tetrahedral molecular geometry

8.63

$FC_S = 6 - 6 = 0$
$FC_{O—} = 6 - 7 = -1$
$FC_{O=} = 6 - 6 = 0$
Net Charge = 0 + 0 + 2(-1) = -2

$FC_B = 3 - 4 = -1$
$FC_F = 7 - 7 = 0$

Net Charge = -1 + 4 (0) = -1

$FC_{Cl} = 7 - 7 = 0$
$FC_O = 6 - 7 = -1$

Net Charge = 1(0) + 1(-1) = -1

$FC_P = 5 - 5 = 0$ $FC_{O=} = 6 - 6 = 0$
$FC_{—O—} = 6 - 6 = 0$ $FC_{—O} = 6 - 7 = -1$

Net Charge = 3(0) + 3(0) + 2(0) + 5(-1) = -5

8.64 a)

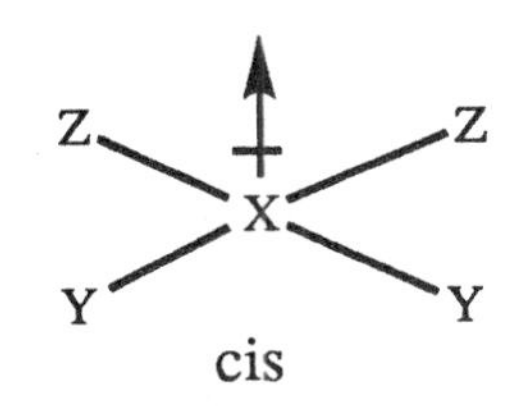

cis

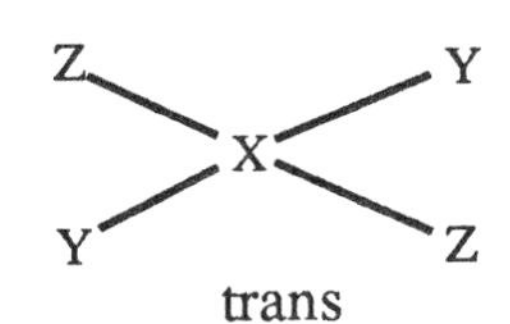

trans

b) The trans form has no molecular dipole moment while the cis form has a dipole moment as shown.

8.65 For your ball-and-stick drawings refer to Figures 8-14 through 8-17 for examples. Then draw ball-and-stick models of orthosilicate (SiO_4^{4-}) and metasilicate ($Si_2O_6^{4-}$) that also reflect their 3-dimensional properties as reflected in Figures 8-17 and 8-18, respectively.

8.66 The Sb atom has 4 sp^3 orbitals formed from 5s and 5p atomic orbitals. One of these sp^3 orbitals holds the lone pair of electrons. The other 3 sp^3 orbitals each overlap with a 3p orbital of a Cl atom to form a Sb—Cl bond. The molecule is a trigonal pyramid. The Cl—Sb—Cl bond angles are probably slightly less than 109.5°. They are not as small as in NCl_3 or PCl_3 because the sp^3 orbitals formed from 5s and 5p orbitals are larger.

8.67

quartz	– silica	mica	– silicate sheets
zircon	– orthosilicate	asbestos	– silicate fibers
jade	– metasilicate		

8.68 InI_3

The In uses sp^2 hybrid orbitals that overlap the I 5p orbitals to form the In—I bonds. The molecule is triangular with bond angles of 120°.

In_2I_6

The In uses sp^3 hybrid orbitals that overlap the I 5p orbitals to form the In—I bonds for the outer I atoms. The In—I bonds with the inner (bridging) I atom are also probably formed by overlap of the sp^3 orbitals of In with 5p orbitals of I. (The In—I—In bond angles are 86.3° and the I—In—I bond angles are 93.88°. The I—In—I, of the outer I, bond angles are 124.79°.)

8.69

	Polyatomic Ion	Formula	Lewis Structure	
a)	bromate	BrO_3^-	:O—Br—O: (with :O: below) $^-$	$FC_{Br} = 7 - 5 = +2$ $FC_O = 6 - 7 = -1$ provisional
			O=Br=O (with :O: below) $^-$	$FC_{Br} = 7 - 7 = 0$ $FC_{O=} = 6 - 6 = 0$ $FC_{O—} = 6 - 7 = -1$ (2 other resonance structures)
b)	cyanide	CN^-	:C—N: $^-$ provisional	$FC_C = 4 - 3 = +1$ $FC_N = 5 - 7 = -2$
			:C=N: $^-$ provisional	$FC_C = 4 - 4 = 0$ $FC_N = 5 - 6 = -1$
			:C—N: $^-$ provisional	$FC_C = 4 - 5 = -1$ $FC_N = 5 - 5 = 0$
			:C≡N: $^-$	$FC_C = 4 - 5 = -1$ $FC_N = 5 - 5 = 0$

(continued)

(8.69 continued)

c) nitrate NO_3^-

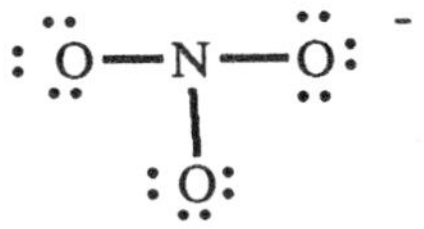

```
 ..        ..   -
: O — N — O :
 ..    |   ..
      : O :
        ..
```

$FC_N = 5 - 3 = +2$
$FC_O = 6 - 7 = -1$
provisional

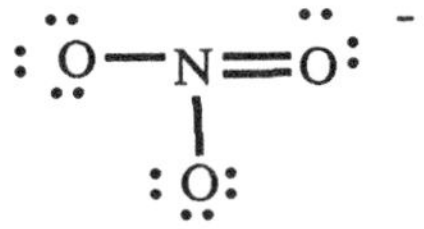

```
 ..        ..   -
: O — N = O :
 ..    |
      : O :
        ..
```

$FC_N = 5 - 4 = +1$
$FC_{O=} = 6 - 6 = 0$
$FC_{O—} = 6 - 7 = -1$
(N limited to 4 valence orbitals)
(2 other resonance structures)

d) nitrite NO_2^-

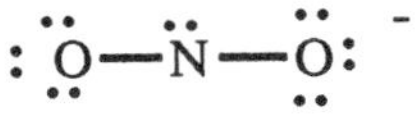

```
 ..   ..   ..   -
: O — N — O :
 ..        ..
```

$FC_N = 5 - 4 = +1$
$FC_O = 6 - 7 = -1$
provisional

```
 ..   ..   ..   -
: O — N = O :
 ..
```

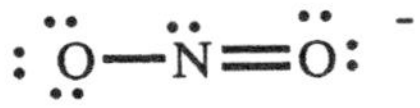

$FC_N = 5 - 5 = 0$
$FC_{O=} = 6 - 6 = 0$
$FC_{O—} = 6 - 7 = -1$
(1 other resonance structure)

e) phosphate PO_4^{3-}

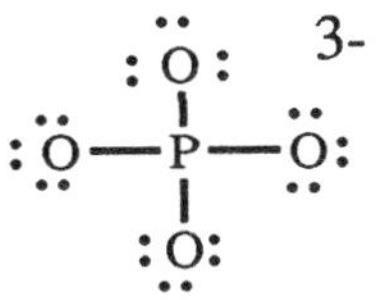

```
       ..      3-
      : O :
 ..     |    ..
: O — P — O :
 ..     |    ..
      : O :
        ..
```

$FC_P = 5 - 4 = +1$
$FC_O = 6 - 7 = -1$
provisional

```
      : O :    3-
 ..     ||   ..
: O — P — O :
 ..     |    ..
      : O :
        ..
```

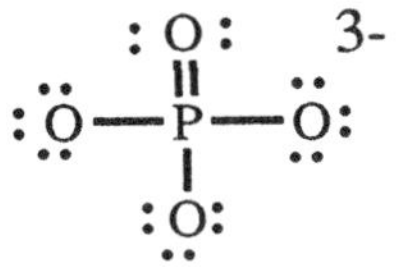

$FC_P = 5 - 5 = 0$
$FC_{O=} = 6 - 6 = 0$
$FC_{O—} = 6 - 7 = -1$
(3 other resonance structures)

f) hydrogen carbonate HCO_3^-

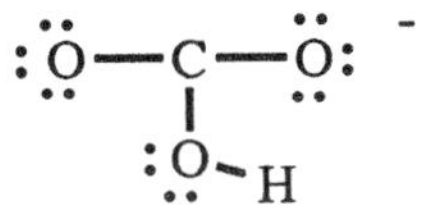

```
 ..        ..   -
: O — C — O :
 ..    |   ..
      : O ~ H
        ..
```

$FC_C = 4 - 3 = +1$
$FC_O = 6 - 7 = -1$
$FC_{—O—} = 6 - 6 = 0$
provisional

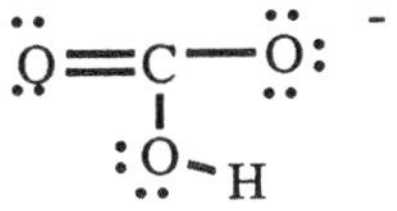

```
 ..       ..   -
 O = C — O :
 ..   |   ..
     : O ~ H
       ..
```

$FC_C = 4 - 4 = 0$
$FC_{O—} = 6 - 7 = -1$
$FC_{—O—} = 6 - 6 = 0$
$FC_{O=} = 6 - 6 = 0$
(1 other resonance structure)

8.70 Electron affinity is the **energy required** when an electron is removed from a negative ion. Electronegativity is a **measure of the tendency** of an atom to draw electrons to itself in a chemical bond.

8.71 $SiCl_4$ tetrahedron

SeF_4 seesaw

CI_4 tetrahedron

$CdCl_4^{2-}$ tetrahedron

XeF_4 square plane

$BeCl_4^{2-}$ tetrahedron

8.72 The hybridization of the C is sp^3 with an orientation of 109.5°. The C—C—C bond angles are 60°. Since the hybridized orbital angles are decreased so much to form the bond angles of the ring of this molecule, the molecule is very unstable and, therefore, very reactive.

8.73 The hybridization of the central atom (N or P) of each of these molecules is sp^3 resulting in an orbital orientation of 109.5°. Each molecule has a lone pair of electrons in one of the sp^3 orbitals which repels the bonded electrons and atoms forcing them closer together and decreasing their bond angles. The sp^3 hybrid orbitals of N (from 2s and 2p atom orbitals) are smaller than the sp^3 of P (from 3s and 3p atom orbitals). Therefore, the lone pair of electrons in NH_3 is held more closely to the N atom and occupies less space. The repulsion of the bonding electrons is less which results in a smaller decrease in bond angles for NH_3 than for PH_3. The large Cl atoms (with 3 lone pairs) repel each other more strongly than H atoms and, therefore, counteract the lone pair repulsion and result in less decrease of the bond angles.

8.74 The orbital overlap picture requested in this question should look very much like that in Figure 8-17 except the Si is replaced by C and the O is replaced by Cl. The orbitals used are the same.

8.75 H_3CNH_2 Draw a ball-and-stick model of the Lewis structure on the left. For examples of ball-and-stick drawings see Figures 8-14 through 8-17 in the textbook.

8.76 $(CH_3)_2CO$

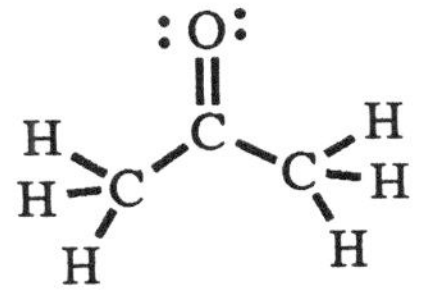

$H_3C-C(=O)-CH_3$

The geometry about the C atom (of CH_3) is tetrahedral and the geometry of the C atom (of C=O) is triangular. The hybridization of the C (of CH_3) is sp^3 and the hybridization of the C (of C=O) is sp^2.

8.77 a) H_2O

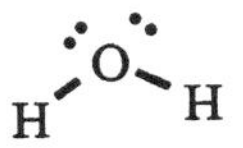

H—Ö—H (two lone pairs on O)

$FC_O = 6 - 6 = 0$

b) C_2H_2

H—C≡C—H

$FC_C = 4 - 4 = 0$

c) HCN

H—C—N̈: (three lone pairs on N)

$FC_C = 4 - 2 = +2$
$FC_N = 5 - 7 = -2$
provisional

H—C≡N:

$FC_C = 4 - 4 = 0$
$FC_N = 5 - 5 = 0$

d) CH_2O

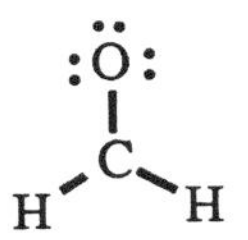

:Ö: — C(H)(H) (single bond)

$FC_C = 4 - 3 = +1$
$FC_O = 6 - 7 = -1$
provisional

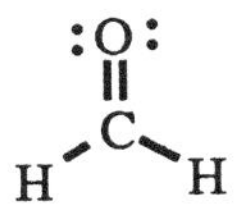

:O: = C(H)(H) (double bond)

$FC_C = 4 - 4 = 0$
$FC_O = 6 - 6 = 0$

e) H_2S

H—S̈—H (two lone pairs on S)

$FC_S = 6 - 6 = 0$

f) H_3CCN

H_3C—C—N̈: (three lone pairs on N)

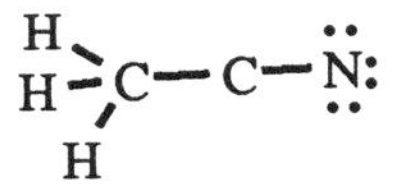

$FC_{H_3C—} = 4 - 4 = 0$
$FC_{—C—} = 4 - 2 = +2$
$FC_N = 5 - 7 = -2$
provisional

H_3C—C≡N:

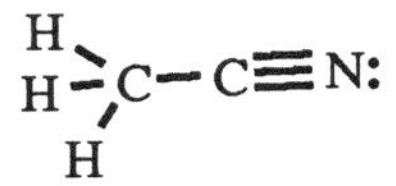

$FC_{H_3C—} = 4 - 4 = 0$
$FC_{—C\equiv} = 4 - 4 = 0$
$FC_N = 5 - 5 = 0$

g) NH_3

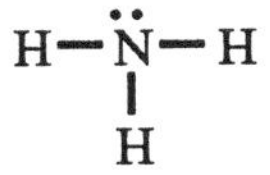

H—N̈—H, with a third H on N

$FC_N = 5 - 5 = 0$

8.78 $(CH_3)_3C^+$

$[C(CH_3)_3]^+$

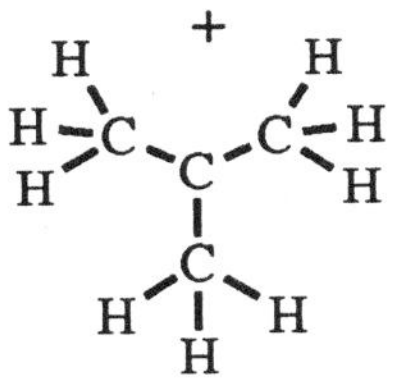

The center C has sp^2 hybrid orbitals which overlap a sp^3 hybrid of each of the outer C atoms to form the C—C bonds. The other sp^3 hybrid orbitals overlap 1s orbitals of H atoms to form C—H bonds. The geometry of the outer C atoms is tetrahedral while that of the center C atom is trigonal planar.

8.79 SF_2

$FC_S = 6 - 6 = 0$
$FC_F = 7 - 7 = 0$

SSF_2

$FC_{S—} = 6 - 7 = -1$
$FC_{—S—} = 6 - 5 = +1$
$FC_F = 7 - 7 = 0$
provisional

$FC_{S=} = 6 - 6 = 0$
$FC_{—S=} = 6 - 6 = 0$
$FC_F = 7 - 7 = 0$

FSSF

$FC_S = 6 - 6 = 0$
$FC_F = 7 - 7 = 0$

F_3SSF

$FC_F = 7 - 7 = 0$
$FC_{—S—} = 6 - 6 = 0$
$FC_{\equiv S—} = 6 - 6 = 0$

SF_4

$FC_S = 6 - 6 = 0$
$FC_F = 7 - 7 = 0$

F_5SSF_5

$FC_F = 7 - 7 = 0$
$FC_S = 6 - 6 = 0$

SF_6

$FC_F = 7 - 7 = 0$
$FC_S = 6 - 6 = 0$

8.80

Molecule	Molecular Geometry	Hybridization of Sulfur
SF_2	bent	sp^3
SSF_2	trigonal plane	sp^2 (inner) none required (outer)
FSSF	bent (for each S)	sp^3 (each S)

(continued)

(8.80 continued)

F_3SSF	seesaw (for F_3S—)	sp^3d
	bent (for —S—F)	sp^3
SF_4	seesaw sp^3d	
F_5SSF_5	octahedron (for each S)	sp^3d^2
SF_6	octahedron	sp^3d^2

8.81 Silica has the empirical formula of SiO_2 and consists of a continuous network of Si—O bonds. Each Si is bonded to 4 O atoms. Each O atom is bonded to 2 Si atoms. Silica has no ionic units. Orthosilicates have discrete SiO_4^{4-} ionic units. Each Si is bonded to 4 outer oxygen atoms. Metasilicates have the empirical formula of SiO_3^{2-} and consist of linear chains or ring systems of Si—O—Si linkages with negative charges. Each Si atom is bonded to 2 outer O atoms and 2 inner O atoms. (Also see Problem 8.65.)

8.82

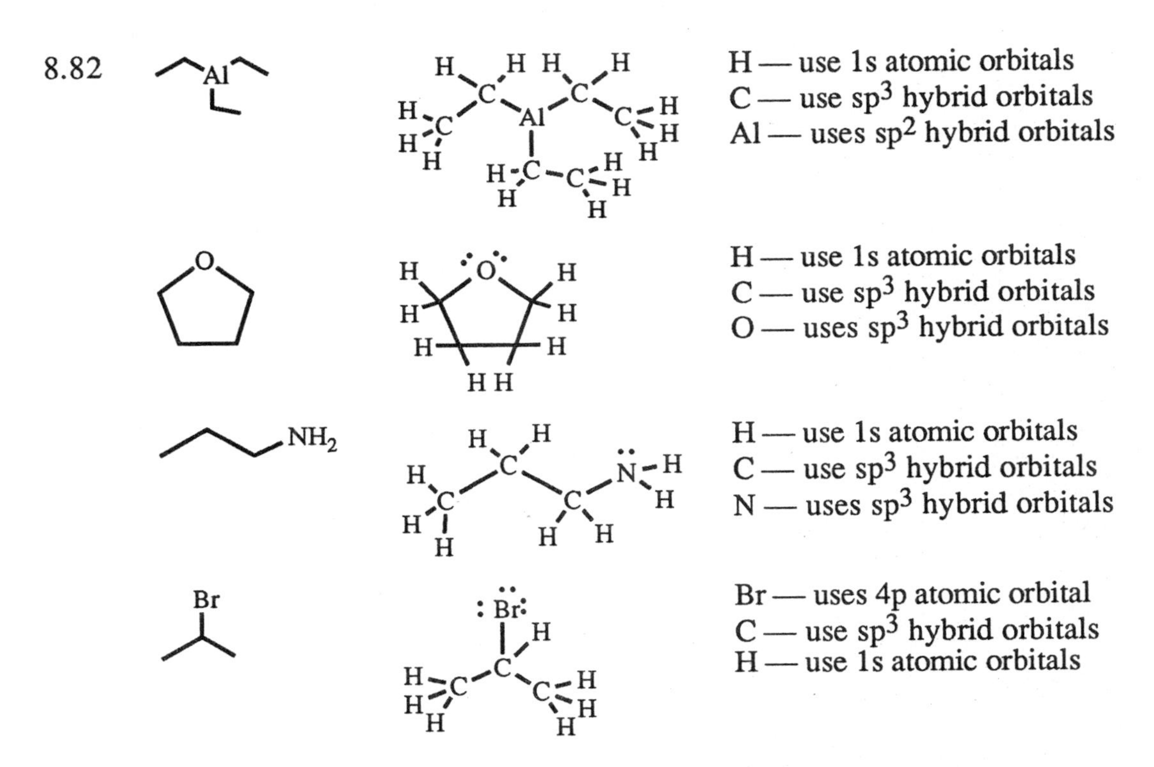

8.83 The hybridization of the central atom O of water molecules is sp^3 resulting in an orbital orientation of 109.5°. Each molecule has a lone pair of electrons in one of the sp^3 orbitals which repels the bonded electrons and atoms forcing them closer together and decreasing their bond angles. The larger atoms (orbitals) of sulfur allow it to bond to two hydrogens in H_2S without hybridizing. The repulsion due to like charges of the bonding hydrogens results in a small increase in bond angles for H_2S, from the 90° of p orbitals to the observed 92.2°.

8.84 cyanate, NCO^-

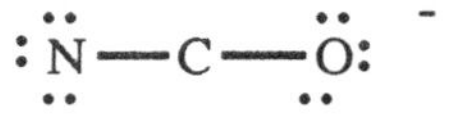

$[:N—C—O:]^-$

$FC_N = 5 - 7 = -2$
$FC_C = 4 - 2 = +2$
$FC_O = 6 - 7 = -1$
provisional

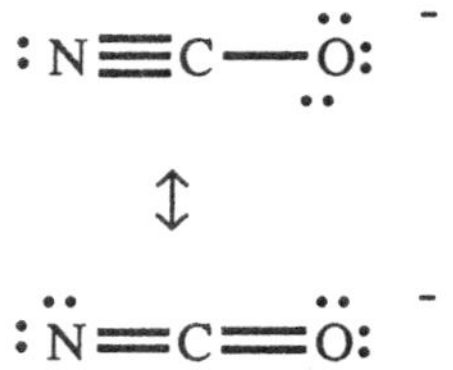

$[:N≡C—O:]^-$

$FC_N = 5 - 5 = 0$
$FC_C = 4 - 4 = 0$
$FC_O = 6 - 7 = -1$

↕

$[:N=C=O:]^-$

$FC_N = 5 - 6 = -1$
$FC_C = 4 - 4 = 0$
$FC_O = 6 - 6 = 0$

isocyanate, CNO^-

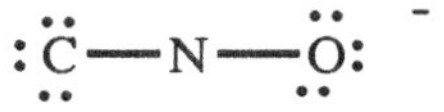

$[:C—N—O:]^-$

$FC_C = 4 - 7 = -3$
$FC_N = 5 - 2 = +3$
$FC_O = 6 - 7 = -1$
provisional

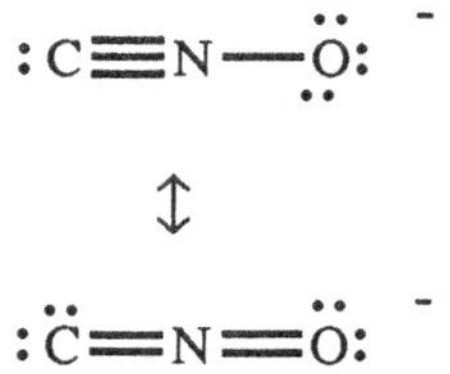

$[:C≡N—O:]^-$

$FC_C = 4 - 5 = -1$
$FC_N = 5 - 4 = +1$
$FC_O = 6 - 7 = -1$

↕

$[:C=N=O:]^-$

$FC_C = 4 - 6 = -2$
$FC_N = 5 - 4 = +1$
$FC_O = 6 - 6 = 0$

CHAPTER 9: CHEMICAL BONDING: MULTIPLE BONDS

9.1 The outer C atoms have sp^3 hybrid orbitals and the center C atom has sp^2 hybrid orbitals. The C—C bonds are formed by overlap of an outer C sp^3 orbital with a center C sp^2 orbital. The C—H bonds are formed by overlap of outer C sp^3 orbitals with H 1s orbitals. The center C atom forms a σ bond with the O atom by end-to-end overlap of an sp^2 orbital of the C with a p orbital of the O. A second C—O bond, a π bond, is formed by the side-by-side overlap of the unused p orbital of the C and a p orbital of the O. Two lone pairs of electrons on the O atom occupy the 2s orbital and the other 2p orbital. The sketches of all the bonding orbitals should include the above information and the structural formula:

See Figures 8-1 through 8-5 for examples of sketches of the various bonding orbitals.

```
        O
 H      ||     H
H-C - C - C
 |          | H
 H          H
```

9.2

```
       H    H H  O    H H  H H   H
  1    |     \/  ||3   \/   \/   | H
CH3O  C  C=C   C   N  C  C   C  C  C=C  C  CH3
      C  C         |     /\   /\  |    |
      || |         H    H H  H H  H    CH3
HO  C  C=C
  2    |    H
       H
```

Atoms	Steric Number	Molecular Geometry	Hybridization
All H	1	-	none, 1s orbital used
O #1 and #2	4	bent	sp^3
O #3	3	linear	none, 2p orbitals used for σ and π bonds
N	4	trigonal pyramid	sp^3
C in benzene ring	3	triangular	sp^2
C in C=C and in C=O	3	triangular	sp^2
All other C	4	tetrahedral	sp^3

9.3 (See Sample Problem 9-1.) The hybridization at both the C and N atoms is sp^2 with "ideal" angles of 120°. The H—C—H angle is less than 120° because the electrons of the π bond occupy space and force the H—C—H angle to be smaller. These π bonds also tend to force the C=N—H to open to greater than 120°, but this is counteracted by the lone pair of electrons on the N atom. The actual angle depends on which of these effects is greater.

9.4

Compound	Formula	Steric no.	Hybridization of central carbons
butane	CH_3—CH_2—CH_2—CH_3	4	sp^3
2-butene	CH_3—CH=CH—CH_3	3	sp^2
2-butyne	CH_3—C≡C—CH_3	2	sp

The end carbon atoms all use sp^3 hybrid orbitals and have tetrahedral geometry. The same is true of the 2 inner C atoms of butane. The 2 inner C atoms of butene use sp^2 hybrid orbitals with triangular geometry to form σ C—C bonds with each other and the outer C atoms. In addition, the two C atoms form a π bond by side-by-side overlap of the unused p orbitals. The 2 inner C atoms of butyne use sp hybrid orbitals with linear geometry to form σ C—C bonds with each other and with the outer C atoms. The unused p orbitals of the inner C atoms form 2 π bonds by side-by-side overlap. The bonding pictures should include the above information. See Figures 8-1 through 8-5 for examples of sketches of the various bonding orbitals.

9.5 H—N < N≡N < C≡N < N—N < Cl—N

H—N is the shortest because the bond is polar and the small 1s orbital of H ($n = 1$) is used in forming the bond. The N≡N and C≡N with $n = 2$ for both N and C are longer than H—N but shorter than N—N and Cl—N because they possess triple bonds. Because the effective nuclear charge of N is greater than C, N≡N is shorter than C≡N. The length of N—N is shorter than Cl—N because $n = 2$ for N but $n =$ 3 for Cl meaning Cl uses larger orbitals for bonding.

9.6 H—F < F—Cl < Cl—O < Cl—Cl < Br—Br

H—F is the shortest because it is a polar bond between an element with $n = 1$ and an element with $n = 2$. F—Cl and Cl—O are both bonds of Cl ($n = 3$) with an element with $n = 2$; the F—Cl bond length would be shorter than the Cl—O because the effective nuclear charge of F is higher than for O and the electronegativity difference for F—Cl is larger than the difference for Cl—O. Finally, we have Cl—Cl with Cl having $n = 3$ and Br—Br with Br having $n = 4$; therefore, Br has larger valence orbitals.

9.7 Table 9-2 N—N < Cl—N < H—N < C≡N < N≡N

Table 9-2	N—N	< Cl—N	< H—N	< C≡N	< N≡N
energy (kJ/mol):	165	315	390	890	942.7

Cl—N is stronger than N—N because of greater electronegativity difference. H—N is stronger than Cl—N because $n = 1$ for H versus $n = 3$ for Cl. C≡N is stronger than H—N because more electrons are shared between C≡N than for H—N. N≡N is stronger than C≡N because the effective nuclear charge is greater for N than for C.

9.8 Carbon monoxide has a very strong chemical bond because it has the maximum of shared electrons (6), a difference in electronegativity, and fairly small principal quantum numbers of $n = 2$ for each atom.

9.9 N_2 + $2\ O_2$ → N_2O_4 (Lewis structure: O–N(=O)–N(=O)–O)

1 N≡N triple bond; 2 O=O double bond; 1 N—N single bond, 2 N—O single bonds, 2 N=O double bonds

$\Delta E_{reaction} = \Sigma\ BE_{(bonds\ broken)} - \Sigma\ BE_{(bonds\ formed)}$
= [(2 mol N≡N)(942.7 kJ/mol N≡N) + (2 mol O=O)(493.6 kJ/mol O=O)]
- [(1 mol N—N)(165 kJ/mol N—N) + (2 mol N—O)(200 kJ/mol N—O) +
(2 mol N=O)(607 kJ/mol N=O)] = 2873 kJ - 1779 kJ = 1094 kJ
This energy change is per mole of N_2O_4.

9.10 For: $2\ HF \rightarrow H_2 + F_2$
ΔE = [(2 mol H—F)(565 kJ/mol H—F)] -
[(1 mol H—H)(432.0 kJ/mol H—H) + (1 mol F—F)(154.8 kJ/mol F—F)]
= 1130 kJ - 586.8 kJ = 543 kJ

For: $2\ HCl \rightarrow H_2 + Cl_2$
ΔE = [(2 mol H—Cl)(428.0 kJ/mol H—Cl)] -
[(1 mol H—H)(432.0 kJ/mol H—H) + (1 mol Cl—Cl)(239.7 kJ/mol Cl—Cl)]
= 856.6.0 kJ - 671.7 kJ = 184.3 kJ

For: $2\ HBr \rightarrow H_2 + Br_2$
ΔE = [(2 mol H—Br)(362.3 kJ/mol H—Br)] -
[(1 mol H—H)(432.0 kJ/mol H—H) + (1 mol Br—Br)(190.0 kJ/mol Br—Br)]
= 724.6 kJ - 622.0 kJ = 102.6 kJ

For: $2\ HI \rightarrow H_2 + I_2$
ΔE = [(2 mol H—I)(294.6 kJ/mol H—I)] -
[(1 mol H—H)(432.0 kJ/mol H—H) + (1 mol I—I)(149.0 kJ/mol I—I)]
= 589.2 kJ - 581.0 kJ = 8.2 kJ

9.11 $2\ CH_3—CH=CH_2 + 2\ NH_3 + 3\ O_2 \rightarrow 2\ H_2C=CH—C\equiv N + 6\ H_2O$

12 C—H single bonds	6 C—H single bonds
2 C—C single bonds	2 C=C double bonds
2 C=C double bonds	2 C—C single bonds
6 N—H single bonds	2 C≡N triple bonds
3 O=O double bonds	12 H—O single bonds

ΔE =[(12 mol C—H)(415 kJ/mol C—H) + (2 mol C—C)(345 kJ/mol C—C)
+ (2 mol C=C)(615 kJ/mol C=C) + (6 mol N—H)(390 kJ/mol N—H)
+ (3 mol O=O)(493.6 kJ/mol O=O)] - [(6 mol C—H)(415 kJ/mol C—H)
+ (2 mol C=C)(615 kJ/mol C=C) + (2 mol C—C)(345 kJ/mol C—C)
+ (2 mol C≡N)(890 kJ/mol C≡N) + (12 mol H—O)(460 kJ/mol H—O)]
= 10721 kJ - 11710 kJ = -989 kJ

9.12

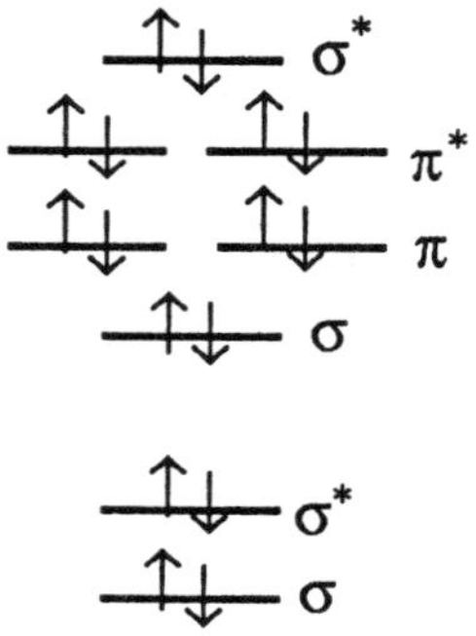

↑↓ σ^*
↑↓ ↑↓ π^*
↑↓ ↑↓ π — orbitals generated by 2p atomic orbitals
↑↓ σ

↑↓ σ^* — orbitals generated by 2s atomic orbitals
↑↓ σ

Ne_2 has 16 valence electrons. The Bond Order is (8 - 8)/2 = 0 as the 8 anti-bonding electrons nullify the 8 bonding electrons resulting in no net bonding

(Ne_2 is unstable).

9.13

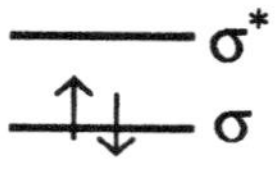

—— σ^*
↑↓ σ — orbitals generated by 3s atomic orbitals

Na_2 has 2 valence electrons. The Bond Order is (2 - 0)/2 = 1 since the 2 valence electrons occupy a bonding orbital and Na_2 is stable.

9.14 a)

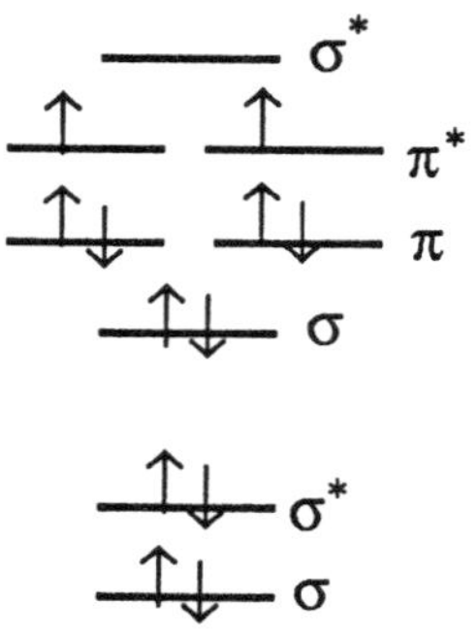

For O_2	For O_2^+
—— σ^*	—— σ^*
↑ ↑ π^*	↑ —— π^*
↑↓ ↑↓ π	↑↓ ↑↓ π
↑↓ σ	↑↓ σ
↑↓ σ^*	↑↓ σ^*
↑↓ σ	↑↓ σ

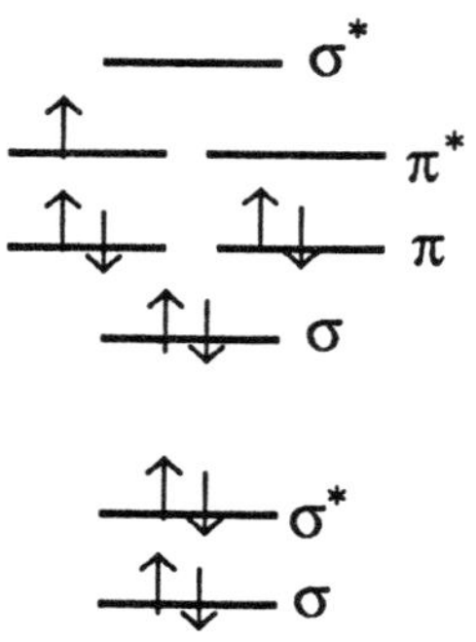

For O_2, Bond Order
= (8 - 4)/2 = 2.
For O_2^+, Bond Order
= (8 - 3)/2 = 2.5.

The O_2^+ has the stronger bond. It has 5 unnullified bonding electrons to only 4 for O_2.

(continued)

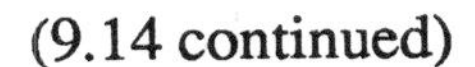

(9.14 continued)

b)

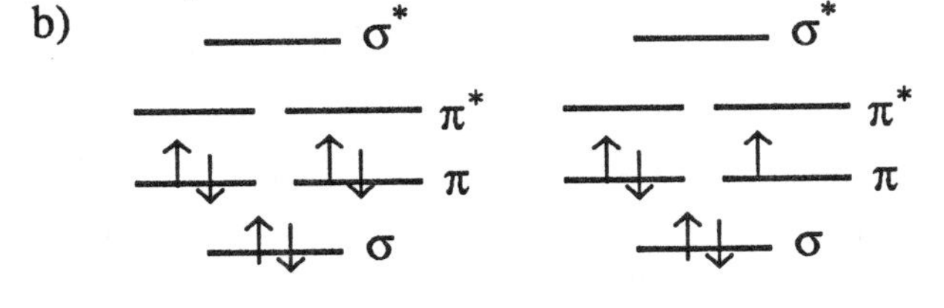

For N_2 For N_2^+

For N_2, Bond Order $= (8 - 2)/2 = 3$.

For N_2^+, Bond Order $= (7 - 2)/2 = 2.5$.

The N_2 has the stronger bond. It has 6 unnullified bonding electrons versus only 5 for N_2^+.

c)

For F_2 For F_2^+

For F_2, Bond Order $= (8 - 6)/2 = 1$.

For F_2^+, Bond Order $= (8 - 5)/2 = 1.5$.

The F_2^+ has the stronger bond. It has 3 unnullified bonding electrons versus only 2 for F_2.

d)

For CN^- For CN

For CN^-, Bond Order $= (8 - 2)/2 = 3$.

For CN, Bond Order $= (7 - 2)/2 = 2.5$.

The CN^- has the stronger bond. It has 6 unnullified bonding electrons versus only 5 for CN.

9.15

orbitals generated by 3p of Cl and 2p of O

orbitals generated by 3s of Cl and 2s of O

For ClO^-

:Cl—O: $^-$

Bond Order $= (8 - 6)/2 = 1$

9.16 The sketches of the shapes of each of the bonding and antibonding orbitals of O_2 needs to include sketches of the 1s + 1s σ orbital and the 2s + 2s σ orbital that would look like the 2s + 2s σ orbital shown in Figure 9-14 in the textbook (the 1s + 1s σ would be smaller). The sketches would also need to include a 1s + 1s σ^* antibonding orbital like that shown in Figure 9-14 and a 2p + 2p π orbital like in Figure 9-18. These are the orbitals that contain electrons. As empty orbitals there will also be σ^* (from 2p orbitals) and π^* orbitals like those shown in Figures 9-14 and 9-15.

9.17 NO: $(\sigma_s)^2\ (\sigma_s^*)^2\ (\sigma_p)^2\ (\pi_x)^2\ (\pi_y)^2\ (\pi_x^*)^1$ B.O. = (8 - 3)/2 = 2.5
NO^+: $(\sigma_s)^2\ (\sigma_s^*)^2\ (\sigma_p)^2\ (\pi_x)^2\ (\pi_y)^2$ B.O. = (8 - 2)/2 = 3
NO^-: $(\sigma_s)^2\ (\sigma_s^*)^2\ (\sigma_p)^2\ (\pi_x)^2\ (\pi_y)^2\ (\pi_x^*)^1\ (\pi_y^*)^1$ B.O. = (8 - 4)/2 = 2
NO has one unpaired electron and NO^- has two unpaired electrons and both, therefore, show magnetism. All of the electrons of NO^+ are paired and, therefore, it is not magnetic.

9.18 :S̤̈—C—S̤̈: provisional Lewis structure
$FC_S = -1$ $FC_C = +2$ $FC_S = -1$

:S̤=C=S̈: Lewis structure
The molecule is linear (steric number is two). The C atom uses sp hybrid orbitals to overlap with a $3p_z$ atomic orbital of each S atom to form the σ C—S bonds. The unused $2p_x$ and $2p_y$ atomic orbitals of C overlap side-by-side with 3p orbitals of each S to form 2 sets of 3 delocalized π orbitals. The 16 valence electrons of the CS_2 molecule are distributed thus: 4 are in the σ bonds, 4 occupy the 3s atomic orbitals on the S atoms, 4 occupy the two delocalized π bonding orbitals, and the last 4 electrons are in the two delocalized π_n non-bonding orbitals. Therefore, 8 valence electrons are in delocalized π orbitals. The 3p orbitals of S atoms overlap less effectively with the 2p orbitals of the C than the 2p orbitals of O. Therefore, the C—S bond is weaker and longer than the C—O bond.

9.19 :Ö̤—C̈—C—C̈—Ö̤: valence electrons = 3 x 4 (from C) + 2 x 6 (from O) = 24 e^-
$FC_O = 6 - 7 = -1$ $FC_{C:} = 4 - 4 = 0$ $FC_C = 4 - 2 = +2$

:Ö=C=C=C=Ö: $FC_O = 6 - 6 = 0$ $FC_C = 4 - 4 = 0$
sp hybrid orbitals are used by the C atoms along with $2p_z$ atomic orbitals of O atoms to form a totally linear molecule. The $2p_x$ and $2p_y$ atomic orbitals of the C and O atoms form 2 sets of delocalized π orbitals. Make the sketches that this question calls for similar to the σ bonding shown in Figure 9-24 (except it needs to include 5 atoms) and, as separate sketches, the π bonding shown in Figure 9-18 for the π bonding in first the y plane, then the z plane (2 of the π bonds will be in each plane).

9.20 NO_2^+ $:\ddot{O}-N-\ddot{O}:^+$

valence electrons
= 1 x 5 (from N) + 2 x 6 (from O) - 1 = 16 e^-
16 e^- - 4 e^- - 12 e^- = 0 e^-

$FC_O = 6 - 7 = -1$
$FC_N = 5 - 2 = +3$

$:O{=}N{=}\ddot{O}:^+$

NO_2^- $:\ddot{O}-\ddot{N}-\ddot{O}:^-$

valence electrons
= 1 x 5 (from N) + 2 x 6 (from O) + 1 = 18 e
18 e^- - 4 e^- - 12 e^- - 2 e^- = 0 e^-

$FC_O = 6 - 7 = -1$
$FC_N = 5 - 4 = +1$

$:\ddot{O}-\ddot{N}{=}\ddot{O}:^- \leftrightarrow :\ddot{O}{=}\ddot{N}-\ddot{O}:^-$

$FC_{O-} = 6 - 7 = -1$ $FC_N = 5 - 5 = 0$ $FC_{O=} = 6 - 6 = 0$

NO_2^+ uses sp hybrid orbitals on N and $2p_z$ orbitals on O to form N—O σ bonds to form a linear molecule (steric number of N is 2). The N in NO_2^- has a steric number of 3 and uses sp^2 hybrid orbitals to overlap with $2p_z$ orbitals on the O to form N—O σ bonds and a bent molecule. The lone pair is in the third sp^2 orbital. The NO_2^- has one set of delocalization π orbitals because only one 2p is left unused on the N atom. The NO_2^+ species can have 2 sets of delocalized π orbitals. The requested sketches need to reflect the details discussed above.

9.21 HN_3 $H-\ddot{N}-\ddot{N}-\ddot{N}:$

valence electrons
= 1 x 1 (from H) + 3 x 5 (from N) = 16 e^-
16 e^- - 6 e^- - 6 e^- - 4 e^- = 0 e^-

$FC_{-N-} = 5 - 4 = +1$
$FC_{-N} = 5 - 7 = -2$

$H-\ddot{N}{=}N{=}\ddot{N}:$

$FC_{-N=} = 5 - 5 = 0$
$FC_{=N=} = 5 - 4 = +1$
$FC_{N=} = 5 - 6 = -1$

A triple bond between the end nitrogen atom and the central nitrogen atom is not possible because the central N is limited to 4 orbitals.

$H-\ddot{N}_1{=}N_2{=}\ddot{N}_3: \leftrightarrow H-\ddot{N}-N{\equiv}N:$

N_1: has sp^2 hybridization and H—N=N has a bond angle of ~ 120°.
N_2: has sp hybridization and N=N=N has a bond angle of 180°.
There is a localized π orbital set for $N_2{=}N_3$ and a delocalized π orbital set for all three of the 3 N atoms.

9.22 CNO^- $:\ddot{C}-N-\ddot{O}:^-$

valence electrons
= 1 x 4 (from C) + 1 x 5 (from N) + 1 x 6 (from O)+ 1
= 16 e^-

$FC_C = 4 - 7 = -3$
$FC_N = 5 - 2 = +3$
$FC_O = 6 - 7 = -1$

16 e^- - 4 e^- - 12 e^- = 0 e^-

(continued)

(9.22 continued)

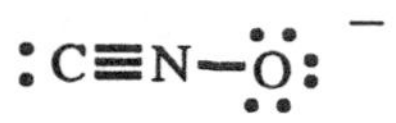

$:C\equiv N-\ddot{\underset{..}{O}}:^-$ $FC_C = 4 - 5 = -1$ $FC_N = 5 - 4 = +1$ $FC_O = 6 - 7 = -1$

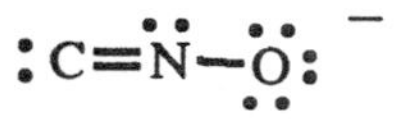

$:C=\ddot{N}-\ddot{\underset{..}{O}}:^-$ $FC_C = 4 - 4 = 0$ $FC_N = 5 - 5 = 0$ $FC_O = 6 - 7 = -1$

The N atom uses sp^2 hybridization to form σ bonds with the C and O atoms and to hold the lone pair of electrons. The steric number of N is 3 and the molecule is bent. A set of delocalized π orbitals extends over all three atoms.

9.23 All three C atoms use sp^2 hybridization to form 3 σ bonds (with H and C and one with O). The geometry of the molecule at each C atom is triangular with bond angles ~ 120°. There is a set of delocalized π orbitals extending over the 3 C atoms and the O atom as each C atom has one unused 2p orbital and the O atom has 1 unused 2p orbital.

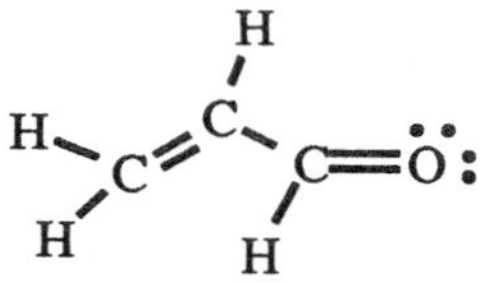

9.24 a) Not stabilized by conjugation

b) Stabilized by delocalization that extends over all the atoms

c) No stabilization

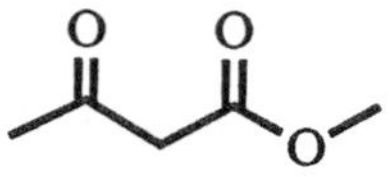

d) Stabilized by delocalization that extends over C atoms 2 through 5

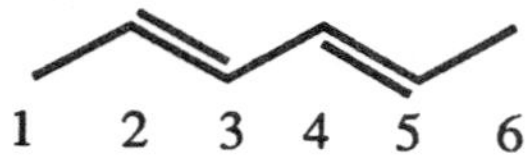

1 2 3 4 5 6

9.25 a) All O's in the OH groups are sp^3 hybridized. The O in the ring is also sp^3 hybridized. The doubly bonded O is not hybridized. The 3 carbon atoms that contain π bonds are sp^2 hybridized. The other 3 C atoms are sp^3 hybridized.
b) In the formula of vitamin C there are 4 delocalized electrons, corresponding to the two π bonds shown in the Lewis structure of the molecule. Delocalization extends over the three sp^2 hybridized C atoms and the unhybridized O atom adjacent to them.

9.26 The C atoms that contain a π bond are sp^2 hybridized. The rest of the C atoms are sp^3 hybridized. There are 25 σ bonds in the molecule, (14 C—H bonds, 10 C—C bonds and 1 C—O bond). There are 3 π bonds. One is localized between carbons at the bottom of the structure given in the textbook and 2 are delocalized over double bonded carbons and the O near the top of the structure as drawn. There are also 2π* orbitals involving the atoms associated with the delocalized π bonds.

9.27 SO_2 :Ö—S̈—Ö: provisional

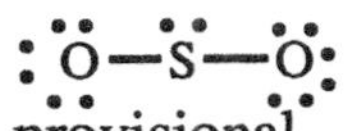

18 valence electrons - 4 e⁻ (in σ bonds) - 12 e⁻ (on O) - 2 e⁻ (on S) = 0 $FC_O = 6 - 7 = -1$ $FC_S = 6 - 4 = +2$

:O=S̈=O:

The steric number of S is 3; the molecule is bent; S uses sp^2 hybridization to form σ bonds with the O atoms by overlap with 2p orbitals of O and to hold a lone pair of electrons. The O—S—O bond angle should be about 120°. The unused 3p orbital on S and the 2p orbitals on the O atoms which are perpendicular to the plane of the molecule overlap to form delocalized π orbitals. There are 4 valence electrons in the 2 σ bonds and 2 valence electrons as a lone pair localized in the sp^2 hybrid orbital on the S. Four electrons are in the 2s orbitals on the outer O atoms and 4 more occupy the O 2p orbitals that lie in the plane of the molecule. This leaves 4 electrons to be placed in the delocalized π orbitals. The antibonding π orbital is empty.

SO_3 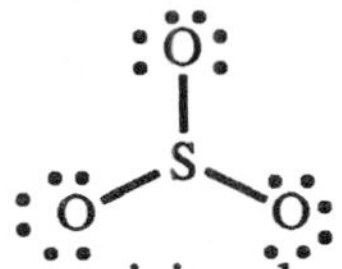 provisional

24 valence electrons - 6 e⁻ (in σ bonds) - 18 e⁻ (on O) = 0

$FC_O = 6 - 7 = -1$ $FC_S = 6 - 3 = +3$

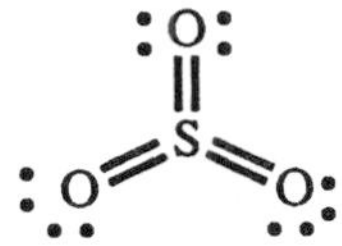

$FC_O = 0$ $FC_S = 0$

The steric number of S is 3; the molecule is triangular; S uses sp^2 hybridization to form σ bonds with the O atoms by overlap with 2p orbitals of O atoms. The O—S—O bond angles should be 120°. The unused 3p orbital on S and the 2p orbitals on the O atoms which are perpendicular to the plane of the molecule overlap to form delocalized π orbitals. There are 6 valence electrons in the 3 σ bonds. There are 6 electrons in the 2s orbitals on the outer O atoms and 6 more occupy the O 2p orbitals that lie in the plane of the molecules. This leaves 6 electrons to be placed in the delocalized π orbitals. The antibonding π orbital is empty.

9.28 :O—P—O—H (2-) provisional

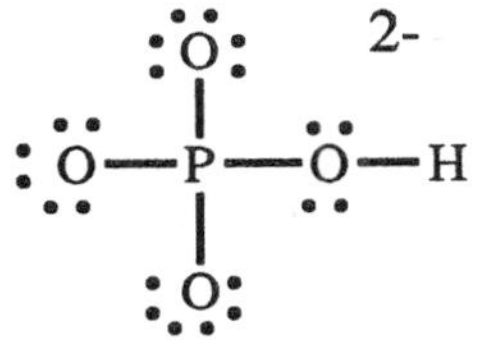

32 valence e⁻ - 10 e⁻ (in σ bonds) - 18 e⁻ (on outer O) - 4 e⁻ (on inner O) = 0

$FC_{O-} = 6 - 7 = -1$ $FC_P = 5 - 4 = +1$
$FC_{-O-} = 6 - 6 = 0$ $FC_H = 1 - 1 = 0$

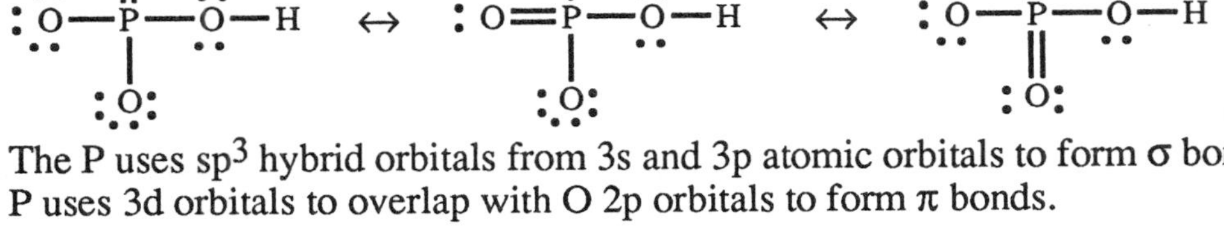

2- ↔ 2- ↔ 2-

The P uses sp^3 hybrid orbitals from 3s and 3p atomic orbitals to form σ bonds. The P uses 3d orbitals to overlap with O 2p orbitals to form π bonds.

9.29 $2-$ 34 valence e^- - 10 e^- (in σ bonds) - 24 e^- (on O) = 0

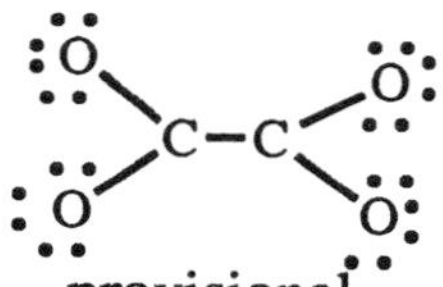

$FC_O = 6 - 7 = -1$ $FC_C = 4 - 3 = +1$

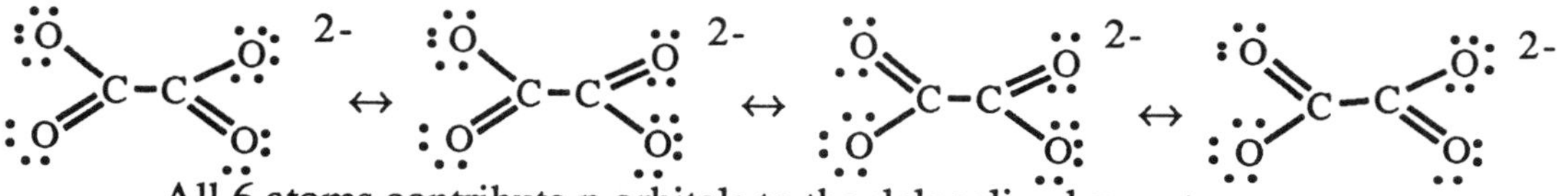

provisional

All 6 atoms contribute p orbitals to the delocalized π system.

9.30 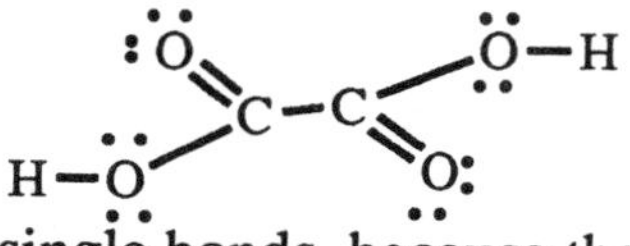 The localized π bonds (O=C) become shorter and the σ bonds (C—O) become longer. The four C—O bonds in the oxalate ion are the same length, between double and single bonds, because the delocalized π system extends over all 6 atoms. But upon adding two protons to form oxalic acid, 2 of the C—O bonds become localized π bonds and become shorter and 2 of the C—O bonds become single bonds and become longer.

9.31 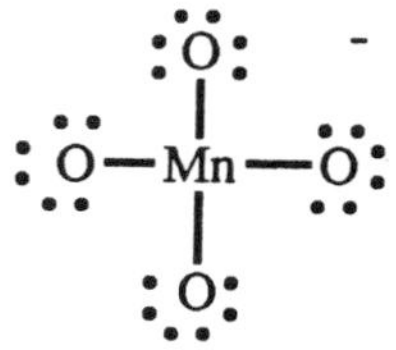 32 valence e^- - 8 e^- (in σ bonds) - 24 e^- (on O) = 0

$FC_{Mn} = 7 - 4 = +3$ $FC_O = 6 - 7 = -1$

provisional

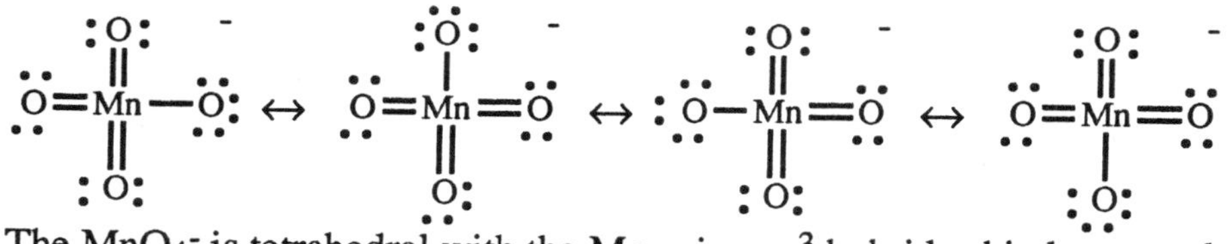

The MnO_4^- is tetrahedral with the Mn using sp^3 hybrid orbitals to overlap 2p orbitals of O to form σ bonds and 3d orbitals to overlap other 2p orbitals of O to form π bonds.

9.32 The p-type semiconductors from the list is: only AlAs doped with Zn
The n-type semiconductors from the list are: InSb doped with Te and Ge doped with P
The undoped semiconductors from the list are: GaP, CdSe, and Ge

9.33 Silicon doped with antimony would be an n-type semiconductor, because Sb has more valence electrons than Si. Its band gap diagram would look like that on the top left portion of Figure 9-36.

9.34 Would dope Si with Al. Al has fewer valence electrons than Si. Its band gap diagram would look like that on the top right portion of Figure 9-36.

9.35 The double bonded compounds have a π bond with electron density above and below the plane of the molecule. In order to rotate about the internuclear axis to convert between the *cis* and *trans* forms of the 1, 2-dichloroethylene, the π bond must be broken and reformed. This would require approximately 279 kJ/mol (615 kJ/mol - 345 kJ/mol, difference in strengths of C=C and C—C). The 1, 2-dichloroethane molecule has only a σ C—C bond with the electron density symmetrical around the internuclear axis. No bond breakage is necessary for rotation about this C—C bond.

9.36 The greater the bond order, the greater the bond energy and the shorter the bond length. The bond length can be measured for any specific bond in any specific compound using such methods as x-ray diffraction. Bond energies can be measured but not as accurately because molecules usually contain more than one kind of bond and it is difficult to control which bonds are broken. It is usually necessary to use several experimental methods and to use averages of close results. Any further measurements that use previous measured results have a built-in uncertainty of these previous measured values. Bond order cannot be measured.

9.37 There are 60 sp^2-hybridized carbon atoms in buckminsterfullerene, leaving one p orbital per atom available to form π orbitals. Thus there are 60 π orbitals, 30 of which are bonding and 30 of which are antibonding. The Lewis structure contains 30 π bonds and has the resonance structures, indicating that all 30 of these orbitals are delocalized.

9.38 Those listed in Tables 9-1 and 9-2 that have both bonds that are longer (142 pm) and stronger (154.8 kJ/mol) than the F—F bond.

F—S	156 pm	285 kJ/mol
F—Si	157 pm	565 kJ/mol

9.39 a) Ge doped with S is an n-type semiconductor.
b) As doped with Si is a p-type semiconductor
c) Si doped with In is a p-type semiconductor.

9.40 allyl alcohol

$$H_2C{=}CH{-}CH_2{-}O{-}H$$

5 C—H bonds
5(415 kJ/mol) = 2075 kJ/mol
1 H—O bond
1(460 kJ/mol) = 460 kJ/mol
1 C=C bond
1(615 kJ/mol) = 615 kJ/mol
1 C—C bond
1(345 kJ/mol) = 345 kJ/mol
1 C—O bond
1(360 kJ/mol) = 360 kJ/mol
3855 kJ/mol

acetone

$$H_3C{-}C({=}O){-}CH_3$$

6 C—H bonds
6(415 kJ/mol) = 2490 kJ/mol
2 C—C bonds
2(345 kJ/mol) = 690 kJ/mol
1 C=O bond
1(750 kJ/mol) = 750 kJ/mol
3930 kJ/mol

The acetone is 75 kJ/mol more stable than the allyl alcohol.

9.41

	NO	CO	CN	NF
σ^*	___	___	___	___
π^*	↑ ___	___ ___	___ ___	↑ ↑
π	↑↓ ↑↓	↑↓ ↑↓	↑↓ ↑	↑↓ ↑↓
σ	↑↓	↑↓	↑↓	↑↓
σ^*	↑↓	↑↓	↑↓	↑↓
σ	↑↓	↑↓	↑↓	↑↓

Stabilized by adding one electron: CN
Stabilized by removing one electron: NO, NF

9.42 The bonding is a three-dimensional lattice of carbon and silicon atoms with each as the center of a tetrahedron and bonded to 2 other carbon atoms and 2 other silicon atoms. The bonding uses sp^3 hybrid orbitals with four σ bonds around each atom. The Si atoms have d orbitals available for formation of delocalized π orbitals, therefore, lessening the energy gap between filled and vacant orbitals. The drawing of an energy band picture showing how one would expect the band gap for SiC to compare with those of diamond and silicon should look like that shown in Figure 9-34. The gap would be smaller for SiC.

9.43

	CN
σ^*	___
π^*	___ ___
π	↑↓ ↑
σ	↑↓
σ^*	↑↓
σ	↑↓

The bonding in CN involves the net effective bonding of one σ bond and two and a half π bonds. These bonds will look like the central σ and π bonds shown in Figure 9-8.

9.44 H_2CCO $H_2C_{(1)}{=}C_{(2)}{=}\ddot{\underset{\cdot\cdot}{O}}$

The $C_{(1)}$ atom uses sp^2 hybridization to form σ bonds with the H 1s orbitals and the $C_{(2)}$ which uses sp hybrid orbitals to form σ bonds with $C_{(1)}$ and the O. The O uses a p atomic orbital to form the σ bond with $C_{(2)}$. The p orbital on $C_{(1)}$ forms a set of π orbitals with the p orbitals similarly aligned on the $C_{(2)}$ and O atoms. You need to make separate sketches of the σ and π bonding systems.

9.45

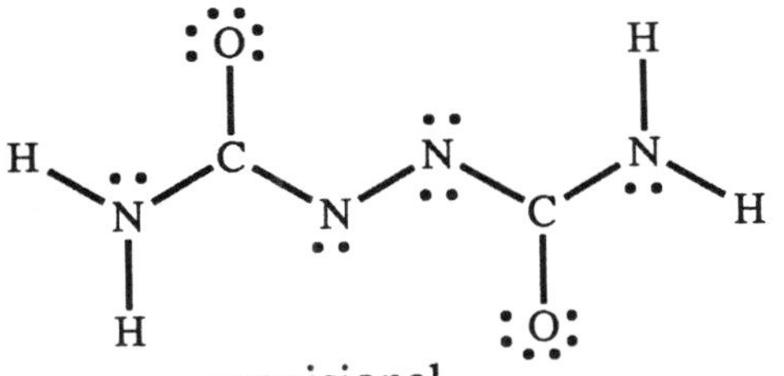

provisional

valence e^-
$= 4\ (1\ e^-$ per H$) + 4\ (5\ e^-$ per N$)$
$+ 2\ (4\ e^-$ per C$) + 2(6\ e^-$ per O$) = 44\ e^-$

$44\ e^- - 22$ (2 per σ bond) $- 12$ (6 per O)
$- 8$ (2 per N) $- 2$ (on an inner N) $= 0$

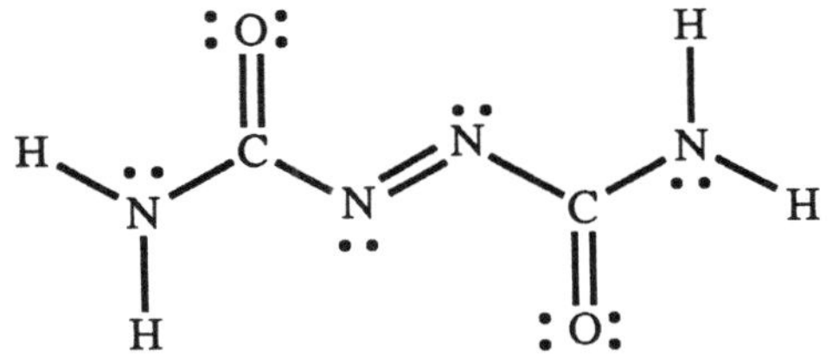

Both C atoms are triangular (bond angles ~ 120°) and the hybridization is sp^2. The two outer N atoms are tetrahedral and hybridization of sp^3. The two inner N's are bent with sp^2 hybridization.

9.46 a) A - sp^3 B - sp^3 C - sp^2 D - sp^3
b) E - 109.5° F - 109.5° G - 120°

c)

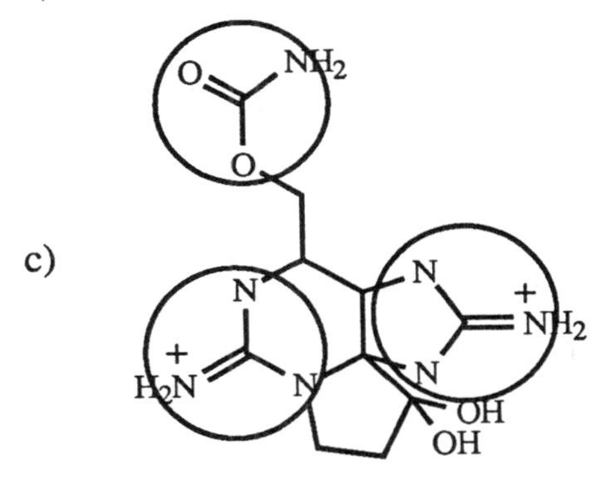

d) 3 π bonds

9.47 ethanol

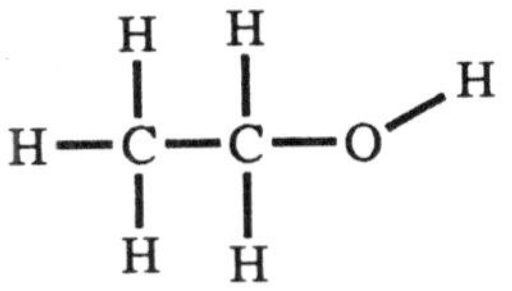

5 C—H bonds	5(415 kJ/mol) = 2075 kJ/mol
1 H—O bond	1(460 kJ/mol) = 460 kJ/mol
1 C—C bond	1(345 kJ/mol) = 345 kJ/mol
1 C—O bond	1(360 kJ/mol) = 360 kJ/mol
	= 3240 kJ/mol

dimethyl ether

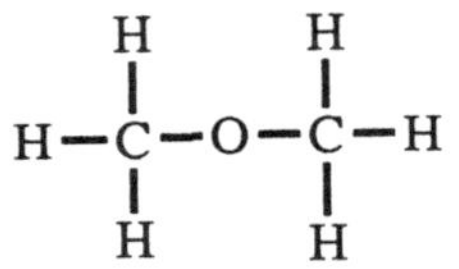

6 C—H bonds	6(415 kJ/mol) = 2490 kJ/mol
2 C—O bonds	2(360 kJ/mol) = 720 kJ/mol
	= 3210 kJ/mol

The ethanol is 30 kJ/mol more stable than the dimethyl ether.

9.48

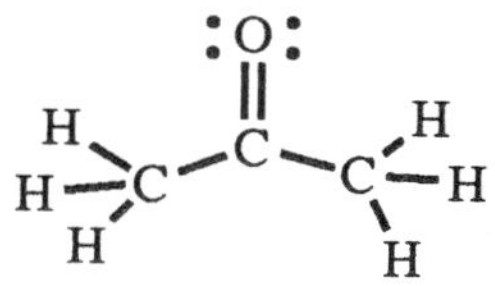

O: uses atomic orbitals
C (outer): sp^3
C (inner): sp^2
H: atomic

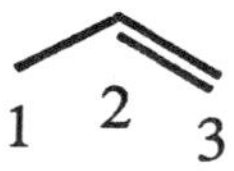

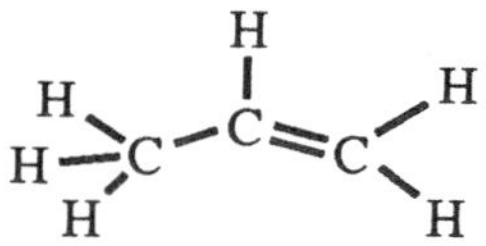

C_1: sp^3
C_2 and C_3: sp^2
H: atomic

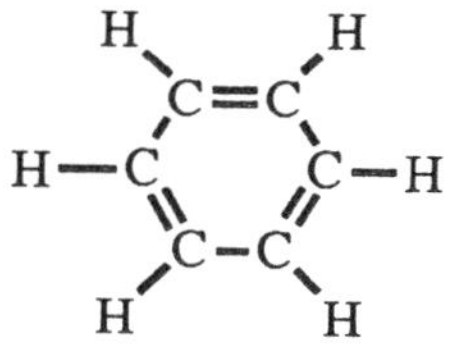

all C: sp^2
H: atomic

N_2H_4

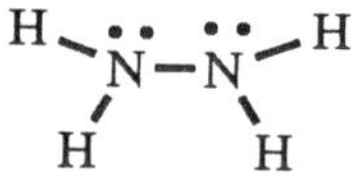

N: sp^3
H: atomic

H_3CNH_2

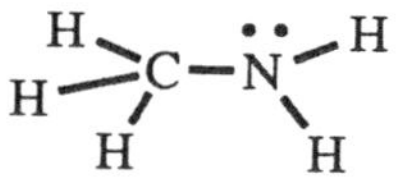

C: sp^3
N: sp^3
H: atomic

9.49 The energy gap between bonding and antibonding orbitals would be associated with the threshold energy. No, they would not all have the same kinetic energy. The bands are made up of a series of orbitals spread over a continuous range of energies. Therefore, the electrons that filled the band have a range of energies, and ejected electrons would also have a range of energies.

9.50

$2\ CH_3\text{–}CH_3$	+	$7\ O_2$	→	4 CO2	+	$6\ H_2O$
2 C—C bonds 12 C—H bonds		7 O=O bonds		8 C=O bonds		12 H—O bonds

$\Delta E = \Sigma\, B.E._{reactants} - \Sigma\, B.E._{products}$

ΔE = [(2 mol C—C)(345 kJ/mol) + (12 mol C—H)(415 kJ/mol) + (7 mol O=O)(493.6 kJ/mol)] - [(8 mol C=O)(805 kJ/mol) + (12 mol H—O)(460 kJ/mol)]
= 9125 kJ - 11,960 kJ = -2835 kJ

$$(-2835\ kJ / 2\ mol\ C_2H_6)\left(\frac{1\ mol\ C_2H_6}{30.08\ g\ C_2H_6}\right) = -47.12\ kJ / g\ C_2H_6$$

$CH_2{=}CH_2$	+	$3\ O_2$	→	$2\ CO_2$	+	$2\ H_2O$
1 C=C bond 4 C—H bonds		3 O=O bonds		4 C=O bonds		4 H—O bonds

ΔE = [(1 mol C=C)(615 kJ/mol) + (4 mol C—H)(415 kJ/mol) + (3 mol O=O)(493.6 kJ/mol)] - [(4 mol C=O)(805 kJ/mol) + (4 mol H—O)(460 kJ/mol)]
= 3756 kJ - 5060 kJ = -1304 kJ

$$(-1304\ kJ / mol\ C_2H_4)\left(\frac{1\ mol\ C_2H_4}{28.06\ g\ C_2H4}\right) = -46.47\ kJ / g\ C_2H_4$$

(continued)

(9.50 continued)

$2\ H—C≡C—H$	+	$5\ O_2$	→	$4\ CO_2$	+	$2\ H_2O$
2 C≡C bonds 4 H—C bonds		5 O=O bonds		8 C=O bonds		4 H—O bonds

ΔE = [(2 mol C≡C)(835 kJ/mol) + (4 mol H—C)(415 kJ/mol) + (5 mol O=O)(493.6 kJ/mol)] - [(8 mol C=O)(805 kJ/mol) + (4 mol H—O)(460 kJ/mol)]
= 5798 kJ - 8280 kJ = -2482 kJ

$$(-2482\text{ kJ} / 2\text{ mol } C_2H_2)\left(\frac{1\text{ mol } C_2H_2}{26.04\text{ g } C_2H_2}\right) = -47.66\text{ kJ / g } C_2H_2$$

This calculation shows that acetylene releases more energy per gram than ethane or ethylene. This is an example of the limitations of using "average" bond energies. Measurements of heats of combustion show that ethane releases the most energy per gram.

9.51 a) CCCO

:C—C—C—O: (with lone pairs), FC +1 0 0 -1

:C=C=C=O:

valence e^- = 3(4 per C) + 1(6 per O) = 18 e^-

18 e^- - 6 e^- (for 3 σ orbitals) - 6 e^- (lone pairs on O) - 6 e^- (lone pair per C) = 0

The 2 middle C's use sp hybridized orbitals to form σ bonds with each other and the end C and O. The end C and O use p atomic orbitals to form σ bonds. The molecule is linear. The 2 sets of p orbitals perpendicular to the σ bonds form 2 sets of π orbitals which extend over the whole molecule.

b) HNCO

H—N—C—O: (with lone pairs), FC 0 -1 +2 -1

H—N=C=O:

valence e^- = 1(1 per H) + 1(5 per N) + 1(4 per C) + 1(6 per O) = 16 e^-

16 e^- - 6 e^- (for 3 σ orbitals) - 6 e^- (lone pairs on O) - 4 e^- (lone pair per N) = 0

The H uses a 1s atomic orbital to form a σ bond to N. The N uses sp^2 hybridized orbitals to form σ bonds to H and C. The C atom uses sp hybridized orbitals to form σ bonds to the N and O. The O atom uses a p atomic orbital to form a σ bond with C. One π system extends over the N, C and O atoms and the other only over the C and O atoms.

c) OCS

:O—C—S: (with lone pairs), FC -1 +2 -1

:O=C=S:

valence e^- = 1(6 per O) + 1(4 per C) + 1(6 per S) = 16 e^-

16 e^- - 4 e^- (for 2 σ orbitals) - 6 e^- (lone pairs on O) - 6 e^- (lone pairs on S) = 0

The bonding is like CO_2. The carbon atom uses sp hybridized orbitals to form σ bonds by overlapping p_z orbitals on O and S atoms. The molecule is linear. Two sets of π orbitals (π_x set and π_y set) are each delocalized over all three atoms.

(continued)

(9.51 continued)

d) HCCCCH

$$\mathrm{H-\ddot{C}-\ddot{C}-\ddot{C}-\ddot{C}-H}$$

FC 0 0 0 0 0 0 valence e^- = 2(1 per H) + 4(4 per C) = 18 e^-

18 e^- - 10 e^- (for 5 σ orbitals) - 8 e^- (lone pair on each C) = 0

$$\mathrm{H-C{\equiv}C-C{\equiv}C-H}$$

FC 0 0 0 0 0 0 The C atoms use sp hybrid orbitals to form σ bonds with other C atoms or with 1s orbitals of H. The unused p orbitals on the C atoms form 2 sets of π orbitals extending over the carbon atoms.

9.52 Potassium has only one valence electron per atom; therefore, there is only one bonding electron per atom in the metal lattice. Iron has 8 valence electrons per atom, all of which occupy bonding orbitals in the metal lattice. These 8 electrons produce stronger attractive forces among the metal atoms than the 1 electron does for potassium. Therefore, iron is harder and melts at a higher temperature than potassium.

9.53 The bond energy for the H—Cl bond was measured for the compound HCl, the only example of the H—Cl bond. The energy for the H—C bond is an average of measured energies for this bond in many different compounds containing this bond. Also, measurements on HCl do not involve any other bonds while most compounds containing H—C bonds have other kinds of bonds present which must be accounted for in the measurement.

9.54 The smaller size of N's and O's valence orbitals results in stronger overlap and a larger stabilization than for P and S which have larger valence orbitals and weaker overlap. However, P and S atoms have d orbitals which can overlap and stabilize long chain molecules.

9.55 a) The bonding in CO is: :C≡O:. The lone pair on the carbon will occupy an sp hybrid orbital when it bonds to a metal. The drawing of the bonding of the sp orbital of CO and the d_{z2} orbital of a metal would look like other σ bonds (such as the one shown in Figure 9-1 in the textbook) except the shapes of the overlapping orbitals would reflect the shapes of the sp and d_{z2} orbitals.

b) The π bonding between the π^* orbital of CO and the d_{xz} orbital of the metal would look like other π bonds such as the one shown in Figure 9-31 in the textbook.

9.56 $2\ C_8H_{18}$ + $25\ O_2$ → $16\ CO_2$ + $18\ H_2O$

14 C—C bonds 36 H—C bonds	25 O=O bonds	32 C=O bonds	36 H—O bonds

$\Delta E = \Sigma\ B.E._{reactants} - \Sigma\ B.E._{products}$

ΔE = [(14 mol C—C)(345 kJ/mol) + (36 mol H—C)(415 kJ/mol) + (25 mol O=O)(493.6 kJ/mol)] - [(32 mol C=O)(805 kJ/mol) + (36 mol H—O)(460 kJ/mol)]

= 32,110 kJ - 42,320 kJ = -10,210 kJ

-10,210 kJ/2 mol isooctane = -5105 kJ/mol isooctane

9.57 The F_2 bond is a single bond; the O_2 bond is a double bond; and the N_2 bond is a triple bond. Each succeeding extra pair of electrons supplies a stronger attraction for the nuclei resulting in a stronger and shorter bond. Looking at the molecular orbital diagrams for these three molecules (See answer for Problem 9.14; a, b and c), the bond orders are: $F_2 = 1$, $O_2 = 2$ and $N_2 = 3$. As the bond order increases, the bond strength increases and the bond length decreases.

9.58

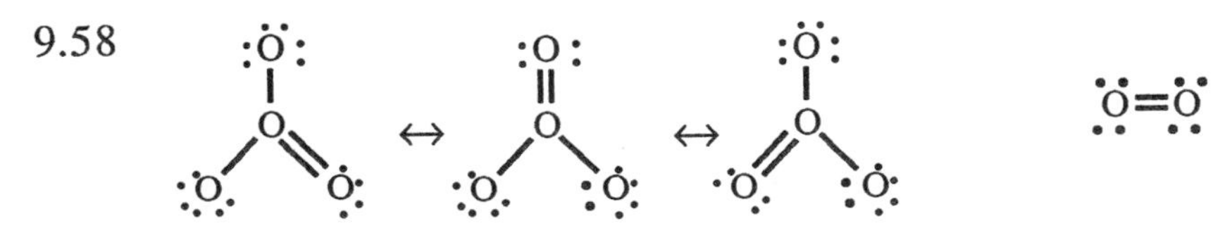

The center O atom forms σ bonds in the molecular plane using sp^2 hybrid orbitals that overlap with 2p orbitals of the three outer oxygen atoms. The 2p orbitals perpendicular to the plane of the molecule could possibly form a π system that extends over the whole molecule which would stabilize the molecule. The distribution of the formal charges suggests that the molecule would be very unstable, especially with a formal charge of +2 for the electronegative O atom. The energy of 2 moles of O=O bonds is (2 mol)(493.6 kJ/mole) = 987.2 kJ. The energy of 1 mole of O_4 is [493.6 kJ + (2)(145 kJ)] = 783.6 kJ. This suggests that the 2 moles of O_2 are 203.6 kJ more stable than one mole of O_4. (This does not consider any stability from the π system of O_4.)

9.59 F—F < Br—Br < Br—Cl < Cl—Cl < Cl—F

F—F	Br—Br	Br—Cl	Cl—Cl	Cl—F
154.8	190.0	216	239.7	249 kJ/mol

The F—F bond is the weakest because the small size and high effective nuclear charge of the fluorine atoms cause decreased overlap of the bonding orbitals and increased repulsion between the non-bonding electrons on the F atoms. The high principal quantum number ($n = 4$) resulting in more diffuse valence orbitals and a larger spread of electron density makes the Br—Br bond weaker than Cl-containing molecules. For the same reason Cl—Cl bond ($n = 3$) is stronger than the Br—Br bond with the Br—Cl bond (with one of each atom) being between. The Cl—F bond has the highest energy because the larger electronegativity difference of F and Cl atoms results in the stability that accompanies a polar bond.

9.60 ClO_4^- ClO_3^- ClO_2^-

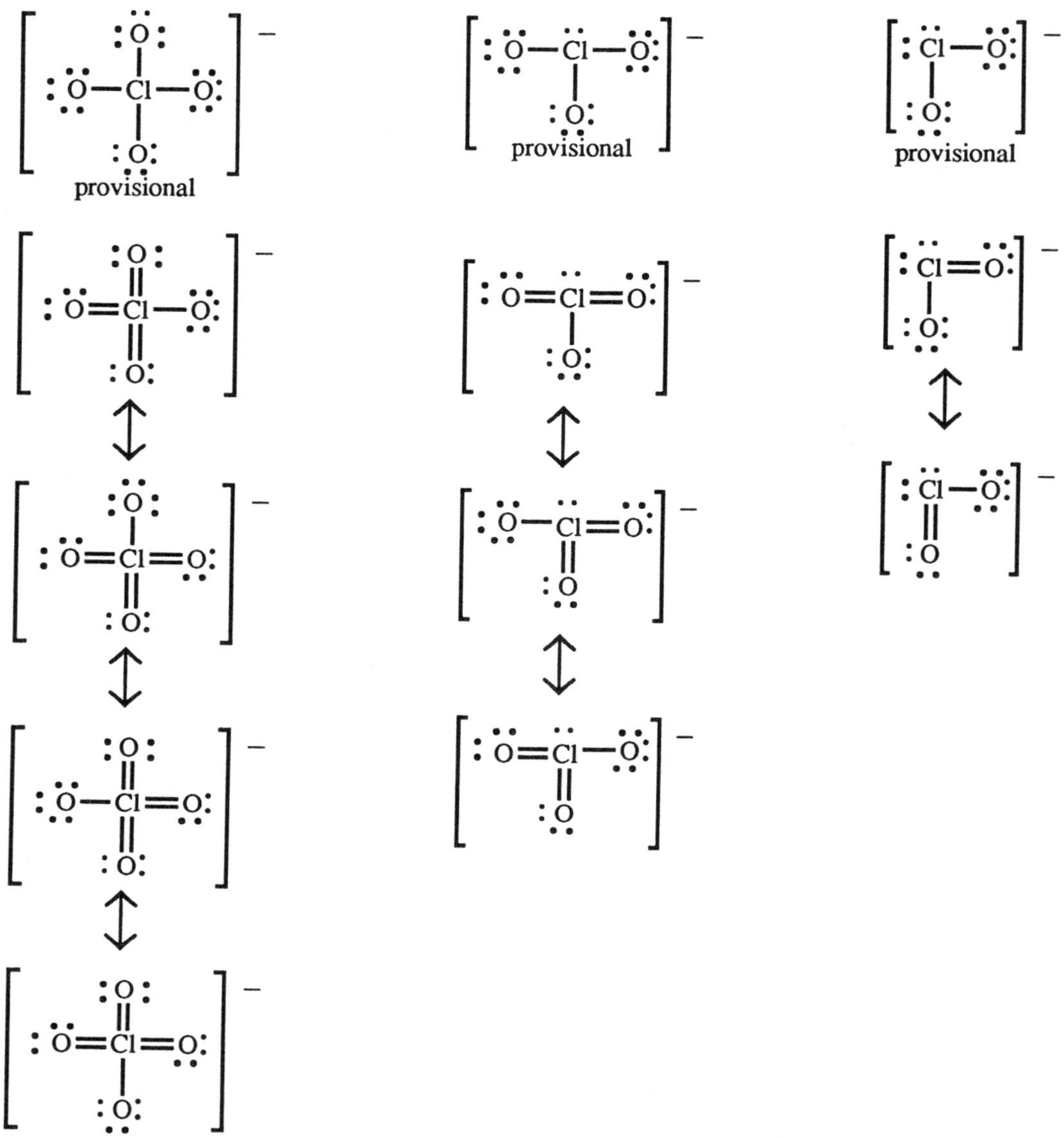

provisional provisional provisional

The Cl atom uses sp^3 hybridization to form σ bonds with the 2p atomic orbitals of O or to hold lone pairs of electrons on chlorine in each of these three ions. Each ion can form π systems using 2p atomic orbitals on the O atoms and 3d orbitals on the Cl. In each case delocalization can occur over all atoms of the ion.

9.61 ClO_2

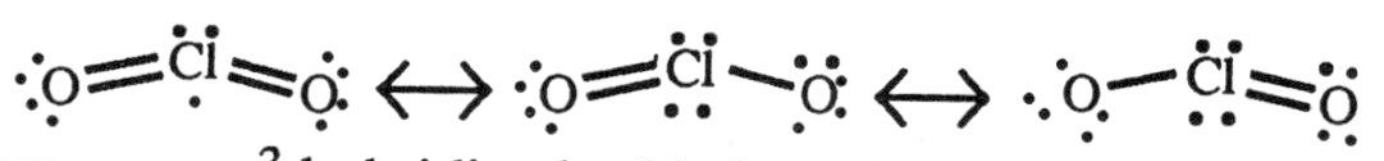

The Cl atom uses sp^3 hybridized orbitals to form σ bonds with the O atoms. The Cl atom uses 3d orbitals to form a π system extending over the 3 atoms. The compound is unusual because having an odd number of valence electrons, it has a lone electron in a π orbital.

9.62 $CO_{(g)}$ + $Cl_{2(g)}$ → $Cl_2CO_{(g)}$

1 C≡O bond   1 Cl—Cl bond   2 Cl—C bonds
1 C=O bond

ΔE = [(1 mol C≡O)(1071 kJ/mol) + (1 mol Cl—Cl)(239.7 kJ/mol)] - [(2 mol Cl—C)(325 kJ/mol) + (1 mol C=O)(750 kJ/mol)] = 1311 kJ - 1400 kJ = -89 kJ

9.63 Four bonds: H—O in H_2O, C=O in CO_2, H—C and C≡N in HCN. The shortest bond is H—O. The longest bond is C=O.

H—O	<	H—C	<	C≡N	<	C=O
96		109		116		120 pm

The small principal quantum number (n = 1) for H makes its bonds shorter. The greater electronegativity difference makes the H—O bond shorter than the H—C bond. The lengths of the other 2 bonds are longer because both atoms involved have n = 2. The C≡N is shorter than the C=O because it has a triple bond versus a double bond.

9.64

H—C	<	H—O	<	C=O	<	C≡N
415		460		805		890 kJ/mol

The C≡N is the strongest bond because the most electrons are shared between the atoms (a triple bond). The C=O is the next strongest bond because 2 pairs of electrons are shared (a double bond). The H—O bond is stronger than the H—C bond because there is a greater electronegativity difference between H and O atoms.

9.65 $2\,(\text{—Si})_n$ + $n\,O_2$ → $2\,(\text{—O—Si})_n$

2 Si—Si bonds   1 O=O bond   4 Si—O bonds

ΔE = [(2 mol Si—Si)(220 kJ/mol) + (1 mol O_2)(493.6 kJ/mol)]
- [(4 mol Si—O)(450 kJ/mol)] = 934 kJ - 1800 kJ = -866 kJ/2 mol Si
= -433 kJ/mol Si

$2\,(\text{—C})_n$ + $n\,O_2$ → $2\,(\text{—O—C})_n$

2 C—C bonds   1 O=O bond   4 C—O bonds

ΔE = [(2 mol C—C)(345 kJ/mol) + (1 mol O_2)(493.6 kJ/mol)]
- [(4 mol C—O)(360 kJ/mol)] = 1184 kJ - 1440 kJ = -256 kJ/2 mol C
= -128 kJ/mol C

9.66

O_2: $(\sigma_s)^2\ (\sigma_s^*)^2\ (\sigma_p)^2\ (\pi_x)^2\ (\pi_y)^2\ (\pi_x^*)^1(\pi_y^*)^1$ B.O. = (8 - 4)/2 = 2

O_2^-: $(\sigma_s)^2\ (\sigma_s^*)^2\ (\sigma_p)^2\ (\pi_x)^2\ (\pi_y)^2\ (\pi_x^*)^2\ (\pi_y^*)^1$ B.O. = (8 - 5)/2 = 1.5

O_2^{2-}: $(\sigma_s)^2\ (\sigma_s^*)^2\ (\sigma_p)^2\ (\pi_x)^2\ (\pi_y)^2\ (\pi_x^*)^2\ (\pi_y^*)^2$ B.O. = (8 - 6)/2 = 1.0

bond order: O_2^{2-} < O_2^- < O_2
1.0   1.5   2

bond energy: O_2^{2-} < O_2^- < O_2

bond length: O_2 < O_2^- < O_2^{2-}

Both O_2 and O_2^- are magnetic. The O_2 has the larger magnetism.

9.67 Increasing band gap: Bi < As < P

Bi: conductor   As: conductor   P: insulator

Both Bi and As have electrical resistivities that fall between what is usually defined as conductors and semiconductors. As is the better conductor.

9.68 In allene the outside carbons are sp^2 hybridized. The central carbon is sp hybridized. The shape defined by three carbons is linear. The shape about each of the outside carbons is trigonal planar. Since the π bonds to the central carbon must use different p orbitals that are 90° to each other, the planar ends of the molecule will be perpendicular to each other. Your sketches must reflect the orbitals used, their shapes, and their orientation to each other. The structure above is provided to help with the understanding of the molecule's shape. The requested sketches need to separately show σ and π bonding systems.

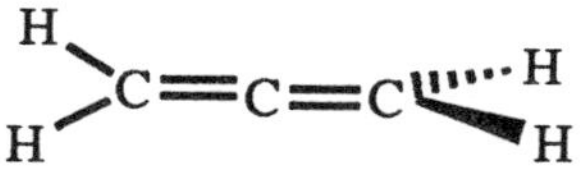

9.69 a) valence e^- = 8(4 per C) + 3(1 per H) + 30 (6 per O) + 5 (5 per N) = 46 e^-

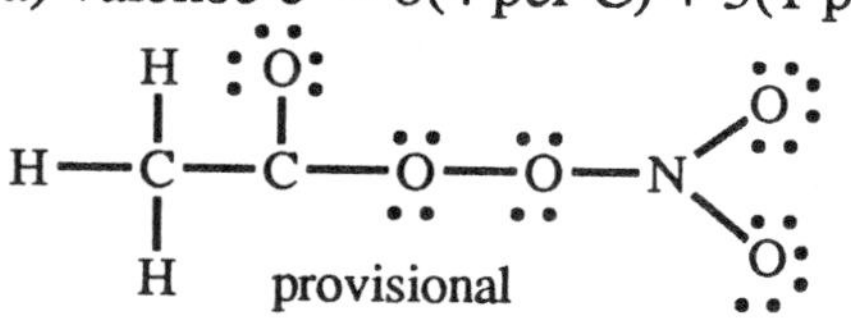

provisional

46 e^- - 20 e^- (10 σ orbitals) - 18 e^- (lone pairs on outer O) - 8 e^- (lone pairs on inner O) = 0

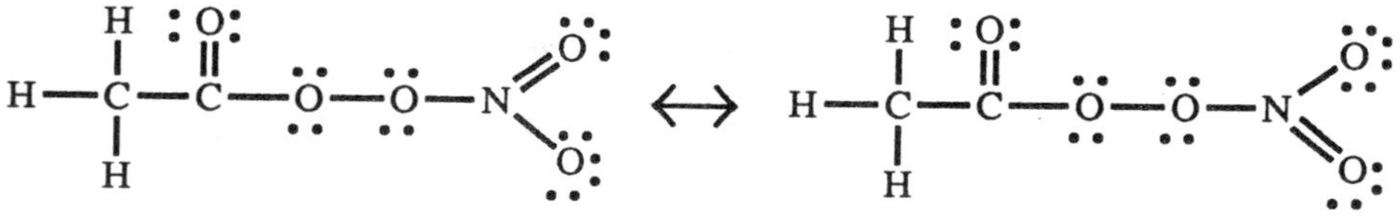

↔

b) The left carbon has sp^3 hybridization. The other carbon has sp^2 hybridization. The oxygen marked with an asterisk in the question has sp^3 hybridization. The nitrogen has sp^2 hybridization.

c) The bond angles around the left carbon atom are 109.5°. The O=C—O bond angle is 120°. The O=N—O bond angle is 120°.

d) There is a set of delocalized π orbitals over O=N—O.

9.70

$H_2C{=}CH_2$	+	HCl	→	$H_3C{-}CH_2Cl$
1 C=C bond		H—Cl bond		1 C—C bond
4 C—H bonds				5 C—H bonds
				1 C—Cl bond

ΔE = [(1 mol C=C)(615 kJ/mol) + (4 mol C—H)(415 kJ/mol) + (1 mol H—Cl) (428.0 kJ/mol)] - [(1 mol C—C)(345 kJ/mol) + (5 mol C—H)(415 kJ/mol) + (1 mol C—Cl)(325 kJ/mol)] = 2703 kJ - 2745 kJ = -42 kJ

$H_2C{=}CH_2$	+	Cl_2	→	$H_2ClC{-}CH_2Cl$
1 C=C bond		Cl—Cl bond		1 C—C bond
4 C—H bonds				2 C—Cl bonds
				4 C—H bonds

ΔE = [(1 mol C=C)(615 kJ/mol) + (4 mol C—H)(415 kJ/mol) + (1 mol Cl—Cl) (239.7 kJ/mol)] - [(1 mol C—C)(345 kJ/mol) + (2 mol C—Cl)(325 kJ/mol) + (4 mol C—H)(415 kJ/mol)] = 2515 kJ - 2655 kJ = -140 kJ

9.71 The bond energy of C—Cl is 325 kJ/mol. The energy per bond is:

$$\left(\frac{325\ \text{kJ}}{\text{mol}}\right)\left(\frac{\text{mol}}{6.02 \times 10^{23}}\right) = 5.40 \times 10^{-22}\ \text{kJ}$$

$E = h\nu$

$$\nu = \frac{E}{h} = \frac{(5.40 \times 10^{-22}\ \text{kJ})(10^{3}\ \text{J / kJ})}{(6.63 \times 10^{-34}\ \text{J s})} = 8.14 \times 10^{14}\ \text{s}^{-1} = 8.14 \times 10^{14}\ \text{Hz}$$

9.72 The C≡C triple bond has one more pair of electrons shared by the C atoms than the C=C double bond. Sharing more electrons strengthens the bond. However, the more electrons shared, the greater the repulsion between the electrons. This spreads the electron density of the π bonds over a wider area and makes them more available for reaction, more reactive.

9.73 The carbon-oxygen bond length in the carbonate ion is between the average length for a C—O single bond and a C=O double bond. Delocalized electrons give the C—O bonds in the carbonate ion properties that are intermediate between single and double bonds.

9.74 I.E. (kJ/mol)

(1503)1314 O:	$1s^2\ 2s^2\ 2p_x^1\ 2p_y^1\ 2p_z^2$
(1314)1164 O_2:	$(\sigma_{1s})^2\ (\sigma_{1s}^*)^2\ (\sigma_{2s})^2\ (\sigma_{2s}^*)^2\ (\sigma_p)^2\ (\pi_x)^2\ (\pi_y)^2\ (\pi_x^*)^1(\pi_y^*)^1$
1402 N:	$1s^2\ 2s^2\ 2p_x^1\ 2p_y^1\ 2p_z^1$
1503 N_2:	$(\sigma_{1s})^2\ (\sigma_{1s}^*)^2\ (\sigma_{2s})^2\ (\sigma_{2s}^*)^2\ (\sigma_p)^2\ (\pi_x)^2\ (\pi_y)^2$

Both N and N_2 have some stability versus O and O_2 since the electron to be removed is a paired bonding electron for N_2 and an electron from a half-filled set of p orbitals with electrons with parallel spins for N. The paired bonding electron of N_2 is most difficult to remove and the unpaired antibonding electron of O_2 is the easiest to remove.

9.75 a) For the N=N=N– portion of the molecule from left to right the first N uses 2p orbitals to the σ bond and the π bond and uses the 2s and a 2p orbital to hold the lone pairs of electrons. The middle N uses sp hybrid orbitals to form the 2 σ bonds and the remaining two 2p orbitals to form π bonds. The third N uses sp^2 hybrid orbitals to form 2 σ bonds and the lone pair of electrons. The remaining 2p orbital is used for the π bond. The 2 N's in the ring use sp^2 hybrid orbitals to each form 3 σ bonds and a p orbital for the lone pairs of electrons. The lone pairs of electrons are part of a delocalized π bond system that involves the entire ring.

b) The bond angle is approximately 109.5°.

c) There are 6 C atoms with sp^3 hybrid orbitals. (4 in the 5-membered ring, 1 attached to this ring and 1 attached to the 6-membered ring).

d) There are 31 σ bonds in AZT.

e) There are 5 π bonds in AZT.

9.76

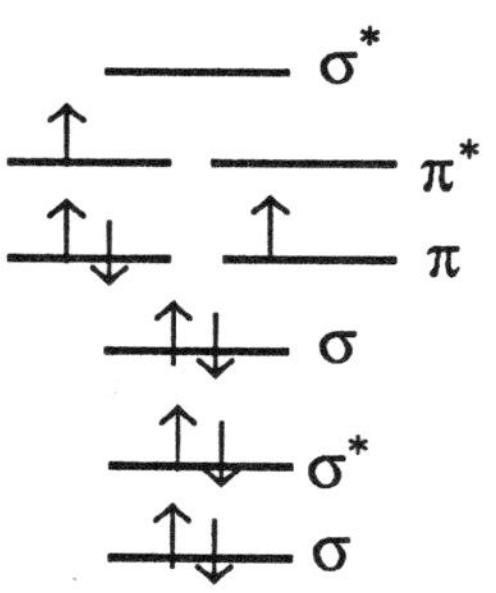

$\underline{\quad\quad}$ σ^*

$\underline{\uparrow}$ $\underline{\quad\quad}$ π^*

$\underline{\uparrow\downarrow}$ $\underline{\uparrow}$ π

$\underline{\uparrow\downarrow}$ σ

$\underline{\uparrow\downarrow}$ σ^*

$\underline{\uparrow\downarrow}$ σ

excited N_2

The N—N bond in this excited state of an N_2 molecule is weaker than the N—N bond in ground state nitrogen. The bond order is (7-3)/2 = 2 versus (8-2)/2 = 3 for ground state nitrogen. The excited electron goes from a bonding to an antibonding orbital thereby decreasing bonding electrons by one and increasing the antibonding electrons by one.

9.77 a) Each carbon atom uses sp^2 hybrid orbitals to form 3 σ bonds and the remaining p orbital for π bonding.

b) Five p orbitals contribute to the π bonding system to form 5 molecular orbitals.

c) There are 4 delocalized electrons.

d)

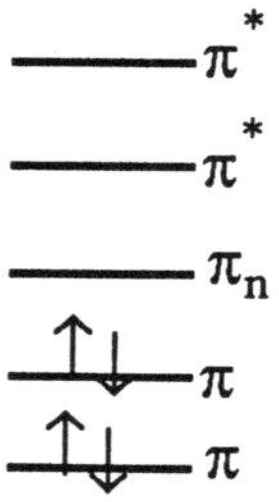

$\underline{\quad\quad}$ π^*

$\underline{\quad\quad}$ π^*

$\underline{\quad\quad}$ π_n

$\underline{\uparrow\downarrow}$ π

$\underline{\uparrow\downarrow}$ π

CHAPTER 10: EFFECTS OF INTERMOLECULAR FORCES

10.1 Your sketch should look like that in Figure 10-3 with the Br_2-Br_2 being replaced by carbon tetrachloride-carbon tetrachloride and the F_2-F_2 being replaced by methane-methane. Carbon tetrachloride has the stronger attraction of the two like Br_2 had the stronger attraction than F_2. The dashed line would remain very nearly at the same level in the two sketches.

10.2 Your molecular pictures should look like those in Figure 10-2 with the following differences: (1) solid Ag unlike solid I_2 is monatomic and will be packed more tightly than the I_2 molecules, (2) Ar gas will be like Cl_2 gas except it is monatomic and (3) Hg liquid will also be monatomic where Br_2 is shown to be diatomic in Figure 10-2.

10.3 a) Intermolecular attractions become less significant when a gas is expanded to a larger volume at constant temperature.
b) Intermolecular attractions become more significant when more gas is forced into the same volume at constant temperature.
c) Intermolecular attractions become more significant when the temperature of the gas is lowered at constant volume.

10.4 a) Molecular volume becomes less significant when a gas is expanded to a larger volume at constant temperature.
b) Molecular volume becomes more significant when more gas is forced into the same volume at constant temperature.
c) The significance of molecular volume remains unchanged when the temperature of the gas is lowered at constant volume.

10.5 $\frac{PV}{nRT} = 1.00$ for an ideal gas.

a) $$\frac{(19.7\text{ atm})(1.20\text{ L})}{(1.00\text{ mol})(0.08206\text{ L}\bullet\text{atm / mol}\bullet\text{K})(313.15\text{ K})} = 0.92$$

$$\frac{1.00 - 0.92}{1.00} \text{ x } 100\% = 8.00\%$$

b) $$\frac{(2.00\text{ atm})\left(189.18\text{ cm}^3 \text{ x } 1\text{ mL / cm}^3\right)(1.00\text{ L / }1000\text{ mL})}{\left(\frac{3.00\text{ g}}{2.016\text{ g / mol}}\right)\left(0.08206\ \frac{\text{L}\bullet\text{atm}}{\text{mol}\bullet\text{K}}\right)(273.15\text{ K})} = 1.13$$

$$\frac{|1.00 - 1.13|}{1.00} \text{ x } 100\% = 13\%$$

10.6 He < Ne < Ar < Xe
The only force between atoms of rare gases is due to the polarizability of their electron clouds. The valence electrons are in atomic orbitals with increasing principal quantum numbers: He (n = 1), Ne (n = 2), Ar (n = 3) and Xe (n = 5). This means the atomic orbitals proceed from small atomic orbitals (n =1) with low polarizability to large atomic orbitals (n = 5) with high polarizability. Low polarizability means weaker intermolecular forces and, therefore, lower boiling points.

10.7 Arranged with easiest to liquefy first: CCl_4 > CF_4 > CH_4
The large Cl atoms in CCl_4 have large atomic orbitals and, therefore, high polarizability, giving CCl_4 strong intermolecular forces and ease of liquefaction (has a high boiling point). The small F atoms in CF_4 have small atomic orbitals, low polarizability, giving CF_4 weaker intermolecular forces and a lower boiling point. The H atoms in CH_4 have no lone pairs of electrons in their smallest atomic orbitals (n = 1) resulting in the weakest intermolecular forces and the lowest boiling point.

10.8

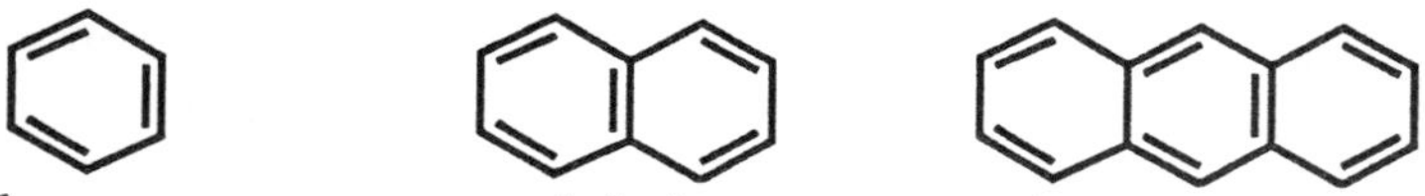

benzene naphthalene anthracene

Benzene is the liquid. Naphthalene is the relatively volatile solid. Anthracene is the less volatile solid. As the ring systems get larger, the polarizability increases and the intermolecular forces increase resulting in higher melting points.

10.9 a) Mg^{2+} has a stronger interaction with water molecules because it has a greater positive charge.
b) Na^{+} has the stronger interaction with water molecules because it has the greater charge density (same charge but smaller size).
c) SO_4^{2-} has the stronger interaction with water molecules because it has 4 O atoms with polarizable lone pairs of electrons versus only 3 O atoms with polarizable lone pairs of electrons on SO_3^{2-}.

10.10 a)

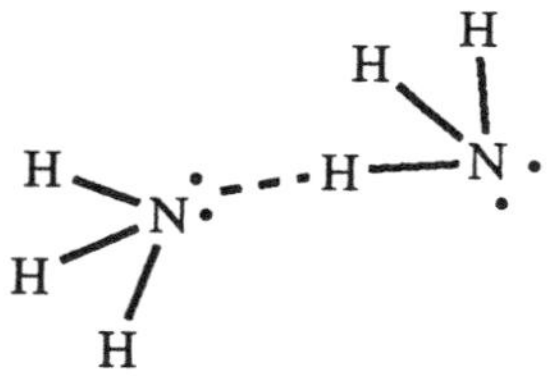

b)

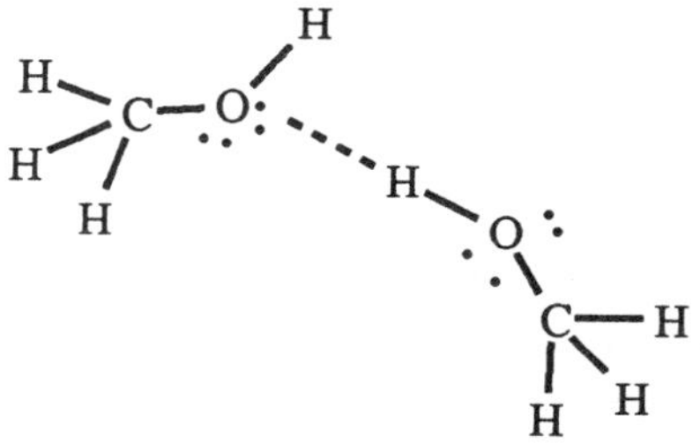

c)

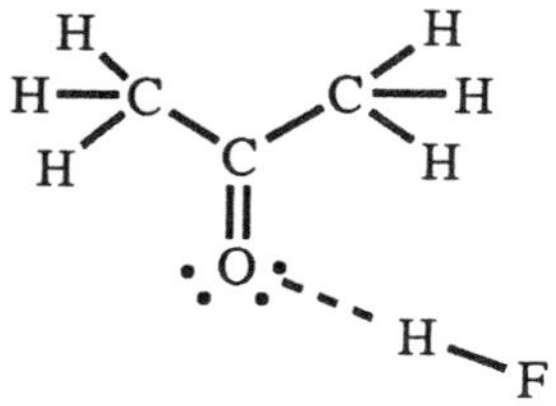

10.11 propane < *n*-pentane < ethanol: boiling points
C_3H_8 C_5H_{12} C_2H_5OH
Dispersion forces are the only intermolecular forces for propane and *n*-pentane. The larger *n*-pentane is more polarizable resulting in stronger intermolecular forces and a higher boiling point than propane. Ethanol not only has dispersion forces but also dipole interactions because of the polar C—O and O—H bonds and the slightly stronger hydrogen bonding between the H atom on the O atom and the lone pairs of electrons on the O atom.

10.12

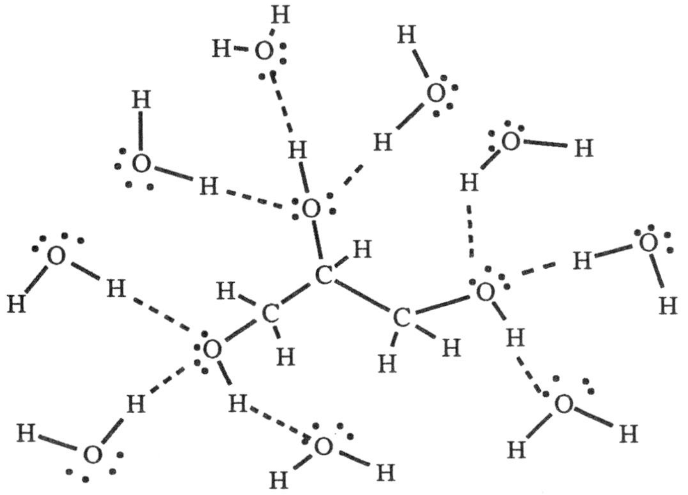

A maximum of 9 hydrogen bonds will be formed between water molecules and one molecule of glycerol.

10.13 a) CH_2Cl_2 will not hydrogen bond.
b) H_2SO_4 will hydrogen bond with molecules of its own kind and other molecules such as H_2O.
c) H_3COCH_3 will hydrogen bond only with other molecules such as H_2O.
d) $H_2NCH_2CO_2H$ will hydrogen bond with molecules of its own kind and other molecules such as H_2O.

10.14 Low viscosity motor oil is preferred for winter because at the lower temperatures of winter the viscosity will increase. If the viscosity is too high when the motor is started, the lubricant will not flow easily until both it and the motor heat up; by then the motor will have been damaged by friction.

10.15 pentane < gasoline < fuel oil
The viscosity increases as the hydrocarbon chains get longer because the tangling between molecules increases and because the cohesive forces (dispersion forces) increase.

10.16 The water forms strong adhesive forces (hydrogen bonds) with oxygen atoms and O—H groups of the silicates in the glass. The water adheres well to the glass and maximizes contact with the glass wall by forming a concave surface. The cohesive forces between mercury atoms are stronger than the adhesive forces between mercury atoms and glass. This results in minimization of the contact with the glass wall of the tube and minimization of the surface area of the mercury (a convex shape).

10.17 Flux must remove those substances to which metals will not adhere. Specifically, any oil or grease and any oxides or other forms of corrosion that prevent a good solder joint.

10.18 a) Grease - dirty: Grease is limited to dispersion forces for intermolecular forces while water uses dipolar forces and hydrogen bonding.
b) Mg^{2+} ions - not dirty: Mg^{2+} ions as charged particles would form adhesive forces with dipolar water molecules.
c) Acetone - not dirty: Acetone would form adhesive forces by hydrogen bonding with water.
d) SiO_2 - not dirty: Water would adhere to the SiO_2 by forming hydrogen bonds with the O atoms of SiO_2.

10.19 Sn - metallic; S_8 - molecular; Se - molecular; SiO_2 - covalent; Na_2SO_4 - ionic

10.20 The two forms of silica have different densities because the efficiency of packing of the atoms is different. The structure of the crystalline quartz places the atoms at constant distances from each other (the bond length). The amorphous form may have some atoms at the same distance but others farther apart because of the lack of symmetry (disordered arrangement). (See Figure 10-20)

10.21 The structure of carborundum has alternating carbon and silicon atoms that are covalently bonded. All atoms are bonded to four other atoms and have a tetrahedral geometry.

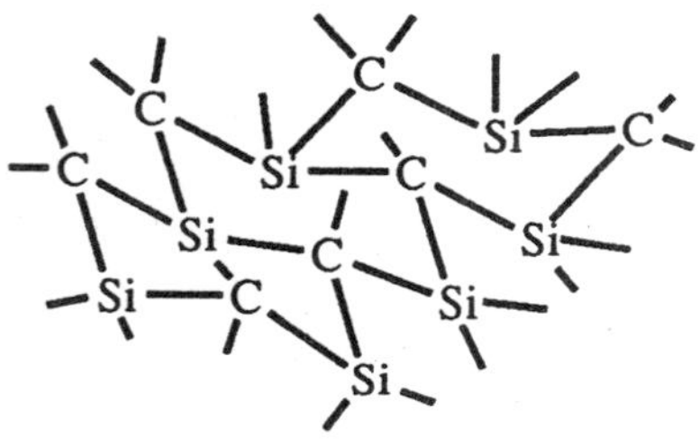

10.22 If the broken lines define the unit cell:
\# of Xenon atoms = 1/8(8 at corners) + 1/1(1 in center) = 2 atoms
\# of Fluorine atoms = 1/4(8 on edges) + 1/1(2 in center) = 4 atoms
Xe_2F_4 formula of compound - XeF_2

10.23 # of Ba atoms = 1/1(2 in interior) = 2 atoms
\# of Cu atoms = 1/8(8 at corners) + 1/4(8 on edges) = 3 atoms
\# of Y atoms = 1/1(1 in center) = 1 atom
\# of O atoms = 1/4(12 on edges) + 1/2(8 on faces) = 7 atoms
$Ba_2Cu_3YO_7$ $YBa_2Cu_3O_7$

10.24 For NaCl draw a crystal pattern like that in Figure 10-25 then isolate the face-centered cubic array of cations.
The nearest neighbors of opposite charge for each ion in NaCl is 6.

10.25 Water is not soluble in gasoline because the major intermolecular forces for water are hydrogen bonds and for gasoline are dispersion forces. Water should not be used to fight a gasoline fire because the gasoline will float on top of the water and continue to burn. Further, if the water flows anywhere, it will spread the fire.

10.26 Acetone can form hydrogen bonds with water and because of its fairly short C chain is miscible with water. Cyclohexane and acetone can share intermolecular forces of the dispersion type and are, therefore, miscible. Acetone can use polar bonding (C=O bond), hydrogen bonding (lone pairs of electrons on O), and dispersion forces (3-carbon bonding framework) and, therefore, can dissolve in different types of compounds. Cyclohexane only uses dispersion forces. Water only uses hydrogen bonding and polar bonding. Therefore, cyclohexane and water are nearly insoluble in each other.

10.27 Solids that form hydrogen bonds, ionic solids and polar solids would be expected to be soluble in liquid ammonia.

10.28 The solubility of O_2 in water at 25°C: (using Henry's Law: $C_i = K_H P_i$)

$C_i = (1.3 \times 10^{-3}\ M/atm)P_i$

We can assume P_i to be 0.21 atm (See Sample Problem 10-9)

$C_i = (1.3 \times 10^{-3}\ M/atm)(0.21\ atm) = 2.73 \times 10^{-4}\ M$

Then 1 L of water contains $(2.73 \times 10^{-4}\ mol/L)(1\ L) = 2.73 \times 10^{-4}$ mol O_2

At 30°C:

$C_i = (0.89 \times 10^{-3}\ M/atm)P_i$ $\quad$ P_i remains 0.21 atm

$C_i = (0.89 \times 10^{-3}\ M/atm)(0.21\ atm) = 1.87 \times 10^{-4}\ M$

Then 1 L of water contains $(1.87 \times 10^{-4}\ mol/L)(1\ L) = 1.87 \times 10^{-4}$ mol O_2

$(2.73 \times 10^{-4}\ mol\ O_2) \div (1.87 \times 10^{-4}\ mol/L) = 1.5$ L of water at 30°C

or $[(1\ L)\ (1.3 \times 10\text{-}3\ M/atm)(P_i)] \div [(0.89 \times 10^{-3}\ M/atm)(P_i)] = 1.5$ L of water at 30°C

Note: The answer is independent of the partial pressure if it remains constant for the two temperatures.

10.29 If the 1.10 atm pressure is the pressure of the CO_2:

$C_i = (34 \times 10^{-3}\ M/atm)(1.10\ atm) = 0.037\ M$

$$\left(0.037\frac{mol}{L}\right)\left(\frac{250\ mL}{1000\ mL/L}\right)\left(\frac{44\ g\ CO_2}{1\ mol}\right) = 0.41\ g\ CO_2$$

10.30 $C_i = K_H P_i$

$0.037\ M = (78 \times 10^{-3}\ M/atm)P_i$

$$P_i = \frac{0.037\ M}{(78 \times 10^{-3}\ M/atm)} = 0.47\ atm$$ If gas space is essentially pure CO_2.

This calculation assumes that the amount of CO_2 that dissolves in the beverage when the beverage is cooled from 25°C to 0°C is not enough to change the concentration in the beverage significantly. The total pressure must include the partial pressure of water.

10.31 The compound is hydrophobic. It will concentrate in the fatty tissue. This makes it dangerous because its concentration builds up in the fatty tissue of an organism through absorption by the organism or by consumption of organisms containing the DDT. Therefore, higher members of the food chain (which humans often eat) acquire the biggest concentration build up.

10.32 a) propionic acid (structure: $CH_3CH_2C(=O)-O-H$) worst surfactant

b) lauryl alcohol (structure: zigzag chain $-O-H$)

c) sodium lauryl sulfate (structure: zigzag chain $-O-S(=O)_2-O^- Na^+$) best surfactant

The best surfactant is the sodium lauryl sulfate because it has a strong hydrophilic end (the sulfate anionic end) and a strong hydrophobic end (the $—C_{12}H_{25}$ end). The propionic acid has a hydrophilic end ($—CO_2H$ end) which can hydrogen bond with several water molecules but its short hydrocarbon end makes that end only weakly hydrophobic. In fact, propionic acid is quite soluble in water. Lauryl alcohol is the opposite with a strong hydrophobic end ($—C_{12}H_{25}$ end) but a weaker hydrophilic end (—O—H end). The molecular pictures called for in this question should show hydrophilic and hydrophobic interactions.

10.33 The drawing asked for in this question should look like one of the horizonal portions of the vesicle shown in Figure 10-37.

10.34 The drawing asked for in this question should look like the horizonal monolayer shown in Figure 10-37 except that the hydrocarbon tail will be pointed toward the nonpolar liquid phase (gasoline) and the polar head group will be protruding above the liquid phase.

10.35 12% by mass ethanol: 12 g ethanol (CH_3CH_2OH) per 88 g H_2O

$$\text{mol of ethanol} = \frac{12\text{ g}}{46\text{ g/mol}} = 0.26\text{ mol} \qquad \text{mol of } H_2O = \frac{88\text{ g}}{18\text{ g/mol}} = 4.89$$

$$\Delta T_f = K_f X_{solute} = (103.2°C)\left(\frac{0.26}{0.26 + 4.89}\right) = 5.2°C$$

0°C - 5.2°C = -5.2°C (freezing point of wine)

10.36 There is not enough information to calculate the boiling point of the wine because Equation 10-3 only applies to a solution containing a nonvolatile solute. Ethanol is volatile. We also do not know the concentration of all dissolved substances such as the sugars.

10.37 a) $$MM = \frac{mRT}{\Pi V} = \frac{(1.00\text{ g})(0.08206\text{ L}\bullet\text{atm/mol}\bullet\text{K})(298\text{ K})}{(64.8\text{ torr})\left(\frac{1\text{ atm}}{760\text{ torr}}\right)(1\text{ L})} = 286.8\text{ g/mol}$$

(The Π has been corrected. If 17.8 torr is used one obtains the value of 1044 g/mol.)

b) $$\text{MM of hydrocarbon portion} = \left(\frac{11\text{ mol C}}{\text{mol}}\right)\left(\frac{12\text{ g}}{\text{mol C}}\right) = \left(\frac{23\text{ mol H}}{\text{mol}}\right)\left(\frac{1\text{ g}}{\text{mol H}}\right)$$

$$= 155\text{ g/mol}$$

MM of polar portion = 287 g/mol - 155 g/mol = 132 g/mol

10.38 Assume the density of water is 1.000 g/mL

$$\text{moles of NaCl} = \frac{2.50\ \text{g}}{58.44\ \text{g/mol}} = 0.04278\ \text{mol NaCl}$$

$$\text{moles of } H_2O = (155\ \text{mL})(1.000\ \text{g/mL}) \div \left(\frac{18.016\ \text{g}}{\text{mol}}\right) = 8.603\ \text{mol } H_2O$$

Using the equation:

$$\Delta T_b = iK_bX_{solute} = (2)(28.9°\text{C})\left(\frac{0.04278}{0.04278 + 8.603}\right) = 0.286°\text{C}$$

boiling point of solution = 100°C + 0.286°C = 100.286°C

10.39 The impure diethyl ether is placed into a distilling flask and pieces of sodium metal are added. The water reacts with the sodium metal to form sodium hydroxide and hydrogen gas. Air is excluded from the distilling system to prevent recontamination and the ether is distilled from the sodium hydroxide.

10.40 TLC could be used to test the compound to see if it would separate into different components. Different stationary and mobile phases (solvents) would be tried. This would result in different intermolecular forces between the stationary phase and the molecules in the sample. The components of the sample would likely interact differently with the stationary phase which would mean they would travel at different speeds and a separation of components would occur. If the components have different colors, the TLC plates would have separated spots of different colors (components).

10.41 95% of 250 g of sample = 237.5 g of $HgCl_2$
5% of 250 g of sample = 12.5 g of impurity
Minimum amount of H_2O at 100°C =

$$(237.5\ \text{g } HgCl_2)\left(\frac{1\ \text{L } H_2O}{380\ \text{g } HgCl_2}\right) = 0.625\ \text{L } H_2O = 625\ \text{mL } H_2O$$

Amount of $HgCl_2$ left in solution at 0°C =

$$(0.625\ \text{L } H_2O)\left(\frac{30\ \text{g } HgCl_2}{1\ \text{L } H_2O}\right) = 18.75\ \text{g } HgCl_2$$

Note: The impurity is more soluble than $HgCl_2$; all 12.5 g will remain in solution.
Amount of $HgCl_2$ precipitated at 0°C = 237.5 g - 18.75 g = 218.8 g
218.8 / 237.5 = 0.92 or 92%

10.42 a) There are 4 components in the sample.
b) There is a larger amount of long-chain polymer molecules in the sample. Long-chain polymer molecules would be too big to enter most of the molecular sieves and, therefore, would move along faster than the other molecules and would be the first to leave the chromatography system. The chromatograph shows that the component requiring the least amount of time to pass through is present in the largest amount.

10.43 Assuming that separation according to polarity means that least polar comes off first and most polar comes off last:
1st - heptane; 2nd - methylpentylether; 3rd (last) - 1-hexanol
Heptane is nonpolar, would have no interaction with the GC column and would come off first. The 1-hexanol with the C—O—H bonding would be the most polar, would interact with the GC column the most and would come off last. The methyl-pentylether with the C—O—C bonding would be intermediate in polarity and would come off between the other two.
Note: All three compounds have about the same size and molecular mass.

10.44 Water will rise the highest and acetone will rise the least. The water hydrogen bonds to the silicate in the glass and is, therefore, attracted most strongly to the glass and the capillary action is strongest. Ethanol forms fewer hydrogen bonds to the silicate and is less strongly attracted. Acetone does not form hydrogen bonds and adheres least to the glass.

10.45 MM = mRT/ΠV (Equation 10-5)

$$MM = \frac{(1.00\text{ g})(0.08206\text{ L atm / mol K})(298\text{ K})}{(1.36\text{ atm})(0.100\text{ L})} = 180\text{ g / mol}$$

10.46 a) metallic b) ionic c) molecular

10.47 C_6H_6 < $C_5H_{11}OH$ < $HOCH_2CH(OH)CH(OH)CH_2OH$
benzene pentanol erythritol
Solubility in water will depend on the ability of the compound to form hydrogen bonds with H_2O. Benzene cannot form any hydrogen bonds with H_2O and is least soluble. Pentanol has only one OH group that can hydrogen bond with H_2O and a 5 carbon chain which would make it only slightly soluble in H_2O. Erythritol has 4 OH groups spread over a 4 carbon chain that allows it to form many hydrogen bonds with many water molecules and be quite soluble in water.

10.48 $\Delta T_f = K_f X_{ethanol}$ (Equation 10-2) $K_f = 103.2°C$

$$X_{ethanol} = \frac{\Delta T_f}{K_f} = \frac{20°C}{103.2°C} = 0.19$$

10.49 Glucose is in the form of molecules in solution. Therefore, a 1 M solution has 1 mole of molecules per liter and a 0.5 M solution has 0.5 moles of molecules per liter. In water, acetic acid is a weak acid that ionizes to a small degree; therefore, a 0.5 M acetic acid solution will have slightly more than 0.5 mole of particles (acetic acid molecules, hydronium ions and acetate ions) and the freezing point will be lower than 0.5 M glucose solution but higher than the freezing point of 1 M glucose (because the acetic acid does not completely ionize). A similar explanation holds for $MgSO_4$ which forms Mg^{2+} ions and SO_4^{2-} ions in aqueous solution. A 0.5 M solution of $MgSO_4$ should be quite close to having 1 mole of particles (Mg^{2+} ions and SO_4^{2-} ions). Because the freezing point is higher than 1 M glucose, there must be some Mg^{2+} - SO_4^{2-} ion pairs which are not separated in solution.

10.50 $\Pi = MRT$ $\quad M = \frac{\Pi}{RT} = \frac{3.03 \text{ atm}}{(0.08206 \text{ L atm / mol K})(300 \text{ K})} = 0.123 \text{ M}$

On this basis H_3PO_4 would be considered a weak acid.

It is only $\left(\frac{0.123 - 0.10}{0.10} \times 100\%\right)$ = about 20% ionized.

10.51 It is not practical to purify NaCl by recrystallization from water because upon cooling the solution from 100°C to 0°C only 2 g of every 28 g (~7 %) are recovered.

10.52 The boiling point of H_2S is lower than the boiling point of H_2O because the intermolecular forces in H_2S are dispersion and dipole-dipole forces while H_2O has the much stronger intermolecular forces of hydrogen bonding that require a higher temperature to overcome. The intermolecular forces in H_2Te are also dispersion and dipole-dipole forces. The dipole-dipole forces are about the same for H_2S and H_2Te, but the dispersion forces of H_2Te are stronger than for H_2S because of the greater polarizability of the electrons on the larger Te atom.

10.53 pentane < butanol < propane-1,3-diol

C_5H_{12} $\quad C_4H_9OH$ $\quad HOCH_2CH_2CH_2OH$

The stronger the intermolecular forces between the molecules of a liquid, the more viscous the liquid. Pentane is least viscous because its intermolecular forces are only weak dispersion forces. Butanol is more viscous because it can hydrogen bond between molecules using the OH group (but only 1 OH per 4 C atoms). The propane-1,3-diol is most viscous because it also forms hydrogen bonds but many more than the butanol (2 OH per 3 C atoms).

10.54 $KCl < C_2H_5OH < C_3H_8$ In order for a solute to be soluble in a solvent, the solute-solute intermolecular forces must be similar to the solvent-solvent intermolecular forces so that solute-solvent intermolecular forces will be similar and there will not be any large energy changes accompanying solution formation. The intermolecular forces for the solvent, cyclohexane, are dispersion forces. The intermolecular forces for C_3H_8 are also dispersion forces and, therefore, C_3H_8 would be quite soluble in cyclohexane. The C_2H_5OH has dispersion forces and dipole forces between molecules, but the most important force is the somewhat stronger hydrogen bonding. Breaking these hydrogen bonds of the solute and replacing them with dispersion forces between solute-solvent is not energetically favorable and, therefore, C_2H_5OH is not very soluble in cyclohexane. The intermolecular forces in KCl are ionic and much stronger than dispersion forces. The only intermolecular forces possible between ions from KCl and cyclohexane are much too weak compared to ionic forces. Therefore, KCl is not soluble in cyclohexane.

10.55 a) slightly greater than 0.1 M, a small amount of ionization

b) slightly less than 0.4 M, $FeCl_{3(aq)} \rightarrow Fe^{3+}_{(aq)} + 3Cl^-_{(aq)}$

c) approximately 0.2 M, $NaOH_{(aq)} \rightarrow Na^+_{(aq)} + OH^-_{(aq)}$

d) approximately 0.3 M, $(NH_4)_2CO_{3(aq)} \rightarrow 2NH_4^+{}_{(aq)} + CO_3^{2-}{}_{(aq)}$

10.56 100 g brackish water consists of 99.5 g H_2O and 0.5 g NaCl

$$0.5 \text{ g NaCl} = \frac{0.5 \text{ g NaCl}}{58.44 \text{ g / mol}} = 8.56 \times 10^{-3} \text{ mol NaCl}$$

mol of ions = 1.7×10^{-2} mol Na^+ and Cl^-

$$(99.5 \text{ g } H_2O)\left(\frac{1.000 \text{ mL}}{1.000 \text{ g}}\right)\left(\frac{1 \text{ L}}{1000 \text{ mL}}\right) = 0.0995 \text{ L}$$

$$\Pi = \left(\frac{1.7 \times 10^{-2} \text{ mol}}{0.0995 \text{ L}}\right)(0.08206 \text{ L atm / mol K})(298 \text{ K}) \quad \text{assuming } 25°C$$

$\Pi = 4$ atm

10.57 0.050 M $KHSO_3$ 0.050 mol $KHSO_3$ per L of solution
Assuming that the density of the solution is 1.00 g/mL, 1 L of solution contains 6g $KHSO_3$ and 994 g H_2O.

$$994 \text{ g } H_2O = \frac{994 \text{ g } H_2O}{18.02 \text{ g / mol}} = 55.2 \text{ mol } H_2O$$

$$0.19°C = i(103.2°C)\left(\frac{0.050 \text{ mol}}{0.050 \text{ mol} + 55.2 \text{ mol}}\right)$$

$0.19°C = i\,(103.2°C)(0.000905) \quad i = 2.03$

$KHSO_{3(s)} \rightarrow K^+_{(aq)} + HSO_3^-{}_{(aq)}$

As can be seen from the calculation, the freezing point depression is about twice what it would be if the $KHSO_3$ did not dissociate. Therefore, it must dissociate into 2 particles (ions) as in equation (b).

10.58 One would expect BHT to be stored in body fat. Although it has an OH group to hydrogen bond with water, it has a large hydrocarbon structure which would make it slightly soluble in water but very soluble in fat.

10.59 Water has intermolecular forces of hydrogen bonding and dipolar forces. Carbon tetrachloride and iodine both have dispersion forces as their intermolecular forces. As a result of this, iodine is more soluble in CCl_4 than in H_2O. When a mixture of H_2O and CCl_4 is shaken and iodine is present, the iodine will collect in the solvent in which it is more soluble: carbon tetrachloride.

10.60 For Groups IV, V and VI the graph of boiling points of the hydrides versus period of the central element is linear within each group for the elements in periods 3, 4 and 5. For Group IV the non-hydrogen bonding hydride from period 2 (CH_4) also lies on the line. For Groups V, VI and VII the hydrogen bonding hydrides from period 2 (NH_3, H_2O and HF) lie well above the line for each of their groups. For Group VII the boiling point for HCl lies somewhat above the line through the boiling points of HBr and HI. This suggests that some hydrogen bonding occurs for HCl.
The molecular picture requested in this question could look like the hydrogen bonding between the HF molecules shown in Figure 10-15 except the F would be replaced by Cl.

10.61 3.0×10^{-3} M glucose < 4.0×10^{-3} M glucose < 3.0×10^{-3} M KBr

$\Pi = MRT$; therefore, $\Pi \propto M$ (or concentration of particles.)

Glucose is present in aqueous solution as molecules. Therefore, 4.0×10^{-3} M glucose has a higher osmotic pressure than 3.0×10^{-3} M glucose because it has a third more molecules in solution (4.0×10^{-3} mol/L versus 3.0×10^{-3} mol/L). The 3.0×10^{-3} M KBr has the highest osmotic pressure because in aqueous solution the KBr dissociates into K^+ ions and Br^- ions. Therefore, the concentration of particles for KBr is approximately 6.0×10^{-3} mol of particles/L.

10.62

molecular crystals	iodine, sulfur	right side of periodic table (Groups V, VI and VII)
metallic crystals	sodium, potassium	left side of periodic table (Groups I, II and transition metals)
covalent crystals	carbon (diamond), silicon	between molecular and metallic (part of Group IV)

10.63 When a fish is placed in a strong salt solution, water is removed from the cells of the fish by osmosis. The fish does not spoil as quickly in this dehydrated condition.

10.64 No, water would not dissolve salts as well if it were linear like CO_2. Instead of having positive and negative ends, it would have both ends positive with negative polarity in the middle. It might efficiently solvate anions but its geometry would make it very difficult to efficiently solvate cations.

10.65

Compound	**Intermolecular Forces**
a) NH_3	dispersion, dipolar, hydrogen bonding
b) Xe	dispersion
c) SF_4	dispersion, dipolar
d) CF_4	dispersion
e) CH_3CO_2H	dispersion, dipolar, hydrogen bonding

10.66 $\Delta T_f = K_f X_{solute}$

$2.30°C = (103.2°C)X_{solute}$

$$X_{solute} = \frac{2.30°C}{103.2°C} = 0.02229$$

$$X_{solute} = 0.02229 = \frac{0.02229 \text{ mol salt ions}}{0.02229 \text{ mol salt ions} + (1.00000 \text{ mol} - 0.02229 \text{ mol})\ H_2O}$$

$$= \frac{0.02229 \text{ mol salt ions}}{0.02229 \text{ mol salt ions} + 0.97771 \text{ mol } H_2O}$$

$$(0.97771 \text{ mol } H_2O)(18.02 \text{ g/mol})\left(\frac{1.000 \text{ mL}}{1.000 \text{ g}}\right)\left(\frac{1 \text{ L}}{1000 \text{ mL}}\right) = 0.0176 \text{ L } H_2O$$

(Assuming volume change from -2.13°C to 15°C is negligible.)

$\Pi = MRT$

$$= \left(\frac{0.02229 \text{ mol salt ions}}{0.0176 \text{ L solution}}\right)(0.08206 \text{ L atm/mol K})(288 \text{ K}) = 29.9 \text{ atm}$$

10.67 1.1 kg NaCl = 1100 g NaCl = (1100 g NaCl)÷(58.44 g/mol) = 18.8 mol NaCl
7.25 kg of ice = 7250 g H_2O = (7250 g H_2O)÷(18.02 g/mol) = 402.3 mol H_2O

$$X_{NaCl} = \left(\frac{18.8 \text{ mol}}{18.8 \text{ mol} + 402.3 \text{ mol}}\right) = 0.0446$$

$\Delta T_f = i\, K_f X_{NaCl} = 2(103.2°C)(0.0446) = 9.2°C$
freezing point = -9.2°C

10.68 Assuming the ionization of vitamin C is negligible in water:

$2.33°C = \Delta T_f = 103.2°C\, X_c \qquad 100 \text{ g } H_2O \text{ x } \frac{1 \text{ mol}}{18.02\text{g}} = 5.549 \text{ mol } H_2O$

$$X_c = \frac{2.33°C}{103.2°C} = 0.02258 = \frac{y}{y + 5.549 \text{ mol}}$$

$(0.02258)(y + 5.549 \text{ mol}) = y \qquad 0.1253 \text{ mol} = y - 0.02258\, y = 0.97742\, y$

$$y = 0.1282 \text{ mol} \qquad MM_c = \frac{22.0 \text{ g}}{0.1282 \text{ mol}} = 172 \text{ g / mol}$$

10.69 a) The CH_3OH has a higher boiling point than CH_3OCH_3 because CH_3OH has hydrogen bonding versus dipolar forces for CH_3OCH_3.

b) The SiO_2 has a higher boiling point than SO_2 because SiO_2 has covalent bonding versus dipolar forces for SO_2.

c) The HF has a higher boiling point than HCl because HF has hydrogen bonding versus dipolar forces (and, perhaps, some weak hydrogen bonding) for HCl.

d) The I_2 has a higher boiling point than Br_2. Although both have dispersion forces as their intermolecular forces, those for I_2 are greater because it is more polarizable due to its larger size.

10.70 The remaining solution will have a lower freezing point because when some of the water is removed from solution as ice, the concentration of solute increases.

$\Delta T_f = K_f X_{ethanol} \qquad 10°C = (103.2°C)X_{ethanol} \qquad X_{ethanol} = \frac{10°C}{103.2°C} = 0.097$

10.71 The intermolecular (interatomic) forces involved here are dispersion forces. Dispersion forces depend on the polarizability of the electrons of the substance. The polarizability is easier for the electron cloud of an extended molecule than for more compact molecules or atoms. Therefore, H_2 is more polarizable than He and has a higher boiling point; methane is more polarizable than Ne and has a higher boiling point.

10.72 $(CH_3)_4C < CH_3CH_2CH_2CH_2CH_3 < CH_3CH_2CH_2CH_2OH < HOCH_2CH_2CH_2OH$
Note that all four compounds are of similar molecular mass (MM = 72-76). The first two compounds have just dispersion forces as their intermolecular forces. The longer chained n-pentane will have stronger dispersion forces than the branched 2,2-dimethylpropane and will also, therefore, be more viscous. The last two compounds have hydrogen bonding as their important intermolecular forces and are, therefore, more viscous than the first two compounds. The propane-1,3-diol forms more hydrogen bonds than the 1-butanol and, therefore, is more viscous.

10.73 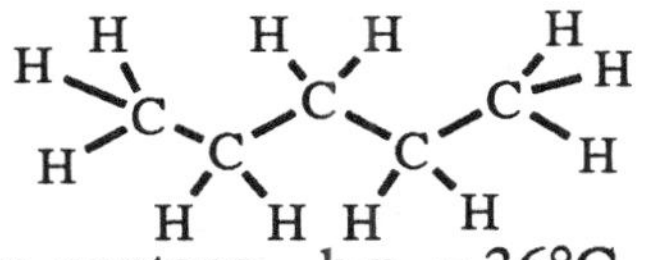 n-pentane b.p. = 36°C

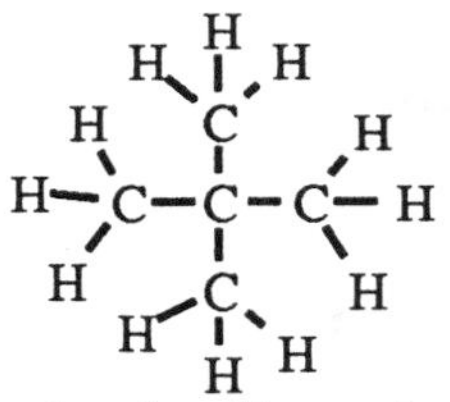

2,2-dimethylpropane b.p. = -10°C

The more compact a molecule is the weaker its dispersion forces are. Therefore, a compact molecule will have a lower boiling point than a molecule of similar mass which is more extended.

10.74

Similarities

hexagonal close-packed	body-centered cubic
ABAB layers	ABAB layers Body-centered and face-centered are similar in that they both are cubic arrays within which atoms are nested.

Differences

hexagonal close packed	body-centered cubic
planar arrangement of sphere within hexagon of six spheres made up usually of atoms or molecules of spherical symmetry	spheres arranged in square planar array
six nearest neighbors of same atom or molecule	made up of 1:1 ionic compounds
empty space minimized	8 nearest neighbors of opposite charged ions maximizes interionic attraction Body-centered and face-centered are different in the locations of the atoms their nearest neighbors.

10.75 Assuming the stationary phase is polar, the most polar dye would move the least distance from the original spot. The dye (black) left at the original spot would be most polar and the purple at the far right the least polar: black > yellow > blue green > orange > purple.

10.76 Refer to Figure 10-20. The regular arrangement of atoms in quartz allows it to be broken into smaller regular crystals of the same structure. The glass is made up of an irregular arrangement of atoms and when it is broken into smaller pieces, they must be irregular also.

10.77 a) Cl_2 would deviate more from ideality than F_2 because it is larger and its dispersion forces are stronger.
b) SnH_4 would deviate more from ideality than CH_4 because it is larger and its dispersion forces are stronger.

10.78 Drinking sea water which is more concentrated in salts than body fluids would result in cells losing water and shrinking. The cell walls are semi-permeable membranes. Water can pass through them but most solutes cannot. Therefore, more water molecules pass out of the cells than pass into the cells and the cell fluid volume decreases causing them to shrink.

10.79 # atoms Ti = 1/1(1 in center) = 1, # atoms Ca = 1/8(8 at corners) = 1,
atoms O = 1/2(6 face centers) = 3, $CaTiO_3$

10.80 Both molecules can form hydrogen bonds. Methyl-2-hydroxybenzoate is likely to form intramolecular hydrogen bonds allowing the compound to melt without breaking hydrogen bonds. Since methyl-4-hydroxybenzoate can only form intermolecular hydrogen bonds, that means hydrogen bonds must be broken in order to melt the compound. This results in a higher melting point.

10.81

Cl Cl
C=C
H H
cis
b.p. = 60°C

Cl H
C=C
H Cl
trans
b.p. = 47°C

Each has two polar C—Cl bonds. Because of the symmetry, the trans molecule is not a polar molecule. The cis form has a dipole moment and the dipole-dipole interactions between the molecules result in the cis form having a higher boiling point.

10.82 In order for a compound to form hydrogen bonds with water it must have a hydrogen atom bonded to a fluorine, oxygen or nitrogen and/or nonbonding electrons on a highly electronegative atom such as F, O or N.

a) CH_4 cannot form hydrogen bonds
b) I_2 cannot form hydrogen bonds
c) HF can form hydrogen bonds with water, has both H bonded to F and nonbonding electrons on F
d) CH_3—O—CH_3 can form hydrogen bonds with water, has nonbonding electrons on O
e) $(CH_3)_3COH$ can form hydrogen bonds with water, has nonbonding electrons on O and H bonded to O

10.83

Compound	Forces to overcome to convert from liquid to gas
a) NH_3	dispersion, dipolar and hydrogen bonding
b) $CHCl_3$	dispersion and dipolar
c) CCl_4	dispersion
d) CO_2	dispersion

10.84

Compound	Type of Crystal
a) HCl	molecular
b) KCl	ionic
c) NH_4NO_3	ionic
d) Mn	metallic
e) Si	covalent

CHAPTER 11: MACROMOLECULES

11.1

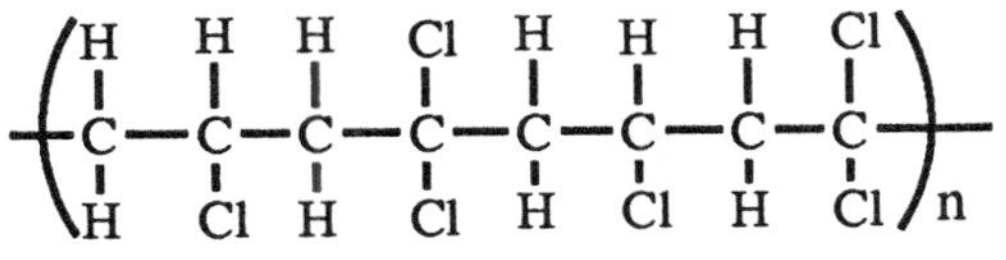

$$\left(CH_2-CHCl-CH_2-CCl_2-CH_2-CHCl-CH_2-CCl_2 \right)_n$$

11.2 acrylonitrile

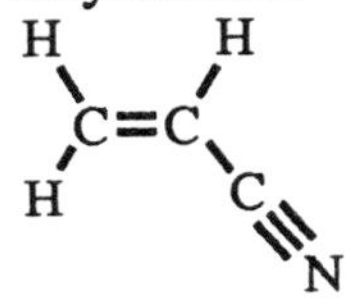

$$H_2C=CH-C\equiv N$$

butadiene

$$H_2C=CH-CH=CH_2$$

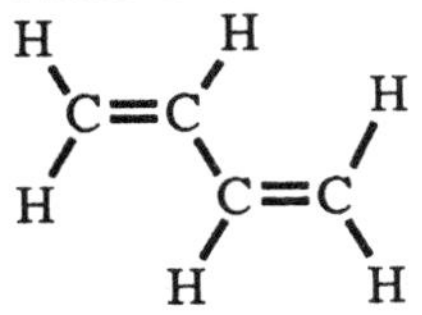

11.3 Initiation: Init.—Init. → 2 Init.•

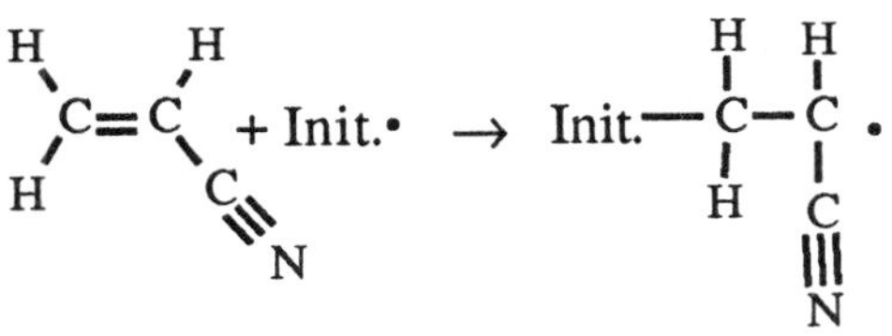

$$H_2C=CH(C\equiv N) + \text{Init.}\bullet \rightarrow \text{Init.}-CH_2-CH(C\equiv N)\bullet$$

Propagation:

$$\text{Init.}-CH_2-CH(C\equiv N)\bullet + H_2C=CH(C\equiv N) \rightarrow \text{Init.}-CH_2-CH(C\equiv N)-CH_2-CH(C\equiv N)\bullet$$

$$\text{Init.}\left(CH_2-CH(C\equiv N)\right)_n-CH_2-CH(C\equiv N)\bullet + H_2C=CH(C\equiv N) \rightarrow \text{Init.}\left(CH_2-CH(C\equiv N)\right)_{n+1}-CH_2-CH(C\equiv N)\bullet$$

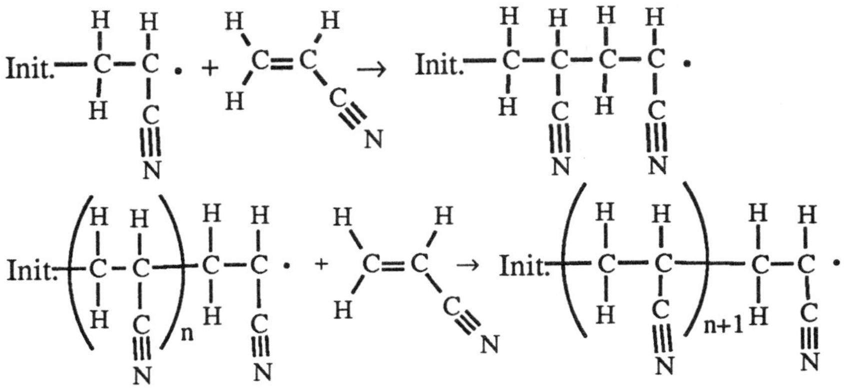

Termination:

$$\text{Init.}\left(CH_2-CH(C\equiv N)\right)_n-CH_2-CH(C\equiv N)\bullet + \text{Init.}\bullet \rightarrow \text{Init.}\left(CH_2-CH(C\equiv N)\right)_n-CH_2-CH(C\equiv N)-\text{Init.}$$

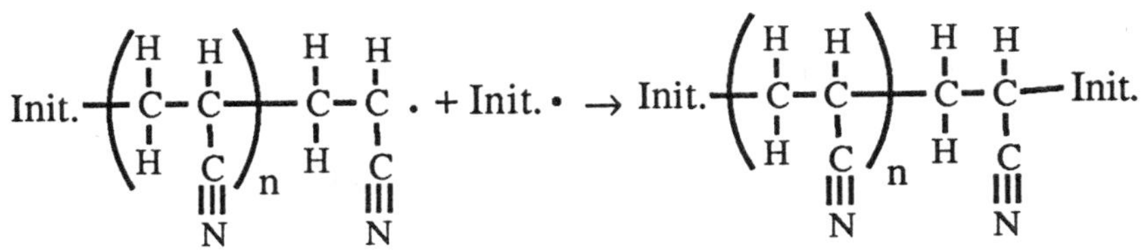

11.4 a) Teflon:

$$-\mathrm{CF_2}-\mathrm{CF_2}-\mathrm{CF_2}-\mathrm{CF_2}-\left(\mathrm{CF_2}-\mathrm{CF_2}\right)_n-\mathrm{CF_2}-\mathrm{CF_2}-\mathrm{CF_2}-\mathrm{CF_2}-$$

b) PVC:

$$-\mathrm{CH_2}-\mathrm{CHCl}-\mathrm{CH_2}-\mathrm{CHCl}-\left(\mathrm{CH_2}-\mathrm{CHCl}\right)_n-\mathrm{CH_2}-\mathrm{CHCl}-\mathrm{CH_2}-\mathrm{CHCl}-$$

c) Styrofoam:

$$-\mathrm{CH_2}-\mathrm{CH(C_6H_5)}-\mathrm{CH_2}-\mathrm{CH(C_6H_5)}-\left(\mathrm{CH_2}-\mathrm{CH(C_6H_5)}\right)_n-\mathrm{CH_2}-\mathrm{CH(C_6H_5)}-\mathrm{CH_2}-\mathrm{CH(C_6H_5)}-$$

11.5 a) $H_2C{=}CH(NO_2)$ nitroethene or nitroethylene

b) $F_2C{=}CFCl$ chlorotrifluoroethene or chlorotrifluoroethylene

11.6 a) alkene, carboxylic acid (O, OH)

b) ketone, amine (NH_2, O)

c) nitrile ($C{\equiv}N$)

11.7 a) ester

b) amine

c) aldehyde

d) alkene (also bromo or bromide)

d) phenol (and phenyl)

11.8 a) 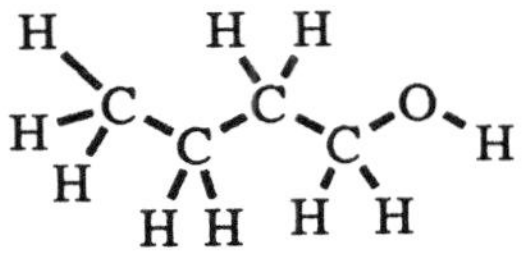 Other isomeric alcohols with the formula $C_4H_{10}O$ are also possible.

b) 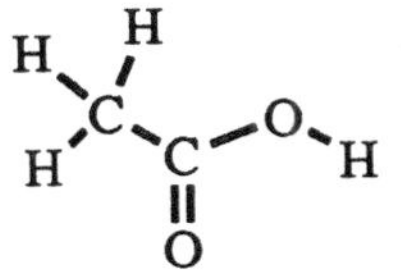 Larger carboxlic acids are also possible answers for this question.

c) 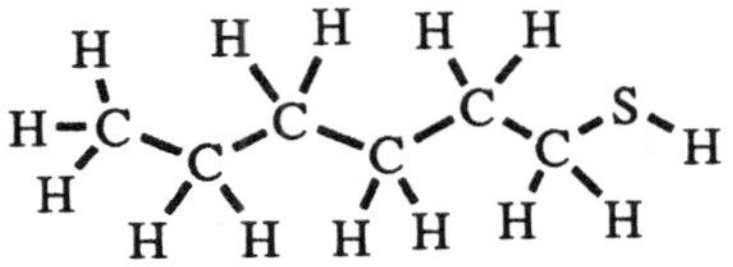 Other isomeric thiols with six carbon atoms are also possible.

d) 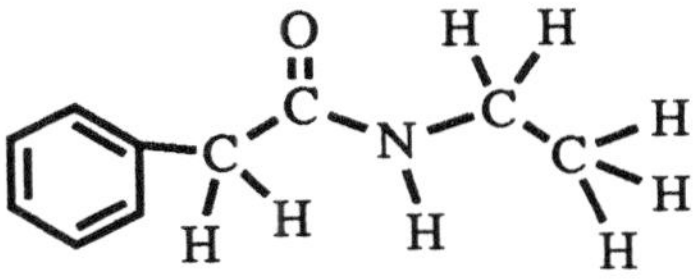 Other isomers and larger molecules are also possible.

11.9 a) 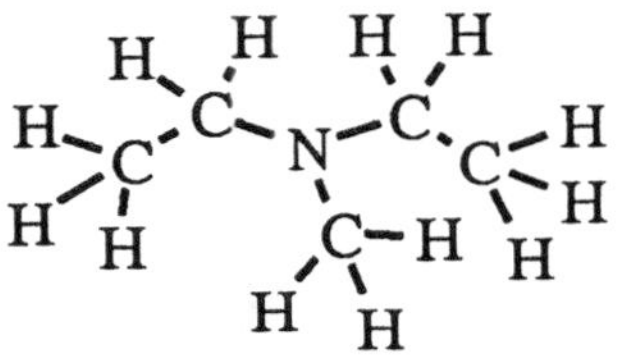 Other isomers are also possible.

b) 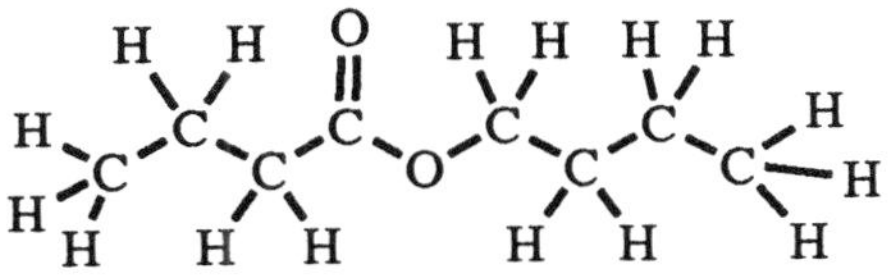 Other isomers are also possible.

c) 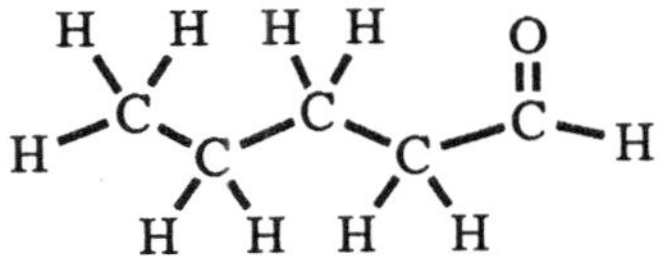 Others are also possible.

d) 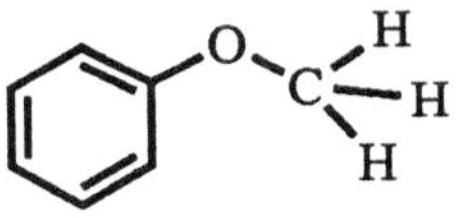 Others are also possible.

11.10 a) 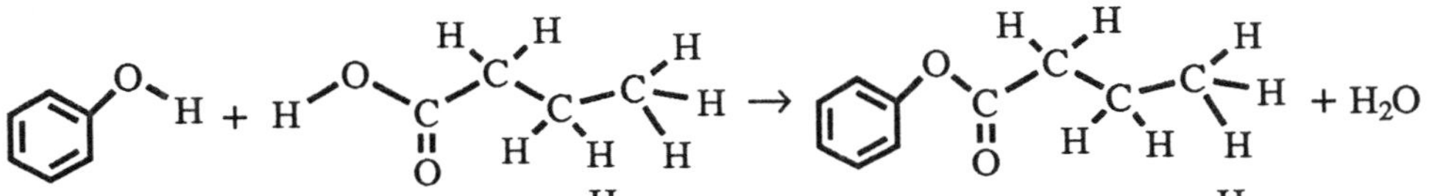 → + H_2O

b) 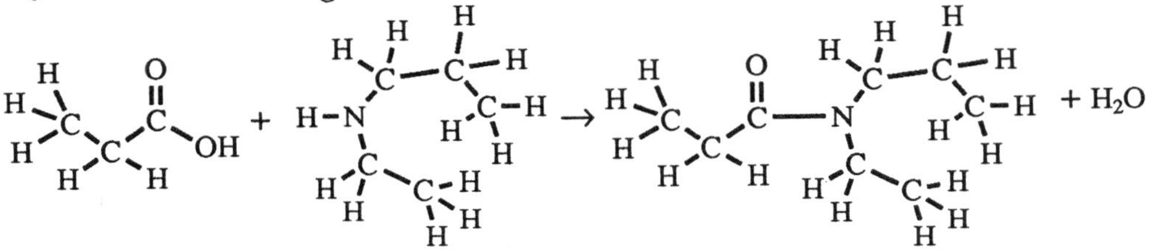 + → + H_2O

(continued)

(11.10 continued)

c) H–O–P(=O)(–O–H)–O–H + 3 H–C(H)(H)–O–H → H–C(H)(H)–O–P(=O)(–O–C(H)(H)–H)–O–C(H)(H)–H + 3 H_2O

11.11 a) (propyl)N(H)(H) + HO–C(=O)–(phenyl) → (propyl)N(H)–C(=O)–(phenyl) + H_2O

b) (methyl)OH + HO–C(=O)–CH_2–(cyclopentyl) → (methyl)O–C(=O)–CH_2–(cyclopentyl) + H_2O

c) 2 (butyl)OH → (butyl)O(butyl) + H_2O

11.12 a) (ethyl)C(=O)–O–(phenyl) + H_2O

b) (cyclopentyl)C(=O)–N(H)–(pentyl) + H_2O

11.13 H–N(H)–$(CH_2)_{10}$–C(=O)–OH

11.14 $-O-\overset{O}{\overset{\|}{C}}-C_6H_4-\overset{O}{\overset{\|}{C}}-O-CH_2-C_6H_{10}-CH_2-O-\overset{O}{\overset{\|}{C}}-C_6H_4-\overset{O}{\overset{\|}{C}}-O-CH_2-C_6H_{10}-CH_2-O-C(=O)-C_6H_4-C(=O)-O-CH_2-C_6H_{10}-CH_2-O-\overset{O}{\overset{\|}{C}}-C_6H_4-\overset{O}{\overset{\|}{C}}-O-CH_2-C_6H_{10}-CH_2-$

11.15 $-CH_2-CH_2-O-CH_2-CH_2-O-CH_2-CH_2-O-$ or $-\!\!\left(CH_2-CH_2-O\right)\!\!_n-$

11.16

This can be extended by further hydrogen bonding with other strands. By rotating some units, more hydrogen bonds can form between these 2 strands.

11.17 The rubber in an automobile tire is more extensively cross-linked than the rubber in surgical gloves. Rubber for tires needs to have greater strength and not be very flexible compared to the rubber in surgical gloves. Cross-linking decreases the flexibility of rubber and increases its strength.

11.18 Plastic wrap is a thermoplastic polymer. It softens (melts) upon heating. It is held together only by dispersion forces. Upon heating, the individual molecules acquire enough kinetic energy to overcome the dispersion forces and it melts. It lacks cross-linking. It has enough branches off the main backbone of the polymer to prevent maximization of dispersion forces resulting in a flexible film that can be stretched slightly.

11.19

item	category of polymer
a) balloon	elastomer
b) rope	fiber
c) camera case	plastic

11.20 Cross-linking decreases the flexibility and increases the strength of a polymer.

11.21 Dioctylphthalate is a common plasticizer that increases the flexibility of polymers when it is added to them.

11.22 a) Tyr-Tyrosine

b) Phe-Phenylalanine

c) Glu-Glutamic acid

(continued)

(11.22 continued)

d) His-Histidine

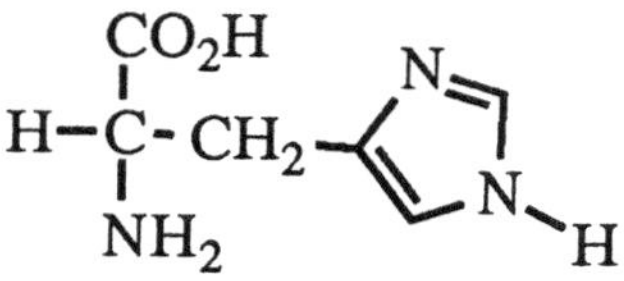

$$H-\underset{NH_2}{\overset{CO_2H}{C}}-CH_2-(\text{imidazole ring, N-H})$$

e) Leu-Leucine

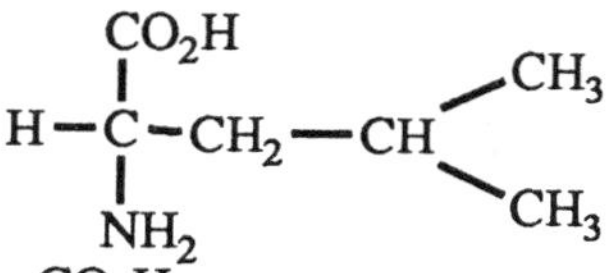

$$H-\underset{NH_2}{\overset{CO_2H}{C}}-CH_2-CH(CH_3)_2$$

f) Pro-Proline

$$H-\overset{CO_2H}{C}(-CH_2-CH_2-CH_2-NH-)$$

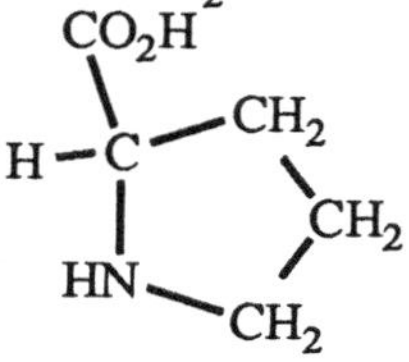

11.23 Assigning each amino acid as hydrophobic or hydrophilic on the basis of its side chain:

a) Tyr - would be hydrophilic because of the hydrophilic side chain.
b) Phe - would be hydrophobic because of the hydrophobic side chain (ring).
c) Glu - would be hydrophilic because of the hydrophilic side chain.
d) His - would be hydrophilic because of the hydrophilic side chain (ring).
e) Leu - would be hydrophobic because of the hydrophobic side chain.
f) Pro - would be hydrophobic because of the hydrophobic side chain (ring).

11.24 a)

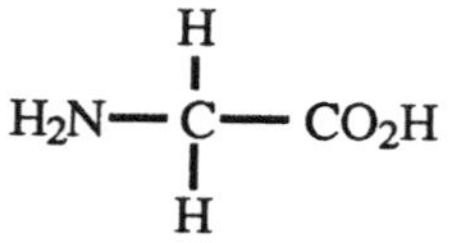

$$H_2N-\underset{H}{\overset{H}{C}}-CO_2H$$ glycine, hydrophobic

b)

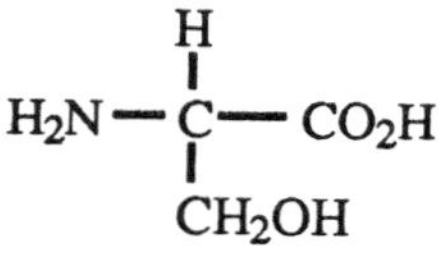

$$H_2N-\underset{CH_2OH}{\overset{H}{C}}-CO_2H$$ serine, hydrophilic

c)

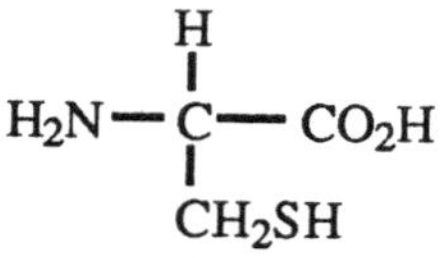

$$H_2N-\underset{CH_2SH}{\overset{H}{C}}-CO_2H$$ cysteine, hydrophilic

d)

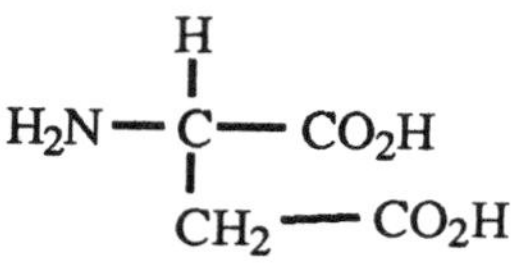

$$H_2N-\underset{CH_2-CO_2H}{\overset{H}{C}}-CO_2H$$ aspartic acid, hydrophilic

e)

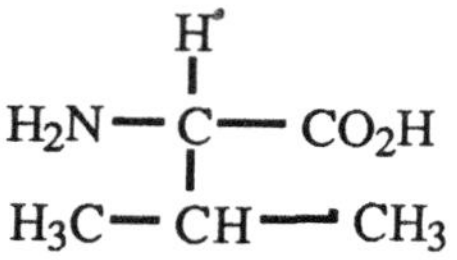

$$H_2N-\underset{H_3C-CH-CH_3}{\overset{H}{C}}-CO_2H$$ valine, hydrophobic

11.25

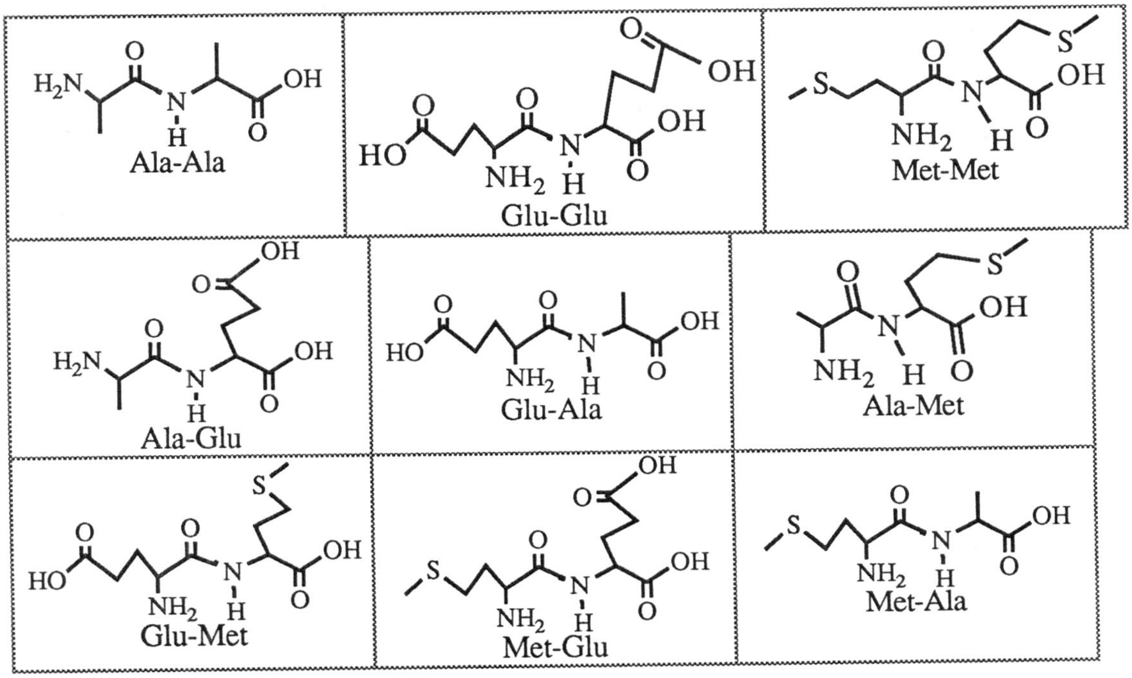

Ala-Ala	Glu-Glu	Met-Met
Ala-Glu	Glu-Ala	Ala-Met
Glu-Met	Met-Glu	Met-Ala

11.26

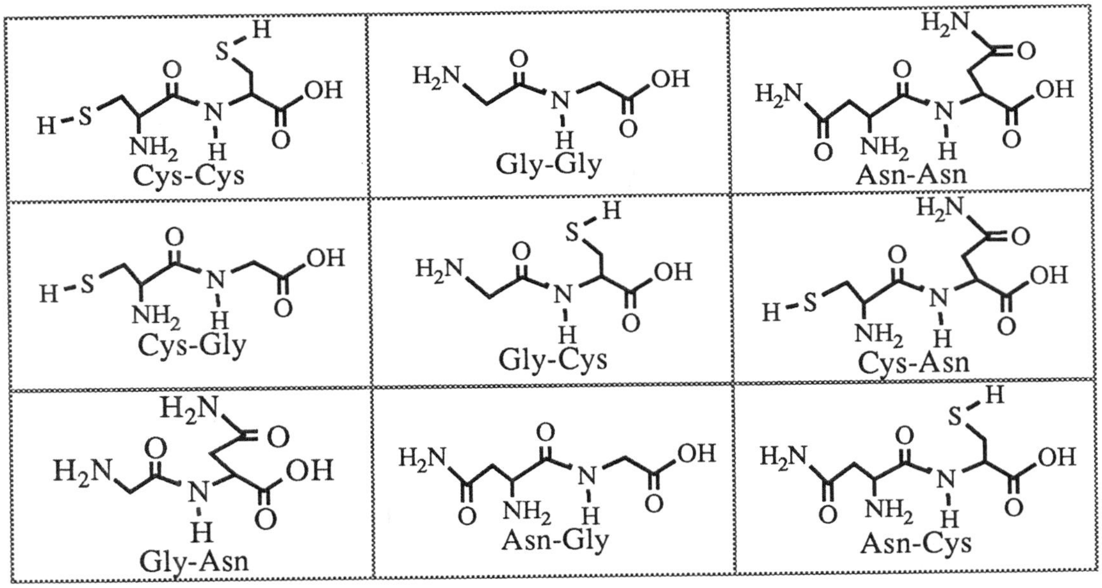

Cys-Cys	Gly-Gly	Asn-Asn
Cys-Gly	Gly-Cys	Cys-Asn
Gly-Asn	Asn-Gly	Asn-Cys

11.27 $\left(-N(H)-C(H)(CH_3)-C(=O)- \right)_n$ C_3H_5NO MM = 71.08 g/empirical unit

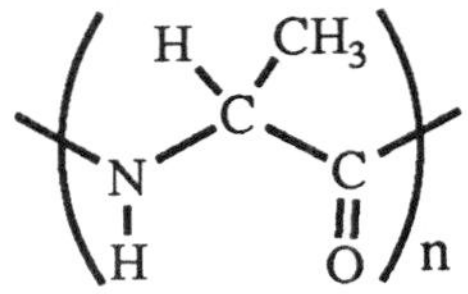

$$\frac{1.20 \times 10^3 \text{ g / mol}}{71.08 \text{ g / unit}} = 16.9 \approx 17 \text{ units / mol or 17 amino acids}$$

An alternate way to obtain the same answer is:
MM of polypeptide - mass of end H - mass of end OH =

$1.20 \times 10^3 - 1 - 17 = 1.18 \times 10^3$ then $\frac{1.18 \times 10^3}{71} = 16.6 = 17$ units or amino acids

11.28 Aspartic acid and phenylalanine are the two amino acids in aspartame.

11.29

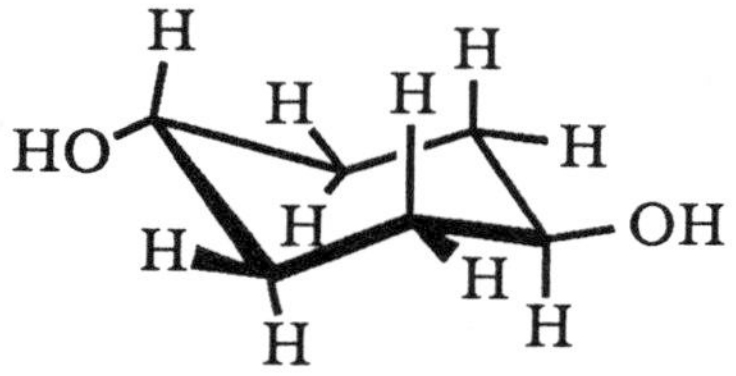

H H H H HO H H OH H H H H

11.30

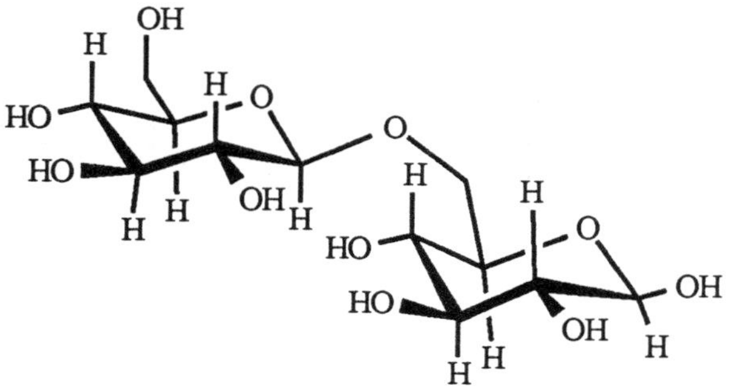

OH H H O HO O HO H H H OH H H H O HO OH HO OH H H H

11.31

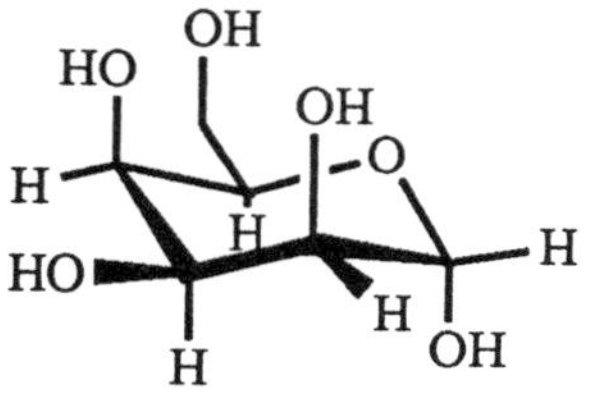

HO OH OH O H H HO H H H OH

α-talose

HO OH OH O H H HO OH H H H

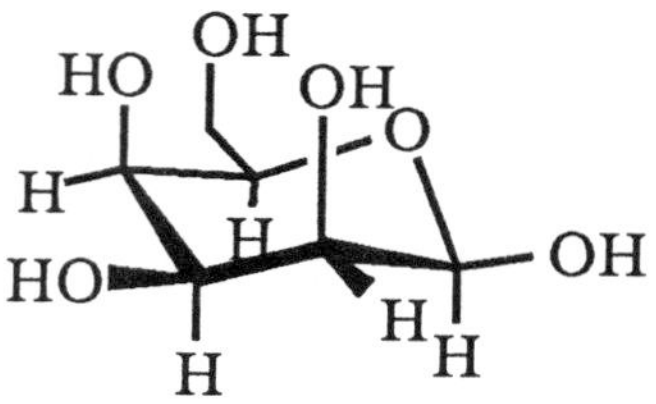

β-talose

11.32 The amylase enzymes present in humans can cleave the α-glucose linkages of glycogen to form α-glucose molecules which are metabolized for energy production. However, the amylase enzymes cannot cleave the β-glucose linkages of cellulose because the geometry of the glycosidic linkages are different. Cows, other ruminants and termites have bacteria in their digestive tracts that break down cellulose.

11.33

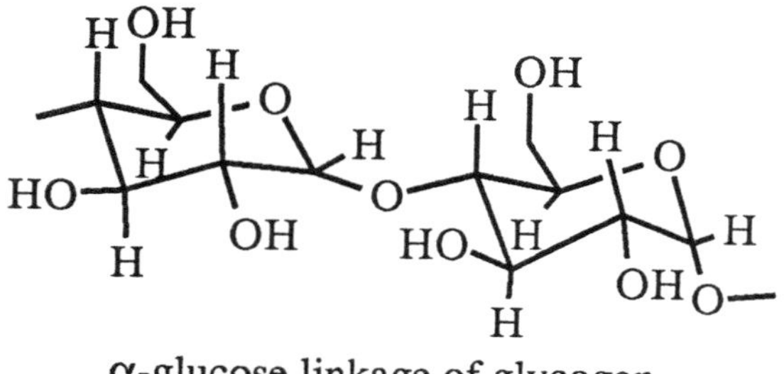

H OH H O H HO H OH O H OH H O HO H H OH O H

α-glucose linkage of glycogen

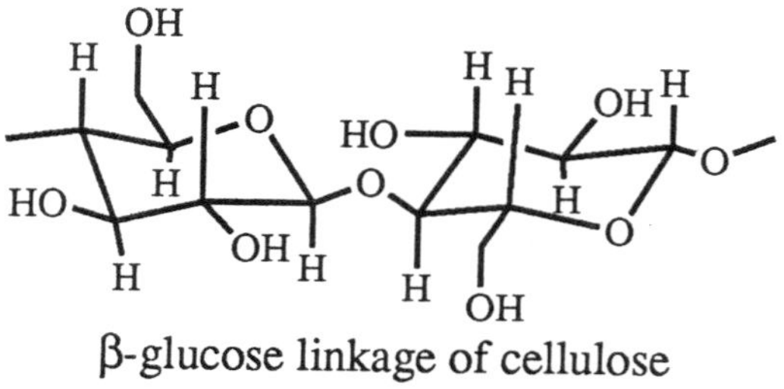

OH H H O HO H OH H H H OH H HO O H O H OH O

β-glucose linkage of cellulose

The glycosidic linkage of glycogen imparts a kink in the structure that can result in the chain coiling upon itself into the granular shape of deposits. The glycosidic linkage of cellulose results in a flat structure giving a long ribbon-like chain of units. The planar arrangement of monomers in cellulose makes it possible for hydrogen bonds to form between chains generating extended packages of ribbons that are similar to a pleated sheet structure.

11.34

	RNA	DNA
Sugar	ribose	deoxyribose
One-ring bases	uracil and cytosine	thymine and cytosine
Two-ring bases	guanine and adenine	guanine and adenine

11.35 DNA sequence: A-A-T-G-C-A-C-T-G
Complementary sequence: T-T-A-C-G-T-G-A-C

11.36 The complete line structure of the segment requested with some carbons being shown via the line-structure notation and hydrogens bonded to carbons not being shown is:

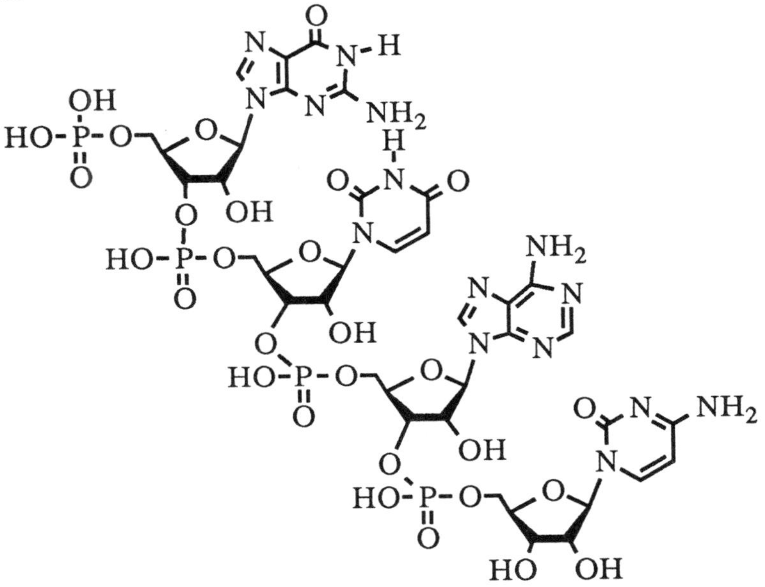

11.37

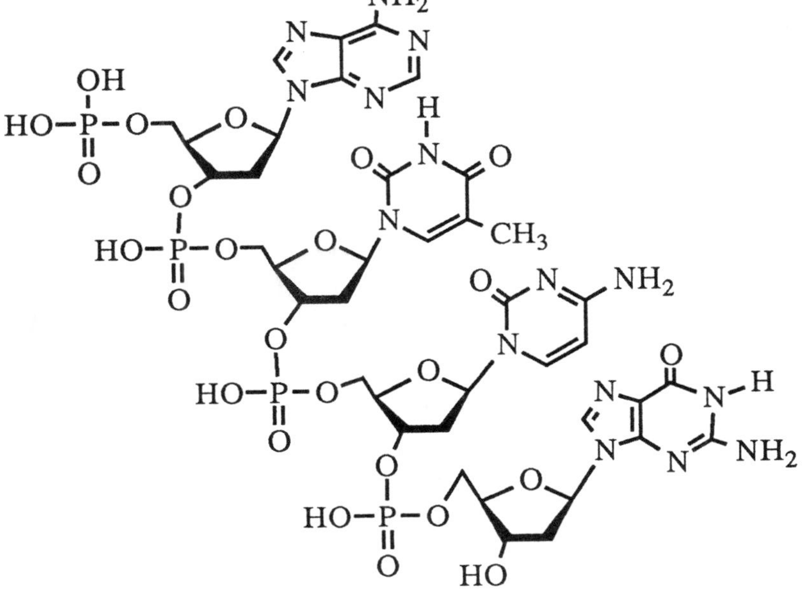

11.38

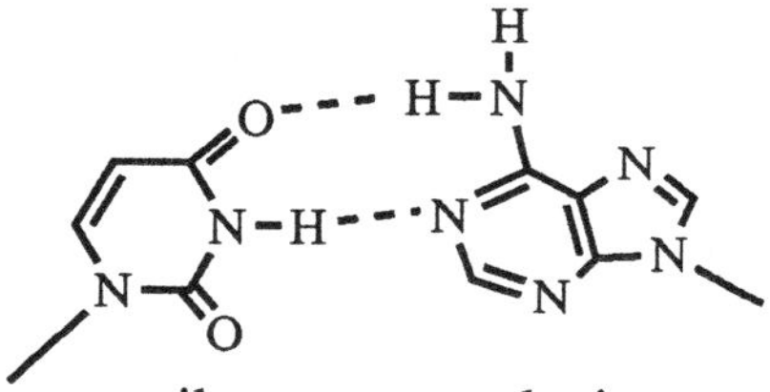

uracil adenine

11.39 DNA sequence: A-T-C Complementary sequence: T-A-G

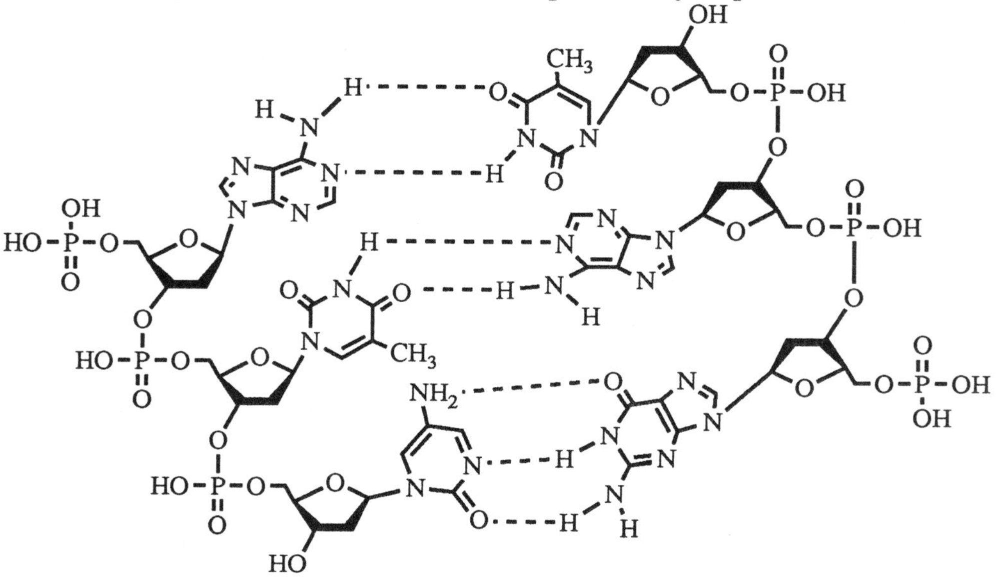

11.40 a) Monomers that contain two identical functional groups are easier and less expensive to produce than monomers with two different groups.
b) The properties of the polymer can be modified by changing the structure of one of the monomers.
c) Bifunctional monomers with different functional groups tend to link into cyclic structures rather than the long chains needed for polymers.

11.41 The nitrogen has sp^2 hybridization in a peptide linkage. By adopting a trigonal planar geometry using sp^2 hybridization, the nitrogen has a p orbital left that may form a delocalized π system with the carbonyl from the acid.

11.42 a) The arrangement would be random, depending on which molecules happened to be in position to form the next bond and therefore the next unit. The probability is determined by the monomer ratio 1:1. But, because of the size and shape of substituents on the ethene monomers (bulky 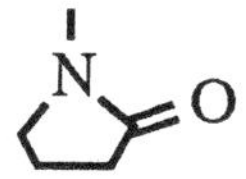 and linear $-O-\underset{\underset{O}{\|}}{C}-CH_3$)

there is a strong likelihood of alternating units because the substituents make it less likely that 2 units of the monomer containing the bulky group would get in position to bond.

(continued)

(11.42 continued)

b) 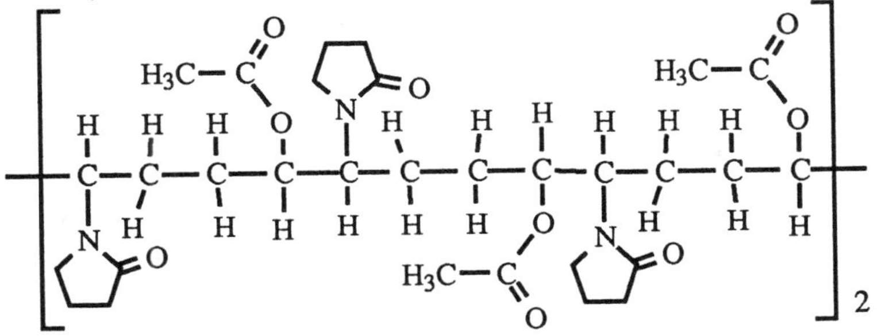 [chemical structure of repeating unit, subscript 2]

c) It forms a thin film and has several sites available for hydrogen bonding by the hair.

11.43 [chemical structure] or [chemical structure]

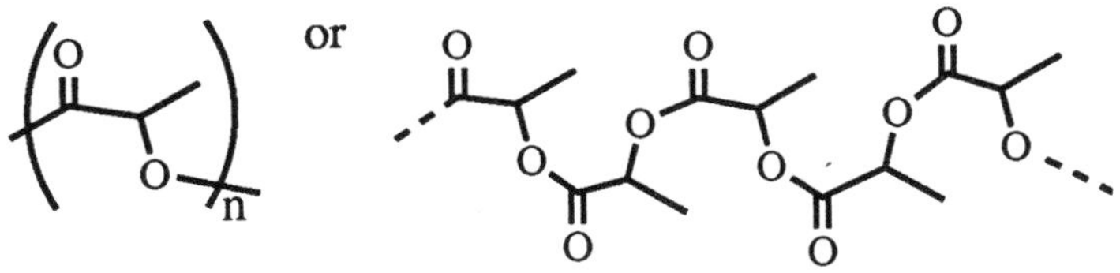

11.44 [chemical structure]

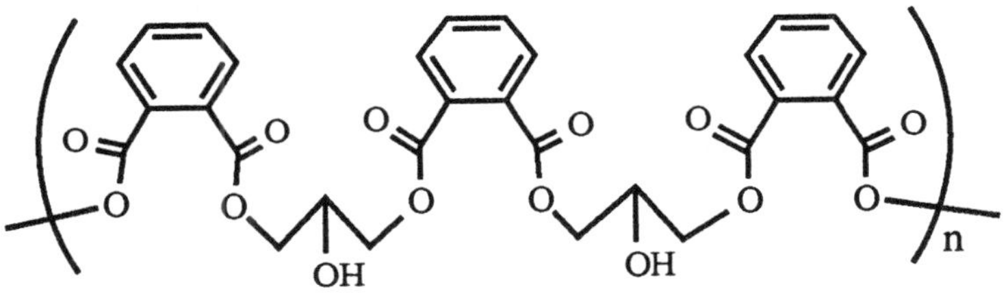

There will undoubtedly be considerable cross-linking with adjacent strands and there will be frequent ring formation.

11.45 [chemical structure] or $\left(C_8H_8\right)_n$ 452 (8 x 12.01 + 8 x 1.008) = 4.71×10^4 g/mol

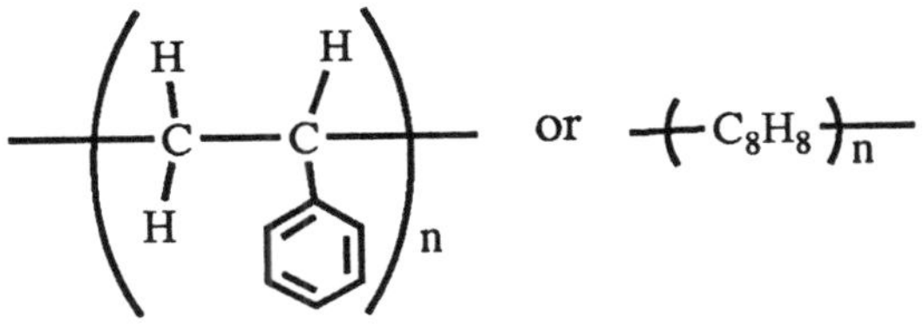

11.46 $20^{50} = 1.1 \times 10^{65} \approx 10^{65}$

11.47 There are 6 molecular building block units in DNA partners, 3 for each strand.
a) Each strand has a pentose sugar which is deoxyribose for DNA.
b) Each strand has a phosphate linkage derived from phosphoric acid.
c) Each strand contains a nitrogen-containing organic base. The possible bases for DNA are: thymine, adenine, guanine and cytosine. These bases are present in certain pairs: guanine with cytosine or thymine with adenine.
This question asks for a molecular picture of DNA partners. The picture in the answer to Question 11.39 is an example of 3 of these DNA partners.

11.48

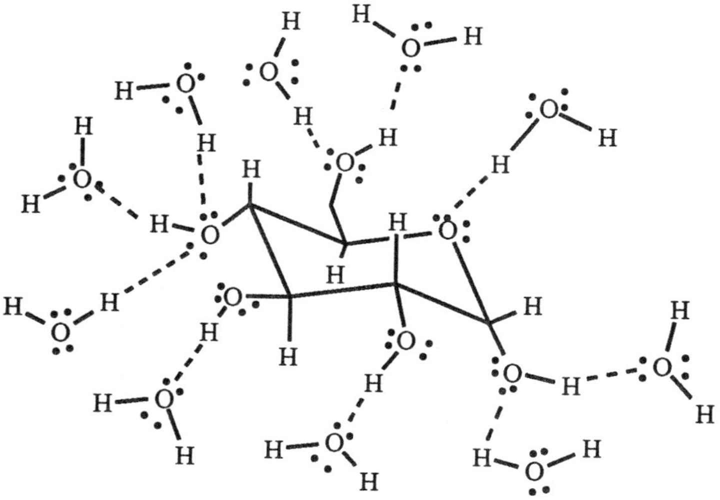

11.49 base 1 = cytosine, base 2 = uracil, base 3 = adenine, base 4 = guanine

11.50 a) Nylon 6 is made from the monomer aminocaproic acid (HO–C(=O)–(CH₂)₅–NH₂)

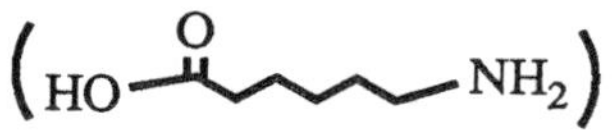

which is formed from caprolactam.

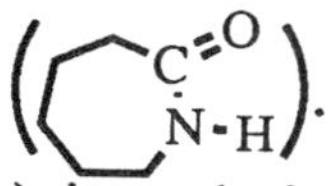

b) PET, poly(ethylene terephthalate), is made from the monomers terephthalic acid (HO–C(=O)–C₆H₄–C(=O)–OH) and ethylene glycol (HO–CH₂CH₂–OH).

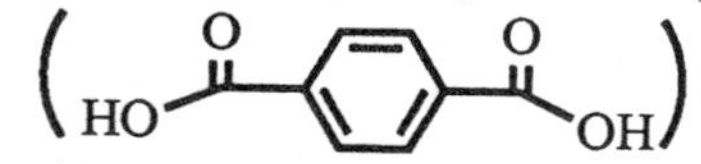

c) Dacron is a fiber form of PET (See part b).

d) PVC, polyvinylchloride, is made from the monomer vinylchloride ($H_2C{=}CHCl$).

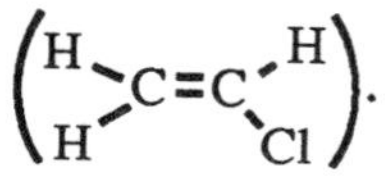

e) Styrofoam is made from the monomer styrene ($H_2C{=}CH{-}C_6H_5$).

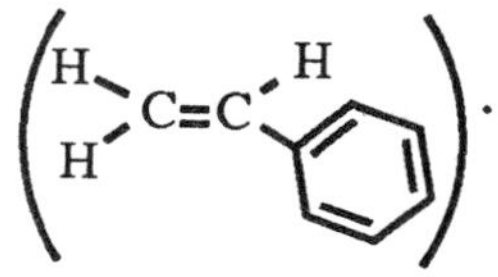

11.51

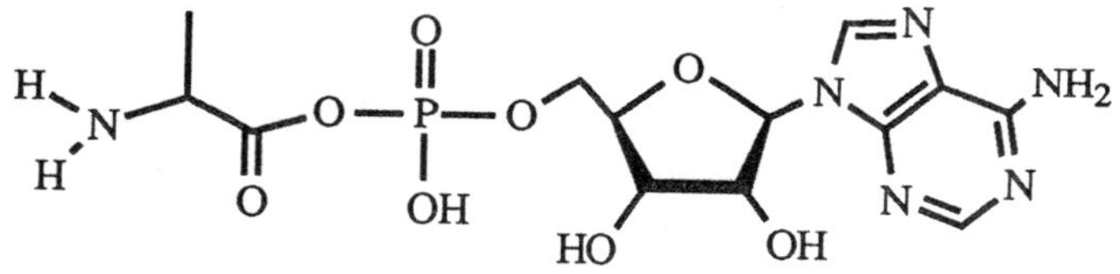

1. Peptidase	c. Hydrolyzes peptide linkages
2. Transmethylase	d. Moves a CH_3 group from one position in a molecule to another
3. Carboxylase	a. Removes carboxyl groups from organic compounds
4. Esterase	b. Catalyzes the formation of esters

11.53 $MM_{enzyme} = 64{,}000$ g/mol
mass of Cu per mole = (64,000 g/mol)(0.0040) = 256 g Cu/mol enzyme

$$(256 \text{ g Cu / mol enzyme})\left(\frac{1 \text{ mol C}}{63.55 \text{ g Cu}}\right) = 4.03 \text{ mol Cu / mol enzyme}$$

Therefore, 4 atoms of Cu are in one molecule of fungal lactase.

11.54 The interactions of the functional groups on the amino acids in the polypeptide chains with water are the most important interactions for the tertiary structure of proteins. The polar interactions with water molecules are maximized and the system is most stabilized when the polypeptide folds into a three-dimensional shape that places its hydrophobic regions inside the overall structure. This folding of the protein is directed and strengthened by a large number of hydrogen bonds. Water forms hydrogen bonds with the protein backbone and with hydrophilic side chains.

11.55

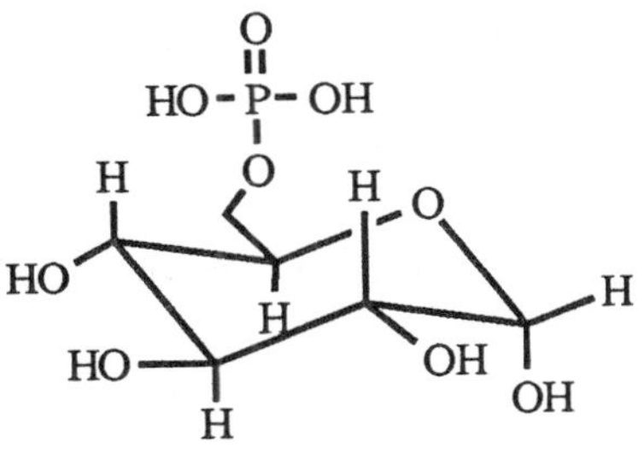

α–glucose phosphate

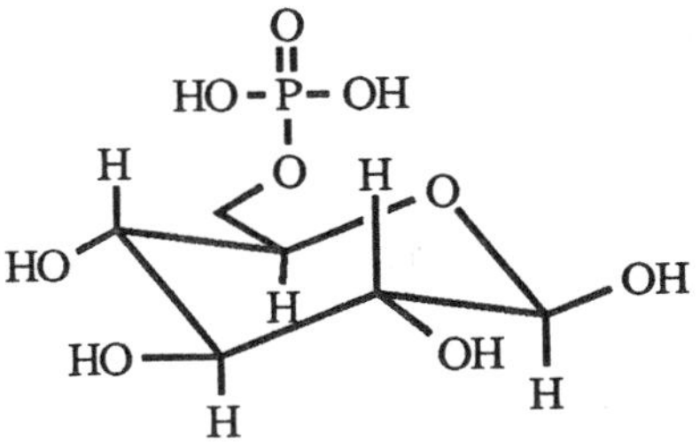

β-glucose phosphate

11.56

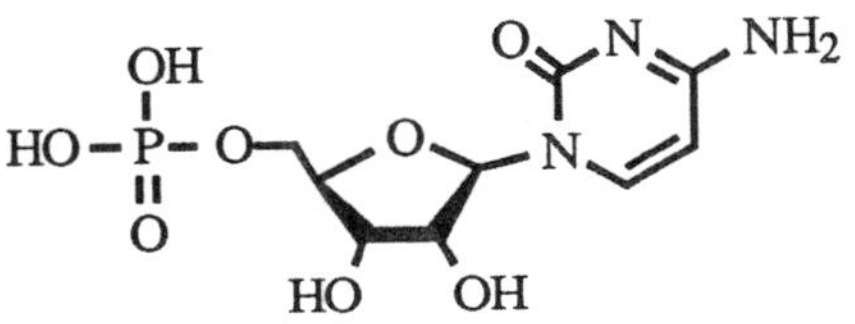

nucleotide
cytidine monophosphate

11.57 Globular proteins have compact, roughly spherical tertiary structures containing folds and grooves. Many have hydrophilic side chains distributed over the outer surface making them soluble in the aqueous environment of the cell. The secondary structures include α-helices and pleated sheets in varying proportions. The unique primary structure of each globular protein leads to a unique distribution of secondary structures and to a specific tertiary structure. Fibrous proteins are the structural components of cells and tissue. The α-helix is a prevalent secondary structure for some fibrous proteins. These molecules are long strands of helical protein that lie with their axes parallel to the axis of the fiber. Because of their compact helical chains, stretching the fibers will stretch and break the relatively weak hydrogen bonds of the α-helix, but as long as none of the stronger covalent bonds are broken, the protein will relax to its original length when released. Other fibrous proteins contain extensive regions of pleated sheets.

Globular proteins carry out most of the work done by cells, including synthesis, transport and energy production. Most globular proteins are enzymes which speed up biochemical reactions. Some are antibodies that protect us from disease. Some transport smaller molecules through the bloodstream such as hemoglobin which carries oxygen in the bloodstream. Some act as hormones. Others are bound in cell

(continued)

(11.57 continued)

membranes where they facilitate passage of nutrients and ions into and out of cells. Fibrous proteins are the cables, girders, bricks, and mortar of organisms. One form called keratins includes wool, hair, skin, fingernails and fur.

11.58 The cytosine-guanine pair forms 3 hydrogen bonds with each other. The adenine-thymine pair forms only 2 hydrogen bonds with each other. As the amount of cytosine-guanine pairs becomes more predominant and the amount of adenine-thymine pairs decreases, the number of hydrogen bonds increases and the melting point of DNA increases.

11.59

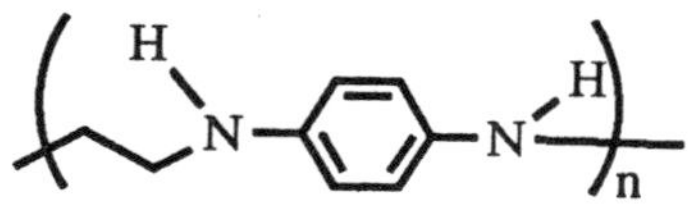

11.60 Both nylon and proteins are polymers formed by condensation of an amine group on one monomer molecule with the carboxylic acid group on another monomer molecule to form amide linkages (also called peptide linkages for proteins). Both are, therefore, polyamides.

Nylon is usually made from 2 monomers which each have 2 amine groups or 2 carboxylic acid groups, although it can be made from monomers that each have an amine group and a carboxylic acid group. Protein is made from monomers (called amino acids) containing an amine group and a carboxylic acid group.

Each type of nylon has carbon chains of different lengths between each pair of functional groups. The 20 amino acids used to form proteins have one carbon atom between the amine group and the carboxylic acid group. This carbon atom has 2 hydrogen atoms bonded to it in glycine. The other 19 amino acids have one hydrogen atom bonded to this carbon and the other bond is to a carbon that may be a part of a ring or a carbon chain with or without functional groups.

11.61

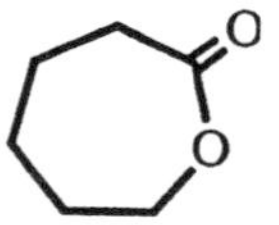

11.62 a)

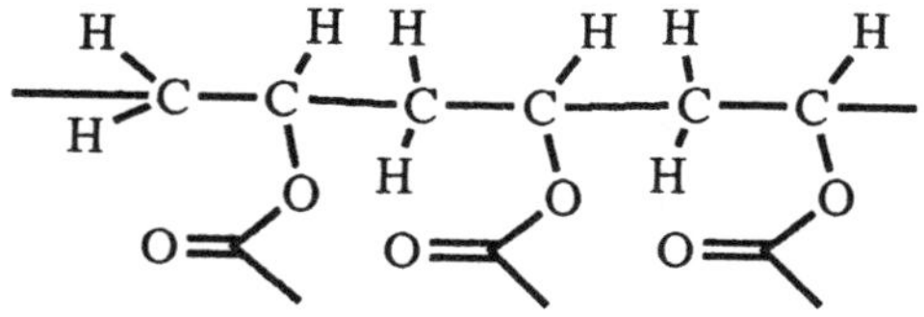

b) Polyvinylacetate is a thermoplastic plastic with a low melting temperature (approximately body temperature). At body temperature the individual molecules have enough kinetic energy to overcome the dispersion forces that hold the molecules together and the gum becomes a viscous liquid that is flexible and easily deformed.

c) Upon cooling, the polymer becomes brittle because the individual molecules do not have enough kinetic energy to overcome the dispersion forces and when force is applied, the polymer does not deform but crumbles easily.

11.63 High-density polyethylene is made up of straight chains of CH_2 units. These linear molecules maximize dispersion forces by lining up in rows that create crystalline regions within the polymer. Low-density polyethylene has chains of CH_2 groups that branch off the main backbone of the polymer. These branches do not allow the polymer molecules to pack close together resulting in an amorphous polymer. The close packed crystalline high-density ethylene therefore has more CH_2 groups per unit volume than the non-close packed amorphous low-density ethylene.

11.64

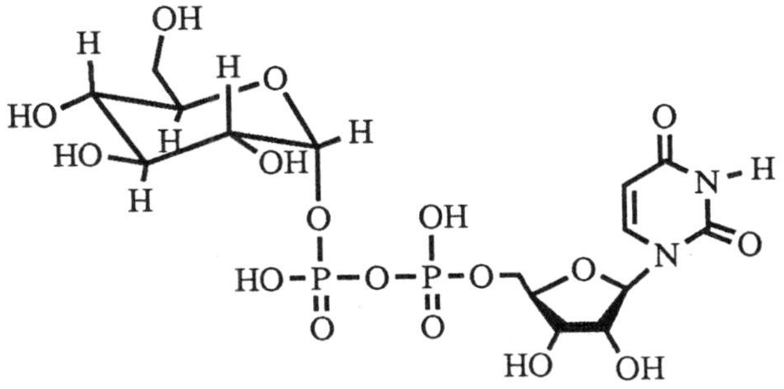

11.65

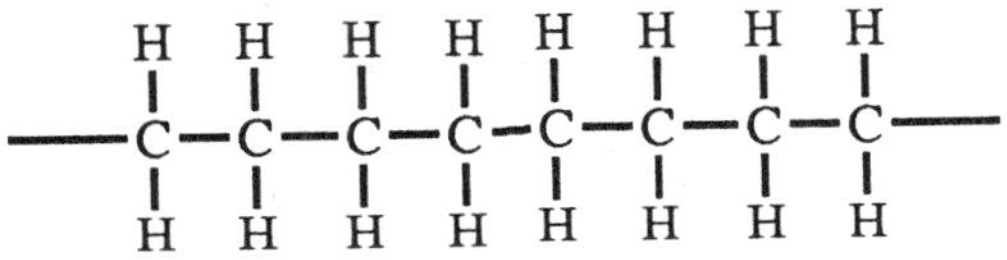

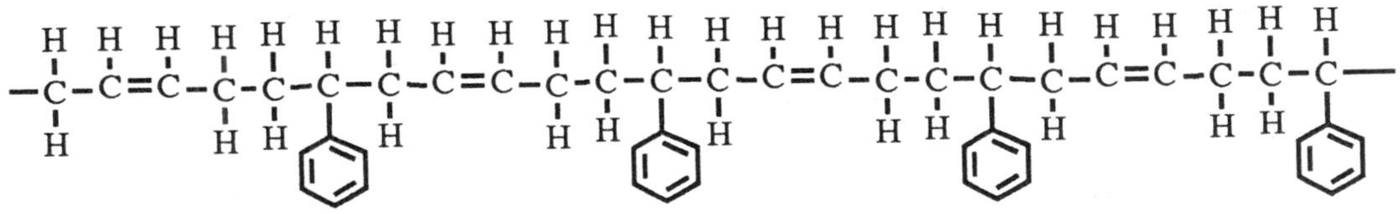

The polyethylene straight chain polymers can maximize dispersion forces by lining up in rows that create crystalline regions within the polymer. This gives strength and rigidity to the polymer. The butadienestyrene copolymer chain cannot line up closely to maximize dispersion forces because the benzene rings take up space and keep portions of the chains apart. Therefore, the butadienestyrene copolymer is less rigid.

11.66 DNA is made up of 2 strands. The bases in each strand form hydrogen bonds with the bases in the other strand. Thymine and adenine pair by forming 2 hydrogen bonds. Cytosine and guanine pair by forming 3 hydrogen bonds. Because of their matching structures these 2 sets of base pairs are said to be complementary. Each base only pairs with its complementary base. Therefore, any sample of DNA will have the same number of cytosine and guanine (their molar ratio equals 1) because they are present in pairs. Any sample of DNA will also have the same number of thymine and adenine (their molar ratio equals 1) because they are present in pairs.

11.67

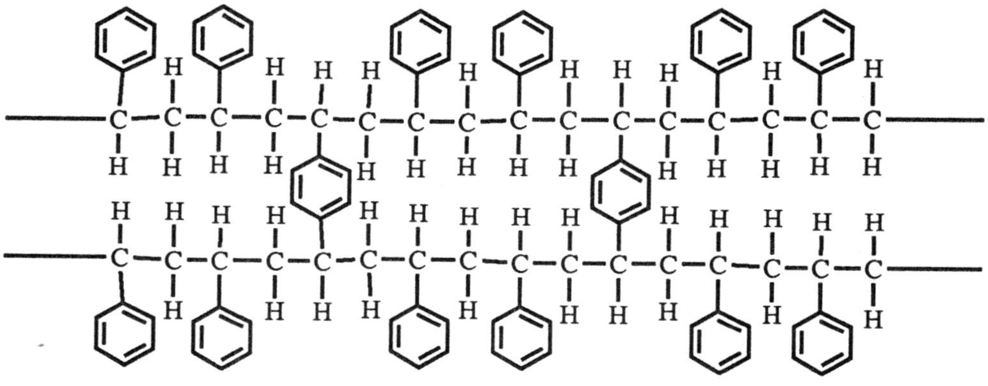

11.68 $+ H_2O \rightarrow$

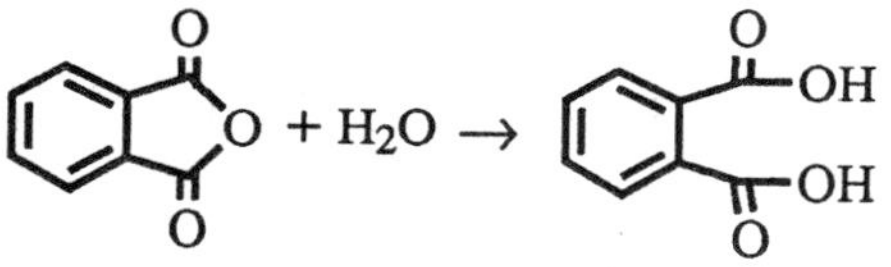

Glyptal

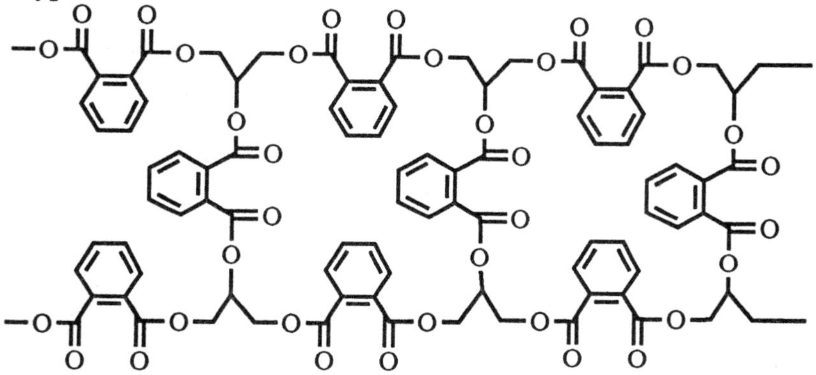

11.69 Amino acids with nonpolar hydrophobic side chains would most likely be found on the inside of a globular protein where they are tucked away from the aqueous environment of the cell. The amino acids with side chains that contain polar function groups are usually on the outer surface of a protein where their interaction with water molecules increases the solubility of the protein.

Amino Acid	Location	Polar or Nonpolar Side Chain	Polar Functional Group
a) Arg	outside	$-CH_2CH_2CH_2NH-C(=NH)-NH_2$	—NH, $-NH_2$, =NH
b) Val	inside	$CH_3-CH-CH_3$	
c) Met	inside	$-CH_2CH_2SCH_3$	
d) Thr	outside	$CH_3-CH-OH$	—OH
e) Asp	outside	$-CH_2CO_2H$	$-CO_2H$

CHAPTER 12: CHEMICAL ENERGIES

12.1 a) A human being is an open system.
b) Coffee in a thermos flask is an isolated system.
c) An ice-cube tray filled with water is a closed system.
d) A helium-filled balloon is an open system.

12.2 a) A can of tomato soup is a closed system because it does not exchange matter with the surroundings but it does exchange energy as it changes to match the temperature of its surroundings.
b) A freezer chest of ice is an isolated system as it does not exchange matter and does not (almost) exchange energy with its surroundings.
c) The Earth is an open system although the exchange of matter is so small that it is almost a closed system.
d) A satellite in orbit is a closed system which exchanges energy. An infinitesimal amount of matter (cosmic dust) may be picked up.

12.3 a) The temperature at which a solution is prepared is not a state variable for its present state.
b) The current temperature of the solution is a state variable.
c) The mass of NaCl in the solution (related to the number of moles) is a state variable.
d) The time when is was prepared is not a state variable.

12.4 a) Height is a state function.
b) Distance traveled in climbing a mountain is a path function not a state function.
c) Energy consumed in climbing the mountain is also a path function not a state function.
d) The gravitational potential energy of a climber is a state function.

12.5 The cylinder's contents, battery, cooling system, engine, the automobile and the human body are possible systems.

12.6 The breathing system, the digestive system, muscles, the liver, the kidneys and the urinary system are some examples.

12.7 a) $(10.0 \text{ g Al})\left(\frac{1 \text{ mol Al}}{26.98 \text{ g Al}}\right) = 0.3706 \text{ mol Al}$

$\Delta T = \frac{q}{nC_p} = \frac{25.0 \text{ J}}{(0.3706 \text{ mol Al})(24.35 \text{ J / mol Al K})}$
$= 2.77 \text{ K} = 2.77°\text{C}$
$T_f = 15.0°\text{C} + \Delta T = 15.0°\text{C} + 2.77°\text{C} = 17.8°\text{C}$

b) $(25.0 \text{ g Al})\left(\frac{1 \text{ mol Al}}{26.98 \text{ g Al}}\right) = 0.9266 \text{ mol Al}$

$\Delta T = \frac{25.0 \text{ J}}{(0.9266 \text{ mol Al})(24.35 \text{ J / mol Al K})} = 1.108 \text{ K}$

$T_f = 295 \text{ K} + \Delta T = 295 \text{ K} + 1.108 \text{ K} = 296 \text{ K}$

c) $(25.0 \text{ g Ag})\left(\frac{1 \text{ mol Ag}}{107.9 \text{ g Ag}}\right) = 0.2317 \text{ mol Ag}$

$\Delta T = \frac{25.0 \text{ J}}{(0.2317 \text{ mol Ag})(25.351 \text{ J / mol Ag K})} = 4.256 \text{ K}$

$T_f = 295 \text{ K} + \Delta T = 295 \text{ K} + 4.256 \text{ K} = 299 \text{ K}$

d) $(25.0 \text{ g } H_2O)\left(\frac{1 \text{ mol } H_2O}{18.02 \text{ g } H_2O}\right) = 1.387 \text{ mol } H_2O$

$\Delta T = \frac{25.0 \text{ J}}{(1.387 \text{ mol } H_2O)(75.291 \text{ J / mol } H_2O \text{ K})} = 0.2394 \text{ K} = 0.3394°\text{C}$

$T_f = 22.0°\text{C} + 0.2394°\text{C} = 22.2°\text{C}$

12.8 For Ag: $(15.0 \text{ g Ag})\left(\frac{1 \text{ mol Ag}}{107.9 \text{ g Ag}}\right) = 0.1390 \text{ mol Ag}$

$\Delta T = 37.6°\text{C} - 24.0°\text{C} = 13.6°\text{C}$ $\qquad C_p$ for Ag = 25.351 J/mol K

$q = nC_p\Delta T = (0.1390 \text{ mol})(25.351 \text{ J/mol K})(13.6 \text{ K}) = 47.9 \text{ J}$

For H_2O: $(25.0 \text{ g } H_2O)\left(\frac{1 \text{ mol } H_2O}{18.02 \text{ g } H_2O}\right) = 1.387 \text{ mol } H_2O$

$\Delta T = 13.6°\text{C} = 13.6 \text{ K}$ $\qquad C_p$ for H_2O = 75.291 J/mol K

$q = (1.387 \text{ mol})(75.291 \text{ J/mol K})(13.6 \text{ K}) = 1420 \text{ J} = 1.42 \times 10^3 \text{ J} = 1.42 \text{ kJ}$

12.9 Let the final temperature = t

$(27.4 \text{ g Ag})\left(\frac{1 \text{ mol Ag}}{107.9 \text{ g Ag}}\right) = 0.2539 \text{ mol Ag}$

$(37.5 \text{ g } H_2O)\left(\frac{1 \text{ mol } H_2O}{18.02 \text{ g } H_2O}\right) = 2.081 \text{ mol } H_2O$

(continued)

(12.9 continued)

Heat lost by hot coin =

(0.2539 mol)(25.351 J/mol K)(373 K - t) = 6.437 J/ K(373 K - t)

Heat gained by H_2O =

(2.081 mol)(75.291 J/mol K)(t - 293.65 K) = 156.7 J/K(t - 293.65 K)

6.437 J/K(373 K - t) = 156.7 J/K(t - 293.65 K)

6.437 J/K(373 K) + 156.7 J/K(293.65 K) = 156.7 t J/K + 6.437 t J/K

2401 J + 46015 J = 163.1 t J/K

48416 J = 163.1 t J/K

t = 48416 J ÷ (163.1 J/K) = 296.8 K

T_f = 296.8 K or 23.6°C

12.10 ΔT = 54.0°C - 5.50°C = 48.5°C = 48.5 K

$$(145 \text{ g } H_2O)\left(\frac{1 \text{ mol } H_2O}{18.02 \text{ g } H_2O}\right) = 8.047 \text{ mol } H_2O$$

q = (8.047 mol)(75.291 J/mol K)(48.5 K) = 2.94 x 10^4 J or 29.4 kJ

12.11 Assuming the atmospheric pressure is 1 atm and using Equation 12-3:

$w = -P\Delta V$ = -(1 atm)(2.5 L - 0) = -2.5 atm L

1 atm = 1.013 x 10^5 Pa = 1.013 x 10^5 kg m^{-1} s^{-2}

1 L = 1000 cm^3 = (1000 cm^3)(1 m/100 cm)^3 = 0.001000 m^3

$$w = (-2.5 \text{ atm L})\left(\frac{1.013 \times 10^5 \text{ kg m}^{-1} \text{ s}^{-2}}{1 \text{ atm}}\right)\left(\frac{0.001000 \text{ m}^3}{1 \text{ L}}\right)$$

$$= -253 \text{ kg m}^2 \text{ s}^{-2} = -2.5 \times 10^2 \text{ J}$$

12.12 $$0.10 \text{ g He}\left(\frac{1 \text{ mol He}}{4.003 \text{ g He}}\right) = 0.0250 \text{ mol He}$$

$\Delta V = V_f - V_i$ = 1.01 L - 1.31 L = -0.30 L

$\Delta T = T_f - T_i$ = 250 K - 320 K = -70 K

q = (0.0250 mol)(20.8 J/mol K)(-70 K) = -36.4 J

$w = -P\Delta V$ Assume P = 1 atm

= -(1 atm)(-0.30 L) = +0.30 atm L

= (0.30 atm L)(1.013 x 10^5 kg m^{-1} s^{-2}/1 atm)(10^{-3} m^3/L) = 30.39 J

$\Delta E = q + w$

= -36.4 J + 30.39 J = -6.0 J or -6 J

12.13 a) Any change in the energy of a system must be counterbalanced by an opposite change in the energy of the surroundings.

b) The total energy of the universe remains unchanged.

c) $\Delta E_{surr} = -\Delta E_{sys}$

$q_{surr} + w_{surr} = -q_{sys} + (-w_{sys})$

12.14 (e) $\Delta E - w$ describes the heat of a chemical reaction under all possible conditions. ΔE and q_v are equal to the heat only under constant volume conditions ($w = 0$). ΔH and q_p are equal to the heat only under constant pressure conditions.

12.15 q for burning of the glucose

$q = (1.7500 \text{ g})(15.57 \text{ kJ/g}) = 27.2475 \text{ kJ}$

$\Delta T = 23.34°C - 21.45°C = 1.89°C = 1.89 \text{ K}$

$$C_{cal} = \frac{q}{\Delta T} = \frac{27.2475 \text{ kJ}}{1.89 \text{ K}} = 14.4 \text{ kJ / K}$$

12.16 $$C_{cal} = \frac{19.75 \text{ kJ}}{4.22 \text{ K}} = 4.68 \text{ kJ / K}$$

For the burning of the methanol:

$q = C_{cal}\Delta T = (4.68 \text{ kJ/K})(8.47 \text{ K}) = 39.64 \text{ kJ}$

Therefore, for methanol:

$$C_{molar} = \left(\frac{39.64 \text{ kJ}}{1.75 \text{ g methanol}}\right)\left(\frac{32.042 \text{ g methanol}}{1 \text{ mol methanol}}\right) = 726 \text{ kJ / mol methanol}$$

12.17 Assume that $C_{cal} \approx C_{H_2O} = n_{H_2O}C_{p,H_2O}$

$$n = \left(\frac{110.0 \text{ g } H_2O}{18.016 \text{ g / mol}}\right) = 6.1057 \text{ mol } H_2O$$

$C_{cal} = C_{H_2O} = (6.1057 \text{ mol})(75.291 \text{ J/mol K}) = 459.7 \text{ J/K}$

$\Delta T = (29.7°C - 22.0°C) = 7.7°C = 7.7 \text{ K}$

$q_{cal} = (459.7 \text{ J/K})(7.7 \text{ K}) = 3.5 \times 10^3 \text{ J}$

$q_{solution} = -q_{cal} = -3.5 \times 10^3 \text{ J}$

$$\left(\frac{-3.5 \times 10^3 \text{ J}}{4.75 \text{ g } CaCl_2}\right)\left(\frac{110.98 \text{ g } CaCl_2}{1 \text{ mol } CaCl_2}\right) = -8.2 \times 10^4 \text{ J / mol } CaCl_2$$

12.18 $q = C_{cal}\Delta T$ $\quad\quad$ $\Delta T = 302.04 \text{ K} - 297.65 \text{ K} = 4.39 \text{ K}$

$q = (7.85 \text{ kJ/K})(4.39 \text{ K}) = 34.46 \text{ kJ} = 34.5 \text{ kJ}$

$$C_{p,caff} = \left(\frac{-34.46 \text{ kJ}}{1.35 \text{ g caff}}\right)\left(\frac{194.2 \text{ g caff}}{1 \text{ mol caff}}\right) = -4.96 \times 10^3 \text{ kJ}$$

12.19 a) $2\ C_7H_6O_2 + 15\ O_2 \rightarrow 14\ CO_2 + 6\ H_2O$

b) $$\left(\frac{35.61 \text{ kJ}}{1.350 \text{ g benzoic acid}}\right)\left(\frac{122.12 \text{ g benzoic acid}}{1 \text{ mol benzoic acid}}\right) = 3221 \text{ kJ / mol benzoic acid}$$

c) $$\left(\frac{3221 \text{ kJ}}{\text{mol benzoic acid}}\right)\left(\frac{2 \text{ mol benzoic acid}}{15 \text{ mol } O_2}\right) = 429.5 \text{ kJ / mol } O_2$$

12.20 Assume densities of all solutions are 1.000 g/mL and the heat capacity as: water, $C_p = 75.291$ J/mol K

Final solution: 100 mL + 100 mL = 200 mL (200 mL)(1.000 g/mL) = 200 g

$$q = nC\Delta T = \left(\frac{200\ g}{18.016\ g/mol}\right)(75.291\ J/mol\ K)(31.8°C - 25.0°C)\left(\frac{K}{°C}\right)$$

$= 5683.6\ J \cong 5.68 \times 10^3\ J = 5.68\ kJ$

$$\left(1.00\frac{mol}{L}\right)(100\ mL)\left(\frac{1\ L}{1000\ mL}\right) = 0.100\ mol$$

$$C_{\text{molar heat of neutralization}} = \frac{5.68\ kJ}{0.100\ mol} = 56.8\ kJ/mol$$

12.21 a) $C_2H_{4(g)} + 3\ O_{2(g)} \rightarrow 2\ CO_{2(g)} + 2\ H_2O_{(l)}$

$\Delta H_{rxn} = [2\ \Delta H^\circ_f(CO_2) + 2\ \Delta H^\circ_f(H_2O)] - [\Delta H^\circ_f(C_2H_4) + 3\ \Delta H^\circ_f(O_2)]$

$$= \left[(2\ mol\ CO_2)\left(\frac{-393.5\ kJ}{mol\ CO_2}\right) + (2\ mol\ H_2O)\left(\frac{-285.8\ kJ}{mol\ H_2O}\right)\right]$$

$$-\left[(1\ mol\ C_2H_4)\left(\frac{52.26\ kJ}{mol\ C_2H_4}\right) + (3\ mol\ O_2)\left(\frac{0\ kJ}{mol\ O_2}\right)\right]$$

$= [(-787.0\ kJ) + (-571.6\ kJ)] - [52.26\ kJ + 0\ kJ] = 1411\ kJ$

b) $2\ NH_{3(g)} \rightarrow N_{2(g)} + 3\ H_{2(g)}$

$\Delta H_{rxn} = [\Delta H^\circ_f(N_2) + 3\ \Delta H^\circ_f(H_2)] - [2\ \Delta H^\circ_f(NH_3)]$

$$= \left[(1\ mol\ N_2)\left(\frac{0\ kJ}{mol\ N_2}\right) + (3\ mol\ H_2)\left(\frac{0\ kJ}{mol\ H_2}\right)\right] - \left[(2\ mol\ NH_3)\left(\frac{-46.1\ kJ}{mol\ NH_3}\right)\right]$$

$= [0\ kJ + 0\ kJ] - [-92.2\ kJ] = 92.2\ kJ$

c) $5\ PbO_{2(s)} + 4\ P_{(s,\ white)} \rightarrow P_4O_{10(s)} + 5\ Pb_{(s)}$

$\Delta H_{rxn} = [\Delta H^\circ_f(P_4O_{10}) + 5\ \Delta H^\circ_f(Pb)] - [5\ \Delta H^\circ_f(PbO_2) + 4\ \Delta H^\circ_f(P)]$

$= [(1\ mol\ P_4O_{10})(-2984.0\ kJ/mol\ P_4O_{10}) + (5\ mol\ Pb)(0\ kJ/mol\ Pb)]$

$- [(5\ mol\ PbO_2)(-277.4\ kJ/mol\ PbO_2) + (4\ mol\ P)(0\ kJ/mol\ P)]$

$= -1597\ kJ$

d) $SiCl_{4(l)} + 2\ H_2O_{(l)} \rightarrow SiO_{2(s)} + HCl_{(g)}$

$\Delta H_{rxn} = [\Delta H^\circ_f(SiO_2) + 4\ \Delta H^\circ_f(HCl)] - [\Delta H^\circ_f(SiCl_4) + 2\ \Delta H^\circ_f(H_2O)]$

$= [(1\ mol\ SiO_2)(-910.94\ kJ/mol\ SiO_2) + (4\ mol\ HCl)(-92.307\ kJ/mol\ HCl)]$

$- [(1\ mol\ SiCl_4)(-687.0\ kJ/mol\ SiCl_4) + (2\ mol\ H_2O)(-285.830\ kJ/mol\ H_2O)]$

$= -21.5\ kJ$

12.22 a) $2\,NH_{3(g)} + 3\,O_{2(g)} + 2\,CH_{4(g)} \rightarrow 2\,HCN_{(g)} + 6\,H_2O_{(g)}$

$\Delta H_{rxn} = [2\,\Delta H^\circ_f(HCN_{(g)}) + 6\,\Delta H^\circ_f(H_2O_{(g)})]$
$\quad - [2\,\Delta H^\circ_f(NH_3) + 3\,\Delta H^\circ_f(O_2) + 2\,\Delta H^\circ_f(CH_4)]$

$$= \left[(2\text{ mol HCN})\left(\frac{135.1\text{ kJ}}{\text{mol HCN}}\right) + (6\text{ mol }H_2O)\left(\frac{-241.8\text{ kJ}}{\text{mol }H_2O}\right)\right]$$

$$-\left[(2\text{ mol }NH_3)\left(\frac{-46.1\text{ kJ}}{\text{mol }NH_3}\right) + (3\text{ mol }O_2)\left(\frac{0\text{ kJ}}{\text{mol }O_2}\right) + (2\text{ mol }CH_4)\left(\frac{-74.81\text{ kJ}}{\text{mol }CH_4}\right)\right]$$

$= [270.2\text{ kJ} + (-1451\text{ kJ})] - [(-92.22\text{ kJ}) + 0\text{ kJ} + (-149.62\text{ kJ})] = -939\text{ kJ}$

b) $2\,Al_{(s)} + 3\,Cl_{2(g)} \rightarrow 2\,AlCl_{3(s)}$

$\Delta H_{rxn} = [2\ \Delta H^\circ_f(AlCl_3)] - [2\ \Delta H^\circ_f(Al) + 3\ \Delta H^\circ_f(Cl_2)]$
$= [(2\text{ mol }AlCl_3)(-704.2\text{ kJ/mol }AlCl_3)] - [(2\text{ mol Al})(0\text{ kJ/mol Al})$
$\quad + (3\text{ mol }Cl_2)\ (0\text{ kJ/mol }Cl_2)] = -1408\text{ kJ} - 0\text{ kJ} = -1408\text{ kJ}$

c) $3\,NO_{2(g)} + H_2O_{(l)} \rightarrow 2\,HNO_{3(aq)} + NO_{(g)}$

$\Delta H_{rxn} = [2\ \Delta H^\circ_f(HNO_3) + \ \Delta H^\circ_f(NO)] - [3\ \Delta H^\circ_f(NO_2) + \ \Delta H^\circ_f(H_2O)]$
$= [(2\text{ mol }HNO_3)(-205.0\text{ kJ/mol }HNO_3) + (1\text{ mol NO})(90.25\text{ kJ/mol NO})]$
$\quad - [(3\text{ mol }NO_2)(33.18\text{ kJ/mol }NO_2) + (1\text{ mol }H_2O)(-285.8\text{ kJ/mol }H_2O)]$
$= [-410.0\text{ kJ} + 90.25\text{ kJ}] - [99.54\text{ kJ} + (-285.8\text{ kJ})]$
$= -133.5\text{ kJ}$

d) $2\,C_2H_{2(g)} + 5\,O_{2(g)} \rightarrow 4\,CO_{2(g)} + 2\,H_2O_{(l)}$

$\Delta H_{rxn} = [4\ \Delta H^\circ_f(CO_2) + 2\ \Delta H^\circ_f(H_2O)] - [2\ \Delta H^\circ_f(C_2H_2) + 5\ \Delta H^\circ_f(O_2)]$
$= [(4\text{ mol }CO_2)(-393.5\text{ kJ/mol }CO_2) + (2\text{ mol }H_2O)(-285.8\text{ kJ/mol }H_2O)]$
$\quad - [(2\text{ mol }C_2H_2)(226.7\text{ kJ/mol }C_2H_2) + (5\text{ mol }O_2)(0\text{ kJ/mol }O_2)]$
$= [(-1574.0\text{ kJ}) + (-571.6\text{ kJ})] - [453.4\text{ kJ} + (0\text{ kJ})]$
$= -2599\text{ kJ}$

12.23 $\Delta E_{rxn} = \Delta H_{rxn} - \Delta(PV)_{rxn}$ Refer to Problem 12.21 for ΔH_{rxn}.
$\Delta(PV) \cong 0$ for condensed phases; $\Delta(PV) = \Delta(nRT)$ for gases;
$\Delta(PV) = RT\Delta n$ if T is constant.
$\Delta E_{rxn} = \Delta H_{rxn} - RT\Delta n$ if T constant Assume T = 298 K

a) $C_2H_{4(g)} + 3\,O_{2(g)} \rightarrow 2\,CO_{2(g)} + 2\,H_2O_{(l)}$ $\Delta n = 2 - 4 = -2$
$\Delta E_{rxn} = \Delta H_{rxn} - (8.314\text{ J/mol K})(298\text{ K})(-2)(\text{kJ}/10^3\text{ J}) = 1411\text{ kJ} + 4.96\text{ kJ}$
$= 1416\text{ kJ}$

b) $2\,NH_{3(g)} \rightarrow N_{2(g)} + 3\,H_{2(g)}$ $\Delta n = 4 - 2 = +2$
$\Delta E_{rxn} = \Delta H_{rxn} - (8.314\text{ J/mol K})(298\text{ K})(+2)(\text{kJ}/10^3\text{ J}) = 92.2\text{ kJ} - 4.96\text{ kJ}$
$= 87.2\text{ kJ}$

c) $5\,PbO_{2(s)} + 4\,P_{(s,\ white)} \rightarrow P_4O_{10(s)} + 5\,Pb_{(s)}$ $\Delta n = 0 - 0 = 0$
$\Delta E_{rxn} = \Delta H_{rxn} - 0 = -1597\text{ kJ} - 0 = -1597\text{ kJ}$

(continued)

(12.23 continued)

d) $SiCl_{4(l)} + 2\ H_2O_{(l)} \rightarrow SiO_{2(s)} + 4\ HCl_{(g)}$

$\Delta E_{rxn} = \Delta H_{rxn} = -(8.314\ J/mol\ K)(298\ K)(+4)(kJ/10^3\ J)$

$= -21.5\ kJ - 9.9\ kJ = -31.4\ kJ)$

12.24 $\Delta E_{rxn} = \Delta H_{rxn} - RT\Delta n$ (See Problem 12.23). Refer to Problem 12.22 for ΔH_{rxn}.
Assume T = 298 K

$2\ NH_{3(g)} + 3\ O_{2(g)} + 2\ CH_{4(g)} \rightarrow 2\ HCN_{(g)} + 6\ H_2O_{(g)}$ $\Delta n = 8 - 7 = +1$

$\Delta E_{rxn} = \Delta H_{rxn} - (8.314\ J/mol\ K)(298\ K)(+1)(kJ/10^3\ J) = -940\ kJ - 2.47\ kJ$

$= -942\ kJ$

Note: Assumption that T = 298 K is questionable since H_2O is in gaseous state.

b) $2\ Al_{(s)} + 3\ Cl_{2(g)} \rightarrow 2\ AlCl_{3(s)}$ $\Delta n = 0 - 3 = -3$

$\Delta E_{rxn} = \Delta H_{rxn} - (8.314\ J/mol\ K)(298\ K)(-3)(kJ/10^3\ J) = -1408\ kJ + 7.43\ kJ$

$= -1401\ kJ$

c) $3\ NO_{2(g)} + H_2O_{(l)} \rightarrow 2\ HNO_{3(aq)} + NO_{(g)}$ $\Delta n = 1 - 3 = -2$

$\Delta E_{rxn} = \Delta H_{rxn} - (8.314\ J/mol\ K)(298\ K)(-2)(kJ/10^3\ J) = -133.5\ kJ + 4.96\ kJ$

$= -128.5\ kJ$

d) $2\ C_2H_{2(g)} + 5\ O_{2(g)} \rightarrow 4\ CO_{2(g)} + 2\ H_2O_{(l)}$ $\Delta n = 4 - 7 = -3$

$\Delta E_{rxn} = \Delta H_{rxn} - (8.314\ J/mol\ K)(298\ K)(-3)(kJ/10^3\ J) = -2599\ kJ + 7.43\ kJ$

$= -2592\ kJ$

12.25 $\Delta H_{rxn} - \Delta E_{rxn} = \Delta(PV)$ $\Delta(PV) = \Delta(nRT)$ for gases

a) 0 because $\Delta(PV) \cong 0$ for condensed phases

b) $\Delta(nRT) = \Delta n\ RT$ $\Delta n = 2 - 0 = 2$

$= (2\ mol)(8.314\ J/mol\ K)(298\ K)$ Assuming temperature = 298 K

$= 4.96 \times 10^3\ J$

c) $C_4H_9OH_{(l)} + 6\ O_{2(g)} \rightarrow 4\ CO_{2(g)} + 5\ H_2O_{(l)}$ Assuming T = 298 K, $\Delta n = 4 - 6 = -2$

$\Delta n\ RT = (-2\ mol)(8.314\ J/mol\ K)(298\ K) = -4.96 \times 10^3\ J$

12.26 a) $4\ C_{(s,\ graphite)} + 5\ H_{2(g)} + 1/2\ O_{2(g)} \rightarrow C_4H_9OH_{(l)}$

b) $2\ Na_{(s)} + C_{(s,\ graphite)} + 3/2\ O_{2(g)} \rightarrow Na_2CO_{3(s)}$

c) $3/2\ O_{2(g)} \rightarrow O_{3(g)}$

d) $3\ Fe_{(s)} + 2\ O_{2(g)} \rightarrow Fe_3O_{4(s)}$

12.27 a) $3\ K_{(s)} + P_{(s,\ white)} + 2\ O_{2(g)} \rightarrow K_3PO_{4(s)}$

b) $2\ C_{(s,\ graphite)} + O_{2(g)} + 2\ H_{2(g)} \rightarrow CH_3CO_2H_{(l)}$

c) $3\ C_{(s,\ graphite)} + 9/2\ H_{2(g)} + 1/2\ N_{2(g)} \rightarrow (CH_3)_3N_{(g)}$

d) $2\ Al_{(s)} + 3/2\ O_{2(g)} \rightarrow Al_2O_{3(s)}$

12.28 a) $(NH_2)_2CO_{(s)} + 3/2\ O_{2(g)} \rightarrow CO_{2(g)} + N_{2(g)} + 2\ H_2O_{(l)}$

or $2\ (NH_2)_2CO_{(s)} + 3\ O_{2(g)} \rightarrow 2\ CO_{2(g)} + 2\ N_{2(g)} + 4\ H_2O_{(l)}$

b) $\dfrac{-632.2\ kJ}{mol\ (NH_2)_2CO} \times \dfrac{2\ mol\ (NH_2)_2CO}{4\ mol\ H_2O} = -316.1\ kJ / mol\ H_2O$

c) $\Delta H_{rxn} = -\ 632.2\ kJ$

$= [\Delta H_f^\circ(CO_2) + \Delta H_f^\circ(N_2) + 2\ \Delta H_f^\circ(H_2O)] - [(\Delta H_f^\circ(NH_2)_2CO) + 3/2\ \Delta H_f^\circ(O_2)]$

$$= \left[(1\ mol\ CO_2)\left(\frac{-393.5\ kJ}{mol\ CO_2}\right) + (1\ mol\ N_2)\left(\frac{0\ kJ}{mol\ N_2}\right) + (2\ mol\ H_2O)\left(\frac{-285.8\ kJ}{mol\ H_2O}\right)\right]$$

$-[(1\ mol\ (NH_2)_2CO)\ \Delta H_f^\circ(NH_2)_2CO + (3/2\ mol\ O_2)(0\ kJ/mol\ O_2)]$

$= [-393.5\ kJ - 571.6\ kJ] - [(1\ mol\ (NH_2)_2CO)(\Delta H_f^\circ(NH_2)_2CO)]$

$(1\ mol\ (NH_2)_2CO)\ \Delta H_f^\circ(NH_2)_2CO = -393.5\ kJ - 571.6\ kJ + 632.2\ kJ = -\ 332.9\ kJ$

$\Delta H_f^\circ(NH_2)_2CO = -332.9\ kJ/mol\ (NH_2)_2CO$

12.29 a) $4\ NH_{3(g)} + 5\ O_{2(g)} \rightarrow 4\ NO_{(g)} + 6\ H_2O_{(l)}$

$\Delta H_{rxn} = [4\ \Delta H_f^\circ(NO) + 6\ \Delta H_f^\circ(H_2O)] - [(4\ \Delta H_f^\circ(NH_3) + 5\ \Delta H_f^\circ(O_2)]$

$$= \left[(4\ mol\ NO)\left(\frac{90.25\ kJ}{mol\ NO}\right) + (6\ mol\ H_2O)\left(\frac{-285.8\ kJ}{mol\ H_2O}\right)\right]$$

$$-\left[(4\ mol\ NH_3)\left(\frac{-46.1\ kJ}{mol\ NH_3}\right) + (5\ mol\ O_2)\left(\frac{0\ kJ}{mol\ O_2}\right)\right]$$

$= [361.0\ kJ + (-\ 1714.8\ kJ] - [-184.4\ kJ + 0\ kJ)] = -1169.4\ kJ$

b) $4\ NH_{3(g)} + 3\ O_{2(g)} \rightarrow 2\ N_{2(g)} + 6\ H_2O_{(l)}$

$\Delta H_{rxn} = [2\ \Delta H_f^\circ(N_2) + 6\ \Delta H_f^\circ(H_2O)] - [(4\ \Delta H_f^\circ(NH_3) + 3\ \Delta H_f^\circ(O_2)]$

$$= \left[(2\ mol\ N_2)\left(\frac{0\ kJ}{mol\ N_2}\right) + (6\ mol\ H_2O)\left(\frac{-285.8\ kJ}{mol\ H_2O}\right)\right]$$

$$-\left[(4\ mol\ NH_3)\left(\frac{-46.1\ kJ}{mol\ NH_3}\right) + (3\ mol\ O_2)\left(\frac{0\ kJ}{mol\ O_2}\right)\right]$$

$= [0\ kJ + (-\ 1714.8\ kJ] - [-184.4\ kJ + 0\ kJ)] = -1530.4\ kJ$

12.30 $\Delta T = 296.36\ K - 298.00 = -1.64\ K$

$$\frac{1.00\ KClO_3}{122.55\ g\ KClO_3\ /\ mol} = 8.16 \times 10^{-3}\ mol \qquad \frac{50.0\ g\ H_2O}{18.02\ g\ H_2O\ /\ mol} = 2.7747\ mol$$

$$q_{H_2O} = (n_{H_2O})(C_{p,H_2O})(\Delta T)$$

$$= (2.7747\ mol)(75.291\ J/mol\ K)(-1.64\ K) = -342.6\ J$$

$$q_{salt} + q_{H_2O} = 0$$

$$q_{salt} = (n_{salt})(\Delta H_{soln}) = -q_{H_2O} = -(-342.6\ J)$$

$$\Delta H_{soln} = \frac{q_{salt}}{n_{salt}} = \frac{342.6\ J}{8.16 \times 10^{-3}\ mol} = 4.20 \times 10^4\ J\ /\ mol = 42.0\ kJ\ /\ mol$$

12.31 a) Solution reaction: $MgSO_{4(s)} \xrightarrow{H_2O} Mg^{2+}_{(aq)} + SO^{2-}_{4(aq)}$

This is the reverse reaction of that given; therefore, ΔH is negative and when $MgSO_4$ dissolves in water, heat is released to the water and absorbed by the water.

b) $q_{H_2O} = -\ q_{salt} = -\ (n_{salt})(\Delta H_{soln})$

$$n_{salt} = \frac{2.55\ g\ MgSO_4}{120.371\ g\ MgSO_4\ /\ mol} = 0.02118\ mol$$

$$q_{H_2O} = -\ (0.02118\ mol)(-91.2\ kJ/mol) = 1.932\ kJ$$

c) moles of H_2O = $(500\ mL)\left(\frac{1.00\ g}{1\ mL}\right)\left(\frac{1\ mol\ H_2O}{18.02\ g}\right) = 27.747\ mol\ H_2O$

$$q_{H_2O} = (n_{H_2O})(C_{p,H_2O})(\Delta T)$$

$$\Delta T = \frac{q_{H_2O}}{(n_{H_2O})(C_{p,H_2O})} = \frac{1.932\ kJ \times 10^3\ J\ /\ kJ}{(27.747\ mol)(75.291\ J\ /\ mol\ K)} = 0.925\ K$$

12.32 The reaction is endothermic (ΔH is positive) which means that the energy needed to counteract the intermolecular forces between the Mg^{2+} and SO_4^{2-} ions and polar water molecules in the solution are greater than the energy produced by the formation of the interionic forces between the Mg^{2+} and SO_4^{2-} ions in the solid salt.

12.33 a) Ethane has stronger dispersion forces between molecules than methane has between its molecules. Therefore, it takes less energy to counteract these forces for methane and cause methane to vaporize resulting in a lower heat of vaporization for methane.

b) Ethanol forms hydrogen bonds between molecules which are much stronger than the dipolar forces between diethyl ether molecules. Therefore, ethanol has a much higher heat of vaporization than diethyl ether.

c) Argon and methane have dispersion forces as their interparticle forces. The polarizability for methane is larger than for argon but the shape of the argon atom allows the argon atoms to approach each other more closely than the methane molecules can approach each other. Dispersion forces decrease rapidly with an increase in distance. Therefore, the forces of attraction for argon are greater than for methane and the heat of fusion for argon is greater.

12.34 $\Delta H_{vap(Ar)} = 6.3$ kJ/mol $\quad \Delta H_{vap(C_2H_6)} = 15.5$ kJ/mol $\quad \Delta H_{vap(Hg)} = 59.0$ kJ/mol

$\Delta E_{vap} = \Delta H_{vap} - RT_{vap} \qquad R = 8.314 \times 10^{-3}$ kJ/mol K

For Ar: $\Delta E_{vap(Ar)} = 6.3$ kJ/mol $- (8.314 \times 10^{-3}$ kJ/mol K$)(87$ K$)$
$= 6.3$ kJ/mol $- 0.72$ kJ/mol $= 5.6$ kJ/mol

$$\% \text{ diff vs } \Delta E = \frac{0.72 \text{ kJ / mol}}{5.6 \text{ kJ / mol}} \times 100\% = 13\%$$

$$\% \text{ diff vs } \Delta H = \frac{0.72 \text{ kJ / mol}}{6.3 \text{ kJ / mol}} \times 100\% = 11\%$$

For ethane: $\Delta E_{vap} = 15.5$ kJ/mol $- (8.314 \times 10^{-3}$ kJ/mol K$)(184$ K$)$
$= 15.5$ kJ/mol $- 1.53$ kJ/mol $= 14.0$ kJ/mol

$$\% \text{ diff vs } \Delta E = \frac{1.53 \text{ kJ / mol}}{14.0 \text{ kJ / mol}} \times 100\% = 10.9\%$$

$$\% \text{ diff vs } \Delta H = \frac{1.53 \text{ kJ / mol}}{15.5 \text{ kJ / mol}} \times 100\% = 9.87\%$$

For mercury: $\Delta E_{vap} = 59.0$ kJ/mol $- (8.314 \times 10^{-3}$ kJ/mol K$)(630$ K$)$
$= 59.0$ kJ/mol $- 5.237$ kJ/mol $= 53.8$ kJ/mol

$$\% \text{ diff vs } \Delta E = \frac{5.237 \text{ kJ / mol}}{53.8 \text{ kJ / mol}} \times 100\% = 9.73\%$$

$$\% \text{ diff vs } \Delta H = \frac{5.237 \text{ kJ / mol}}{59.0 \text{ kJ / mol}} \times 100\% = 8.88\%$$

Argon has the biggest percentage difference between ΔE_{vap} and ΔH_{vap}.

12.35 a) $\Delta H = q_p = nC_p\Delta T \qquad n = \frac{2.50 \text{ g}}{18.02 \text{ g / mol}} = 0.1387$ mol

$\Delta T = 100.0°C - 37.5°C = 62.5°C = 62.5$ K

$\Delta H = q_p = (0.1387$ mol$)(75.291$ J/mol K$)(62.5$ K$) = 653$ J or 0.653 kJ

b) $q_p = n\Delta H_{vap} + nC_p\Delta T$

$= (0.1387$ mol$)(40.79$ kJ/mol$)(10^3$ J/kJ$) + 653$ J

$= 5658$ J $+ 653$ J $= 6.31 \times 10^3$ J or 6.31 kJ

12.36 $\Delta E = q + w$

The 1 g of gasoline provides the same amount of energy in each case. The automobile engine is doing less work while idling; therefore, more heat must be removed during idling.

12.37 $q_{Cu} + q_{ice} = 0$

$$q_{Cu} = nC_p\Delta T = \left(\frac{12.7 \text{ g}}{63.55 \text{ g / mol}}\right)(24.435 \text{ J / mol K})(0°\text{C} - 200°\text{C})\left(\frac{\text{K}}{°\text{C}}\right) = -976.6 \text{ J}$$

$q_{ice} = -q_{Cu} = 976.6$ J $= n\Delta H_{fus} = n\ (6.01$ kJ/mol$)(10^3$ J/kJ$)$

$$n = \frac{976.6 \text{ J}}{(6.01 \text{ kJ / mol})(10^3 \text{ J/kJ})} = 0.1625 \text{ mol}$$

$m_{ice} = (0.1625$ mol$)(18.02$ g/mol$) = 2.93$ g of ice will melt

12.38 a) Your body could be the system. Its mass and temperature and chemical makeup would be needed to define it. It would release heat to the water.

b) One could define the system as the Bunsen burner, the methane, the air and the products of combustion. The system would release heat.

c) The system would be the contents of the thermos flask. The concentrations of acid and base, the equation for the reaction, the amount of solutions, the amount of products and the temperature would be needed to define the system. The reaction releases heat. The system would approximate an isolated system and would neither absorb nor release heat.

12.39 $C_{(s,\ graphite)} + O_{2(g)} \rightarrow CO_{2(g)}$ $\Delta H = -\ 393.51$ kJ/mol

$C_{60} + 60\ O_{2(g)} \rightarrow 60\ CO_{2(g)}$

$\Delta H^\circ_f(C_{60})$ = probably positive

Therefore, the ΔH for combustion of buckminsterfullerene equals 60 ΔH of combustion of graphite minus the positive ΔH°_f for C_{60}. Therefore, the value for buckminsterfullerene will be more negative. The heat of combustion per gram will be greater for buckminsterfullerene (by (1/60)(1/12) of the ΔH°_f of buckminsterfullerene).

12.40 a) $N_2 + O_2 \rightarrow 2\ NO$ balanced

$\Delta H_{rxn} = [2\ \Delta H^\circ_f(NO)] - [\Delta H^\circ_f(N_2) + \Delta H^\circ_f(O_2)]$

$= [(2\text{ mol } NO)(90.25\text{ kJ/mol } NO)]$

$- [(1\text{ mol } N_2)(0\text{ kJ/mol } N_2) + (1\text{ mol } O_2)(0\text{ kJ/mol } O_2)]$

$= [180.5\text{ kJ}] - [0\text{ kJ} + 0\text{ kJ}] = 180.5\text{ kJ}$

$(180.5\text{ kJ/mol } N_2)(1\text{ mol } N_2/2\text{ mol N}) = 90.2\text{ kJ/mol N}$

b) $N_2O + O_2 \rightarrow NO$

$2\ N_2O + O_2 \rightarrow 4\ NO$

$\Delta H_{rxn} = [4\ \Delta H^\circ_f(NO)] - [2\ \Delta H^\circ_f(N_2O) + \Delta H^\circ_f(O_2)]$

$= [(4\text{ mol } NO)(90.25\text{ kJ/mol } NO)]$

$- [(2\text{ mol } N_2O)(82.05\text{ kJ/mol } N_2O) + (1\text{ mol } O_2)(0\text{ kJ/mol } O_2)]$

$= [361.0\text{ kJ}] - [164.1\text{ kJ} + 0\text{ kJ}] = 196.9\text{ kJ}$

$(196.9\text{ kJ/2 mol } N_2O)(\text{mol } N_2O/2\text{ mol N}) = 49.2\text{ kJ/mol N}$

c) $NO + O_2 \rightarrow NO_2$

$2\ NO + O_2 \rightarrow 2\ NO_2$

$\Delta H_{rxn} = [2\ \Delta H^\circ_f(NO_2)] - [2\ \Delta H^\circ_f(NO) + \Delta H^\circ_f(O_2)]$

$= [(2\text{ mol } NO_2)(33.18\text{ kJ/mol } NO_2)]$

$- [(2\text{ mol } NO)(90.25\text{ kJ/mol } NO) + (1\text{ mol } O_2)(0\text{ kJ/mol } O_2)]$

$= [66.36\text{ kJ}] - [180.50\text{ kJ} + 0\text{ kJ}] = -114.1\text{ kJ}$

$(-114.1\text{ kJ/2 mol } NO)(1\text{ mol } NO/\text{mol N}) = -57.0\text{ kJ/mol N}$

(continued)

(12.40 continued)

d) $NO_2 + O_2 \rightarrow N_2O_5$

$4\ NO_2 + O_2 \rightarrow 2\ N_2O_5$

$\Delta H_{rxn} = [2\ \Delta H^\circ_f(N_2O_5)] - [4\ \Delta H^\circ_f(NO_2) + \Delta H^\circ_f(O_2)]$

$= [(2\text{ mol } N_2O_5)(-43.1\text{ kJ/mol } N_2O_5)]$

$- [(4\text{ mol } NO_2)(33.18\text{ kJ/mol } NO_2) + (1\text{ mol } O_2)(0\text{ kJ/mol } O_2)]$

$= [-86.2\text{ kJ}] - [132.72\text{ kJ} + 0\text{ kJ}] = -218.9\text{ kJ}$

$(-218.9\text{ kJ/4 mol } NO_2)(1\text{ mol } NO_2/1\text{ mol N}) = -54.7\text{ kJ/mol N}$

12.41 Any living organism will die if made into an isolated thermodynamic system. Even before dying, its performance will change from the normal.

12.42 a) C_p has units of J/mol K and $C_{calorimeter}$ has units of J/K.

$$\text{b) } (115\text{ mL})\left(\frac{1.00\text{ g}}{\text{mL}}\right)\left(\frac{1\text{ mol } H_2O}{18.02\text{ g } H_2O}\right)\left(\frac{75.291\text{ J}}{\text{mol K}}\right) = 480\text{ J/K}$$

12.43 Prefer classification as physical change. The process can be reversed by a physical process; that is, the water can be allowed to evaporate (or be boiled off) to recover the original substance, $NaCl_{(s)}$.

12.44 Let T = the final temperature of the system.

$q_{H_2O} = -q_{Cu}$

$nC_{p,H_2O}\Delta T_{H_2O} = -nC_{p,Cu}\Delta T_{Cu}$

$$(200\text{ mL})\left(\frac{1.00\text{ g}}{\text{mL}}\right)\left(\frac{\text{mol } H_2O}{18.02\text{ g } H_2O}\right)\left(\frac{75.291\text{ J}}{\text{mol K}}\right)(T - 278\text{ K})$$

$$= -(9.50\text{ g})\left(\frac{\text{mol Cu}}{63.55\text{ g Cu}}\right)\left(\frac{24.435\text{ J}}{\text{mol K}}\right)(T - 473\text{ K})$$

$(835.6\text{ J} + 3.653\text{ J})\ T = 1728\text{ J K} + 232{,}300\text{ J K}$

$(839.3\text{ J})\ T = 2.340 \times 10^5\text{ J K}$

$T = 278.8\text{ K} = 5.7°C$

12.45 a) $C_6H_{12}O_{6(s)} + 6\ O_{2(g)} \rightarrow 6\ H_2O_{(l)} + 6\ CO_{2(g)}$

b) $$\left(\frac{-15.7\ \text{kJ}}{1.00\ \text{g glucose}}\right)\left(\frac{180.156\ \text{g glucose}}{\text{mol glucose}}\right) = -2.83 \times 10^3\ \text{kJ / mol glucose}$$

c) $\Delta H_{rxn} = [6\,\Delta H_f^\circ(H_2O_{(l)}) + 6\,\Delta H_f^\circ(CO_2)] - [\Delta H_f^\circ(C_6H_{12}O_6) + 6\,\Delta H_f^\circ(O_2)]$

$$-2.83 \times 10^3\ \text{kJ} = \left[(6\ \text{mol}\ H_2O_{(l)})\left(\frac{-285.8\ \text{kJ}}{\text{mol}\ H_2O_{(l)}}\right) + (6\ \text{mol}\ CO_2)\left(\frac{-393.5\ \text{kJ}}{\text{mol}\ CO_2}\right)\right]$$

$$-\left[(1\ \text{mol}\ C_6H_{12}O_6)\Delta H_f^\circ(C_6H_{12}O_6) + (6\ \text{mol}\ O_2)\left(\frac{0\ \text{kJ}}{\text{mol}\ O_2}\right)\right]$$

$-2.83 \times 10^3\ \text{kJ} =$

$[-1714.8\ \text{kJ} - 2361.0\ \text{kJ}] - [(1\ \text{mol}\ C_6H_{12}O_6)\ \Delta H_f^\circ(C_6H_{12}O_6) + 0\ \text{kJ}]$

$-2.83 \times 10^3\ \text{kJ} + [4075.8\ \text{kJ}] = -\ (1\ \text{mol}\ C_6H_{12}O_6)\ \Delta H_f^\circ(C_6H_{12}O_6)$

$\Delta H_f^\circ(C_6H_{12}O_6) = -1.25 \times 10^3\ \text{kJ/ mol}\ C_6H_{12}O_6$

12.46 From Problem 12.45: 15.7 kJ/1.00 g glucose

$$\left(\frac{220\ \text{kJ}}{1\ \text{km}}\right)\left(\frac{1.00\ \text{g glucose}}{15.7\ \text{kJ}}\right)\left(\frac{100\ \text{g cereal}}{35\ \text{g glucose}}\right)\left(\frac{100}{30}\right) = 133\ \text{g cereal / km}$$

12.47 $\Delta E = q + w = q - P\Delta V = nC_p\Delta T - P(V_f - V_i)$

$$= \left(\frac{1.25\ \text{g}}{132.92\ \text{g / mol}}\right)\left(\frac{80.7\ \text{J}}{\text{mol K}}\right)(20^\circ\text{C} - 50^\circ\text{C})\left(\frac{\text{K}}{^\circ\text{C}}\right)$$

$$-(1\ \text{atm})(248\ \text{mL} - 274\ \text{mL})\left(\frac{\text{cm}^3}{\text{mL}}\right)$$

$$= -22.767\ \text{J} + 26\ \text{atm cm}^3\left(\frac{1\ \text{m}}{10^2\ \text{cm}}\right)^3\left(\frac{1.013 \times 10^5\ \text{kg m}^{-1}\ \text{s}^{-2}}{1\ \text{atm}}\right)$$

$= -22.767\ \text{J} + 2.6338\ \text{kg m}^2\ \text{s}^{-2} = -22.767\ \text{J} + 2.6338\ \text{J} = -20.1\ \text{J}$

$\Delta H = \Delta E + \Delta(nRT);\ \Delta(nRT) = R\Delta(nT) = RT\Delta n + Rn\Delta T$ $\quad\quad \Delta n = 0$

$\Delta T = 20^\circ\text{C} - 50^\circ\text{C} = -\ 30^\circ\text{C} = -\ 30\ \text{K}$

$$= -20.1\ \text{J} + nR\Delta T = -20.1\ \text{J} + \left(\frac{1.25\ \text{g}}{132.92\ \text{g / mol}}\right)(8.314\ \text{J / mol K})(-30\ \text{K})$$

$= -20.1\ \text{J} - 2.3\ \text{J} = -22.4\ \text{J}$

12.48 $q = nC_{p,H_2O}\Delta T$

$\Delta T = (50°C - 25°C) = 25°C = 25\ K$

$$q = (40\ L)\left(\frac{1000\ g}{1\ L}\right)\left(\frac{1\ mol\ H_2O}{18.02\ g\ H_2O}\right)\left(\frac{75.291\ J}{mol\ K}\right)(25\ K)\left(\frac{100}{80}\right)$$

$= 5.22 \times 10^6$ J (heat needed)

$$5.22 \times 10^6\ J = N\frac{hc}{\lambda} =$$

$$\frac{(6.626 \times 10^{-34} J\ s)(2.998 \times 10^8 m\ s^{-1})}{(500\ nm)(10^{-9}\ m / nm)}(6.022 \times 10^{23} / \text{mol phot's})(\#\ \text{mol phot's})$$

mol photons = 21.8

12.49 Let T = final temperature $\qquad q_{spoon} = - q_{coffee}$

$nC_{p,spoon}\Delta T = - nC_{pH_2O}\Delta T$

$$\left(\frac{99\ g\ Ag}{107.9\ Ag / mol}\right)\left(\frac{25.351\ J}{mol\ K}\right)(T - 280\ K)$$

$$= -(200\ mL\ H_2O)\left(\frac{1.00\ g}{mL}\right)\left(\frac{1\ mol}{18.02\ g\ H_2O}\right)\left(\frac{75.291\ J}{mol\ K}\right)(T - 350\ K)$$

$(23.26\ J/K)\ T - 6512.8\ J = - (835.64\ J/K)\ T + 292473.4\ J$

$2.98986 \times 10^5\ J = (858.9\ J/K)\ T$

$T = 348.1\ K$

An aluminum spoon of the same mass (99 g) would be (99 g/27 g mol^{-1}) 3.7 moles while the silver spoon would be (99 g/107.9 g mol^{-1}) 0.92 moles. C_p times number of moles yields:

For Ag: (25.351 J/mol K)(0.92 mol) = 23 J/K

For Al: (24.35 J/mol K)(3.7 mol) = 90 J/K

Therefore, the final temperature of the coffee would be lower for the aluminum spoon.

12.50 $CH_{4(g)} + 2\ O_{2(g)} \rightarrow CO_{2(g)} + 2\ H_2O_{(g)}$

$\Delta H_{rxn} = [\ \Delta H^\circ_f(CO_2) + 2\ \Delta H^\circ_f(H_2O_{(g)})\ ] - [\ \Delta H^\circ_f(CH_4) + 2\ \Delta H^\circ_f(O_2)]$

$$= \left[(1\ mol\ CO_2)\left(\frac{-393.5\ kJ}{mol\ CO_2}\right) + (2\ mol\ H_2O_{(g)})\left(\frac{-241.8\ kJ}{mol\ H_2O_{(g)}}\right)\right]$$

$$- \left[(1\ mol\ CH_4)\left(\frac{-74.81\ kJ}{mol\ CH_4}\right) + (2\ mol\ O_2)\left(\frac{0\ kJ}{mol\ O_2}\right)\right]$$

$= [-393.5\ kJ - 483.6\ kJ] - [\ 74.81\ kJ + 0\ kJ] = -802.3\ kJ/mol\ CH_4$

$\Delta T = 25°C - 15°C = 10°C = 10\ K \qquad$ Assume P = 1.00 atm.

(continued)

(12.50 continued)

size of room = (3.0 m)(5.0 m)(4.0 m) = $(60\ m^3)(10^2\ cm/m)^3(L/10^3\ cm^3) = 6.0 \times 10^4$ L

air volume at 15°C (288 K) = 6.0×10^4 L

N_2 volume at 15°C (288 K) = $(0.78)(6.0 \times 10^4\ L) = 4.68 \times 10^4$ L

O_2 volume at 15°C (288 K) = $(0.22)(6.0 \times 10^4\ L) = 1.32 \times 10^4$ L

This assumes that all of the air originally in the room at 15°C is heated and that none escapes on the expansion accompanying increase of temperature to 25°C.

$$\text{mol } O_2 = n_{O_2} = \frac{PV}{RT} = \frac{(1.00\ \text{atm})(1.32 \times 10^4\ \text{L})}{(0.08206\ \text{L atm / mol K})(288\ \text{K})} = 558.5\ \text{mol } O_2$$

$$\text{mol } N_2 = n_{N_2} = \frac{PV}{RT} = \frac{(1.00\ \text{atm})(4.68 \times 10^4\ \text{L})}{(0.08206\ \text{L atm / mol K})(288\ \text{K})} = 1980\ \text{mol } N_2$$

heat needed to warm O_2

$= n_{O_2}C_{p,O_2}\Delta T = (558.5\ \text{mol})(29.355\ \text{J/mol K})(10\ \text{K}) = 1.639 \times 10^5$ J

heat needed to warm N_2

$$= n_{N_2}C_{p,N_2}\Delta T = (1980\ \text{mol})\left(\frac{29.125\ \text{J}}{\text{mol K}}\right)(10\ \text{K}) = 5.767 \times 10^5\ \text{J}$$

total heat needed = 7.406×10^5 J = 7.406×10^2 kJ

$$\text{moles of } CH_4 \text{ needed} = (7.406 \times 10^2\ \text{kJ})\left(\frac{\text{mol } CH_4}{802.3\ \text{kJ}}\right) = 0.9231\ \text{mol } CH_4$$

$$\text{mass of } CH_4 \text{ needed} = (0.9231\ \text{mol } CH_4)\left(\frac{16.042\ \text{g } CH_4}{\text{mol } CH_4}\right) = 14.8\ \text{g } CH_4 \text{ or } 15\ \text{g } CH_4$$

12.51 $w_{sys} = -P_{ext}\Delta V_{sys}$

$w_{gas} = -(4.00\ \text{atm})(20.0\ \text{L} - 30.0\ \text{L})$

$$= (+40.0\ \text{atm L})\left(\frac{1000\ \text{cm}^3}{\text{L}}\right)\left(\frac{1\ \text{m}}{10^2\ \text{cm}}\right)^3\left(\frac{1.013 \times 10^5\ \text{kg m}^{-1}\ \text{s}^{-2}}{\text{atm}}\right)$$

$= 4052\ \text{kg m}^2\ \text{s}^{-2} = 4052\ \text{J} \cong 4.05$ kJ

$q_{gas} = nCp\Delta T \quad \Delta T = 0 \therefore q_{gas} = 0$

$\Delta E_{gas} = q_{gas} + w_{gas} = 0 + 4.05\ \text{kJ} = 4.05$ kJ or 4.05×10^3 J

12.52 The heat supplied to a monatomic gas results in an increase in temperature in the form of an increase in its translational (motion) energy. For diatomic gases the heat supplied not only increases translational energy but additional heat is used to increase molecular rotational and vibrational motions.

12.53 a) $2\ SO_{2(g)} + O_{2(g)} \rightarrow 2\ SO_{3(g)}$

$\Delta H_{rxn} = [2\ \Delta H^\circ_f(SO_3)] - [2\ \Delta H^\circ_f(SO_2) + \Delta H^\circ_f(O_2)]$

$$= \left[(2 \text{ mol } SO_3)\left(\frac{-454.51 \text{ kJ}}{\text{mol } SO_3}\right)\right] - \left[(2 \text{ mol } SO_2)\left(\frac{-296.83 \text{ kJ}}{\text{mol } SO_2}\right) + (1 \text{ mol } O_2)\left(\frac{0 \text{ kJ}}{\text{mol } O_2}\right)\right]$$

$= [-909.02 \text{ kJ}] - [-593.66 \text{ kJ} + 0 \text{ kJ}] = -315.36 \text{ kJ}$

b) $2\ NO_{2(g)} \rightarrow N_2O_{4(g)}$

$\Delta H_{rxn} = [\Delta H^\circ_f(N_2O_4)] - [2\ \Delta H^\circ_f(NO_2)]$

$$= \left[(1 \text{ mol } N_2O_4)\left(\frac{9.16 \text{ kJ}}{\text{mol } N_2O_4}\right)\right] - \left[(2 \text{ mol } NO_2)\left(\frac{33.18 \text{ kJ}}{\text{mol } NO_2}\right)\right]$$

$= [9.16 \text{ kJ}] - [66.36 \text{ kJ}] = -\ 57.20 \text{ kJ}$

c) $Fe_2O_{3(s)} + 2\ Al_{(s)} \rightarrow Al_2O_{3(s)} + 2\ Fe_{(s)}$

$\Delta H_{rxn} = [\Delta H^\circ_f(Al_2O_3) + 2\ \Delta H^\circ_f(Fe)] - [\Delta H^\circ_f(Fe_2O_3) + 2\ \Delta H^\circ_f(Al)]$

$$= \left[(1 \text{ mol } Al_2O_3)\left(\frac{-1675.7 \text{ kJ}}{\text{mol } Al_2O_3}\right) + (2 \text{ mol Fe})\left(\frac{0 \text{ kJ}}{\text{mol Fe}}\right)\right] -$$

$$\left[(1 \text{ mol } Fe_2O_3)\left(\frac{-824.2 \text{ kJ}}{\text{mol } NO_2}\right) + (2 \text{ mol Al})\left(\frac{0 \text{ kJ}}{\text{mol Al}}\right)\right]$$

$= [-1675.7 \text{ kJ} + 0 \text{ kJ}] - [-824.2 \text{ kJ} + 0 \text{ kJ}] = -851.5 \text{ kJ}$

12.54 $\Delta H_{H_2O} = q_{H_2O} = nC_{p,H_2O}\Delta T =$

$$\left(\frac{21.5 \text{ g } H_2O}{18.02 \text{ g } H_2O/\text{mol}}\right)\left(\frac{75.291 \text{ J}}{\text{mol K}}\right)(21.5°C - 15.5°C)\left(\frac{K}{°C}\right) = 539 \text{ J}$$

$\Delta H_{coin} = q_{coin} = nC_{p,coin}\Delta T = (15.5 \text{ g/MM})(C_{p,coin})(21.5°C - 100°C)(K/°C)$

$\Delta H_{coin} = -\ 539 \text{ J} = -\ 1216.75\ (C_p/MM)\text{g K}$

$C_p/MM = 0.443 \text{ J/g K}$

for silver: $\dfrac{C_p}{MM} = \dfrac{25.351 \text{ J/mol K}}{107.9 \text{ g/mol}} = 0.235 \text{ J/g K}$

for nickel: $\dfrac{C_p}{MM} = \dfrac{26.07 \text{ J/mol K}}{58.70 \text{ g/mol}} = 0.444 \text{ J/g K}$

The coin is a counterfeit nickel copy.

12.55 $(1 \text{ km})\left(\dfrac{1 \text{ L}}{6.0 \text{ km}}\right)\left(\dfrac{10^3 \text{ mL}}{1 \text{ L}}\right)\left(\dfrac{0.68 \text{ g}}{\text{mL}}\right)\left(\dfrac{48 \text{ kJ}}{\text{g}}\right) = 5.4 \times 10^3 \text{ kJ}$

12.56 The heat absorbed by the droplet is equal to ΔE because $w = 0$.

12.57 a) $2\ NH_{3(g)} + 3\ O_{2(g)} + 2\ CH_{4(g)} \rightarrow 2\ HCN_{(g)} + 6\ H_2O_{(g)}$

$$\Delta H_{rxn} = [2\,\Delta H_f^\circ(HCN_{(g)}) + 6\,\Delta H_f^\circ(H_2O_{(g)})] - [2\,\Delta H_f^\circ(NH_3) + 3\,\Delta H_f^\circ(O_2) + 2\,\Delta H_f^\circ(CH_4)]$$

$$= \left[(2\text{ mol HCN})\left(\frac{135.1\text{ kJ}}{\text{mol HCN}}\right) + (6\text{ mol }H_2O)\left(\frac{-241.8\text{ kJ}}{\text{mol }H_2O}\right)\right]$$

$$-\left[(2\text{ mol }NH_3)\left(\frac{-46.1\text{ kJ}}{\text{mol }NH_3}\right) + (3\text{ mol }O_2)\left(\frac{0\text{ kJ}}{\text{mol }O_2}\right) + (2\text{ mol }CH_4)\left(\frac{-74.81\text{ kJ}}{\text{mol }CH_4}\right)\right]$$

$$= [270.2\text{ kJ} + (-1450.8\text{ kJ})] - [(-92.2\text{ kJ}) + 0\text{ kJ} + (-149.62\text{ kJ})] = -938.8\text{ kJ}$$

b) $2\ C_2H_{2(g)} + 5\ O_{2(g)} \rightarrow 4\ CO_{2(g)} + 2\ H_2O_{(g)}$

$$\Delta H_{rxn} = [4\ \Delta H_f^\circ(CO_2) + 2\ \Delta H_f^\circ(H_2O_{(g)})] - [2\ \Delta H_f^\circ(C_2H_2) + 5\ \Delta H_f^\circ(O_2)]$$

$$= \left[(4\text{ mol }CO_2)\left(\frac{-393.5\text{ kJ}}{\text{mol }CO_2}\right) + (2\text{ mol }H_2O)\left(\frac{-241.8\text{ kJ}}{\text{mol }H_2O}\right)\right]$$

$$-\left[(2\text{ mol }C_2H_2)\left(\frac{226.7\text{ kJ}}{\text{mol }C_2H_2}\right) + (5\text{ mol }O_2)\left(\frac{0\text{ kJ}}{\text{mol }O_2}\right)\right]$$

$$= [(-1574.0\text{ kJ}) + (-483.6\text{ kJ})] - [453.4\text{ kJ} + 0\text{ kJ}] = -2511.0\text{ kJ}$$

c) $C_2H_{4(g)} + O_{3(g)} \rightarrow CH_3CHO_{(g)} + O_{2(g)}$

$$\Delta H_{rxn} = [\,\Delta H_f^\circ(CH_3CHO) + \Delta H_f^\circ(O_2)] - [\,\Delta H_f^\circ(C_2H_4) + \Delta H_f^\circ(O_3)]$$

$$= \left[(1\text{ mol }CH_3CHO)\left(\frac{-166.2\text{ kJ}}{\text{mol }CH_3CHO}\right) + (1\text{ mol }O_2)\left(\frac{0\text{ kJ}}{\text{mol }O_2}\right)\right]$$

$$-\left[(1\text{ mol }C_2H_4)\left(\frac{52.3\text{ kJ}}{\text{mol }C_2H_4}\right) + (1\text{ mol }O_3)\left(\frac{142.7\text{ kJ}}{\text{mol }O_3}\right)\right]$$

$$= [-166.2\text{ kJ} + 0\text{ kJ}] - [52.3\text{ kJ} + 142.7\text{ kJ}] = -361.2\text{ kJ}$$

12.58 a) $q = nC_{p,H_2O}\Delta T \qquad \Delta T = (30°C - 20°C) = 10°C = 10\text{ K}$

$$q = (155\text{ m}^3)\left(\frac{10^2\text{ cm}}{\text{m}}\right)^3\left(\frac{1.00\text{ g}}{\text{cm}^3}\right)\left(\frac{1\text{ mol }H_2O}{18.02\text{ g }H_2O}\right)\left(\frac{75.291\text{ J}}{\text{mol K}}\right)(10\text{ K})$$

$$= 6.476 \times 10^9\text{ J}$$

(continued)

(12.58 continued)

b) $(6.476 \times 10^9 \text{ J})\left(\frac{\text{kJ}}{10^3 \text{ J}}\right)\left(\frac{\text{mol } CH_4}{803 \text{ kJ}}\right)\left(\frac{16.043 \text{ g } CH_4}{\text{mol } CH_4}\right)\left(\frac{100}{80}\right)$

$= 1.6 \times 10^5$ g of CH_4

12.59 Heat to convert $H_2O = nC_{p,H_2O}\Delta T + n\Delta H_{vap}$ $\Delta T = (100°C - 25°C) = 75°C = 75$ K

$$q = \left(\frac{250 \text{ g } H_2O}{18.02 \text{ g } H_2O/\text{mol}}\right)\left(\frac{75.291 \text{ J}}{\text{mol K}}\right)(75 \text{ K}) + \left(\frac{250 \text{ g } H_2O}{18.02 \text{ g } H_2O/\text{mol}}\right)\left(\frac{40.79 \text{ kJ}}{\text{mol}}\right)$$

$= 7.834 \times 10^4 \text{ J} + 565.9 \text{ kJ} = 7.834 \times 10^1 \text{ kJ} + 565.9 \text{ kJ} = 644 \text{ kJ}$

mass of CH_4 = $(644 \text{ kJ})\left(\frac{\text{mol } CH_4}{803 \text{ kJ}}\right)\left(\frac{16.043 \text{ g } CH_4}{\text{mol } CH_4}\right) = 12.9 \text{ g } CH_4$

12.60 $2\, ClF_{3(g)} + 2\, NH_{3(g)} \rightarrow N_{2(g)} + 6\, HF_{(g)} + Cl_{2(g)}$

$\Delta H_{rxn} = [\Delta H_f^\circ(N_2) + 6\ \Delta H_f^\circ(HF) + \Delta H_f^\circ Cl_2] - [2\ \Delta H_f^\circ(ClF_3) + 2\ \Delta H_f^\circ(NH_3)]$

-1196 kJ =

$$\left[(1 \text{ mol } N_2)\left(\frac{0 \text{ kJ}}{\text{mol } N_2}\right) + (6 \text{ mol HF})\left(\frac{-271.1 \text{ kJ}}{\text{mol HF}}\right) + (1 \text{ mol } Cl_2)\left(\frac{0 \text{ kJ}}{\text{mol } Cl_2}\right)\right]$$

$$-\left[(2 \text{ mol } ClF_3)\Delta H_f^\circ(ClF_3) + (2 \text{ mol } NH_3)\left(\frac{-46.1 \text{ kJ}}{\text{mol } NH_3}\right)\right]$$

$- 1196 \text{ kJ} = [0 \text{ kJ} + (-1626.6 \text{ kJ}) + 0 \text{ kJ})] - [(2 \text{ mol } ClF_3)\, \Delta H_f^\circ(ClF_3) + (- 92.2 \text{ kJ})]$

$(2 \text{ mol } ClF_3)\, \Delta H_f^\circ(ClF_3) = 1196 \text{ kJ} + [-1626.6 \text{ kJ}] + [92.2 \text{ kJ}] = -338.4 \text{ kJ}$

$\Delta H_f^\circ(ClF_3) = -169.2$ kJ/mol

12.61 $H_2O_{(l,\,298\,K)} \rightarrow H_2O_{(g,\,298\,K)}$ $\Delta H_{rxn} = [\,\Delta H^\circ_f(H_2O_{(g)})] - [\,\Delta H^\circ_f(H_2O_{(l)})] =$

$$\left[(1 \text{ mol } H_2O_{(g)})\left(\frac{-241.8 \text{ kJ}}{\text{mol } H_2O_{(g)}}\right)\right] - \left[(1 \text{ mol } H_2O_{(l)})\left(\frac{-285.8 \text{ kJ}}{\text{mol } H_2O_{(l)}}\right)\right]$$

$= -\ 241.8 \text{ kJ} + 285.8 \text{ kJ} = 44.0 \text{ kJ}$ $\Delta H_{vap} = 40.79 \text{ kJ/mol}$

$H_2O(l,\ 373\ K)$ —(+ 40.79 kJ/mol)→ $H_2O(g,\ 373\ K)$

$H_2O(l,\ 298\ K)$ → $H_2O(l,\ 373\ K)$: + 5.65 kJ/mol = $(C_{p,\,H_2O(l)})(75\ K)$

$H_2O(g,\ 373\ K)$ → $H_2O(g,\ 298\ K)$: - 2.74 kJ/mol = $(C_{p,\,H_2O(g)})(-75\ K)$

$H_2O(l,\ 298\ K)$ —(+ 44.0 kJ/mol)→ $H_2O(g,\ 298\ K)$

44.0 kJ/mol ≅ 5.65 kJ/mol + 40.79 kJ/mol - 2.74 kJ/mol = 43.7 kJ/mol
The ΔH_{vap} is for the steam and water at 100°C (373 K) and 1 atm. The ΔH_{rxn} is for steam and water at standard conditions of 25°C (298 K) and 1 atm. Note that the 2 paths for the change give the same enthalpy change to two significant digits.

12.62 a) $I_{2(s)} \rightarrow I_{2(g)}$

b) $\frac{1}{2} I_{2(s)} \rightarrow I_{(g)}$

c) $2\ C_{(s,\,graphite)} + \frac{3}{2} H_{2(g)} + \frac{1}{2} Cl_{2(g)} \rightarrow C_2H_3Cl_{(g)}$

d) $Na_2SO_{4(s)} \rightarrow 2\ Na^+_{(aq)} + SO_{4(aq)}^{2-}$

12.63 $$C_{p,\,Pb} = \frac{100.0 \text{ J}}{(52.5 \text{ g})(299.6 \text{ K} - 280.0 \text{ K})}\left(\frac{207.2 \text{ g Pb}}{1 \text{ mol Pb}}\right) = 20.1 \text{ J / mol K}$$

12.64 a) $$\Delta H_{combustion} = -(1694 \text{ kJ / mol } (CH_3)_2NNH_2)\left(\frac{1 \text{ mol } (CH_3)_2NNH_2}{60.10 \text{ g } (CH_3)_2NNH_2}\right)$$

$= -28.19 \text{ kJ/g } (CH_3)_2NNH_2$

b) $$\Delta H_{combustion} = -(726 \text{ kJ / mol } CH_3OH)\left(\frac{1 \text{ mol } CH_3OH}{32.04 \text{ g } CH_3OH}\right)$$

$= -22.7 \text{ kJ/g } CH_3OH$

(continued)

(12.64 continued)

c) $\Delta H_{combustion} = -(5500 \text{ kJ / mol } C_8H_{18})\left(\frac{1 \text{ mol } C_8H_{18}}{114.22 \text{ g } C_8H_{18}}\right)$

$= -48.15 \text{ kJ/g } C_8H_{18}$

Octane has the largest energy content per gram and methanol has the smallest energy content per gram.

12.65 $P_4S_{3(s)} + 8\ O_{2(g)} \rightarrow P_4O_{10(s)} + 3\ SO_{2(g)} \qquad \Delta H_{rxn} = -\ 3677 \text{ kJ}$

$-\ 3677 \text{ kJ} = \Delta H_{rxn} = [\Delta H^\circ_f(P_4O_{10}) + 3\ \Delta H^\circ_f(SO_2)] - [\Delta H^\circ_f(P_4S_3) + 8\ \Delta H^\circ_f(O_2)]$

$$= \left[(1 \text{ mol } P_4O_{10})\left(\frac{-2984.0 \text{ kJ}}{\text{mol } P_4O_{10}}\right) + (3 \text{ mol } SO_2)\left(\frac{-296.8 \text{ kJ}}{\text{mol } SO_2}\right)\right]$$

$$-\left[\Delta H^\circ_f(P_4S_3) + (8 \text{ mol } O_2)\left(\frac{0 \text{ kJ}}{\text{mol } O_2}\right)\right]$$

$-3677 \text{ kJ} = -2984.0 \text{ kJ} + (-890.4 \text{ kJ}) - \Delta H^\circ_f(P_4S_3) + 0$

$\Delta H^\circ_f(P_4S_3) = -197 \text{ kJ}$

12.66 The kinetic energy of the automobile is transformed into work on the brakes. Some of that is transformed into heat by friction and is passed on to the tires where it is transformed into more heat by the friction of the tire surface with the road surface.

12.67 The process would appear to be deposition (the opposite of sublimation).
a) ΔH_{sys} would be (-) as heat is released (exothermic process) as a gas is transformed into a solid.

b) ΔE_{surr} would be (+) as it would have the opposite sign of ΔE_{sys} which would be approximately the same value as ΔH_{sys}.

c) $\Delta E_{universe}$ would equal zero (0); the total energy of the universe is conserved.

12.68 The process would appear to be a reaction of a diatomic gas becoming a monatomic gas at constant V.
a) $w_{sys} = 0$ because this is like a constant-volume calorimeter, $\Delta V_{sys} = 0$.

b) q_{sys} would be (+) since energy must be added to the system to break bonds.

c) ΔE_{surr} would be (-) since it has the opposite sign of ΔE_{sys} which would be (+).

12.69 The process is an expansion at constant temperature.
a) w_{sys} is (-) for an expansion: [$w_{sys} = -\ P_{ext}\Delta V_{sys}$, ΔV_{sys} is (+)].

b) ΔE_{surr} would be (+) because its sign is opposite ΔE_{sys} which is (-).

c) $q_{sys} = 0$ because this is a constant temperature process.

12.70 $4\ PH_{3(g)} + 8\ O_{2(g)} \rightarrow P_4O_{10(s)} + 6\ H_2O_{(g)}$

$\Delta H_{combustion} = [\Delta H_f^\circ(P_4O_{10}) + 6\ \Delta H_f^\circ(H_2O_{(g)})] - [4\ \Delta H_f^\circ(PH_3) + 8\ \Delta H_f^\circ(O_2)]$

$$= \left[(1 \text{ mol } P_4O_{10})\left(\frac{-2984.0 \text{ kJ}}{\text{mol } P_4O_{10}}\right) + (6 \text{ mol } H_2O)\left(\frac{-241.8 \text{ kJ}}{\text{mol } H_2O}\right)\right]$$

$$- \left[(4 \text{ mol } PH_3)\left(\frac{5.4 \text{ kJ}}{\text{mol } PH_3}\right) + (8 \text{ mol } O_2)\left(\frac{0 \text{ kJ}}{\text{mol } O_2}\right)\right]$$

$= (-2984.0 - 1450.8) - (21.6 + 0) = -4456.4$ kJ

12.71 $C_{20}H_{32}O_{2(s)} + 27\ O_{2(g)} \rightarrow 20\ CO_{2(g)} + 16\ H_2O_{(l)}$

$\Delta H_{combustion} = [20\ \Delta H_f^\circ(CO_2) + 16\ \Delta H_f^\circ(H_2O_{(l)})]$

$- [\Delta H_f^\circ(C_{20}H_{32}O_2) + 27\ \Delta H_f^\circ(O_2)]$

$$= \left[(20 \text{ mol } CO_2)\left(\frac{-393.5 \text{ kJ}}{\text{mol } CO_2}\right) + (16 \text{ mol } H_2O)\left(\frac{-285.8 \text{ kJ}}{\text{mol } H_2O}\right)\right]$$

$$- \left[(1 \text{ mol } C_{20}H_{32}O_2)\left(\frac{-636 \text{ kJ}}{\text{mol } C_{20}H_{32}O_2}\right) + (27 \text{ mol } O_2)\left(\frac{0 \text{ kJ}}{\text{mol } O_2}\right)\right]$$

$= -1.18 \times 10^4$ kJ

heat needed to warm bear flesh =

$(500 \text{ kg})(10^3 \text{ g/kg})(4.18 \text{ J g}^{-1} \text{ K}^{-1})(25°C - 5°C)(K/°C) = 4.18 \times 10^7 \text{ J}$

$$(4.18 \times 10^7 \text{ J})\left(\frac{\text{kJ}}{10^3 \text{ J}}\right)\left(\frac{1 \text{ mol } C_{20}H_{32}O_2}{1.18 \times 10^4 \text{ kJ}}\right)\left(\frac{304.46 \text{ g } C_{20}H_{32}O_2}{1 \text{ mol } C_{20}H_{32}O_2}\right)$$

$= 1080 \text{ g } C_{20}H_{32}O_2$ or 1.08 kg

12.72 $2\ N_2H_{4(l)} + N_2O_{4(g)} \rightarrow 3\ N_{2(g)} + 4\ H_2O_{(g)}$

a) $\Delta H_{rxn} = [3\ \Delta H_f^\circ(N_2) + 4\ \Delta H_f^\circ(H_2O_{(g)})] - [2\ \Delta H_f^\circ(N_2H_4) + \Delta H_f^\circ(N_2O_4)]$

$$= \left[(3 \text{ mol } N_2)\left(\frac{0 \text{ kJ}}{\text{mol } N_2}\right) + (4 \text{ mol } H_2O)\left(\frac{-241.8 \text{ kJ}}{\text{mol } H_2O}\right)\right]$$

$$- \left[(2 \text{ mol } N_2H_4)\left(\frac{50.6 \text{ kJ}}{\text{mol } N_2H_4}\right) + (1 \text{ mol } N_2O_4)\left(\frac{9.16 \text{ kJ}}{\text{mol } N_2O_4}\right)\right]$$

$= [0 \text{ kJ} + (-967.2 \text{ kJ})] - [101.2 \text{ kJ} + 9.16 \text{ kJ}] = -1077.6$ kJ

$(\Delta H_{rxn})/2 \text{ mol } N_2H_4 = -538.8 \text{ kJ mol}^{-1}$

(continued)

(12.72 continued)

b) $N_2H_{4(l)} + O_{2(g)} \rightarrow N_{2(g)} + 2\ H_2O_{(g)}$

$\Delta H_{rxn} = [\ \Delta H^\circ_f(N_2) + 2\ \Delta H^\circ_f(H_2O_{(g)})] - [\ \Delta H^\circ_f(N_2H_4) + \Delta H^\circ_f(O_2)]$

$$= \left[(1\text{ mol }N_2)\left(\frac{0\text{ kJ}}{\text{mol }N_2}\right) + (2\text{ mol }H_2O)\left(\frac{-241.8\text{ kJ}}{\text{mol }H_2O}\right)\right]$$

$$-\left[(1\text{ mol }N_2H_4)\left(\frac{50.6\text{ kJ}}{\text{mol }N_2H_4}\right) + (1\text{ mol }O_2)\left(\frac{0\text{ kJ}}{\text{mol }O_2}\right)\right]$$

$= [\ 0\text{ kJ} + (-483.6\text{ kJ})] - [50.6\text{ kJ} + 0\text{ kJ}] = -534.2\text{ kJ mol}^{-1}$

The reaction with O_2 gives off less heat per mole of hydrazine because the hydrazine reaction with dinitrogentetroxide results in the formation of more (1.5 moles vs. 1.0 moles) of the highly stable N_2.

12.73 a) The drawing of this figure after work has been done on the system would need to reflect a $-\Delta V$. Therefore, the piston would need to be moved in to keep pressure constant.

b) Pressure will increase and the piston will move out as work is done by the system.

12.74 $\Delta H_{H_2O} = -\Delta H_{metal}$

$n_{H_2O}C_{p,\ H_2O}\ \Delta T_{H_2O} = -n_{metal}C_{p,\ metal}\Delta T_{metal}$ where C_p is per gram

$$(80.0\text{ g }H_2O)\left(\frac{4.184\text{ J}}{(\text{g }H_2O)\text{ K}}\right)(28.4°C - 24.8C)\left(\frac{K}{°C}\right)$$

$$= -(44.0\text{ g metal})\left(C_{p,\ metal}\right)(28.4C - 100.0°C)\left(\frac{K}{°C}\right)$$

$1205\text{ J} = (+\ 3150.4\text{ g K})\ C_{p,\ metal}$

$C_{p,\ metal} = (0.382\text{ J/g K})$

The metal was Cu (0.385 J/g K).

CHAPTER 13: SPONTANEITY OF CHEMICAL PROCESSES

13.1 a) The ordered arrangement of sand particles in the sand castle is destroyed as the ocean waves wash them away into a disordered state.

b) The secretary expends energy as she orders (straightens) the objects on the boss' desk. Her body becomes disordered as it produces CO_2 and H_2O from carbohydrates and/or fat and the CO_2 is exhaled.

c) The sticks become disordered as they drop to the floor and land in a disorganized manner.

d) The engine is ordered as the skilled mechanic works to reassemble the engine. The mechanic's body becomes disordered as it produces energy and reaction products from the metabolism of carbohydrates and/or fat.

13.2 a) As water and acetone dissolve in each other to form a liquid solution both water molecules and acetone molecules become disordered as they are dispersed.

b) As wood burns, the ordered arrangement of C, O and H atoms in the cellulose polymers become disordered as they are converted to the gaseous molecules CO_2 and H_2O which pass off into the air.

c) The iodine molecules become more ordered as they are converted into crystals of I_2. The larger system becomes more disordered because heat is passed off into the surroundings.

13.3 The organized molecules of ink in the drop become disorganized in the beaker of water. After the drop enters the water, collisions with H_2O molecules cause the ink molecules to become separated and to become homogeneously distributed throughout the beaker of water.

13.4 When Humpty-Dumpty fell and broke into many small pieces, he became disorganized. In order to become organized and whole again, much work would have to be done to collect and reorganize the pieces. This was a difficult, if not impossible, task.

13.5 a) The air confined inside the tire has some order. Upon puncturing, the air molecules escape and become disordered as they spread throughout the atmosphere.

b) The ordered molecules of perfume in the bottle escape out into the air of the room and become disordered as they spread throughout the room.

13.6 a) $n_{CO_2} = (27.5 \text{ g})/(44.01 \text{ g / mol}) = 0.6249 \text{ mol}$

$$q_{CO_2} = +(n_{CO_2})(\Delta H_{subl}) = +(0.6249 \text{ mol})(25.2 \text{ kJ / mol})\left(\frac{10^3 \text{ J}}{\text{kJ}}\right) = 1.575 \times 10^4 \text{ J}$$

$$\Delta S_{CO_2} = \frac{1.575 \times 10^4 \text{ J}}{195 \text{ K}} = 80.8 \text{ J/K}$$

b) $q_{room} = -q_{CO_2} = 1.575 \times 10^4 \text{ J}$

$$\Delta S_{room} = \frac{-1.575 \times 10^4 \text{ J}}{299.6 \text{ K}} = -52.6 \text{ J / K}$$

c) $\Delta S_{overall} = \Delta S_{room} + \Delta S_{CO_2} = -52.6 \text{ J/K} + 80.8 \text{ J/K} = 28.2 \text{ J/K}$

13.7 a) As the dry ice sublimes (changes from solid CO_2 to gaseous CO_2), its entropy increases and some of the water freezes to become ice with a decrease in its entropy.

b) $$q_{CO_2} = +(n_{CO_2})(\Delta H_{subl}) = \left(\frac{12.5 \text{ g } CO_2}{44.01 \text{ g } CO_2\text{/mole}}\right)\left(\frac{25.2 \text{ k J}}{\text{mol}}\right) = 7.157 \text{ kJ}$$

$$\Delta S_{CO_2} = \left(\frac{7.157 \text{ kJ}}{195 \text{ K}}\right)\left(\frac{10^3 \text{ J}}{\text{kJ}}\right) = 36.7 \text{ J/K}$$

$q_{H_2O} = -q_{CO_2} = -7.157 \text{ kJ}$

$$\Delta S_{H_2O} = \frac{-7.157 \text{ kJ}}{273.15 \text{ K}}\left(\frac{10^3 \text{ J}}{\text{kJ}}\right) = -26.2 \text{ J/K}$$

$\Delta S_{overall} = \Delta S_{CO_2} + \Delta S_{H_2O} = 36.7 \text{ J/K} - 26.2 \text{ J/K} = 10.5 \text{ J/K}$

13.8

	ΔS_{system}	$\Delta S_{surroundings}$
a)	ice (-)	surrounding air (+)
b)	coffee (-)	surrounding air and cup (+)
c)	popsicle (+)	surrounding air and table (-)

13.9 $$q_{H_2O} = -n_{H_2O}\Delta H_{vap} = -\left(\frac{15.5 \text{ g } H_2O}{18.02 \text{ g } H_2O\text{/mol}}\right)\left(\frac{40.79 \text{ kJ}}{\text{mol}}\right) = -35.09 \text{ kJ}$$

$$\Delta S_{H_2O} = \frac{-35.09 \text{ kJ}}{373.15 \text{ K}}\left(\frac{10^3 \text{ J}}{\text{kJ}}\right) = -94.0 \text{ J/K}$$

The entropy change of the surroundings will be positive and greater in value than 94.0 J/K because $\Delta S_{universe} > 0$.

13.10 $q_{vap} = + n\,\Delta H_{vap}$ if n = 1 $q_{vap} = \Delta H_{vap}$ $\Delta S_{vap} = \Delta H_{vap}/T_{vap}$

a) $\Delta S_{O_2} = \dfrac{9.8\ \text{kJ/mol}}{90\ \text{K}}\left(\dfrac{10^3\ \text{J}}{\text{kJ}}\right) = 1.1 \times 10^2\ \text{J/mol K}$

b) $\Delta S_{C_2H_6} = \dfrac{15.5\ \text{kJ/mol}}{184\ \text{K}}\left(\dfrac{10^3\ \text{J}}{\text{kJ}}\right) = 84.2\ \text{J/mol K}$

c) $\Delta S_{C_6H_6} = \dfrac{31.0\ \text{kJ/mol}}{353\ \text{K}}\left(\dfrac{10^3\ \text{J}}{\text{kJ}}\right) = 87.8\ \text{J/mol K}$

d) $\Delta S_{Hg} = \dfrac{59.0\ \text{kJ/mol}}{630\ \text{K}}\left(\dfrac{10^3\ \text{J}}{\text{kJ}}\right) = 93.7\ \text{J/mol K}$

13.11 a) The soft drink is the system; ΔS is negative (cooling decreases disorder).

b) The air is the system; ΔS is negative (concentration is increased).

c) The juice concentrate is the system; ΔS is positive (concentration is decreased).

13.12 a) Cl_2 has the larger molar entropy at 298 K because Cl_2 is a gas and Br_2 is a liquid at 298 K.

b) Pt has the larger molar entropy because Ni and Pt have similar structure but Pt has the larger molar mass.

c) C_8H_{18} has the larger molar entropy because both are liquid but C_8H_{18} is a larger molecule (has more atoms) than C_5H_{12}.

d) SiH_4 has the larger molar entropy because while both are gases, both have the same number of atoms and both have similar structures, SiH_4 has a larger molar mass.

13.13 a) HgS would have the larger molar entropy because HgO and HgS have similar structures but HgS has the larger molar mass.

b) The $MgCl_2$ in aqueous solution would have the larger molar entropy because it dissociates into 3 ions in solution while NaCl only dissociates into 2 ions in solution.

c) Br_2 would have the larger molar entropy because it is a liquid while I_2 is a solid.

13.14 Br_2 is more disordered than mercury because Br_2 is diatomic and, therefore, has a greater amount of intramolecular disorder. Water is the most highly ordered of the three in the liquid state because the strong hydrogen bonding between water molecules tends to restrict their movements and orientation.

13.15 $S^{\circ}_{He} = 126.15$ J/mol K   Per mole of atoms = 126.15 J/mol K

$S^{\circ}_{H_2} = 130.684$ J/mol K   Per mole of atoms =

$$\left(\frac{130.684\ \text{J}}{\text{mol K}}\right)\left(\frac{1\ \text{molecule}}{2\ \text{atoms}}\right) = 65.34\ \text{J/mol atoms K}$$

$S^{\circ}_{CH_4} = 186.26$ J/mol K   Per mole of atoms =

$$\left(\frac{186.26\ \text{J}}{\text{mol K}}\right)\left(\frac{1\ \text{molecule}}{5\ \text{atoms}}\right) = 37.25\ \text{J/mol atoms K}$$

$S^{\circ}_{C_2H_6} = 229.60$ J/mol K   Per mole of atoms =

$$\left(\frac{229.60\ \text{J}}{\text{mol K}}\right)\left(\frac{1\ \text{molecule}}{8\ \text{atoms}}\right) = 28.70\ \text{J/mol atoms K}$$

$S^{\circ}_{C_3H_6} = 226.9$ J/mol K   Per mole of atoms =

$$\left(\frac{226.9\ \text{J}}{\text{mol K}}\right)\left(\frac{1\ \text{molecule}}{9\ \text{atoms}}\right) = 25.21\ \text{J/mol atoms K}$$

The tabulated values of S° do not include a value for C_3H_8. The trend for CH_4 and C_2H_6 and C_3H_6 suggest the S° for C_3H_8 would be 260 - 270 J/mol K, resulting in 23 - 24.5 J/mol atoms K. These values show the trend that as the number of bonded atoms increases the entropy increases per mole of molecules. This is to be expected since larger molecules have more ways to arrange their atoms in space. Thus, they have greater intermolecular disorder. However, the above data also shows that as atoms are bonded the order increases for each mole of atoms. This is to be expected since free atoms have less order than bonded atoms.

13.16 a) $2\ H_{2(g)} + O_{2(g)} \rightarrow 2\ H_2O_{(l)}$

$\Delta S^{\circ}_{rxn} = [2\ S^{\circ}(H_2O_{(l)})] - [\ 2\ S^{\circ}(H_2) + S^{\circ}(O_2)]$

$= [(2\ \text{mol}\ H_2O_{(l)})(69.91\ \text{J/mol}\ H_2O\ \text{K})] -$

$[(2\ \text{mol}\ H_2)(130.68\ \text{J/ mol}\ H_2\ \text{K})] + (1\ \text{mol}\ O_2)(205.14\ \text{J/ mol}\ O_2\ \text{K})]$

$= -326.68$ J/K

b) $C_{(s)} + 2\ H_{2(g)} \rightarrow CH_{4(g)}$ Assuming $C_{(s)}$ is graphite form:

$\Delta S^{\circ}_{rxn} = [S^{\circ}(CH_4)] - [S^{\circ}(C_{(s,\ graphite)}) + 2\ S^{\circ}(H_2)]$

$$= \left[(1\ \text{mol}\ CH_4)\left(\frac{186.264\ \text{J/K}}{\text{mol}\ CH_4}\right)\right]$$

$$-\left[\left(1\ \text{mol}\ C_{(s,\ graphite)}\right)\left(\frac{5.74\ \text{J/K}}{\text{mol}\ C_{(s)}}\right) + (2\ \text{mol}\ H_2)\left(\frac{130.68\ \text{J/K}}{\text{mol}\ H_2}\right)\right] = -80.84\ \text{J/K}$$

(continued)

(13.16 continued)

c) $C_2H_5OH_{(l)} + 3\ O_{2(g)} \rightarrow 2\ CO_{2(g)} + 3\ H_2O_{(l)}$

$$\Delta S^{\circ}_{rxn} = [2\ S^{\circ}(CO_2) + 3\ S^{\circ}(H_2O_{(l)})] - [S^{\circ}(C_2H_5OH) + 3\ S^{\circ}(O_2)]$$

$$= \left[(2 \text{ mol } CO_2)\left(\frac{213.74 \text{ J/K}}{\text{mol } CO_2}\right) + (3 \text{ mol } H_2O_{(l)})\left(\frac{69.91 \text{ J/K}}{\text{mol } H_2O}\right)\right]$$

$$- \left[(1 \text{ mol } C_2H_5OH)\left(\frac{160.7 \text{ J/K}}{\text{mol } C_2H_5OH}\right) + (3 \text{ mol } O_2)\left(\frac{205.14 \text{ J/K}}{\text{mol } O_2}\right)\right] = -138.9 \text{ J/K}$$

d) $Fe_2O_{3(s)} + 2\ Al_{(s)} \rightarrow Al_2O_{3(s)} + 2\ Fe_{(s)}$

$$\Delta S^{\circ}_{rxn} = [S^{\circ}(Al_2O_3) + 2\ S^{\circ}(Fe)] - [S^{\circ}(Fe_2O_3) + 2\ S^{\circ}(Al)]$$

$$= \left[(1 \text{ mol } Al_2O_3)\left(\frac{50.92 \text{ J/K}}{\text{mol } Al_2O_3}\right) + (2 \text{ mol Fe})\left(\frac{27.28 \text{ J/K}}{\text{mol Fe}}\right)\right]$$

$$- \left[(1 \text{ mol } Fe_2O_3)\left(\frac{87.40 \text{ J/K}}{\text{mol } Fe_2O_3}\right) + (2 \text{ mol Al})\left(\frac{28.33 \text{ J/K}}{\text{mol Al}}\right)\right] = -38.58 \text{ J/K}$$

13.17 a) $N_{2(g)} + 3\ H_{2(g)} \rightarrow 2\ NH_{3(g)}$

$$\Delta S^{\circ}_{rxn} = [2\ S^{\circ}(NH_3)] - [S^{\circ}(N_2) + 3\ S^{\circ}(H_2)]$$

$$= \left[(2 \text{ mol } NH_3)\left(\frac{192.45 \text{ J/K}}{\text{mol } NH_3}\right)\right]$$

$$- \left[(1 \text{ mol } N_2)\left(\frac{191.61 \text{ J/K}}{\text{mol } N_2}\right) + (3 \text{ mol } H_2)\left(\frac{130.684 \text{ J/K}}{\text{mol } H_2}\right)\right] = -198.76 \text{ J/K}$$

b) $3\ O_{2(g)} \rightarrow 2\ O_{3(g)}$

$$\Delta S^{\circ}_{rxn} = [2\ S^{\circ}(O_3)] - [3\ S^{\circ}(O_2)]$$

$$= \left[(2 \text{ mol } O_3)\left(\frac{238.93 \text{ J/K}}{\text{mol } O_3}\right)\right] - \left[(3 \text{ mol } O_2)\left(\frac{205.138 \text{ J/K}}{\text{mol } O_2}\right)\right] = -137.55 \text{ J/K}$$

c) $PbO_{2(s)} + 2\ Ni_{(s)} \rightarrow Pb_{(s)} + 2\ NiO_{(s)}$

$$\Delta S^{\circ}_{rxn} = [S^{\circ}(Pb) + 2\ S^{\circ}(NiO)] - [S^{\circ}(PbO_2) + 2\ S^{\circ}(Ni)]$$

$$= \left[(1 \text{ mol Pb})\left(\frac{64.81 \text{ J/K}}{\text{mol Pb}}\right) + (2 \text{ mol NiO})\left(\frac{37.99 \text{ J/K}}{\text{mol NiO}}\right)\right]$$

$$- \left[(1 \text{ mol } PbO_2)\left(\frac{68.6 \text{ J/K}}{\text{mol } PO_2}\right) + (2 \text{ mol Ni})\left(\frac{29.87 \text{ J/K}}{\text{mol Ni}}\right)\right] = 12.5 \text{ J/K}$$

(continued)

(13.17 continued)

d) $C_2H_{4(g)} + 3\ O_{2(g)} \rightarrow 2\ CO_{2(g)} + 2\ H_2O_{(l)}$

$\Delta S^{\circ}_{rxn} = [2\ S°(CO_2) + 2\ S°(H_2O_{(l)})] - [S°(C_2H_4) + 3\ S°(O_2)]$

$$= \left[(2 \text{ mol } CO_2)\left(\frac{213.74 \text{ J/K}}{\text{mol } CO_2}\right) + (2 \text{ mol } H_2O)\left(\frac{69.91 \text{ J/K}}{\text{mol } H_2O}\right)\right]$$

$$- \left[(1 \text{ mol } C_2H_4)\left(\frac{219.56 \text{ J/K}}{\text{mol } C_2H_4}\right) + (3 \text{ mol } O_2)\left(\frac{205.14 \text{ J/K}}{\text{mol } O_2}\right)\right] = -267.68 \text{ J/K}$$

13.18 a) Disorder decreases due to the intermolecular interactions present in the water formed. The magnitude is large due to the considerable organization in water.

b) Disorder decreases due to the increased bonding formed. Magnitude is small due to few bonds being formed.

c) Disorder decreases due to the intermolecular interactions present in the water formed. The magnitude is large due to the considerable organization in water.

d) Disorder decreased due to stronger bonds present in Al_2O_3. The magnitude is small due to the small difference between the bonding in Fe_2O_3 (which has very little covalent properties) and Al_2O_3 (which has some covalent properties).

13.19 a) Disorder decreases due to increased number of bonds. The magnitude reflects the increased bonds.

b) Disorder decreases due to increased number of bonds. The magnitude reflects the increased bonds.

c) Disorder increases slightly due to the formation of the smaller NiO compared to PbO_2.

d) Disorder increases by a considerable amount due to the large amount of intermolecular interactions present in water.

13.20 a) 1.00 mol H_2 at p = 5.0 atm (Assume T = 298 K)

$S_{(p = 5.0 \text{ atm})} = S° - R \ln P = 130.68$ J/mol K - (8.314 J/mol K) ln (5.0)
= 130.68 J/mol K - 13.38 J/mol K = 117 J/mol K
or for 1 mole °S = 117 J/K

b) 0.25 mol of propane gas at p = 0.10 atm (Assume T = 298 K)

$S°_{(propane)} = 270$ J/mol K (from CRC)
$S_{(p = 5.0 \text{ atm})} = S° - R \ln P = 270$ J/mol K - (8.314 J/mol K) ln (0.10)
= 270 J/mol K + 19.1 J/mol K) = 289 J/mol K
For 0.25 mol: S = (289 J/mol K)(0.25 mol) = 72 J/K

c) 1.00 mol mixture of N_2 (p = 125 atm) and of H_2 (p = 375 atm)
Total pressure = 500 atm

$$n_{N_2} = (1.00 \text{ mol})\left(\frac{125 \text{ atm}}{500 \text{ atm}}\right) = 0.250 \text{ mol}$$

$$n_{H_2} = (1.00 \text{ mol})\left(\frac{375 \text{ atm}}{500 \text{ atm}}\right) = 0.750 \text{ mol}$$

(continued)

(13.20 continued)

For N_2: $S_{(p = 125\ atm)} = S° - R \ln P$ = 191.61 J/mol K - (8.314 J/mol K) ln (125 atm)
= 191.61 J/mol K - 40.14 J/mol K = 151.47 J/mol K
For 0.250 moles of N_2: S = (151.47 J/mol K)(0.250 mol) = 37.87 J/K
for H_2: $S_{(p = 375\ atm)} = S° - R \ln P$ = 130.68 J/mol K - (8.314 J/mol K) ln (375 atm)
= 130.68 J/mol K - 49.28 J/mol K = 81.40 J/mol K
For 0.750 moles of H_2: S = (81.40 J/mol K)(0.750 mol) = 61.05 J/K
Assuming this mixture to be an ideal mixture, the entropy of the mixture equals the sum of the entropies of the individual gases.

$S_{mixture} = S_{N_2} + S_{H_2}$ = 37.87 J/K + 61.05 J/K = 98.92 J/K

13.21 a)

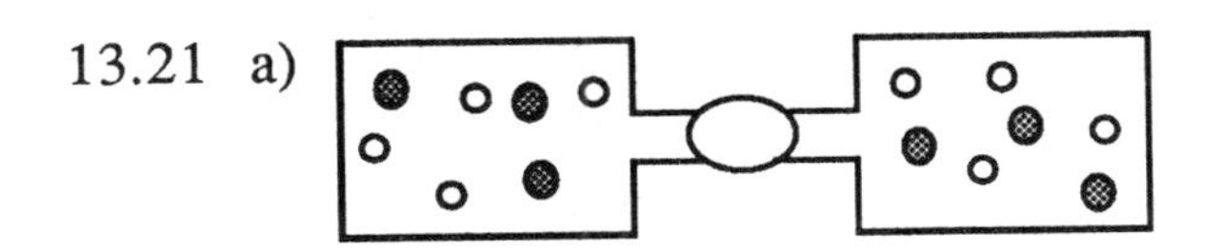

b)

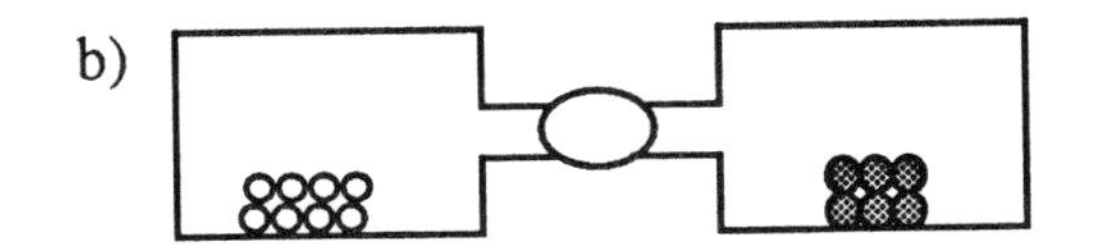

13.22 a) For any spontaneous process at constant T and P, the free energy of the system decreases.

b) When T increases during a process, the free energy of a system always decreases if ΔS is positive.

c) $\Delta G_{sys} < 0$ for any spontaneous process under constant T and P.

d) ΔG = ΔH - Δ(TS)

13.23 a) $2\ H_{2(g)} + O_{2(g)} \rightarrow 2\ H_2O_{(l)}$

$\Delta G^{\circ}_{rxn} = [2\ \Delta G°(H_2O_{(l)})] - [2\ \Delta G°(H_2) + \Delta G°(O_2)]$
$= [(2\text{ mol } H_2O_{(l)})(-237.13\text{ kJ/mol } H_2O)] -$
$[(2\text{ mol } H_2)(0.00\text{ kJ/mol } H_2)] + (1\text{ mol } O_2)(0.00\text{ kJ/mol } O_2)]$
= -474.26 kJ

b) $C_{(s)} + 2\ H_{2(g)} \rightarrow CH_{4(g)}$ Assuming $C_{(s)}$ is graphite form:

$\Delta G^{\circ}_{rxn} = [\Delta G°(CH_4)] - [\Delta G°(C_{(s,\ graphite)}) + 2\ \Delta G°(H_2)]$

$$= \left[(1\text{ mol } CH_4)\left(\frac{-50.72\text{ kJ}}{\text{mol } CH_4}\right)\right]$$

$$-\left[\left(1\text{ mol } C_{(s,\ graphite)}\right)\left(\frac{0.00\text{ kJ}}{\text{mol } C_{(s)}}\right) + (2\text{ mol } H_2)\left(\frac{0.00\text{ kJ}}{\text{mol } H_2}\right)\right] = -50.72\text{ kJ}$$

(continued)

(13.23 continued)

c) $C_2H_5OH_{(l)} + 3\ O_{2(g)} \rightarrow 2\ CO_{2(g)} + 3\ H_2O_{(l)}$

$\Delta G^{\circ}_{rxn} = [2\ \Delta G^{\circ}(CO_2) + 3\ \Delta G^{\circ}(H_2O_{(l)})] - [\Delta G^{\circ}(C_2H_5OH) + 3\ \Delta G^{\circ}(O_2)]$

$$= \left[(2 \text{ mol } CO_2)\left(\frac{-394.36 \text{ kJ}}{\text{mol } CO_2}\right) + \left(3 \text{ mol } H_2O_{(l)}\right)\left(\frac{-237.13 \text{ kJ}}{\text{mol } H_2O}\right)\right]$$

$$-\left[(1 \text{ mol } C_2H_5OH)\left(\frac{-174.78 \text{ kJ}}{\text{mol } C_2H_5OH}\right) + (3 \text{ mol } O_2)\left(\frac{0.00 \text{ kJ}}{\text{mol } O_2}\right)\right] = -1{,}325.3 \text{ kJ}$$

d) $Fe_2O_{3(s)} + 2\ Al_{(s)} \rightarrow Al_2O_{3(s)} + 2\ Fe_{(s)}$

$\Delta G^{\circ}_{rxn} = [\Delta G^{\circ}(Al_2O_3) + 2\ \Delta G^{\circ}(Fe)] - [\Delta G^{\circ}(Fe_2O_3) + 2\ \Delta G^{\circ}(Al)]$

$$= \left[(1 \text{ mol } Al_2O_3)\left(\frac{-1582.3 \text{ kJ}}{\text{mol } Al_2O_3}\right) + (2 \text{ mol Fe})\left(\frac{0.0 \text{ kJ}}{\text{mol Fe}}\right)\right]$$

$$-\left[(1 \text{ mol } Fe_2O_3)\left(\frac{-742.2 \text{ kJ}}{\text{mol } Fe_2O_3}\right) + (2 \text{ mol Al})\left(\frac{0.00 \text{ kJ}}{\text{mol Al}}\right)\right] = -840.1 \text{ kJ}$$

13.24 a) $N_{2(g)} + 3\ H_{2(g)} \rightarrow 2\ NH_{3(g)}$

$\Delta G^{\circ}_{rxn} = [2\ \Delta G^{\circ}(NH_3)] - [\Delta G^{\circ}(N_2) + 3\ \Delta G^{\circ}(H_2)]$

$$= \left[(2 \text{ mol } NH_3)\left(\frac{-16.45 \text{ kJ}}{\text{mol } NH_3}\right)\right]$$

$$-\left[(1 \text{ mol } N_2)\left(\frac{0.00 \text{ kJ}}{\text{mol } N_2}\right) - (3 \text{ mol } H_2)\left(\frac{0.00 \text{ kJ}}{\text{mol } H_2}\right)\right] = -32.90 \text{ kJ}$$

b) $3\ O_{2(g)} \rightarrow 2\ O_{3(g)}$

$\Delta G^{\circ}_{rxn} = [2\ \Delta G^{\circ}(O_3)] - [3\ \Delta G^{\circ}(O_2)]$

$$= \left[(2 \text{ mol } O_3)\left(\frac{163.2 \text{ kJ}}{\text{mol } O_3}\right)\right] - \left[(3 \text{ mol } O_2)\left(\frac{0.00 \text{ kJ}}{\text{mol } O_2}\right)\right] = 326.4 \text{ kJ}$$

c) $PbO_{2(s)} + 2\ Ni_{(s)} \rightarrow Pb_{(s)} + 2\ NiO_{(s)}$

$\Delta G^{\circ}_{rxn} = [\Delta G^{\circ}(Pb) + 2\ \Delta G^{\circ}(NiO)] - [\Delta G^{\circ}(PbO_2) + 2\ \Delta G^{\circ}(Ni)]$

$$= \left[(1 \text{ mol Pb})\left(\frac{0.00 \text{ kJ}}{\text{mol Pb}}\right) + (2 \text{ mol NiO})\left(\frac{-211.7 \text{ kJ}}{\text{mol NiO}}\right)\right]$$

$$-\left[(1 \text{ mol } PbO_2)\left(\frac{-217.33 \text{ kJ}}{\text{mol } PO_2}\right) + (2 \text{ mol Ni})\left(\frac{0.00 \text{ kJ}}{\text{mol Ni}}\right)\right] = -206.1 \text{ kJ}$$

(continued)

(13.24 continued)

d) $C_2H_{4(g)} + 3\ O_{2(g)} \rightarrow 2\ CO_{2(g)} + 2\ H_2O_{(l)}$

$$\Delta G^{\circ}_{rxn} = [2\ \Delta G^{\circ}(CO_2) + 2\ \Delta G^{\circ}(H_2O_{(l)})] - [\Delta G^{\circ}(C_2H_4) + 3\ \Delta G^{\circ}(O_2)]$$

$$= \left[(2 \text{ mol } CO_2)\left(\frac{-394.359 \text{ kJ}}{\text{mol } CO_2}\right) + (2 \text{ mol } H_2O)\left(\frac{-237.129 \text{ kJ}}{\text{mol } H_2O}\right)\right]$$

$$-\left[(1 \text{ mol } C_2H_4)\left(\frac{68.15 \text{ kJ}}{\text{mol } C_2H_4}\right) + (3 \text{ mol } O_2)\left(\frac{0.00 \text{ kJ}}{\text{mol } O_2}\right)\right] = -1{,}331.13 \text{ kJ}$$

13.25 We can estimate free energy changes at temperatures other than 298 using the equation $\Delta G_{rxn} = \Delta H^{\circ} + T\Delta S^{\circ}$. ΔS°'s were calculated in Problem 13.17. ΔH°'s will need to be calculated before we can answer this question.

a) $N_{2(g)} + 3\ H_{2(g)} \rightarrow 2\ NH_{3(g)}$

$$\Delta H^{\circ}_{rxn} = [2\ \Delta H^{\circ}(NH_3)] - [\Delta H^{\circ}(N_2) + 3\ \Delta H^{\circ}(H_2)]$$

$$= \left[(2 \text{ mol } NH_3)\left(\frac{-46.11 \text{ kJ}}{\text{mol } NH_3}\right)\right]$$

$$-\left[(1 \text{ mol } N_2)\left(\frac{0.00 \text{ kJ}}{\text{mol } N_2}\right) - (3 \text{ mol } H_2)\left(\frac{0.00 \text{ kJ}}{\text{mol } H_2}\right)\right] = -92.22 \text{ kJ}$$

$$\Delta G^{\circ}_{rxn,425} = \Delta H^{\circ}_{rxn} - T\Delta S^{\circ}_{rxn} = -92.22 \text{ kJ} - [425 \text{ K } (-198.76 \text{ J/K})(10^{-3} \text{ kJ/J})] = -7.7 \text{ kJ}$$

b) $3\ O_{2(g)} \rightarrow 2\ O_{3(g)}$

$$\Delta H^{\circ}_{rxn} = [2\ \Delta H^{\circ}(O_3)] - [3\ \Delta H^{\circ}(O_2)]$$

$$= \left[(2 \text{ mol } O_3)\left(\frac{142.7 \text{ kJ}}{\text{mol } O_3}\right)\right] - \left[(3 \text{ mol } O_2)\left(\frac{0.00 \text{ kJ}}{\text{mol } O_2}\right)\right] = 285.4 \text{ kJ}$$

$$\Delta G^{\circ}_{rxn,425} = \Delta H^{\circ}_{rxn} - T\Delta S^{\circ}_{rxn} = 285.4 \text{ kJ} - [425 \text{ K } (-137.55 \text{ J/K})(10^{-3} \text{ kJ/J})] = 343.9 \text{ kJ}$$

c) $PbO_{2(s)} + 2\ Ni_{(s)} \rightarrow Pb_{(s)} + 2\ NiO_{(s)}$

$$\Delta H^{\circ}_{rxn} = [\Delta H^{\circ}(Pb) + 2\ \Delta H^{\circ}(NiO)] - [\Delta H^{\circ}(PbO_2) + 2\ \Delta H^{\circ}(Ni)]$$

$$= \left[(1 \text{ mol Pb})\left(\frac{0.00 \text{ kJ}}{\text{mol Pb}}\right) + (2 \text{ mol NiO})\left(\frac{-239.7 \text{ kJ}}{\text{mol NiO}}\right)\right]$$

$$-\left[(1 \text{ mol } PbO_2)\left(\frac{-277.4 \text{ kJ}}{\text{mol } PO_2}\right) + (2 \text{ mol Ni})\left(\frac{0.00 \text{ kJ}}{\text{mol Ni}}\right)\right] = -202.0 \text{ kJ}$$

$$\Delta G^{\circ}_{rxn,425} = \Delta H^{\circ}_{rxn} - T\Delta S^{\circ}_{rxn} = -202.0 \text{ kJ} - [425 \text{ K } (12.5 \text{ J/K})(10^{-3} \text{ kJ/J})] = -207.3 \text{ kJ}$$

(continued)

(13.25 continued)

d) $C_2H_{4(g)} + 3\ O_{2(g)} \rightarrow 2\ CO_{2(g)} + 2\ H_2O_{(l)}$

$$\Delta H^\circ_{rxn} = [2\ \Delta H^\circ(CO_2) + 2\ \Delta H^\circ(H_2O_{(l)})] - [\Delta H^\circ(C_2H_4) + 3\ \Delta H^\circ(O_2)]$$

$$= \left[(2 \text{ mol } CO_2)\left(\frac{-393.509 \text{ kJ}}{\text{mol } CO_2}\right) + (2 \text{ mol } H_2O)\left(\frac{-285.83 \text{ kJ}}{\text{mol } H_2O}\right)\right]$$

$$- \left[(1 \text{ mol } C_2H_4)\left(\frac{52.26 \text{ kJ}}{\text{mol } C_2H_4}\right) + (3 \text{ mol } O_2)\left(\frac{0.00 \text{ kJ}}{\text{mol } O_2}\right)\right] = -1410.94 \text{ kJ}$$

$$\Delta G^\circ_{rxn,425} = \Delta H^\circ_{rxn} - T\Delta S^\circ_{rxn} = -1410.94 \text{ kJ} - [425 \text{ K } (-267.68 \text{ J/K})(10^{-3} \text{ kJ/J})]$$

$$= -1297 \text{ kJ}$$

13.26 For gases: $\Delta G_{rxn} = \Delta G^\circ_{rxn} + RT \ln Q$, values for ΔG°_{rxn} are from Problem 13.24.

a) $N_{2(g)} + 3\ H_{2(g)} \rightarrow 2\ NH_{3(g)}$

$$\Delta G_{rxn} = \Delta G^\circ_{rxn} + RT \ln\left[\frac{p^2(NH_3)}{p(N_2)p^3(H_2)}\right]$$

$$= -32.90 \text{ kJ} + \left(\frac{8.314 \text{ J}}{\text{mol K}}\right)(298 \text{ K})\ln\left(\frac{(0.25)^2}{(0.25)(0.25)^3}\right)$$

$$= (-32.90 \text{ kJ}) + [(2.47757 \times 10^3 \text{ J})(2.7726)10^{-3} \text{ kJ/J}] = -26.03 \text{ kJ}$$

b) $3\ O_{2(g)} \rightarrow 2\ O_{3(g)}$

$$\Delta G_{rxn} = \Delta G^\circ_{rxn} + RT \ln\left[\frac{p^2(O_3)}{p^3(O_2)}\right] = 326.4 \text{ kJ} + (8.314 \times 298 \text{ K})\ln\left(\frac{(0.25)^2}{(0.25)^3}\right)$$

$$= 326.4 \text{ kJ} + (2.47757 \times 10^3 \text{ J})(1.3863)(10^{-3} \text{ kJ/J}) = 329.8 \text{ kJ}$$

c) $PbO_{2(s)} + 2\ Ni_{(s)} \rightarrow Pb_{(s)} + 2\ NiO_{(s)}$

$$\Delta G_{rxn} = \Delta G^\circ_{rxn} + RT \ln\left[\frac{(1)(1)^2}{(1)(1)^2}\right]$$

$$= -206.1 \text{ kJ} + (2.4776 \times 10^3 \text{ J})(\ln 1)(10^{-3} \text{ kJ/J}) = -206.1 \text{ kJ} + 0 = -206.1 \text{ kJ}$$

d) $C_2H_{4(g)} + 3\ O_{2(g)} \rightarrow 2\ CO_{2(g)} + 2\ H_2O_{(l)}$

$$\Delta G_{rxn} = \Delta G^\circ_{rxn} + RT \ln\left[\frac{(p)^2(CO_2)}{p(C_2H_4)p^3(O_2)}\right]$$

$$= -1331.1 \text{ kJ} + \left[(2.4776 \times 10^3 \text{ J}) \ln\left(\frac{(0.25)^2}{(0.25)(0.25)^3}\right)(10^{-3} \text{ kJ/J})\right]$$

$$= -1331.1 \text{ kJ} + (2.4776 \times 10^3 \text{ J})(\ln 16)(10^{-3} \text{ kJ/J}) = -1324.2 \text{ kJ}$$

13.27 $2\ NH_{3(g)} + 2\ O_{2(g)} \rightarrow NH_4NO_{3(s)} + H_2O_{(l)}$

$$\Delta H^\circ_{rxn} = [\Delta H^\circ_f(NH_4NO_3) + \Delta H^\circ_f(H_2O_{(l)})] - [2\ \Delta H^\circ_f(NH_3) + 2\ \Delta H^\circ_f(O_2)]$$

$$= \left[(1 \text{ mol } NH_4NO_3)\left(\frac{-365.56 \text{ kJ}}{\text{mol } NH_4NO_3}\right) + (1 \text{ mol } H_2O)\left(\frac{-285.83 \text{ kJ}}{\text{mol } H_2O}\right)\right]$$

$$-\left[(2 \text{ mol } NH_3)\left(\frac{-46.11 \text{ kJ}}{\text{mol } NH_3}\right) + (2 \text{ mol } O_2)\left(\frac{0 \text{ kJ}}{\text{mol } O_2}\right)\right] = -559.17 \text{ kJ}$$

$$\Delta S^\circ_{rxn} = [S^\circ(NH_4NO_3) + S^\circ(H_2O_{(l)})] - [2\ S^\circ(NH_3) + 2\ S^\circ(O_2)]$$

$$= \left[(1 \text{ mol } NH_4NO_3)\left(\frac{151.08 \text{ J/K}}{\text{mol } NH_4NO_3}\right) + (1 \text{ mol } H_2O_{(l)})\left(\frac{69.91 \text{ J/K}}{\text{mol } H_2O}\right)\right]$$

$$-\left[(2 \text{ mol } NH_3)\left(\frac{192.45 \text{ J/K}}{\text{mol } NH_3}\right) + (2 \text{ mol } O_2)\left(\frac{205.14 \text{ J/K}}{\text{mol } O_2}\right)\right] = -574.19 \text{ J/K}$$

$$\Delta G^\circ_{rxn} = [\Delta G^\circ_f(NH_4NO_3) + \Delta G^\circ_f(H_2O_{(l)})] - [2\ \Delta G^\circ_f(NH_3) + 2\ \Delta G^\circ_f(O_2)]$$

$$= \left[(1 \text{ mol } NH_4NO_3)\left(\frac{-183.87 \text{ kJ}}{\text{mol } NH_4NO_3}\right) + (1 \text{ mol } H_2O_{(l)})\left(\frac{-237.13 \text{ kJ}}{\text{mol } H_2O}\right)\right]$$

$$-\left[(2 \text{ mol } NH_3)\left(\frac{-16.45 \text{ kJ}}{\text{mol } NH_3}\right) + (2 \text{ mol } O_2)\left(\frac{0.00 \text{ kJ}}{\text{mol } O_2}\right)\right] = -388.10 \text{ kJ}$$

or $\Delta G^\circ_{rxn} = \Delta H^\circ_{rxn} - T\Delta S^\circ_{rxn} = -559.17 \text{ kJ} - (298 \text{ K})(-574.19 \text{ J/K})(10^{-3} \text{ kJ/J})$

$= -388.06 \text{ kJ}$

13.28 $CO_{2(g)} + 2\ NH_{3(g)} \rightarrow (NH_2)_2CO_{(s)} + H_2O_{(l)}$ (Assume $H_2O_{(l)}$)

$$\Delta H^\circ_{rxn} = [\Delta H^\circ_f((NH_2)_2CO) + \Delta H^\circ_f(H_2O_{(l)})] - [\Delta H^\circ_f(CO_2) + 2\ \Delta H^\circ_f(NH_3)]$$

$$= \left[(1 \text{ mol } (NH_2)_2CO)\left(\frac{-334 \text{ kJ}}{\text{mol } (NH_2)_2CO}\right) + (1 \text{ mol } H_2O_{(l)})\left(\frac{-285.83 \text{ kJ}}{\text{mol } H_2O}\right)\right]$$

$$-\left[(1 \text{ mol } CO_2)\left(\frac{-393.51 \text{ kJ}}{\text{mol } CO_2}\right) + (2 \text{ mol } NH_3)\left(\frac{-46.11 \text{ kJ}}{\text{mol } NH_3}\right)\right] = -134 \text{kJ}$$

$$\Delta S^\circ_{rxn} = [S^\circ((NH_2)_2CO) + S^\circ(H_2O_{(l)})] - [S^\circ(CO_2) + 2\ S^\circ(NH_3)]$$

$$= \left[(1 \text{ mol } (NH_2)_2CO)\left(\frac{105 \text{ J/K}}{\text{mol } (NH_2)_2CO}\right) + (1 \text{ mol } H_2O_{(l)})\left(\frac{69.91 \text{ J/K}}{1 \text{ mol } H_2O}\right)\right]$$

$$-\left[(1 \text{ mol } CO_2)\left(\frac{213.74 \text{ J/K}}{\text{mol } CO_2}\right) + (2 \text{ mol } NH_3)\left(\frac{192.45 \text{ J/K}}{1 \text{ mol } NH_3}\right)\right] = -424 \text{ J/K}$$

(continued)

(13.28 continued)

$$\Delta G^\circ_{rxn} = [\Delta G^\circ_f((NH_2)_2CO) + \Delta G^\circ_f(H_2O_{(l)})] - [\Delta G^\circ_f(CO_2) + 2\,\Delta G^\circ_f(NH_3)]$$

$$= \left[(1 \text{ mol } (NH_2)_2CO)\left(\frac{-198 \text{ kJ}}{\text{mol } (NH_2)_2CO}\right) + (1 \text{ mol } H_2O_{(l)})\left(\frac{-237.13 \text{ kJ}}{1 \text{ mol } H_2O}\right)\right]$$

$$-\left[(1 \text{ mol } CO_2)\left(\frac{-394.36 \text{ kJ}}{\text{mol } CO_2}\right) + (2 \text{ mol } NH_3)\left(\frac{-16.45 \text{ kJ}}{1 \text{ mol } NH_3}\right)\right] = -7.9 \text{ kJ}$$

or

$$\Delta G^\circ_{rxn} = \Delta H^\circ_{rxn} - T\,\Delta S^\circ_{rxn} = -134 \text{ kJ} -(298 \text{ K})(-424 \text{ J/K})(10^{-3} \text{ kJ/J}) = -7.6 \text{ kJ}$$

13.29 Ammonia is a gas at standard conditions while both urea and ammonium nitrate are solids. All three are quite water soluble. Some of the gaseous ammonia can be lost during placement in the field and after placement before use by the plants. An aqueous solution of ammonia is a weak base. A solution with too high a concentration of ammonia not only increases the percentage lost as gas but raises the pH of the soil. Neither the dissolved molecular urea nor the ionic ammonium nitrate raises the pH of the soil.

13.30 Your diagram should look like most other phase diagrams; e.g. that of ammonia in Sample Problem 13.10, except the placement of the pressure and temperature values will be shifted. The triple point will be across from 0.0015 atm and above 54 K. The boiling point will be shown on the liquid/vapor line at 1 atm and 90 K. Label the diagram as oxygen and show where the solid, liquid, and vapor phases exist.

13.31 a) As the CO_2 gas is compressed from 1.00 to 50.0 atm at a constant T of 298 K, it liquefies at ~6 - 6.5 atm and remains a liquid during the rest of the compression to 50.0 atm.

b) As the N_2 gas is compressed from 1.00 to 50.0 atm at constant T of 298 K, it remains a gas and the volume decreases to ~1/50 of original volume.

c) As the N_2 is cooled from 298 to 50 K at a constant P of 1.00 atm, it liquefies at 77 K and then freezes at 63 K.

d) As the CO_2 is cooled from 298 to 50 K at a constant P of 1.00 atm, it condenses directly to the solid phase at 195 K.

13.32 Because of the second law of thermodynamics excess heat must be generated by a system any time a process is driven in a nonspontaneous direction. Heat spontaneously flows from a warmer to a cooler reservoir. Heat pumps force the heat to flow in a nonspontaneous direction (from a cold to a warmer reservoir) and, therefore, the heat pumps must convert some energy into heat.

13.33 At cellular conditions, the pressure is not 1 atm and the concentrations are not 1 M.

13.34 The spontaneous process of the 2 kg mass falling creates the energy to produce the work to perform the nonspontaneous process of lifting the 1 kg mass. Excess heat is produced from the energy used to overcome the friction of the rope and pulley.

13.35 Reactions

a) acetyl phosphate + $H_2O \rightarrow$ acetic acid + H_3PO_4 $\quad \Delta G° = -41.8$ kJ

b) glutamic acid + $NH_3 \rightarrow$ glutamine + H_2O $\quad \Delta G° = +14$ kJ

Net: glutamic acid + NH_3 + acetyl phosphate $\rightarrow$ glutamine + acetic acid + H_3PO_4

$\Delta G^{\circ}_{rxn} = \Delta G^{\circ}_{rxn(a)} + \Delta G^{\circ}_{rxn(b)} = -41.8 \text{ kJ} + (+14. \text{ kJ}) = -28 \text{ kJ}$

13.36 When the cell burns fat or carbohydrates, it stores the energy as (is paid with) ATP. When the cell needs energy for its processes, it converts (spends) ATP which becomes ADP.

13.37 No, this will not work. Moving heat from the cooler refrigerator to the warmer surroundings (the kitchen) is a nonspontaneous process. Therefore, not only is the kitchen made warmer by the heat removed from the refrigerator, but according to the second law of thermodynamics, excess heat must be generated by the system while driving this nonspontaneous process.

13.38 The molecules of H_2 gas and Cl_2 gas are made up of 2 atoms of the same element while HCl gas has 1 atom of 2 different elements.

13.39 a) $\Delta E_{universe} = 0$

b) $\Delta E_{teaspoon}$ is positive

c) $\Delta S_{universe}$ is positive.

d) ΔS_{water} is negative

e) $q_{teaspoon}$ increases.

13.40 a) $\Delta E_{universe} = 0$

b) ΔE_{pie} is negative

c) $\Delta S_{universe}$ is positive.

d) ΔS_{pie} is negative

e) q_{pie} decreases.

13.41 0.50 mol H_2O (liquid, 298 K) < 1.0 mol H_2O (liquid, 298 K) < 1.0 mol H_2O (liquid, 373 K) < 1.0 mol H_2O (gas, 373 K, 1 atm) < 1.0 mol H_2O (gas, 373 K, 0.1 atm)

13.42 1.00 g Br_2 (liquid, 331.9 K, 1.00 atm) < 1.00 g Br_2 (gas, 331.9 K, 1.00 atm) < 1.00 g Br_2 (gas, 331.9 K, 0.10 atm) < 1.00 g Br atoms (gas, 331.9 K, 0.10 atm)

13.43 a) $q_{H_2O} = (n_{H_2O})(\Delta H_{fusion}) = \left(\frac{150\ g\ H_2O}{18.02\ g\ H_2O/mol}\right)\left(-6.01\frac{kJ}{mol}\right) = -50.03\ kJ$

$$\Delta S_{H_2O} = \left(\frac{-50.03\ kJ}{273\ K}\right)\left(\frac{10^3\ J}{kJ}\right) = -183\ J/K$$

b) $q_{freezer} = -\ q_{H_2O} = 50.03\ kJ$

$$\Delta S_{freezer} = \left(\frac{50.03\ kJ}{253\ K}\right)\left(\frac{10^3\ J}{kJ}\right) = 198\ J/K$$

$\Delta S_{universe} = \Delta S_{H_2O} + \Delta S_{freezer} = -183\ J/K + 198\ J/K = 15\ J/K$

c) Yes, an additional entropy change would occur for the universe as the tray of ice was cooled to -20°C because the freezer would generate excess heat and, therefore, more disorder.

13.44 a) $CH_{4(g)} + H_2O_{(g)} \rightarrow CO(g) + 3\ H_{2(g)}$

$\Delta G^{\circ}_{rxn} = [\Delta G^{\circ}(CO) + 3\ \Delta G^{\circ}(H_2)] - [\Delta G^{\circ}(CH_4) + \Delta G^{\circ}(H_2O)]$

$= [-137.168\ kJ + (3 \times 0.00)] - [(-50.72\ kJ) + (-228.72\ kJ)] = 142.27\ kJ$

b) $\Delta H^{\circ}_{rxn} = [\Delta H^{\circ}(CO) + 3\ \Delta G^{\circ}(H_2)] - [\Delta H^{\circ}(CH_4) + \Delta H^{\circ}(H_2O)]$

$= [(-110.525\ kJ + 3(0.00)] - [-74.81\ kJ) + (-241.818\ kJ)] = 206.10\ kJ$

$\Delta S^{\circ}_{rxn} = [S^{\circ}(CO) + 3\ S^{\circ}(H_2)] - [S^{\circ}(CH_4) + S^{\circ}(H_2O)]$

$= [197.674\ J/K + 3(130.684\ J/K)] - [186.264\ J/K + 188.825\ J/K]$
$= 214.637\ J/K$

$\Delta G^{\circ}_{rxn,\ 1300} = \Delta H^{\circ} - T\Delta S^{\circ} = 206.10\ kJ - (1300\ K)(214.637\ J/K)(10^{-3}\ kJ/J)$
$= -72.9\ kJ$

13.45 $Al_2O_3 + 3\ H_2 \rightarrow 2\ Al + 3\ H_2O_{(l)}$

$\Delta H^{\circ}_{rxn} = [2\ \Delta H^{\circ}(Al) + 3\ \Delta H^{\circ}(H_2O_{(l)})] - [\Delta H^{\circ}(Al_2O_3) + 3\ \Delta H^{\circ}(H_2)]$

$\Delta H^{\circ}_{rxn} = [0.00 + 3(-285.830\ kJ] - [-1675.7\ kJ + 0.00] = 818.2\ kJ$

$\Delta G^{\circ}_{rxn} = [2\ \Delta G^{\circ}(Al) + 3\ \Delta G^{\circ}(H_2O_{(l)})] - [\Delta G^{\circ}(Al_2O_3) + 3\ \Delta G^{\circ}(H_2)]$

$\Delta G^{\circ}_{rxn} = [0.00 + 3\ (-237.129\ kJ)] - [-1582.3\ kJ + 0.00] = 870.9\ kJ$

$\Delta S^{\circ}_{rxn} = [2\ S^{\circ}(Al) + 3\ S^{\circ}(H_2O_{(l)})] - [S^{\circ}(Al_2O_3) + 3\ S^{\circ}(H_2)]$

$\Delta S^{\circ}_{rxn} = [(2 \times 28.33\ J/K) + 3(69.91)] - [+50.92\ J/K + 3(130.684\ J/K)] = -176.58\ J/K$

a) As indicated by the positive value for ΔG°_{rxn}, the reaction is not spontaneous.

b) As indicated by the positive value for ΔH°_{rxn}, the reaction absorbs heat.

c) As indicated by the negative value for ΔS°_{rxn}, the products are more ordered than the reactants.

13.46 $3\ Fe + 4\ H_2O_{(l)} \rightarrow Fe_3O_4 + 4\ H_2$

a) $\Delta G^\circ_{rxn} = [\Delta G^\circ(Fe_3O_4) + 4\ \Delta G^\circ(H_2)] - [3\ \Delta G^\circ(Fe) + 4\ \Delta G^\circ(H_2O)]$

$= [(-1015.4\ kJ) + 0.00] - [-(0.00) + 4(-237.129\ kJ)] = -66.88\ kJ$

The reaction is spontaneous.

b) $\Delta H^\circ_{rxn} = [\Delta H^\circ(Fe_3O_4) + 4\ \Delta H^\circ(H_2)] - [3\ \Delta H^\circ(Fe) + 4\ \Delta H^\circ(H_2O)]$

$= [(-1118.4\ kJ) + 0.00] - [-(0.00) + 4(-285.83\ kJ)] = +24.92\ kJ$

The reaction absorbs heat.

c) $\Delta S^\circ_{rxn} = [S^\circ(Fe_3O_4) + 4\ S^\circ(H_2)] - [3\ S^\circ(Fe) + 4\ S^\circ(H_2O)]$

$= [(146.4\ J/K) + 4(130.684\ J/K)] - [3(27.28\ J/K) + 4(69.91\ J/K)]$

$= 307.66\ J/K$ The products are less ordered than the reactants.

13.47 $C_{16}H_{32}O_2 + 23\ O_2 + 130\ ADP + 130\ H_3PO_4 \rightarrow 16\ CO_2 + 146\ H_2O + 130\ ATP$

$C_{16}H_{32}O_2 + 23\ O_2 \rightarrow 16\ CO_2 + 16\ H_2O$ $\Delta G^\circ = -\ 9790\ kJ$

$130\ ADP + 130\ H_3PO_4 \rightarrow 130\ H_2O + 130\ ATP$

$\Delta G^\circ = (130\ mol\ ATP)(+\ 30.6\ kJ/mol) = +\ 3978\ kJ$

$\Delta G^\circ_{overall} = -\ 9790\ kJ + 3978\ kJ = -\ 5812\ kJ$

$$\text{efficiency} = \left(\frac{3978\ kJ}{3978\ kJ\ +\ 9790\ kJ}\right) x\ 100\% = 40.6\%$$

13.48 Each mole of palmitic acid produces 5812 kJ of heat in excess of what is used to convert ADP to ATP.

$$q_{H_2O} = (n_{H_2O})(\Delta H_{vap}) = \left(\frac{75\ g}{18.02\ g/mol\ H_2O}\right)(40.79\ kJ/mol) = 170\ kJ$$

$$(170\ kJ)\left(\frac{1\ mol\ palmitic\ acid}{5812\ kJ}\right)\left(\frac{256.42\ g\ palmitic\ acid}{1\ mol\ palmitic\ acid}\right) = 7.5\ g\ palmitic\ acid$$

13.49 $N_2 + O_2 + \text{Lightning} \rightarrow 2\ NO$ ΔG would be positive because lightning is needed to supply the energy to drive this nonspontaneous reaction.

Luciferin $\rightarrow$ Dehydroluciferin + Light ΔG could be negative because an enzyme catalyzes this reaction; could be positive if energy from metabolism drives the reaction; probably positive because firefly exerts control over "lightning".

13.50 $$q_{H_2O} = n_{H_2O}\Delta H_{vap} = \left(\frac{1\ g}{18.02\ g/mol}\right)(40.79\ kJ/mol) = 2.26\ kJ$$

$$\Delta S_{H_2O} = \frac{2.26\ kJ}{(23.5\ +\ 273.15)K}\left(\frac{10^3\ J}{kJ}\right) = +7.62\ J/K$$

$$\Delta S_{body} = \frac{-2.26\ kJ}{(37.5\ +\ 273.15)K}\left(\frac{10^3\ J}{kJ}\right) = -7.28\ J/K$$

$\Delta S_{universe} = +7.62\ J/K - 7.28\ J/K = 0.34\ J/K$

13.51 a) $2\ ClO_2^-{}_{(aq)} + O_{2(g)} \rightarrow 2\ ClO_3^-{}_{(aq)}$

$\Delta S^\circ_{rxn} = [2\ S^\circ(ClO_3^-{}_{(aq)})] - [2\ S^\circ(ClO_2^-{}_{(aq)}) + S^\circ(O_2)]$

$$= \left[(2 \text{ mol } ClO^-_{3(aq)})\left(\frac{162 \text{ J/K}}{\text{mol } ClO_3^-}\right)\right]$$

$$-\left[(2 \text{ mol } ClO_2^-)\left(\frac{101 \text{ J/K}}{\text{mol } ClO_2^-}\right) + (1 \text{ mol } O_2)\left(\frac{205.138 \text{ J/K}}{\text{mol } O_2}\right)\right] = -83 \text{ J/K}$$

b) $4\ FeCl_{3(s)} + 3\ O_{2(g)} \rightarrow 2\ Fe_2O_{3(s)} + 6\ Cl_{2(g)}$

$\Delta S^\circ_{rxn} = [2\ S^\circ(Fe_2O_3) + 6\ S^\circ(Cl_2)] - [4\ S^\circ(FeCl_3) + 3\ S^\circ(O_2)]$

$$= \left[(2 \text{ mol } Fe_2O_3)\left(\frac{87.40 \text{ J/K}}{\text{mol } Fe_2O_3}\right) + (6 \text{ mol } Cl_2)\left(\frac{223.066 \text{ J/K}}{\text{mol } Cl_2}\right)\right]$$

$$-\left[(4 \text{ mol } FeCl_3)\left(\frac{142.3 \text{ J/K}}{\text{mol } FeCl_3}\right) + (3 \text{ mol } O_2)\left(\frac{205.138 \text{ J/K}}{\text{mol } O_2}\right)\right] = 328.6 \text{ J/K}$$

c) $3\ N_2H_{4(l)} + 4\ O_{3(g)} \rightarrow 6\ NO_{(g)} + 6\ H_2O_{(l)}$

$\Delta S^\circ_{rxn} = [6\ S^\circ(NO) + 6\ S^\circ(H_2O_{(l)})] - [3\ S^\circ(N_2H_4) + 4\ S^\circ(O_3)]$

$$= \left[(6 \text{ mol } NO)\left(\frac{210.761 \text{ J/K}}{\text{mol } NO}\right) + (6 \text{ mol } H_2O_{(l)})\left(\frac{69.91 \text{ J/K}}{\text{mol } H_2O}\right)\right]$$

$$-\left[(3 \text{ mol } N_2H_4)\left(\frac{121.21 \text{ J/K}}{\text{mol } N_2H_4}\right) + (4 \text{ mol } O_3)\left(\frac{238.93 \text{ J/K}}{\text{mol } O_3}\right)\right] = -478.37 \text{ J/K}$$

13.52 a) $2\ ClO_2^-{}_{(aq)} + O_{2(g)} \rightarrow 2\ ClO_3^-{}_{(aq)}$

$\Delta G^\circ_{rxn} = [2\ \Delta G^\circ(ClO_3^-{}_{(aq)})] - [2\ \Delta G^\circ(ClO_2^-{}_{(aq)}) + \Delta G^\circ(O_2)]$

$= [2(-3 \text{ kJ})] - [2(17 \text{ kJ}) + 1(0.00)] = -40 \text{ kJ}$

b) $4\ FeCl_{3(s)} + 3\ O_{2(g)} \rightarrow 2\ Fe_2O_{3(s)} + 6\ Cl_{2(g)}$

$\Delta G^\circ_{rxn} = [2\ \Delta G^\circ(Fe_2O_3) + 6\ \Delta G^\circ(Cl_2)] - [4\ \Delta G^\circ(FeCl_3) + 3\ \Delta G^\circ(O_2)]$

$= [2(-742.2 \text{ kJ}) + 6(0.00)] - [4(-334.00 \text{ kJ}) + 3(0.00)] = -148.4 \text{ kJ}$

c) $3\ N_2H_{4(l)} + 4\ O_{3(g)} \rightarrow 6\ NO_{(g)} + 6\ H_2O_{(l)}$

$\Delta G^\circ_{rxn} = [6\ \Delta G^\circ(NO) + 6\ \Delta G^\circ(H_2O_{(l)})] - [3\ \Delta G^\circ(N_2H_4) + 4\ \Delta G^\circ(O_3)]$

$= [6(86.55) + 6(-237.129)] - [3(149.34) + 4(163.2)] = -2004.3 \text{ kJ}$

13.53 a) $2\ ClO_2^-{}_{(aq)} + O_{2(g)} \rightarrow 2\ ClO_3^-{}_{(aq)}$

$\Delta H^\circ_{rxn} = [2\ \Delta H^\circ(ClO_3^-{}_{(aq)})] - [2\ \Delta H^\circ(ClO_2^-{}_{(aq)}) + \Delta H^\circ(O_2)]$

$=[2(-104)] - [2(-67) + (0.00)] = -74$ kJ

$\Delta G^\circ_{350} = \Delta H^\circ - T\Delta S^\circ$ (Use ΔS° from Problem 13.51)

$= -74$ kJ $- [(350\ K)(-83\ J/K)(10^{-3}\ kJ/J)] = -45$ kJ

b) $4\ FeCl_{3(s)} + 3\ O_{2(g)} \rightarrow 2\ Fe_2O_{3(s)} + 6\ Cl_{2(g)}$

$\Delta H^\circ_{rxn} = [2\ \Delta H^\circ(Fe_2O_3) + 6\ \Delta H^\circ(Cl_2)] - [4\ \Delta H^\circ(FeCl_3) + 3\ \Delta H^\circ(O_2)]$

$= [2(-824.2) + 6(0.00)] - [4(-399.49) + 3(0.00)] = -50.44$ kJ

$\Delta G^\circ_{350} = \Delta H^\circ - T\Delta S^\circ = -50.44$ kJ $- (350\ K)(328.6\ J/K)(10^{-3}\ kJ/J) = -165$ kJ

c) $3\ N_2H_{4(l)} + 4\ O_{3(g)} \rightarrow 6\ NO_{(g)} + 6\ H_2O_{(l)}$

$\Delta H^\circ_{rxn} = [6\ \Delta H^\circ(NO) + 6\ \Delta H^\circ(H_2O_{(l)})] - [3\ \Delta H^\circ(N_2H_4) + 4\ \Delta H^\circ(O_3)]$

$= [6(90.25) + 6(-285.83)] - [3(50.63) + 4(142.7)] = -1896$ kJ

$\Delta G^\circ_{350} = \Delta H^\circ - T\Delta S^\circ = -1896$ kJ $- (350\ K)(-478.37\ J/K)(10^{-3}\ kJ/J) = -1729$ kJ

13.54 a) q_{sys} is negative for deposition (reverse of sublimation)

b) ΔS_{sys} is negative because the system has become more ordered.

c) $\Delta E_{universe}$ is 0; conservation of energy.

13.55 a) W_{sys} is 0 because the system is constant volume.

b) q_{sys} is positive because system requires energy to separate the molecules into their constituent atoms (ΔS is positive).

c) ΔS_{surr} is negative because the surroundings supply energy to the system.

13.56 The four most important nitrogen-containing chemicals are ammonia (NH_3), nitric acid (HNO_3), urea [$(NH_2)_2CO$], and ammonium nitrate (NH_4NO_3). All are used in the fertilizer industry as feedstocks or fertilizers or both.

13.57 a) $P_{4\beta}$ has a more ordered crystalline structure; q is negative; therefore, ΔS is negative.

b) ΔS is negative for this process because the process results in a more ordered system. ΔH is negative because the process is spontaneous (ΔG is negative) at -77°C.

13.58 $2\ H_2S_{(g)} + O_{2(g)} \rightarrow 2\ H_2O_{(l)} + 2\ S_{(s)}$

$\Delta G^\circ_{rxn,H_2S} = [2\ \Delta G^\circ(H_2O) + 2\ \Delta G^\circ(S)] - [2\ \Delta G^\circ(H_2S) + 6\ \Delta G^\circ(O_2)]$

$= [2(-237.129) + 2(0.00)] - [2(-33.56) + 6(0.00)]$

$= -407$ kJ

$6\ CO_{2(g)} + 6\ H_2O_{(l)} \rightarrow C_6H_{12}O_{6(s)} + 6\ O_{2(g)}$

In order for these to be coupled in a spontaneous reaction $\Delta G^\circ_{rxn,H_2S}$ will need to be repeated sufficient times to yield a negative value when added to $\Delta G^\circ_{rxn,\ glucose}$.

[(-407 kJ) x ?] + [2870 kJ] = - ? => 7

$[2\ H_2S_{(g)} + O_{2(g)} \rightarrow 2\ H_2O_{(l)} + 2\ S_{(s)}] \times 8$ $\Delta G^\circ = 8 \times (-407$ kJ)

$6\ CO_{2(g)} + 6\ H_2O_{(l)} \rightarrow C_6H_{12}O_{6(s)} + 6\ O_{2(g)}$ $\Delta G^\circ = 2870$ kJ

Net: $16\ H_2S_{(g)} + 6\ CO_{2(g)} + 2\ O_{2(g)} \rightarrow 10\ H_2O_{(l)} + 16\ S_{(s)} + C_6H_{12}O_{6(s)}$

$\Delta G^\circ = -386$ kJ

13.59 a) The sign of ΔH is - and of ΔG is -. The NO_2 molecule contains an unpaired electron and readily dimerizes to form N_2O_4; therefore, ΔG is negative. $\Delta G = \Delta H - T\Delta S$. For the reaction $2\ NO_2 \rightarrow N_2O_4$, ΔS is negative. Therefore, in order that ΔG be negative, ΔH must be negative and of greater magnitude than $T\Delta S$.

b) The sign of ΔH is + and of ΔG is +. The gold requires an input of energy (+ q) to be melted; therefore, ΔH is positive. The melting is **nonspontaneous**; therefore, ΔG is positive.

c) The sign of ΔH is - and of ΔG is -. This is a typical **exothermic** combustion that is **spontaneous**; therefore, ΔH is negative and ΔG is negative.

13.60 a) The sketch showing how the system appears when it reaches its final state will show the piston raised to give a volume of 2.2 L and the bromine molecules well separated.

b) $\Delta H^\circ_{vap} = \Delta H^\circ_{(g)}(Br_2) - \Delta H^\circ_{(l)}(Br_2)$

$= (30.907$ kJ/mol$) - 0.00 = 30.907$ kJ/mol

$q = n[Cp_{(l)}\ \Delta T_{(l)} + \Delta H^\circ_{vap}]$

$= 0.08$ mole$[(75.7$ J/mol K$)(10^{-3}$ kJ/J$)(332$ K $- 273$ K$) + (30.907$ kJ/mol$)]$

$q = 2.8$ kJ or 3 kJ

$w = -p\Delta V = -1.0$ atm x 2.2 L x 10^{-3} m^3/L

x 1.013×10^5 kg m^{-1} s^{-2} atm^{-1}

x 1 J kg^{-1} m^{-2} s^2 = -2.2×10^2 J

$\Delta E = q + w = 2.8$ kJ $- 0.22$ kJ $= 2.6$ kJ

$\Delta H = q = 2.8$ kJ

$\Delta S = q/T = 2.8$ kJ/332 K $= 8.4$ J/K

13.61 The process of breaking the bonds in oxygen (or in sulfur) and forming oxide (or sulfide) bonds with other elements are spontaneous (ΔG negative) because the oxide (or sulfide) bonds are stronger than the bonds in oxygen (or sulfur). The bond in nitrogen is stronger than a nitride bond and, therefore, nitrogen exists as a pure element rather than being bonded to other elements.

13.62 a) $\Delta G° = \Delta H° - T\Delta S° = \Delta H° - (298\ K)\Delta S°$

$= 5.65\ kJ/mol - (298\ K)(28.9\ J/mol\ K)(10^{-3}\ kJ/J) = -2.96\ kJ/mol$

b) $\Delta G°_T = 0 = \Delta H_T - T\Delta S_T = \Delta H° - T\Delta S°$ Assuming $\Delta H° \approx \Delta H_T$ and $\Delta S° \approx \Delta S_T$

$0 = 5.65\ kJ/mol - T\ (28.9\ J/mol)(10^{-3}\ kJ/J)$

$T(2.89 \times 10^{-2}\ kJ/mol\ K) = 5.65\ kJ/mol$

$T = 196\ K$

13.63 The K^+ amd Ca^{2+} ions are similar in size and mass. The Cl^- ion is both larger and of greater mass than the O^{2-} ion which could account for KCl having a greater disorder.

13.64 $3\ O_{2(g)} \rightarrow 2\ O_{3(g)}$

a) $\Delta H°_{rxn} = [2\ \Delta H°(O_3)] - [3\ \Delta H°(O_2)]$

$\Delta H°_{rxn} = [2(142.7)] - [3(0.00)] = 285.4\ kJ$

$\Delta S°_{rxn} = [2\ S°(O_3)] - [3\ S°(O_2)]$

$= [2(238.93)] - [3(205.138)] = -137.6\ J/K$

$\Delta G°_{rxn} = [2\ \Delta G°(O_3)] - [3\ \Delta G°(O_2)]$

$= [2(163.2)] - [3(0.00)] = 326.4\ kJ$

or $\Delta G° = \Delta H - T\Delta S = 285.4 - (298)(-0.1376) = 326.4\ kJ$

b) The signs of $\Delta H°$ and $\Delta S°$ indicate that there is no temperature at which this reaction becomes spontaneous (ΔG negative) at 1 atm pressure.

c) $$\Delta G_{rxn} = \Delta G°_{rxn} + RT \ln Q = \Delta G°_{rxn} + RT \ln\left[\frac{p^2(O_3)}{(0.20)^3}\right] = 0$$

$$0 = (326.4 \times 10^3\ J) + (8.314\ J/mol\ K)(298\ K)\ln\left[\frac{p^2(O_3)}{(0.20)^3}\right]$$

$$\frac{-326.4 \times 10^3}{8.314 \times 298} = \ln\left[\frac{p^2(O_3)}{(0.20)^3}\right] = -131.7 \qquad \frac{p^2(O_3)}{(0.20)^3} = 6.1 \times 10^{-58}$$

$p^2(O_3) = 4.9 \times 10^{-60}$ $\quad$ $p(O_3) = 2.2 \times 10^{-30}$ atm

d) The ozone layer is thick but very dilute. The ozone layer lies between 20 and 35 km above the surface of the Earth. However, the production of ozone in the ozone layer is not spontaneous if only oxygen to ozone is considered. The reaction is driven by the absorption of photons in the wavelength region between 180 and 240 nm.

13.65 a) The water flow is spontaneous. The turning of the turbine and the generating of the electricity is nonspontaneous.

b) The burning of the gasoline is spontaneous. The movement of the engine and the movement of the water uphill are nonspontaneous.

13.66 a) $C_{(s, coal)} + H_2O_{(g)} \rightarrow CO_{(g)} + H_{2(g)}$

$\Delta H^\circ_{rxn} = [1\ \Delta H^\circ(CO) + 1\ \Delta H^\circ(H_2)] - [1\ \Delta H^\circ(C) + 1\ \Delta H^\circ(H_2O)]$
$= [-110.525\ kJ + 0.00] - [(0.00) + (-241.818\ kJ)] = 131.293\ kJ$

$\Delta S^\circ_{rxn} = [1\ S^\circ(CO)] - [1\ S^\circ(H_2)] - [1\ S^\circ(C) + 1\ S^\circ(H_2O)]$
$= [(197.647\ J/K) + (130.684\ J/K)] - [(5.740\ J/K) + (188.825\ J/K)]$
$= 133.793\ J/K$

$\Delta G^\circ_{rxn,\ T} = \Delta H^\circ_{rxn,\ 298\ K} - T\Delta S^\circ_{rxn,\ 298\ K}$

$\Delta G^\circ_{rxn,\ 473\ K} = (131.293\ kJ) - (473\ K)(133.793\ J/K)(10^{-3}\ kJ/J) = 68.0\ kJ$

b) $C_{(s, coal)} + O_{2(g)} \rightarrow CO_{2(g)}$

$\Delta H^\circ_{rxn} = [1\ \Delta H^\circ(CO_2)] - [1\ \Delta H^\circ(C) + 1\ \Delta H^\circ(O_2)]$
$= (-393.509\ kJ) - (0.00 + 0.00) = -393.509\ kJ$

$\Delta S^\circ_{rxn} = [1\ S^\circ(CO_2)] - [1\ S^\circ(C) + 1\ S^\circ(O_2)]$
$= (213.74\ J/K) - [(5.740\ J/K) + (205.138\ J/K)] = 2.86\ J/K$

$\Delta G^\circ_{rxn,\ 473\ K} = (-393.509\ kJ) - (473\ K)(2.86\ J/K)(10^{-3}\ kJ/J) = -395\ kJ$

c) $C_{(s, coal)} + 1/2\ O_{2(g)} \rightarrow CO_{2(g)}$

$\Delta H^\circ_{rxn} = [1\ \Delta H^\circ(CO)] - [1\ \Delta H^\circ(C) + 1/2\ \Delta H^\circ(O_2)]$
$= (-110.525\ kJ) - [0.00 + 1/2(0.00)] = -110.525\ kJ$

$\Delta S^\circ_{rxn} = [1\ S^\circ(CO)] - [1\ S^\circ(C) + 1/2 S^\circ(O_2)]$
$= (197.674\ J/K) - [5.740\ J/K + 1/2(205.138\ J/K)] = 89.365\ J/K$

$\Delta G^\circ_{rxn,\ 473\ K} = (-110.525\ kJ) - (473\ K)(89.365\ J/K)(10^{-3}\ kJ/J) = -153\ kJ$

d) $CO_{(g)} + H_2O_{(g)} \rightarrow CO_{2(g)} + H_{2(g)}$

$\Delta H^\circ_{rxn} = [1\ \Delta H^\circ(CO_2) + 1\ \Delta H^\circ(H_2)] - [1\ \Delta H^\circ(CO) + 1\ \Delta H^\circ(H_2O)]$
$= [(-393.509\ kJ) + (0.00)] - [-110.525\ kJ + (-241.818\ kJ)]$
$= -41.166\ kJ$

$\Delta S^\circ_{rxn} = [1\ S^\circ(CO_2)] - [1\ S^\circ(H_2)] - [1\ S^\circ(CO) + 1\ S^\circ(H_2O)]$
$= [(213.74\ J/K) + (130.684\ J/K)] - [(197.674\ J/K) + (188.825\ J/K)]$
$= -42.07\ J/K$

$\Delta G^\circ_{rxn,\ 473\ K} = (-41.166\ kJ) - (473\ K)(-42.07\ J/K)(10^{-3}\ kJ/J) = -21.27\ kJ$

13.67 $3\ NO_{2(g)} + H_2O_{(l)} \rightarrow 2\ HNO_{3(g)} + NO_{(g)}$

a) $\Delta G° = [\Delta G°(HNO_3) + 1\ \Delta G°(NO)] - [3\ \Delta G°(NO_2) + 1\ \Delta G°(H_2O)]$

$= [2(-74.72\ kJ) + 1(86.55\ kJ)] - [3(51.31\ kJ) + 1(-237.129\ kJ)]$

$= 20.31\ kJ$ Not thermodynamically feasible at standard conditions.

b) $\Delta G = \Delta G° + RT \ln Q$

$$\Delta G = (20.31\ kJ) + (8.314\ J/mol\ K)(298K)(10^{-3}\ kJ / J)\ln\frac{(10^{-6})^3}{(10^{-6})^3}$$

$= 20.31\ kJ + 2.478\ kJ \ln 1$

$= 20.31\ kJ + 0 = 20.31\ kJ$ Not thermodynamically feasible.

If $H_2O_{(g)}$ were used instead of $H_2O_{(l)}$ in this problem:

a) $\Delta G° = 11.90\ kJ$ b) $\Delta G° = 46.13\ kJ$

It is still not thermodynamically feasible but at higher pressures (~123 atm) it could be.

13.68 This is not drawn to scale.

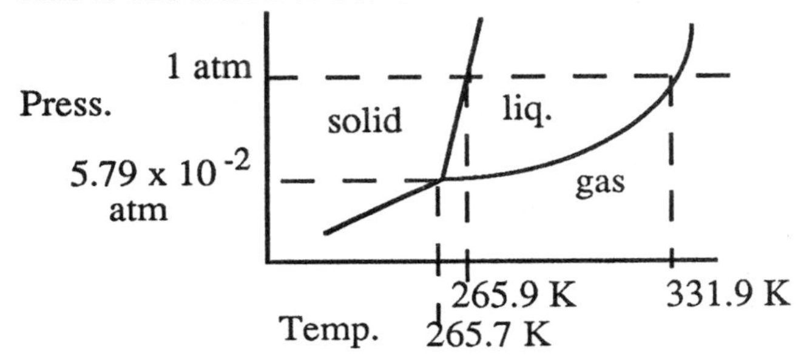

13.69 a) The sample will condense to a liquid at 331.9 K and freeze at 265.9 K if maintained at 1 atm pressure. Therefore, at 250 K and 1.00 atm the sample will be a solid. This can be shown by drawing a horizontal line at 1 atm between T = 400 K and T = 250 K.

b) The temperature of 265.8 K is between the temperature of the triple point and the normal freezing point. A vertical line drawn at 265.8 K between the pressures of 1.00×10^{-3} atm and 1.00×10^{3} atm will start in the vapor phase, pass through the liquid phase, and terminate in the solid phase.

c) A horizontal line drawn at $P = 2.00 \times 10^{-2}$ atm will pass below the triple point. Therefore, a sample heated from 250 to 400 K at a constant $P = 2.00 \times 10^{-2}$ atm will start as a solid and pass directly into the vapor phase (sublime).

13.70 From Section 13.6

Glucose = $C_6H_{12}O_6$ $\Delta G° = -2870$ kJ/mol

Palmitic acid = $C_{15}H_{31}CO_2H$ $\Delta G° = -9790$ kJ/mol

$C_6H_{12}O_6$ MM = 180 g/mol

$C_{15}H_{31}CO_2H$ MM = 256 g/mol

(continued)

(13.70 continued)

$$\Delta G°(C_6H_{12}O_6) = \frac{-2870\ \text{kJ/mol}}{180\ \text{g/mol}} = -15.9\ \text{kJ/g}$$

$$\Delta G°(C_{15}H_{31}CO_2H) = \frac{-9790\ \text{kJ/mol}}{256\ \text{g/mol}} = -38.2\ \text{kJ/g}$$

Palmitic acid is the better energy source on a per gram basis and on a per mole basis. An inspection of the formula of glucose indicates that it is already partially oxygenated.

13.71 Methane would be expected to have the highest S° because it has the most atoms (5 vs. 4 for ammonia and 3 for water). Methane may have a lower absolute entropy because it has 4 hydrogen atoms located symmetrically around the carbon atom. The other molecules have hydrogen atoms and lone pairs of electrons located around the central atom.

13.72 $$q_{H_2O} = +(25.0\ \text{g}\ H_2O)\left(\frac{1\ \text{mol}\ H_2O}{18.0\ \text{g}\ H_2O}\right)(40.79\ \text{kJ/mol}) = +56.65\ \text{kJ}$$

$$\Delta S_{H_2O} = \left(\frac{56.65\ \text{kJ}}{(100+273)\text{K}}\right)\left(\frac{10^3\ \text{J}}{\text{kJ}}\right) = +152\ \text{J/K} = \Delta S_{sys}$$

$$\Delta S_{hotplate} = \left(\frac{-56.65\ \text{kJ}}{(300+273)\text{K}}\right)\left(\frac{10^3\ \text{J}}{\text{kJ}}\right) = -98.9\ \text{J/K} = \Delta S_{surr}$$

$$\Delta S_{universe} = \Delta S_{sys} + \Delta S_{surr} = 152\ \text{J/K} + (-98.9\ \text{J/K}) = 53\ \text{J/K}$$

13.73 $CCl_{4(l)} + 5\ O_{2(g)} \rightarrow CO_{2(g)} + 4\ ClO_{2(g)}$

ΔH°_{rxn} = (1 mol CO_2) $\Delta H°(CO_2)$ + (4 mol ClO_2) $\Delta H°ClO_2$ - [(1 mol CCl_4) $\Delta H°(CCl_4)$ + (5 mol O_2) $\Delta H°(O_2)$]= (1 mol CO_2)(-393.509 kJ/mol) + (4 mol ClO_2) (102.5 kJ/mol) - [(1 mol CCl_4)(-135.44 kJ/mol) + (5 mol O_2)(0.0 kJ/mol)]
= -393.509 kJ + 410.0 kJ + 135.44 kJ + 0.0 kJ = 151.9 kJ

ΔG°_{rxn} = (1 mol CO_2) $\Delta G°(CO_2)$ + (4 mol ClO_2) $\Delta G°ClO_2$ - [(1 mol CCl_4) $\Delta G°(CCl_4)$ + (5 mol O_2) $\Delta G°(O_2)$]
= (1 mol CO_2)(-394.359 kJ/mol) + (4 mol ClO_2)(120.5 kJ/mol) - [(1 mol CCl_4) (-65.21 kJ/mol) + (5 mol O_2)(0 kJ/mol)]
= -394.359 kJ + 482.0 kJ + 65.21 kJ + 0.0 kJ = 152.9 kJ

(continued)

(13.73 continued)

$CS_{2(l)} + 3\ O_{2(g)} \rightarrow CO_{2(g)} + 2\ SO_{2(g)}$

$\Delta H^\circ_{rxn} = (1\ \text{mol}\ CO_2)\ \Delta H^\circ(CO_2) + (2\ \text{mol}\ SO_2)\ \Delta H^\circ(SO_2) - [(1\ \text{mol}\ CS_2)\ \Delta H^\circ(CS_2)$
$+ (3\ \text{mol}\ O_2)\ \Delta H^\circ(O_2)] = (1\ \text{mol}\ CO_2)(-393.509\ \text{kJ/mol})$
$+ (2\ \text{mol}\ SO_2)(-296.830\ \text{kJ/mol}) - [(1\ \text{mol}\ CS_2)(89.70\ \text{kJ/mol}) + (3\ \text{mol}\ O_2)(0)]$
$= -393.509\ \text{kJ} - 593.66\ \text{kJ} - 89.70\ \text{kJ} + 0\ \text{kJ} = -1076.87\ \text{kJ}$

$\Delta G^\circ_{rxn} = (1\ \text{mol}\ CO_2)\ \Delta G^\circ(CO_2) + (2\ \text{mol}\ SO_2)\ \Delta G^\circ(SO_2) - [(1\ \text{mol}\ CS_2)\ \Delta G^\circ(CS_2)$
$+ (3\ \text{mol}\ O_2)\ \Delta G^\circ(O_2)]$
$= (1\ \text{mol}\ CO_2)(-394.359\ \text{kJ/mol}) + (2\ \text{mol}\ SO_2)(-300.194\ \text{kJ/mol}) - [(1\ \text{mol}\ CS_2)$
$(65.27\ \text{kJ/mol}) + (3\ \text{mol}\ O_2)(0)]$
$= -394.359\ \text{kJ} - 600.388\ \text{kJ} - 65.27\ \text{kJ} + 0.0\ \text{kJ} = -1060.02\ \text{kJ}$

Special precautions against fires would be recommended for industrial plants using carbon disulfide. The combustion reaction is spontaneous ($\Delta G^\circ = -1060$ kJ) and a great amount of heat is produced ($\Delta H^\circ = -1077$ kJ) which would keep the fire going and might well ignite other flammable substances. The combustion of carbon tetrachloride is nonspontaneous and is endothermic.

13.74 The structure of diamond is a 3-dimensional network of carbon atoms bonded to each other. The structure of graphite is a 2-dimensional network of carbon atoms that are stacked in layers that may slide by each other. Therefore, graphite has greater disorder and has a greater molar entropy. The molar entropy of fullerene is probably larger than graphite because individual molecules contain 60 carbon atoms while a layer of carbon atoms in graphite may contain many more carbon atoms. Also the layers in graphite can move in 2-dimensions while the C_{60} individual molecules could move in 3-dimensions.

13.75 a) $q_v = \Delta E$ b) $q_p = \Delta H$ c) $q_T = T\Delta S$

13.76 The stored, potential energy of the uranium creates a heat flow that generates steam which drives turbines to generate electricity. Electricity is an organized motion of electrons in the same direction in a wire. Producing this order among the electrons requires a decrease in entropy that must be paid for by creating greater disorder somewhere else. Much energy used for thermal generation of electrical power is wasted by being dumped into the surroundings as thermal pollution.

13.77 $NaCl_{(s)} \rightarrow Na^+_{(aq)} + Cl^-_{(aq)}$

a) $\Delta G^\circ_{rxn} =$
$(1\ \text{mol}\ Na^+_{(aq)})\ \Delta G^\circ_f(Na^+_{(aq)}) + (1\ \text{mol}\ Cl^-_{(aq)})\ \Delta G^\circ_f(Cl^-_{(aq)}) - (1\ \text{mol}\ NaCl_{(s)})$
$\Delta G^\circ_f(NaCl_{(s)}) = (1\ \text{mol})(-261.905\ \text{kJ/mol}) + (1\ \text{mol})(-131.228\ \text{kJ/mol}) -$
$(1\ \text{mol})(-384.138\ \text{kJ/mol}) = -8.995\ \text{kJ}$

This reaction is spontaneous; ΔG is negative.

(continued)

(13.77 continued)

b) ΔH°_{rxn} = (1 mol $Na^+_{(aq)}$) $\Delta H^\circ_f(Na^+_{(aq)})$ + (1 mol $Cl^-_{(aq)}$) $\Delta H^\circ_f(Cl^-_{(aq)})$ - (1 mol $NaCl_{(s)}$) $\Delta H^\circ_f(NaCl_{(s)})$

= (1 mol)(-240.12 kJ/mol) + (1 mol)(-167.59 kJ/mol) - (1 mol)(-411.153 kJ/mol)

= +3.443 kJ

This reaction does not release energy; it is endothermic (ΔH is positive.)

c) ΔS°_{rxn} =

(1 mol $Na^+_{(aq)}$) $S^\circ(Na^+_{(aq)})$ + (1 mol $Cl^-_{(aq)}$) $S^\circ(Cl^-_{(aq)})$ - (1 mol $NaCl_{(s)}$) $S^\circ(NaCl_{(s)})$

= (1 mol)(59.0 J/K) + (1 mol)(56.5 J/K) - (1 mol)(72.13 J/K) = 43.4 J/K

No, the amount of disorder of the system increases; ΔS is positive.

13.78 $AgCl_{(s)} \rightarrow Ag^+_{(aq)} + Cl^-_{(aq)}$

a) ΔG°_{rxn} = (1 mol $Ag^+_{(aq)}$) (77.107 kJ/mol) + (1 mol $Cl^-_{(aq)}$) (-131.228 kJ/mol) - (1 mol $AgCl_{(s)}$) (-109.789 kJ/mol) = 55.668 kJ

The reaction is not spontaneous; ΔG is positive.

b) ΔH°_{rxn} = (1 mol $Ag^+_{(aq)}$) (105.579 kJ/mol) + (1 mol $Cl^-_{(aq)}$) (-167.59 kJ/mol) - (1 mol $AgCl_{(s)}$) (-127.068 kJ/mol) = 65.06 kJ

The reaction does not release energy; it is endothermic (ΔH is positive).

c) ΔS°_{rxn} = (1 mol $Ag^+_{(aq)}$) (72.68 J/mol K) + (1 mol $Cl^-_{(aq)}$) (56.5 J/mol K) - (1 mol $AgCl_{(s)}$) (96.2 J/mol K) = 33.0 J/K

The amount of disorder increases; ΔS is positive.

Table salt is soluble (its dissolution reaction is spontaneous) while silver chloride is not soluble (its dissolution reaction is not spontaneous).

$\Delta G^\circ_{rxn} = \Delta H^\circ_{rxn} - T\Delta S^\circ_{rxn}$

There is a slight difference in the $-T\Delta S^\circ_{rxn}$ terms at 298.15 K

NaCl: $-T\Delta S^\circ_{rxn}$ = -12.9 kJ AgCl: $-T\Delta S^\circ_{rxn}$ = -9.84 kJ

But ΔH°_{rxn} for AgCl dissolution is much more positive (+65.06 kJ) than for NaCl dissolution (+3.443 kJ).

CHAPTER 14: MECHANISMS OF CHEMICAL REACTIONS

14.1 a) Filling each cup with coffee from the urn.
b) The ringing up of the items by the checker.
c) Making payment.
d) The passing through the airplane door.

14.2 a) $Cl_2 \xrightarrow{h\nu} 2Cl$

b) $NO + Cl_2 \rightarrow NOCl + Cl$

c) $2\ NO + Cl_2 \rightarrow 2\ NOCl$

14.3 a) $\xrightarrow{h\nu}$ [ ] → +

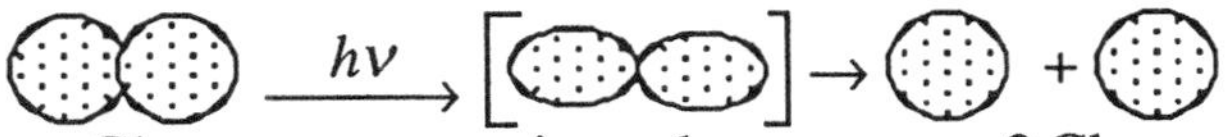

Cl_2 activated 2 Cl

b)

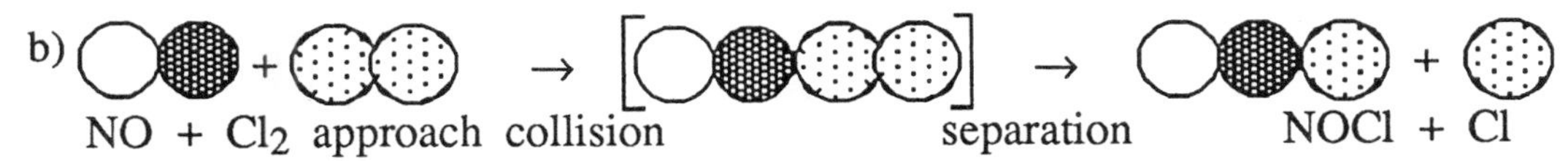

NO + Cl_2 approach collision separation NOCl + Cl

c) + + → [ ] → 2

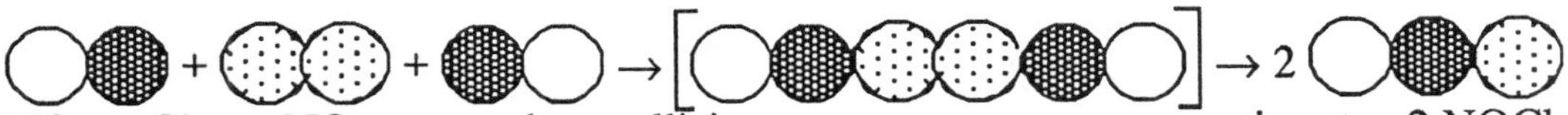

NO + Cl_2 + NO approach collision separation 2 NOCl

14.4 a) $Cl_2 \xrightarrow{h\nu} 2\ Cl$

$NO + Cl \rightarrow NOCl$

b) $NO + Cl_2 \rightarrow NOCl + Cl$

$NO + Cl \rightarrow NOCl$

c) $2\,NO + Cl_2 \rightarrow 2\,NOCl$

14.5 The intermediate for the decompositon of ozone is oxygen atoms (O).

14.6 $\xrightarrow{h\nu}$ +

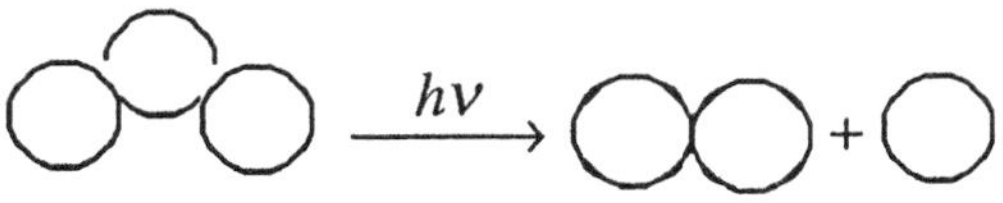

$O_3 \xrightarrow{h\nu} O_2 + O$

+ → +

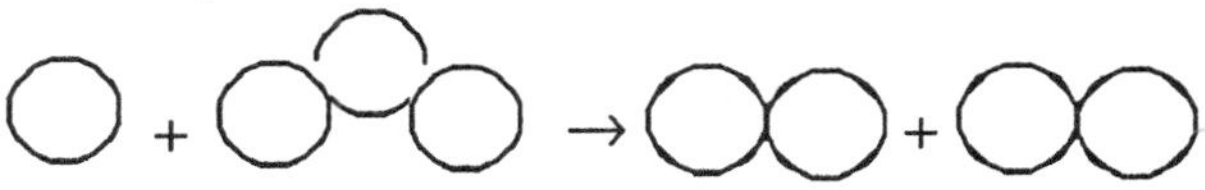

$O + O_3 \rightarrow O_2 + O_2$

14.7 $2\ NO + Cl_2 \rightarrow 2\ NOCl$

a) $\text{Rate} = -\frac{\Delta[Cl_2]}{\Delta t}$

b) $\frac{1}{2}\frac{\Delta[NOCl]}{\Delta t} = -\frac{\Delta[Cl_2]}{\Delta t}$, i.e., the rate of NOCl formation is twice the rate of Cl_2 disappearance.

c) The rate of NOCl formation = 2 x rate of Cl_2 disappearance = 2 x 47 M s^{-1} = 94 M s^{-1}

14.8 a) mol/sec, mol/min, M/sec, M/min. molecules/sec, molecules/min, mol/hr, M/hr, molecules/hr, atm/sec, atm/min

b) Assuming reaction: $2\ O_3 \rightarrow 3\ O_2$

$\text{Rate} = \frac{1}{3}\frac{\Delta[O_2]}{\Delta t}$

c) $-\frac{1}{2}\frac{\Delta[O_3]}{\Delta t} = \frac{1}{3}\frac{\Delta[O_2]}{\Delta t}$

14.9 $2\ NO + Cl_2 \rightarrow 2\ NOCl$

a)

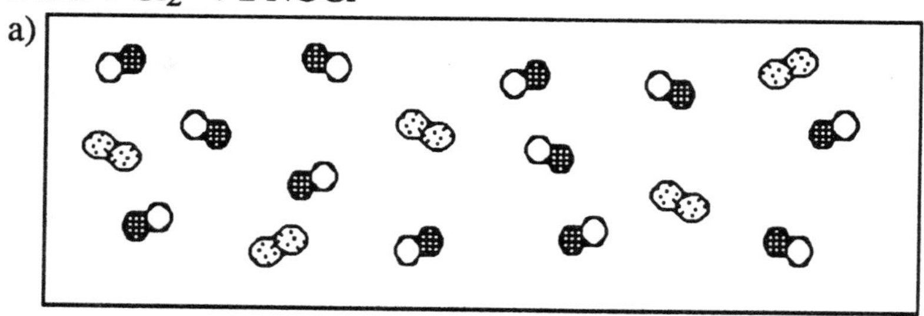

b)

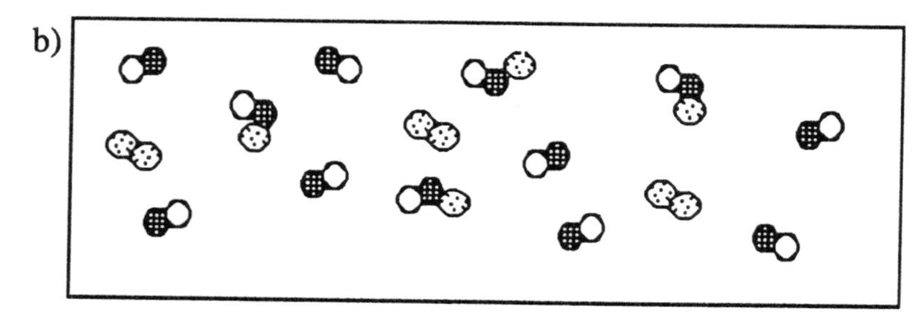

c)

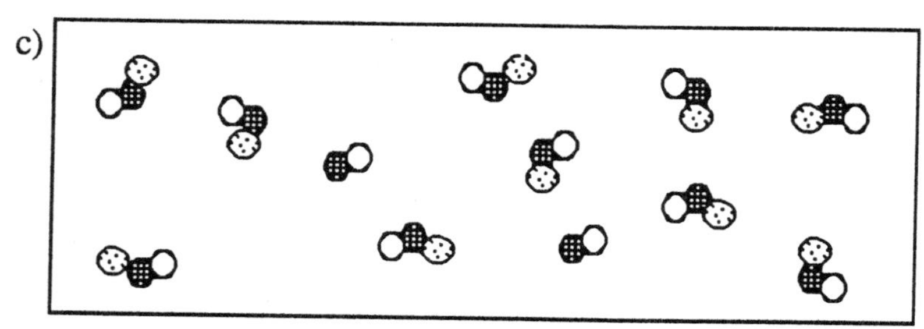

14.10 $CaCO_{3(s)} \rightarrow CaO_{(s)} + CO_{2(g)}$ $\qquad$ 550°C = 823 K

moles of total CO_2 produced =

$$n = \frac{PV}{RT} = \frac{(0.15 \text{ atm})(5.0 \text{ L})}{(0.08206 \text{ L atm mol}^{-1} \text{ K}^{-1})(823 \text{ K})} = 0.0111 \text{ moles}$$

a) $\dfrac{\Delta[CO_2]}{\Delta t} = \dfrac{0.0111 \text{ moles}}{5 \text{ min}} = 2.2 \times 10^{-3}$ mol/min

b) moles of $CaCO_3$ decomposed = moles of CO_2 formed = 0.011 moles

14.11 a)

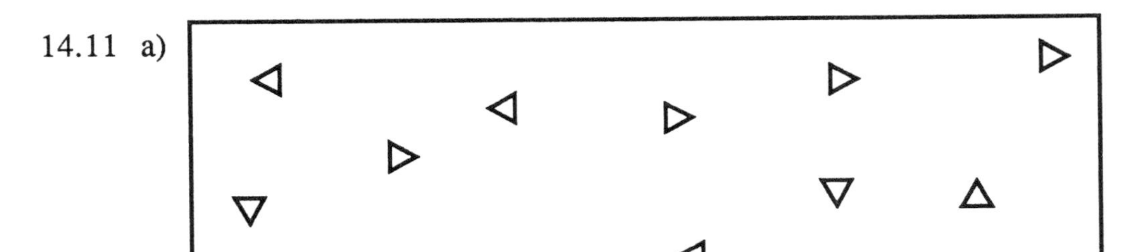

Cyclopropane (C_3H_6) before any isomerization.

b)

Propene (C_3H_6) and cyclopropane (C_3H_6) after the sample in part a has isomerized for 20 minutes.

14.12 $N_2 + 3\,H_2 \rightarrow 2\,NH_3$

a) Rate of reaction of H_2 =

$$-\frac{\Delta[H_2]}{\Delta t} = -\frac{(0.25 \text{ mol L}^{-1} - 0.50 \text{ mol L}^{-1})}{(30 \text{ s} - 0 \text{ s})} = 8.3 \times 10^{-3} \text{ mol L}^{-1} \text{ s}^{-1}$$

b) $-\dfrac{1}{3}\dfrac{\Delta[H_2]}{\Delta t} = \dfrac{1}{2}\dfrac{\Delta[NH_3]}{\Delta t}$

$$\frac{\Delta[NH_3]}{\Delta t} = -\frac{2}{3}\frac{\Delta[H_2]}{\Delta t} = \left(\frac{2}{3}\right)8.3 \times 10^{-3} \text{ mol L}^{-1} \text{ s}^{-1} = 5.5 \times 10^{-3} \text{ mol L}^{-1} \text{ s}^{-1}$$

c) 0.25 mol/L of H_2 consumed in 30 seconds;
1/3(0.25 mol/L) of N_2 consumed in 30 seconds = 0.083 mol/L of N_2 consumed;
original concentration of N_2 - amount consumed = concentration after 30 seconds;
1.25 mol/L N_2 - 0.083 mol/L N_2 = 1.17 mol/L N_2 (after 30 seconds).

14.13 A single-step elementary reaction would require 4 molecules to collide while having proper orientation. The chances of this occurring are nil.

14.14 rate = k [kernels]

6 kernels/5 s = k [150] $k = 8 \times 10^{-3}\ s^{-1}$

a) rate = $(8 \times 10^{-3}\ s^{-1})$[100 kernels] = $\frac{0.8 \text{ kernels}}{s} = \frac{4 \text{ kernels}}{5 \text{ s}}$

b) The average "popping time" per kernel has changed from 0.83 s/kernel to 1.25 s/kernel.

The average "popping time" (time for an event to occur) is the reciprocal of the rate. The rate is proportional to the concentration (number) of kernels. The concentration for (a) is 2/3 the original concentration and the rate is 2/3 the original rate. The average time per event for (a) is 3/2 (reciprocal of 2/3) the original average time per event; 1.25 vs. 0.83.

14.15 Rate = k [red][white]

$\frac{4 \text{ objects}}{2 \text{ min}} = k\ [20\ \text{objects}_r][10\ \text{objects}_w];\ k = \frac{4 \text{ objects}}{2\ \text{min}(20\ \text{objects}_r)(10\ \text{objects}_w)}$

$= 0.01\ \text{objects}^{-1}\ \text{min}^{-1}$

a) rate = $(0.01\ \text{objects}^{-1}\ \text{min}^{-1})[10\ \text{objects}_r][10\ \text{objects}_w]$

= 1 object/min = 2 objects/2 min

The concentration of red balls was halved; therefore, half as many pairs are formed: (4 pairs/2 min)(1/2) = 2 pairs/2 min.

b) rate = $(0.01\ \text{objects}^{-1}\ \text{min}^{-1})[20\ \text{objects}_r][15\ \text{objects}_w]$

= 3 objects/min = 6 objects/2 min

The concentration of red balls was kept the same (compared to original) and the concentration of white balls was increased by a factor of 1.5:
(4 pairs/2 min)(1.5) = 6 pairs/2min

c) rate = $(0.01\ \text{objects}^{-1}\ \text{min}^{-1})[10\ \text{objects}_r][20\ \text{objects}_w]$

= 2 objects/min = 4 objects/2 min

The concentration of red balls was halved and the concentration of white balls was doubled (compared to original): (4 pairs/2 min)(1/2)(2) = 4 pairs/2 min

d) rate = $(0.01\ \text{object}^{-1}\ \text{min}^{-1})[40\ \text{objects}_r][20\ \text{objects}_w]$

= 8 objects/min = 16 objects/2 min

The concentration of red balls was doubled and the concentration of white balls was doubled (compared to original): (4 pairs/2 min)(2)(2) = 16 pairs/2 min
This assumes that cramming 60 balls in the machine does not exceed the capacity of the machine to keep all the balls in motion. If all of the balls are not kept in motion (some lying in the bottom of the machine), the rate would be less.

e) When the number of balls is increased (in the same volume) there will be more collisions and, therefore, more collisions with the correct orientation for reaction. The reaction rate will be greater. If the number of balls is decreased, the opposite is true and the rate will be smaller.

14.16 a) $2\ C + AB \rightarrow AC + BC$ rate = $k[C]^2[AB]$

b) $2\ AB \rightarrow A_2 + 2\ B$ rate = $k[AB]^2$

c) $C + AB \rightarrow BC + A$ rate = $k[C][AB]$

d) $AB \rightarrow A + B$ rate = $k[AB]$

14.17 a) (concentration)(time)$^{-1}$ = k (concentration)^2(concentration)

$$k = \frac{(\text{concentration})(\text{time})^{-1}}{(\text{concentration})^2(\text{concentration})} = (\text{concentration})^{-2}(\text{time})^{-1}$$

b) (concentration)(time)$^{-1}$ = k (concentration)^2

$$k = \frac{(\text{concentration})(\text{time})^{-1}}{(\text{concentration})^2} = (\text{concentration})^{-1}(\text{time})^{-1}$$

c) (concentration)(time)$^{-1}$ = k (concentration)(concentration)

$$k = \frac{(\text{concentration})(\text{time})^{-1}}{(\text{concentration})(\text{concentration})} = (\text{concentration})^{-1}(\text{time})^{-1}$$

d) (concentration)(time)$^{-1}$ = k (concentration)

$$k = \frac{(\text{concentration})(\text{time})^{-1}}{(\text{concentration})} = \text{time}^{-1}$$

14.18 a)

$2\ C + AB \rightarrow AC + BC$ (rate determining)
$AC + AB \rightarrow A_2 + BC$ (fast)
$2\ C + 2\ AB \rightarrow A_2 + 2\ BC$ (overall reaction)

b)

$2\ AB \rightarrow A_2 + 2\ B$ (rate determining)
$C + B \rightarrow BC$ (twice) (fast)
$2\ AB + 2\ C \rightarrow A_2 + 2\ BC$ (overall reaction)

c)

$C + AB \rightarrow BC + A$ (twice)(rate determining)
$A + A \rightarrow A_2$ (fast)
$2\ C + 2\ AB \rightarrow 2\ BC + A_2$ (overall reaction)

d)

$AB \rightarrow A + B$ (twice)(rate determining)
$B + C \rightarrow BC$ (twice)(fast)
$A + A \rightarrow A_2$ (fast)
$2\ C + 2\ AB \rightarrow 2\ BC + A_2$ (overall reaction)

14.19 a) rate = $k[H_2]^2$ rate ratio = $\frac{k\,[3\,x]^2}{k\,[x]^2} = 9$ The rate would increase 9 times.

b) rate = $k[H_2]^0$ rate ratio = $\frac{k\,[3\,x]^0}{k\,[x]^0} = 1$

The rate would not change because it is independent of the concentration of H_2.

c) rate = $k[H_2]^{3/2}$ rate ratio = $\frac{k\,[3\,x]^{3/2}}{k\,[x]^{3/2}} = [3]^{3/2} = 5.196$

The rate would increase by 5.196 times.

14.20 Flask B will react at the same rate as flask A. The O_2 concentration (number of molecules) has been doubled and, therefore, the number of collisions of oxygen molecules will be increased. There would have been double the number of effective collisions between O_2 and NO_2 if the number of NO_2 had not changed, but the number of NO_2 was halved which halves the possible effective collisions. Therefore, the number of effective collisions in flask B is the same as in flask A.

$$\text{rate ratio} = \frac{\text{rate (B)}}{\text{rate (A)}} = \frac{k\,[12][3]}{k\,[6][6]} = 1$$

14.21 $2\,N_2O_{5(g)} \rightarrow 4\,NO_{2(g)} + O_{2(g)}$

Time (s)	0	200	400	600	800
$[N_2O_5]$(atm)	2.50	2.22	1.96	1.73	1.53
ln $[N_2O_5]$	0.916	0.798	0.673	0.548	0.425
$1/[N_2O_5]$	0.400	0.450	0.510	0.578	0.654

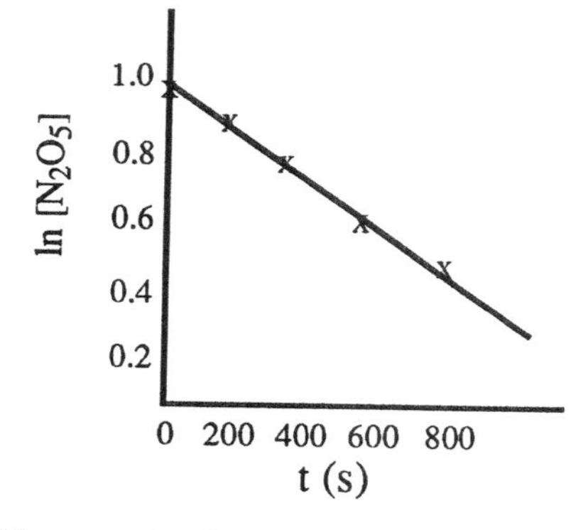

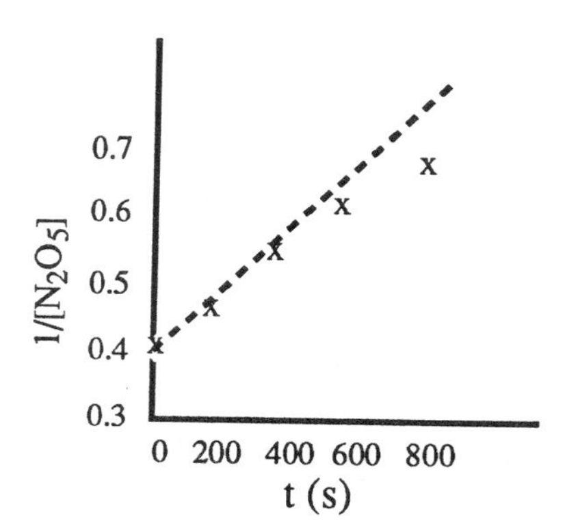

The rate is first order with respect to N_2O_5; rate = $k[N_2O_5]$. The rate constant, k, equals the – slope for the graph of $\ln[N_2O_5]$ vs. t.

$$\text{From graph:} -\text{slope} = -\frac{\Delta y}{\Delta x} = -\frac{0.673 - 0.890}{400\text{ s} - 0\text{ s}} = +5.43 \times 10^{-4}\ \text{s}^{-1}$$

14.22

Time (days)	0	4.0	8.0	12.0	16.0
Mass $^{131}I(\mu g)$	12.0	8.48	6.0	4.24	3.0
ln [I]	2.48	2.14	1.79	1.44	1.10
1/[I]	0.0833	0.118	0.167	0.236	0.333

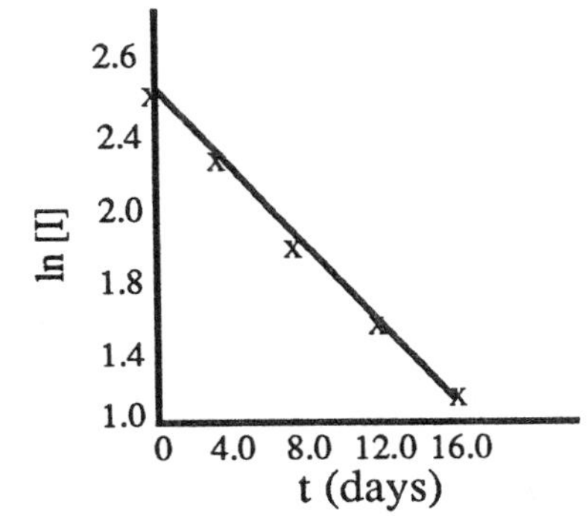

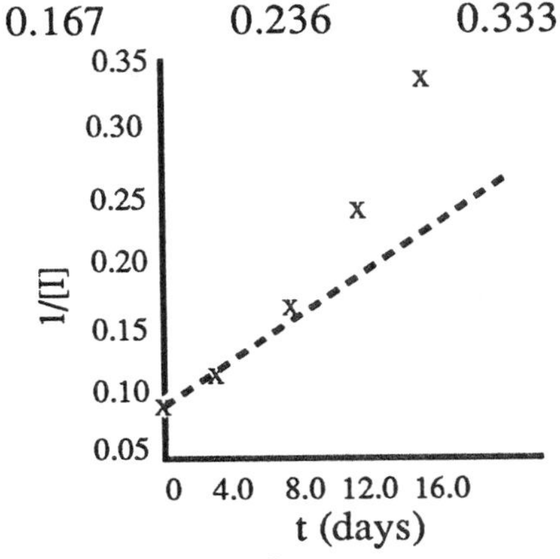

a) The reaction is first order with respect to ^{131}I; rate = k [^{131}I]. Note: It is not necessary to draw the graphs. Comparing 0 and 8 hours, or 4 and 12 hours or 8 and 16 hours it is obvious that the half-life for this decomposition is 8 days and independent of the starting concentration. Therefore, the rate is first order with respect to ^{131}I.

b) $k = \frac{\ln 2}{t_{1/2}} = \frac{0.6931}{8.0 \text{ days}} = 0.087 \text{ days}^{-1}$

c) $kt = \ln\frac{[A]_0}{[A]}$ $\qquad (0.087 \text{ days}^{-1})(32 \text{ days}) = \ln\frac{[A]_0}{[A]} = 2.78$

$\frac{[A]_0}{[A]} = e^{2.78} = 16. \qquad \frac{[A]_0}{16.18} = [A] = \frac{12\ \mu g}{16} = 0.75 \mu g$

Note: Using half-lifes, 1.5 μg after 24.0 days and 0.75 μg after 32.0 days.

d) $t = \frac{\ln\frac{[A]_0}{[A]}}{k} = \frac{\ln\left(\frac{12.0\ \mu g}{1.2\ \mu g}\right)}{0.087 \text{ days}^{-1}} = 26.6 \text{ days}$

14.23 Rate = k [NOBr]^2 = 25 L mol^{-1} min^{-1} [NOBr]^2

a) $\frac{1}{[A]} - \frac{1}{[A]_0} = kt = 25 \text{ L mol}^{-1} \text{ min}^{-1} t$

$\frac{1}{0.010 \text{ M}} - \frac{1}{0.025 \text{ M}} = 25 \text{ L mol}^{-1} \text{ min}^{-1} t$

$60 \text{ M}^{-1} = 60 \text{ L mol}^{-1} = 25 \text{ L mol}^{-1} \text{ min}^{-1} t$

$t = \frac{60 \text{ L mol}^{-1} \text{ min}^{-1}}{25 \text{ L mol}^{-1} \text{ min}^{-1}} = 2.4 \text{ min}$

b) $\frac{1}{[A]} = kt + \frac{1}{[A]_0} = (25 \text{ L mol}^{-1} \text{ min}^{-1})(100 \text{ min}) + \frac{1}{0.025 \text{ M}}$

$1/[A] = 2500 \text{ L mol}^{-1} + 40 \text{ M}^{-1} = 2540 \text{ M}^{-1}$

$[A] = 0.00039 \text{ M}$

14.24 a) Rate = k [A]

The order of the reaction with respect to B is zero because the rate does not change when the concentration of B is changed. The order of the reaction with respect to A is first because comparison of concentrations of A for 0 s and 30 s or 10 s and 40 s or 20 s and 50 s shows a half-life of 30 s that is independent of starting concentration.

b) k = ln 2/30 s = 0.6931/30 s = 0.023 s^{-1}

14.25 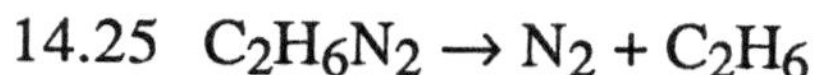
$C_2H_6N_2 \rightarrow N_2 + C_2H_6$

Time (s)	0	100	150	200	250	300
Azomethane (mM)	7.94	6.15	5.40	4.75	4.20	3.69
ln [Azomethane]	2.07	1.82	1.69	1.56	1.44	1.31
1/[Azomethane]	0.126	0.163	0.185	0.211	0.238	0.271

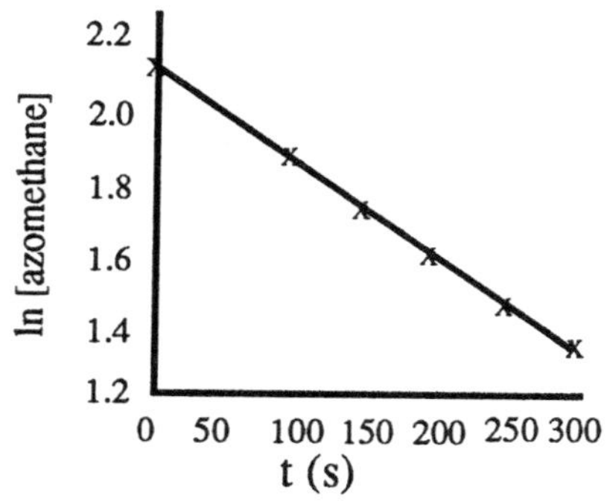

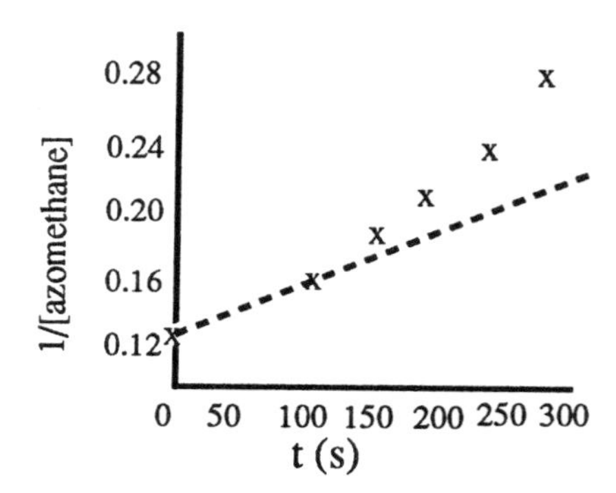

The rate of reaction is first order.

k = – slope = – Δy/Δx = – (1.31 - 1.82) / (300 s - 100 s) = 0.00255 s^{-1}

14.26

Time (s)	0	200	400	600	800
[A](M)	2.50	2.22	1.96	1.73	1.53
ln [A]	0.916	0.798	0.673	0.548	0.425
1/[A]	0.400	0.450	0.510	0.578	0.654

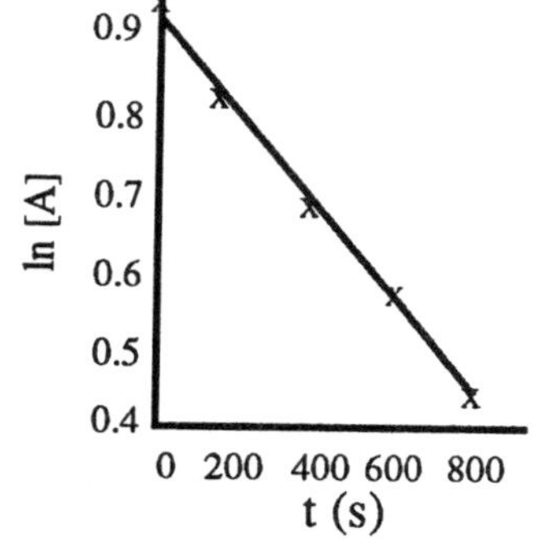

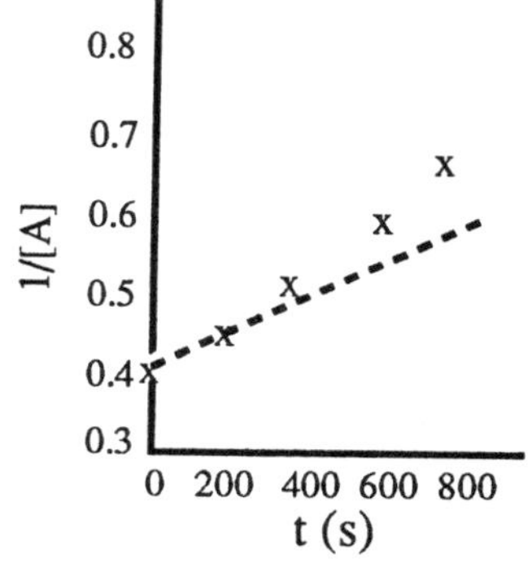

The rate of reaction is first order with respect to [A]; rate = k[A]. The rate constant, k, equals – slope for graph of ln [A] vs. t.

k = – slope = – Δy/Δx = – (0.548 - 0.916) / (600 s - 0 s) = 6.13 x 10^{-4} s^{-1}

14.27 In a system where one of the products of a fast, reversible reaction is being used as a reactant for a second reaction, the forward and reverse rates for the fast, reversible reaction will not ever be exactly equal. The forward and reverse rates are equal only when the concentrations of the reactant(s) and product(s) fulfill a certain relationship. Because one of the products is constantly being used by the second reaction, this relationship will be deficient in the concentration of that product. Therefore, the forward rate will be slightly faster as it tries to replace the product being used by the second reaction.

14.28 a) $Cl_2 \rightleftarrows 2\ Cl$ (fast, reversible)

$Cl + CO \rightarrow COCl$ (slow, rate determining)

$COCl + Cl_2 \rightarrow COCl_2 + Cl$ (fast)

$Cl_2 \rightarrow 2\ Cl$

$2\ Cl \rightarrow Cl_2$

$Cl + CO \rightarrow COCl$

$COCl + Cl_2 \rightarrow COCl_2 + Cl$

Net reaction: $CO + Cl_2 \rightarrow COCl_2$

b) $Cl_2 \underset{k_{-1}}{\overset{k_1}{\rightleftarrows}} 2\ Cl$

$k_1[Cl_2] = k_{-1}[Cl]^2$ $\quad [Cl]^2 = \frac{k_1}{k_{-1}}[Cl_2]$ $\quad [Cl] = \left(\frac{k_1}{k_{-1}}\right)^{1/2}[Cl_2]^{1/2}$

$$\text{rate} = k_2[Cl][CO] = k_2\left(\frac{k_1}{k_{-1}}\right)^{1/2}[Cl_2]^{1/2}[CO] = k[Cl_2]^{1/2}[CO]$$

c) Both Cl and COCl are reactive intermediates.

14.29 $O_3 + O_2 \rightleftarrows O_5$ (fast, reversible)

$O_5 \rightarrow 2\ O_2 + O$ (slow, rate determining)

a) $O_3 + O \rightarrow 2\ O_2$ (fast)

b) $k_1\ [O_3]\ [O_2] = k_{-1}\ [O_5]$

$[O_5] = k_1/k_{-1}[O_3][O_2]$

$$\text{rate} = k_2\ [O_5] = k_2\frac{k_1}{k_{-1}}[O_3][O_2] = k[O_3][O_2]$$

c) The mechanism requires O_5 to decompose by two different processes; a fast one that returns it to the starting materials and a slow one that leads to the product.

14.30 a) $k_1[NO]^2 = k_{-1}[N_2O_2]$ $\quad [N_2O_2] = (k_1/k_{-1})[NO]^2$

rate $= k_2[N_2O_2][O_2] = k_2\ (k_1/k_{-1})[NO]^2[O_2] = k[NO]^2[O_2]$

b) Yes, the rate is third-order overall, second-order in NO and first-order in O_2.

(continued)

(14.30 continued)

c) 2 $\cdot\ddot{N}{=}\ddot{O}{:}$ → $:\ddot{N}{=}\ddot{O}{-}\ddot{N}{=}\ddot{O}{:}$ or $:\ddot{O}{=}\ddot{N}{-}\ddot{N}{=}\ddot{O}{:}$ The second possible intermediate is the most reasonable since it is logical that the next step of the mechanism is the collision between oxygen and the N—N to form an intermediate that will have two oxygens bonded to each nitrogen, either before or after the N—N bond breaks.

14.31 a) $Cl_2 + h\nu \underset{k_{-1}}{\overset{k_1}{\rightleftharpoons}} 2\ Cl$ (fast, reversible)

$NO + Cl \xrightarrow{k_2} NOCl$ (rate determining)

$$k_1[Cl_2] = k_{-1}[Cl]^2 \qquad [Cl]^2 = \left(\frac{k_1}{k_{-1}}\right)[Cl_2] \qquad [Cl] = \left(\frac{k_1}{k_{-1}}\right)^{1/2}[Cl_2]^{1/2}$$

$$\text{rate} = k_2[NO]\,[Cl] = k_2\left(\frac{k_1}{k_{-1}}\right)^{1/2}[NO][Cl_2]^{1/2} = k[NO][Cl_2]^{1/2}$$

b) $NO + Cl_2 \underset{k_{-1}}{\overset{k_1}{\rightleftharpoons}} NOCl + Cl$ (rev.) $NO + Cl \xrightarrow{k_2} NOCl$ (rate determining)

$$k_1[NO][Cl_2] = k_{-1}[NOCl][Cl] \qquad [Cl] = \left(\frac{k_1}{k_{-1}}\right)\frac{[NO]\,[Cl_2]}{[NOCl]}$$

$$\text{rate} = k_2[NO]\,[Cl] = k_2\left(\frac{k_1}{k_{-1}}\right)[NO]\frac{[NO]\,[Cl_2]}{[NOCl]} = k\frac{[NO]^2[Cl_2]}{[NOCl]}$$

14.32 b) $2\ AB \underset{k_{-1}}{\overset{k_1}{\rightleftharpoons}} A_2 + 2\ B$ (reversible) $C + B \xrightarrow{k_2} BC$ (rate determining)

$$k_1[AB]^2 = k_{-1}[A_2]\,[B]^2 \qquad [B]^2 = \left(\frac{k_1}{k_{-1}}\right)\frac{[AB]^2}{[A_2]} \qquad [B] = \left(\frac{k_1}{k_{-1}}\right)^{1/2}\frac{[AB]}{[A_2]^{1/2}}$$

$$\text{rate} = k_2[C][B] = k_2\left(\frac{k_1}{k_{-1}}\right)^{1/2}[C]\frac{[AB]}{[A_2]^{1/2}} = k\frac{[C][AB]}{[A_2]^{1/2}}$$

c) $C + AB \underset{k_{-1}}{\overset{k_1}{\rightleftharpoons}} BC + A$ (reversible) $A + A \xrightarrow{k_2} A_2$ (rate determining)

$$k_1[C][AB] = k_{-1}[BC][A] \qquad [A] = \left(\frac{k_1}{k_{-1}}\right)\frac{[C]\,[AB]}{[BC]} \qquad [A]^2 = \left(\frac{k_1}{k_{-1}}\right)^2\frac{[C]^2[AB]^2}{[BC]^2}$$

$$\text{rate} = k_2[A]^2 = k_2\left(\frac{k_1}{k_{-1}}\right)^2\frac{[C]^2[AB]^2}{[BC]^2} = k\frac{[C]^2[AB]^2}{[BC]^2}$$

(continued)

(14.32 continued)

d) $AB \underset{k_{-1}}{\overset{k_1}{\rightleftarrows}} A + B$ (reversible) $\quad B + C \xrightarrow{k_2} BC$ (rate determining)

$k_1[AB] = k_{-1}[A][B] \quad [B] = \left(\frac{k_1}{k_{-1}}\right)\frac{[AB]}{[A]}$

$\text{rate} = k_2[B][C] = k_2\left(\frac{k_1}{k_{-1}}\right)[C]\ \frac{[AB]}{[A]} = k\frac{[C][AB]}{[A]}$

14.33 If the activation energy (E_a) is equal to zero, the rate constant is independent of temperature (i.e., it does not change when the temperature changes).
$k = A\ e^{-E_a/RT} = A\ e^{-0/RT} \doteq A$
A zero E_a means that no bonds are distorted or broken during the reaction.

14.34 $\Delta E_a(f) = -\Delta E_a(r)$ — The energy change for the forward reaction (ΔE) would be equal to the negative of E_a ($-E_a$) for the reverse reaction.

14.35 a) ΔE_{rxn} = A - C; false; ΔE_{rxn} = C - A (Reaction is exothermic.)

b) ΔE_{rxn} = B - C; false

c) E_a (forward) = E_a (reverse); false; E_a (forward) = B - A and E_a (reverse) = B - C

d) true

e) E_a (forward) = B - C; false

f) true

14.36 a) $NO_2 + NO_2 \rightarrow N_2O_4$

b)

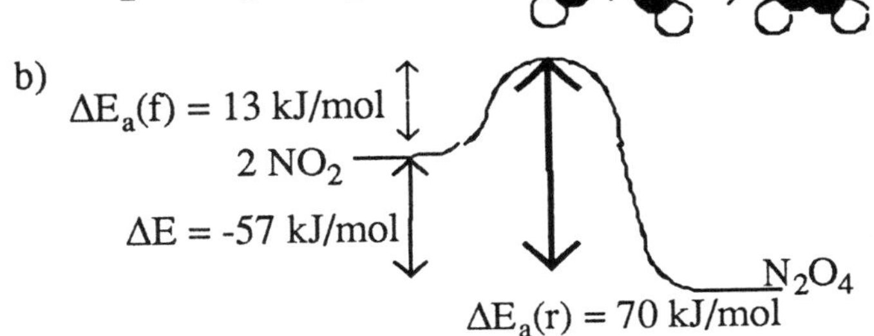

14.37

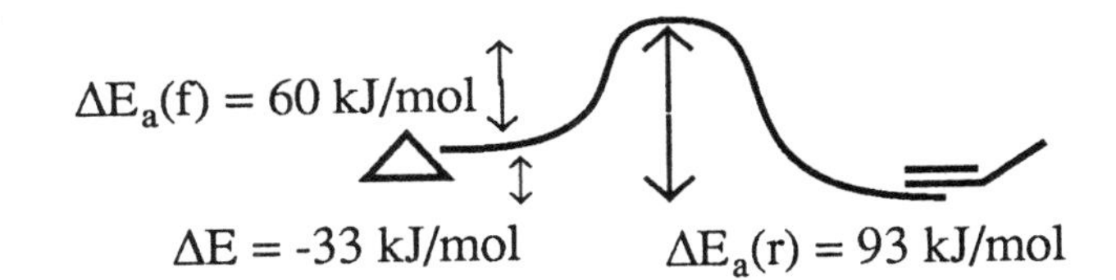

14.38 $E_a = R\ln\left(\frac{k_2}{k_1}\right) \Big/ \left(\frac{1}{T_1} - \frac{1}{T_2}\right)$ (Eqn. 14-8)

$$E_a = \frac{\left(8.314 \text{ J mol}^{-1} \text{ K}^{-1}\right)\left[\ln\left(\frac{1.9 \times 10^2/\text{min}}{39.6/\text{min}}\right)\right]}{((1/278) - (1/301))}$$

$$= \frac{\left(8.314 \text{ J mol}^{-1} \text{ K}^{-1}\right)[\ln(4.798)]}{\left(3.597 \times 10^{-3} \text{ K}^{-1} - 3.322 \times 10^{-3} \text{ K}^{-1}\right)} = \frac{\left(8.314 \text{ J mol}^{-1} \text{ K}^{-1}\right)(1.568)}{\left(2.75 \times 10^{-4} \text{ K}^{-1}\right)}$$

$$= 4.74 \times 10^4 \text{ J/mol or } 47.4 \text{ kJ/mol}$$

14.39 An intermediate appears as a product of an early step and is used as a reactant in a later step. A catalyst is used as a reactant in an early step and is regenerated as a product in a later step of the mechanism.

14.40 a) Pd metal is a heterogeneous catlayst.

b) The molecular picture would be similar to that shown in Figure 14-26. The molecular hydrogen would attach to the surface of the palladium. The concentration of the hydrogen on the surface of the catalyst would be very high and the hydrogen to hydrogen bonds would be weakened or broken. The ethylene would also attach to the surface of the catalyst. There would then be materials migration, reaction of bound materials, and escape of product (desorption).

14.41 a) A and B are O_2 and Br^- (or Br^- and O_2)

b) $2\,H_2O_{2(aq)} \rightarrow 2\,H_2O_{(l)} + O_{2(g)}$

$$\Delta H_{rxn} = \left(2 \text{ mol } H_2O_{(l)}\right)\left(\frac{-285.830 \text{ kJ}}{\text{mol } H_2O_{(l)}}\right) + (1 \text{ mol } O_2)\left(\frac{0.0 \text{ kJ}}{\text{mol } O_2}\right)$$

$$- \left[\left(2 \text{ mol } H_2O_{2(aq)}\right)\left(\frac{-187.78 \text{ kJ}}{\text{mol } H_2O_{2(aq)}}\right)\right] = -196.1 \text{ kJ}$$

Note: $\Delta E = \Delta H$ assumed and $\Delta H_{H_2O_2(l)}$ used for $\Delta H_{H_2O_2(aq)}$

$H_2O_{2(aq)} + Br^-_{(aq)} \rightarrow H_2O_{(l)} + BrO^-_{(aq)}$ (First step in catalyzed mechanism.)

$$\Delta H_{int} = \left(1 \text{ mol } H_2O_{(l)}\right)\left(\frac{-285.830 \text{ kJ}}{\text{mol } H_2O_{(l)}}\right) + \left(1 \text{ mol } BrO^-_{(aq)}\right)\left(\frac{-94.1 \text{ kJ}}{\text{mol } BrO^-_{(aq)}}\right)$$

$$- \left[\left(1 \text{ mol } H_2O_{2(aq)}\right)\left(\frac{-187.78 \text{ kJ}}{\text{mol } H_2O_{2(aq)}}\right) + \left(1 \text{ mol } Br^-_{(aq)}\right)\left(\frac{-121.55 \text{ kJ}}{\text{mol } Br^-_{(aq)}}\right)\right] = -70.60 \text{ kJ}$$

(continued)

(14.41 continued)

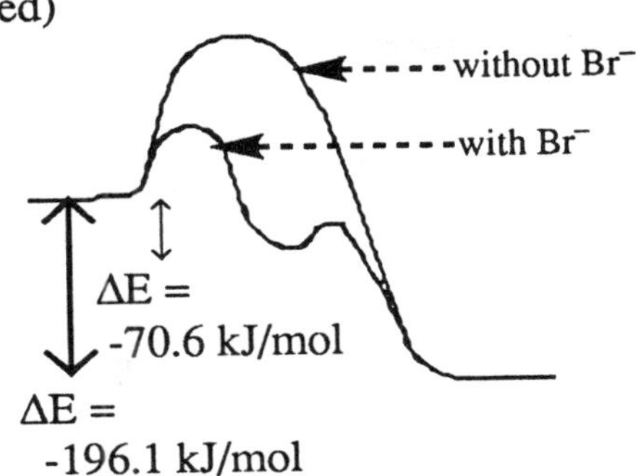

14.42 Ozone decomposition to oxygen is an exothermic process. The energy released can help supply energy for activation for further reaction of O_3. The process of O_3 production from O_2 is endothermic. Energy must be supplied for the process to continue.

14.43 Enzymes are proteins with many different shapes. They have different grooves or pockets which have shape and size similar to the shape and size of the specific reactant molecule that they catalyze. Only that molecule tends to fit in the groove or pocket. During binding of the reactant molecule in the enzyme's groove or pocket, the enzyme often changes shape to match the reactant molecule's shape even more closely in order to bind it more tightly.
Your diagrams of a hypothetical enzyme need to show the shape of a groove or pocket that will fit the shape of the square but not the triangle.

14.44 You may make your drawing similar to Figures 14-28 and 14-29 except yours should be more like a cartoon and needs to have starch in place of sucrose in one set of frames and cellulose in place of starch in another set of frames.

14.45 Rate = k $[CO][NO_2]$

a) $\text{Rate} = k\left[\dfrac{0.5\text{ mol}}{2.0\text{ L}}\right]\left[\dfrac{0.5\text{ mol}}{2.0\text{ L}}\right] = 0.06\text{ k mol}^2\text{ L}^{-2}$

b) $\text{Rate} = k\left[\dfrac{0.5\text{ mol}}{1.0\text{ L}}\right]\left[\dfrac{0.5\text{ mol}}{1.0\text{ L}}\right] = 0.25\text{ k mol}^2\text{ L}^{-2}$

c) $\text{Rate} = k\left[\dfrac{0.1\text{ mol}}{1.0\text{ L}}\right]\left[\dfrac{2.0\text{ mol}}{1.0\text{ L}}\right] = 0.20\text{ k mol}^2\text{ L}^{-2}$

The fastest rate would be for the conditions for (b). The rate for (b) is faster than the rate for (a) because the concentrations for (b) are each twice the concentrations for (a). The rate for (b) is faster than the rate for (c) because the concentration of CO is 5 times greater for (b) than for (c) and the concentration of NO_2 for (b) is 1/4 that for (c). Therefore, (b) is 5/4 faster than (c).

14.46 $O_2 + CO \rightarrow CO_2 + O$

a) $O + CO \rightarrow CO_2$

b) Your molecular pictures of the first step should be similar to the drawings in Figure 14-12 with O_2 in place of O_3 and CO in place of NO. The drawings for the second step would involve the collision of O and CO to form CO_2.

14.47 a) Rate $= \frac{\Delta[C_6H_6]}{\Delta T} = -\frac{1}{3}\frac{\Delta[C_2H_2]}{\Delta T}$

b) Not enough information is given to find the rate law. Experiments would need to be performed in which the starting concentrations of acetylene are varied and either the formation of benzene vs. time or the disappearance of acetylene vs. time is measured. Comparison of the change of starting concentrations of acetylene with the changes of rates of reaction would determine the order of the reaction with respect to acetylene. Then substitution of the starting concentration of acetylene and the rate (for any of these experiments) into the rate expression allows solving for k.

14.48 a) First-order with respect to N_2O_5 and first-order overall.

b) Second-order with respect to NO, first-order with respect to H_2, and third-order overall.

c) First-order with respect to $CHCl_3$, half-order with respect to Cl_2, and three halves-order overall.

d) Zero-order with respect to N_2, zero-order with respect to H_2 and zero-order overall.

e) Zero-order with respect to glucose, zero-order with respect to ATP, first-order with respect to enzyme and first-order overall.

14.49 $A \rightarrow B + C$ rate $= k[A]^n$

rate of flask 1 = $rate_1 = k[5\ A]^n$

rate of flask 2 = $rate_2 = 4\ rate_1 = k[10\ A]^n$

$$\frac{rate_2}{rate_1} = \frac{4\ rate_1}{rate_1} = 4 = \frac{k[10\ A]^n}{k[5\ A]^n} = [2]^n \qquad n = 2$$

The rate-determining step depends on a collision between 2 molecules of A. Doubling the concentration of A increases the number of collisions by $[2]^2 = 4$ times; that increases the reaction rate by 4 times.

14.50 relative rate $= -1/4\frac{\Delta[NH_3]}{\Delta t} = -1/5\frac{\Delta[O_2]}{\Delta t} = 1/4\frac{\Delta[NO]}{\Delta t} = 1/6\frac{\Delta[H_2O]}{\Delta t}$

$$-1/4\frac{\Delta[NH_3]}{\Delta t} = 1/4\frac{\Delta[NO]}{\Delta t}$$

$$\frac{\Delta[NH_3]}{\Delta t} = -\frac{\Delta[NO]}{\Delta t} = -1.5 \times 10^3 \text{ mol L}^{-1} \text{ s}^{-1}$$

$$-1/5\frac{\Delta[O_2]}{\Delta t} = 1/4\frac{\Delta[NO]}{\Delta t}$$

$$\frac{\Delta[O_2]}{\Delta t} = -5/4\frac{\Delta[NO]}{\Delta t} = -(5/4)(1.5 \times 10^3 \text{ mol L}^{-1} \text{ s}^{-1}) = -1.9 \times 10^3 \text{ mol L}^{-1} \text{ s}^{-1}$$

$$1/6\frac{\Delta[H_2O]}{\Delta t} = 1/4\frac{\Delta[NO]}{\Delta t}$$

$$\frac{\Delta[H_2O]}{\Delta t} = 6/4\frac{\Delta[NO]}{\Delta t} = (3/2)(1.5 \times 10^3 \text{ mol L}^{-1} \text{ s}^{-1}) = 2.3 \times 10^3 \text{ mol L}^{-1} \text{ s}^{-1}$$

14.51 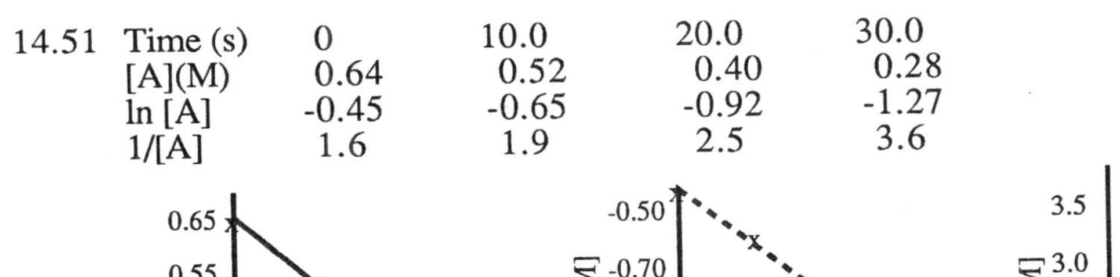

Time (s)	0	10.0	20.0	30.0
[A](M)	0.64	0.52	0.40	0.28
ln [A]	-0.45	-0.65	-0.92	-1.27
1/[A]	1.6	1.9	2.5	3.6

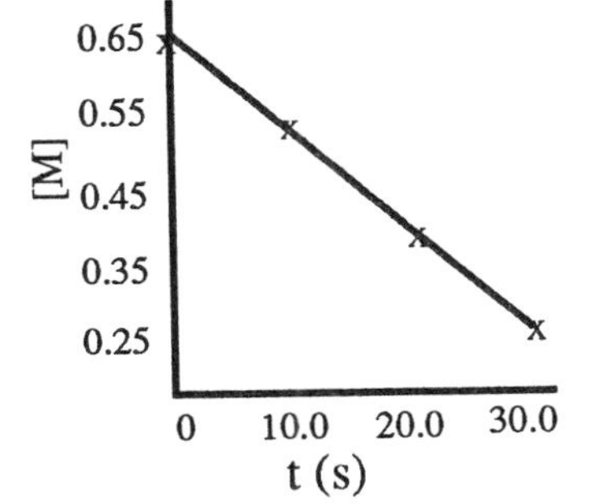

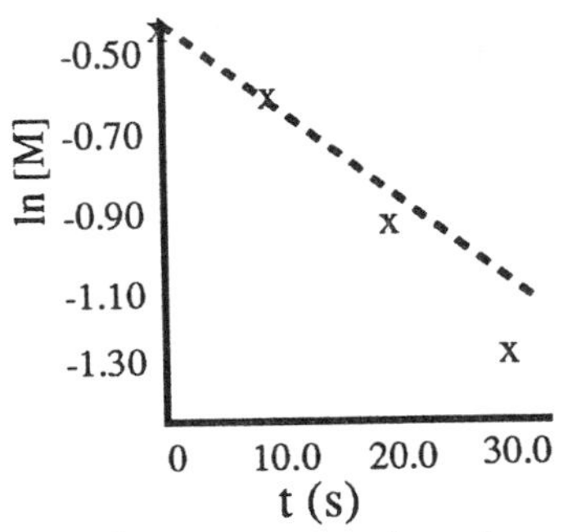

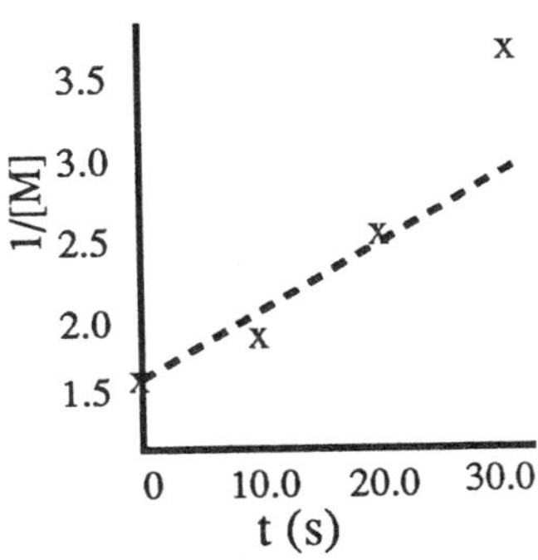

The reaction is zero-order. (Note: the concentration decreases by 0.12 M every ten seconds; therefore, the rate is independent of concentration.)

14.52 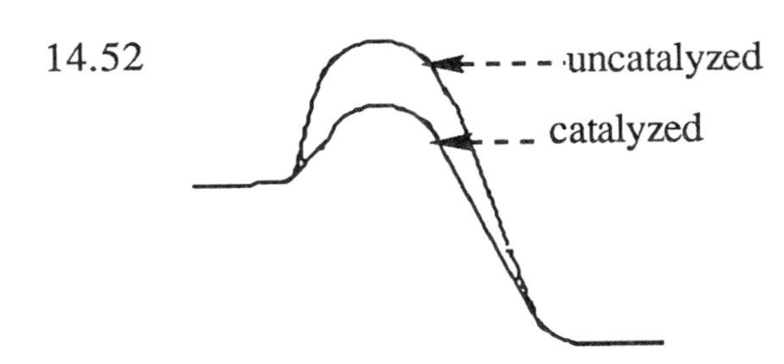

The enzyme lowers the activation energy by its action on the bond to be broken. The enzyme binds with the reactant, subtle changes in the structure of the enzyme distort the reactant and make the reactant's bonds easier to break. The reaction between the enzyme and the reactant depends upon the shape of the reactant.

14.53 $k = Ae^{-E_a/RT}$ (Arrhenius equation) $21°C = 294$ K

$rate_{urease} = A\,e^{-(46\ kJ/mol)/RT}$ $rate_{uncatalyzed} = A\,e^{-(125\ kJ/mol)/RT}$

$$\frac{rate_{urease}}{rate_{uncat.}} = \frac{A\,e^{-(46\ kJ/mol)/RT}}{A\,e^{-(125\ kJ/mol)/RT}} = e^{\left(\frac{-46\ kJ/mol}{RT} + \frac{125\ kJ/mol}{RT}\right)} = e^{\frac{79\ kJ/mol}{RT}}$$

$$= e^{\frac{79\ kJ/mol}{(8.314\ J/mol\ K)\left(\frac{kJ}{10^3\ J}\right)(294\ K)}} = e^{32.32} = 1.1 \times 10^{14}$$

14.54 $2\,O_3 \rightarrow 3\,O_2$ $3\,O_2 \rightarrow 2\,O_3$
In both of these reactions O is formed as an intermediate and then used in a later step.

14.55 a) A catalyst lowers the activation energy for the reaction. This speeds up the reaction because more molecules possess enough energy to react.

b) An increase in temperature increases the average energy of the molecules that in turn increases the proportion of molecules with enough energy to exceed the activation energy.

c) An increase in concentration increases the number of collisions that have the correct orientation and enough energy to react.

14.56 A spontaneous chemical reaction occurs under specified conditions (i.e., the products are thermodynamically more stable than the reactants under these conditions). ΔG is negative for such a reaction. Speed of a chemical reaction is the rate at which molecules of reactants become molecules of products. For a chemical reaction to be useful, it must be spontaneous under economical conditions and must proceed at a reasonable speed.

14.57 a)

Time (hr)	0	2	3	5	7	9
conc (M)	0.500	0.300	0.250	0.188	0.150	0.125
ln [ ]	-0.693	-1.20	-1.39	-1.67	-1.90	-2.08
1/[ ]	2.00	3.33	4.00	5.32	6.67	8.00

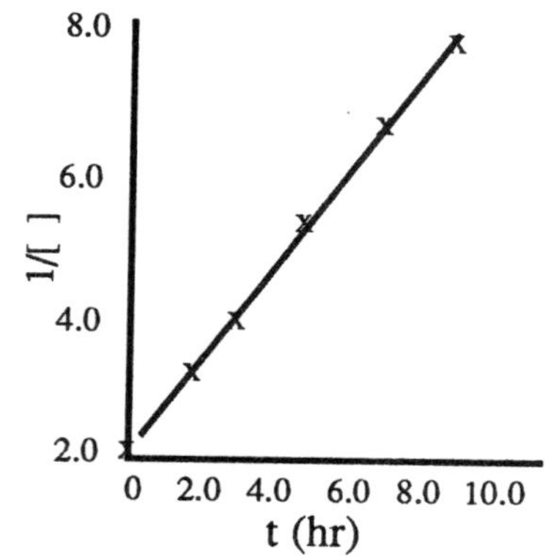

Rate = $k[NH_4NCO]^2$

b) Rate constant (k) at 50°C equals the slope of line;

$$k_2 = \frac{\Delta y}{\Delta x} = \frac{8.00\ M^{-1} - 2.00\ M^{-1}}{9\ hr - 0\ hr} = \frac{6.00}{9} M^{-1}\ hr^{-1} = 0.67\ M^{-1}\ hr^{-1}\ (\text{at } 50°C)$$

c) 25°C $k_1 = \frac{(1/0.300\ M) - (1/0.500\ M)}{6\ hr} = \frac{3.33\ M^{-1} - 2.00\ M^{-1}}{6\ hr} = 0.22\ M^{-1}\ hr^{-1}$

$$E_a = \frac{R \ln\left(\frac{k_2}{k_1}\right)}{\left(\frac{1}{T_1} - \frac{1}{T_2}\right)} = \frac{(8.314\ J\ mol^{-1}\ K^{-1})\left[\ln\left(\frac{0.67\ M^{-1}\ hr^{-1}}{0.22\ M^{-1}\ hr^{-1}}\right)\right]}{\left(\frac{1}{298\ K} - \frac{1}{323\ K}\right)}$$

$$= \frac{(8.314\ J\ mol^{-1}\ K^{-1})[\ln(3.05)]}{(3.356 \times 10^{-3}\ K^{-1} - 3.096 \times 10^{-3}\ K^{-1})} = \frac{9.271\ J\ mol^{-1}\ K^{-1}}{2.60 \times 10^{-4}\ K^{-1}}$$

$$= 3.57 \times 10^4\ J\ mol^{-1} = 35.7\ kJ\ mol^{-1}$$

14.58 a) true b) cannot tell if true c) true d) false e) false
f) true [The odds against a termolecular reaction (three-body collision) are high. The possibility of a four-body collision is nil.]
g) true (If there is an activation energy, it will be positive. Almost all reactions have an activation energy. If there was no activation energy, the reaction would occur whenever N_2 and H_2 were mixed without any special conditions such as the Haber process.)

14.59 a) false (The reaction is first-order, not fourth-order.)

b) false (One needs the rate constant for 2 different temperatures to evaluate the activation energy.)

c) true (This assumes that A, the geometric factor, does not change over this temperature range. If A changes, the truth of the statement depends on how A changes.)

d) true (This step is first-order.)

14.60 $\ln[A] - \ln[A_0] = \ln\left[\frac{[A]}{[A_0]}\right] = -kt \qquad A = 0.10\ A_0$

$$\ln\left[\frac{[0.10\ A_0]}{[A_0]}\right] = -(1.73 \times 10^{-2}\ s^{-1})t$$

$$t = \frac{\ln\left[\frac{[0.10\ A_0]}{[A_0]}\right]}{-(1.73 \times 10^{-2}\ s^{-1})} = 133\ s\ (\text{or } 2.2\ \text{min})$$

14.61

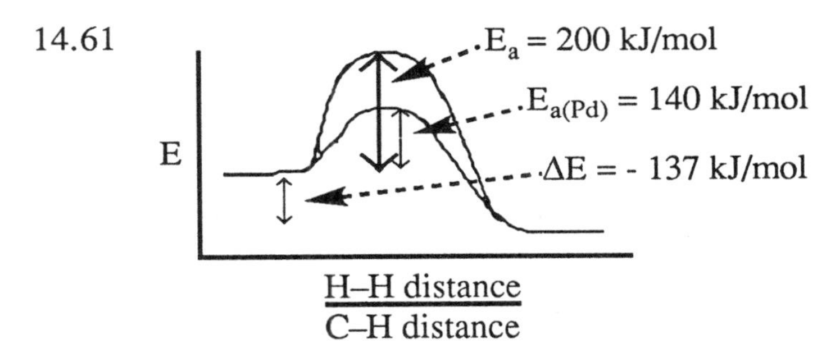

14.62 a) $A + B \underset{}{\overset{k_1}{\rightleftarrows}} C$

$C + D \xrightarrow{k_2} A + E$

$E + B \xrightarrow{k_3} 2\ F$

$2\ B + D \rightarrow 2\ F$

b) $A + B \underset{k_{-1}}{\overset{k_1}{\rightleftarrows}} C$

$k_1[A][B] = k_{-1}[C] \qquad [C] = (k_1/k_{-1})[A][B]$

$\text{rate} = k_2[C][D] = k_2(k_1/k_{-1})[A][B][D] = k[A][B][D]$

c) Catalyst — A

Intermediates — C and E

14.63 a) $NO_2 + O_3 \rightarrow NO_3 + O_2$ (rate-determining step)

$NO_2 + NO_3 \rightarrow N_2O_5$

b) Your drawing of a molecular picture should be similar to Figure 14-12 for the first step except you have NO_2 instead of NO. The drawing of a molecular picture for the second step would involve the combining of NO_2 and NO_3 to form N_2O_5.

c)

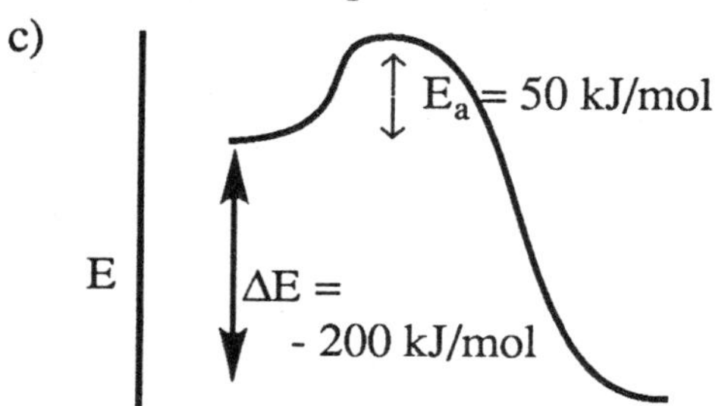

14.64 While a graphical method can be used to determine the order with respect to O_2, note that the concentration of O_2 halves every 3.0 seconds (Experiment A). Therefore, the reaction is first-order with respect to O_2 and:

$$k_{obs\ A} = \frac{\ln 2}{t_{1/2}} = \frac{0.6931}{3.05} = 0.23\ s^{-1}$$

Note the time required for the concentration of NO to halve (Experiment B) is not constant. Therefore, the reaction is not first-order with respect to NO.

Time (s)	0	100	200	300	400
[NO](10^{-4} M)	4.1	2.05	1.43	1.02	0.82
1/[NO](10^4 M^{-1})	0.24	0.49	0.70	0.98	1.2

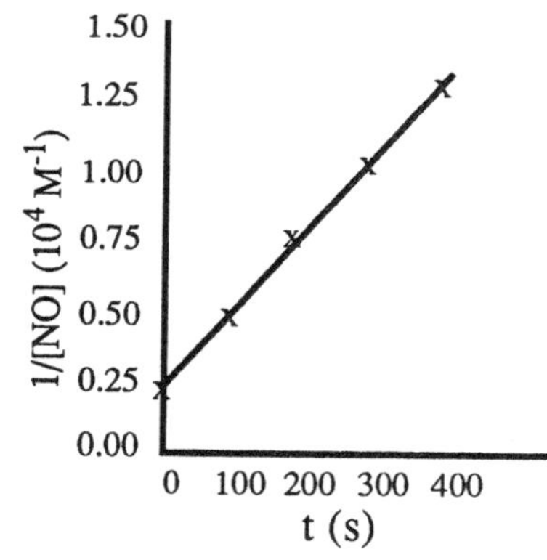

Therefore, the reaction is second-order with respect to NO.

$$k_{obs\ B} = \text{slope} = \frac{\Delta x}{\Delta y} = \frac{0.24 \times 10^4\ M^{-1}}{100\ s} = 24\ M^{-1}\ s^{-1}$$

Rate = $k[O_2][NO]^2$

For experiment A: $k_{obs\ A} = 0.23\ s^{-1} = k[NO]^2$

For experiment B: $k_{obs\ B} = 24\ M^{-1}\ s^{-1} = k[O_2]$

$$k = \frac{k_{obs\ A}}{[NO]^2} = \frac{0.23\ s^{-1}}{(9.63 \times 10^{-3}\ M)^2} = 2.5 \times 10^3\ M^{-2}\ s^{-1} \quad \text{(experiment A)}$$

or $$k = \frac{k_{obs\ B}}{[O_2]} = \frac{24\ M^{-1}\ s^{-1}}{(9.75 \times 10^{-3}\ M)} = 2.5 \times 10^3\ M^{-2}\ s^{-1} \quad \text{(experiment B)}$$

14.65

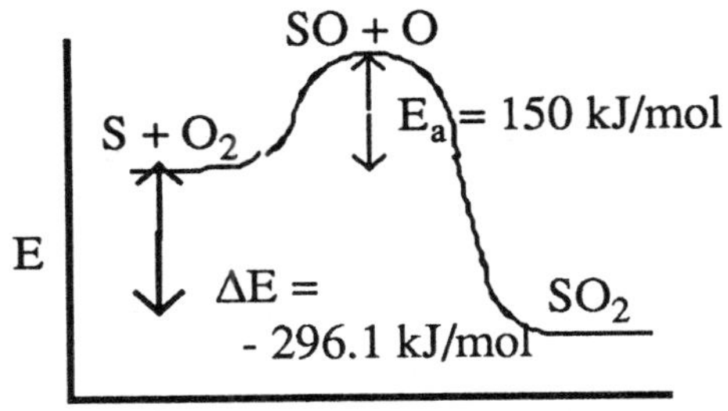

SO + O

$S + O_2$ — E_a = 150 kJ/mol

E

ΔE = - 296.1 kJ/mol — SO_2

14.66 a) $[Co(NH_3)_5H_2O]^{3+} \underset{k_{-1}}{\overset{k_1}{\rightleftharpoons}} [Co(NH_3)_5]^{3+} + H_2O$ (rapid, reversible)

$[Co(NH_3)_5]^{3+} + Cl^- \xrightarrow{k_2} [Co(NH_3)_5Cl]^{2+}$ (slow, rate-determining)

b) $k_1[[Co(NH_3)_5H_2O]^{3+}] = k_{-1}[[Co(NH_3)_5]^{3+}][H_2O]$

$$[[Co(NH_3)_5]^{3+}] = \frac{k_1}{k_{-1}} \frac{[[Co(NH_3)_5H_2O]^{3+}]}{[H_2O]}$$

$$\text{rate} = k_2[[Co(NH_3)_5]^{3+}][Cl^-]$$

$$= k_2 \frac{k_1}{k_{-1}} \frac{[[Co(NH_3)_5H_2O]^{3+}][Cl^-]}{[H_2O]} = k\frac{[[Co(NH_3)_5H_2O]^{3+}][Cl^-]}{[H_2O]}$$

c) Yes, this observation is consistent with the proposed mechanism. The observed kinetics would be for $[Co(NH_3)_5H_2O]^{3+}$ which is first-order. The reaction has been "flooded" with excess water and HCl (HCl is 1000 times Co complex concentration and water is about 55,000 times Co complex concentration)

14.67 $\ln\left[\frac{A}{A_0}\right] = -kt \qquad t = \frac{\ln\left[\frac{A}{A_0}\right]}{-k} = \frac{\ln\left[\frac{A}{A_0}\right]}{-5.5 \times 10^{-4}\ s^{-1}}$

10% decomposed
$A = 0.90\ A_0$

$$t = \frac{\ln\left(\frac{0.90\ A_0}{A_0}\right)}{-5.5 \times 10^{-4}\ s^{-1}} = 1.9 \times 10^2\ s \quad \text{(or 3.2 min)}$$

50% decomposed
$A = 0.50\ A_0$

$$t = \frac{\ln\left(\frac{0.50\ A_0}{A_0}\right)}{-5.5 \times 10^{-4}\ s^{-1}} = 1.3 \times 10^3\ s \quad \text{(or 21 min)}$$

(Note: this is $t_{1/2} = \frac{\ln 2}{5.5 \times 10^{-4}\ s^{-1}} = 1.26 \times 10^3$ s)

(continued)

(14.67 continued)

99.9% decomposed

$A = 0.001\ A_0$

$$t = \frac{\ln\left(\frac{0.001\ A_0}{A_0}\right)}{-5.5 \times 10^{-4}\ s^{-1}} = 1.3 \times 10^4\ s$$

(or 209 min) (or 3.5 hr)

14.68 Rate = $(3.5 \times 10^{-2}\ M^{-1}\ s^{-1})[Cl_2][H_2S]$

The reaction is "flooded" with Cl_2 and, therefore, the $[Cl_2]$ remains essentially constant. The reaction can be treated as first-order in H_2S with $k_{obs} = k[Cl_2]$ $= (3.5 \times 10^{-2}\ M^{-1}\ s^{-1})(0.035\ M) = 1.2 \times 10^{-3}\ s^{-1}$.

$$\ln\left(\frac{[H_2S]}{[H_2S]_0}\right) = -k_{obs}t = \left(-1.2\ 10^{-3}\ s^{-1}\right)t$$

a) for 200 seconds:

$$\ln\left(\frac{[H_2S]}{[H_2S]_0}\right) = -1.2\ 10^{-3}\ s^{-1}(200\ s) = -0.24 \qquad \frac{[H_2S]}{[H_2S]_0} = e^{-0.24} = 0.787$$

$[H_2S] = 0.787[H_2S]_0 = (0.787)(5.0 \times 10^{-5}\ M) = 3.9 \times 10^{-5}\ M$

b) $$\ln\left(\frac{[H_2S]}{[H_2S]_0}\right) = \ln\left(\frac{1.0 \times 10^{-5}\ M}{5.0 \times 10^{-5}\ M}\right) = -(1.2 \times 10^{-3}\ s^{-1})t$$

$$t = \frac{\ln\left(\frac{1.0 \times 10^{-5}\ M}{5.0 \times 10^{-5}\ M}\right)}{-1.2 \times 10^{-3}\ s^{-1}} = 1.3 \times 10^{3}\ s \quad \text{(or 22 min)}$$

14.69 a) Rate = $k[H_2][X_2]$ b) Rate = $k[X_2]$ c) $X_2 \underset{k_{-1}}{\overset{k_1}{\rightleftharpoons}} X + X$

$$k_1[X_2] = k_{-1}[X]^2 \qquad [X] = \left(\frac{k_1}{k_{-1}}\right)^{1/2}[X_2]^{1/2}$$

$$\text{Rate} = k_2[X][H_2] = k_2\left(\frac{k_1}{k_{-1}}\right)^{1/2}[X_2]^{1/2}[H_2] = k[X_2]^{1/2}[H_2]$$

d) $X_2 + X_2 \underset{k_{-1}}{\overset{k_1}{\rightleftharpoons}} X_3 + X$

$$k_1[X_2]^2 = k_{-1}[X_3][X] \qquad [X] = \frac{k_1}{k_{-1}}\frac{[X_2]^2}{[X_3]}$$

$$\text{Rate} = k_2[X][H_2] = k_2\left(\frac{k_1}{k_{-1}}\right)\frac{[X_2]^2}{[X_3]} = k\frac{[X_2]^2}{[X_3]}$$

This rate law cannot be tested experimentally because it contains the intermediate X_3.

14.70 a) $N_2O_5 \underset{k_{-1}}{\overset{k_1}{\rightleftarrows}} NO_2 + NO_3$ (fast decomposition) (used twice)

$N_2O_5 \rightleftarrows NO_2 + NO_3$

$NO_2 + NO_3 \xrightarrow{k_2} NO + NO_2 + O_2$ (slow)

$NO + NO_3 \rightarrow 2\ NO_2$ (fast)

$2\ N_2O_5 \rightarrow 4\ NO_2 + O_2$ (overall)

b) $k_1[N_2O_5] = k_{-1}[NO_2][NO_3]$

$$[NO_2][NO_3] = \frac{k_1}{k_{-1}}[N_2O_5]$$

$$\text{Rate} = k_2[NO_2][NO_3] = k_2\left(\frac{k_1}{k_{-1}}\right)[N_2O_5] = k[N_2O_5]$$

c) Being first-order does not prove whether step 2 or step 1 is rate-determining because the rate law is first-order in N_2O_5 if either step is rate-determining.

14.71 a) Assume the rate (and, therefore, the rate constant) at 25°C (298 K) is double the

$$E_a = \frac{R\ln\left(\frac{k_2}{k_1}\right)}{\left(\frac{1}{T_1} - \frac{1}{T_2}\right)} = \frac{8.314\ J/mol\ K\ \ln\ (2)}{\left(\frac{1}{288\ K} - \frac{1}{298\ K}\right)} = \frac{5.7628\ J/mol\ K}{(1.165 \times 10^{-4}\ K^{-1})}$$

$= 4.95 \times 10^4$ J/mol or 49.5 kJ/mol

b) $\ln\left(\frac{k_2}{k_1}\right) = \frac{E_a}{R}\left(\frac{1}{T_1} - \frac{1}{T_2}\right)$ $T_1 = 20°C = 293\ K$ $T_2 = 25°C = 298\ K$

$k_1 = 1/10 = 0.1$ $k_2 = 1/t$

$$\ln\left(\frac{1/t}{0.1\ min^{-1}}\right) = \frac{4.95 \times 10^4\ J/mol}{8.314\ J/mol\ K}\left(\frac{1}{293\ K} - \frac{1}{298\ K}\right)$$

$$= 5.9538 \times 10^3\ K\ (5.726 \times 10^{-5}\ K^{-1}) = 0.3409$$

$$\ln\left(\frac{1\ min}{t}\right) - \ln(0.1) = 0.3409$$

$$\ln\left(\frac{1\ min}{t}\right) = 0.3409 + \ln(0.1) = (0.3409) - 2.3026 = -1.9617$$

$$\left(\frac{1\ min}{t}\right) = 0.1406 \qquad t = 7.1\ min$$

14.72 An automobile catalytic converter has 3 purposes for which it functions as a catalyst. It catalyzes the oxidation of any unburned hydrocarbons to CO_2 and H_2O. It catalyzes the oxidation of poisonous CO to CO_2 and it catalyzes the conversion of nitrogen oxides (NO and NO_2) to N_2 and O_2.

14.73 Initially the rate of formation of product increases with the increase of the concentration of reactant. As the concentration of reactant is increased, more and more active sites on the enzyme are occupied and more and more product is formed. At some point all active sites are occupied and the rate of formation of product becomes constant as reactant molecules can only bind to the active sites after reaction occurs and the product is released.

14.74 $2NO \underset{k_{-1}}{\overset{k_1}{\rightleftarrows}} N_2O_2$ (fast)

$N_2O_2 + H_2 \xrightarrow{k_2} N_2O + H_2O$ (slow)

$N_2O + H_2 \rightarrow N_2 + H_2O$ (fast)

a) $2\ NO + 2\ H_2 \rightarrow N_2 + 2\ H_2O$ (overall stoichiometry)

b) $k_1[NO]^2 = k_{-1}[N_2O_2]$ $[N_2O_2] = \frac{k_1}{k_{-1}}[NO]^2$

$$\text{Rate} = k_2[N_2O_2][H_2] = k_2\left(\frac{k_1}{k_{-1}}\right)[NO]^2[H_2] = k[NO]^2[H_2]$$

14.75 $CF_2Cl_2 \xrightarrow{h\nu} CF_2Cl + Cl$, $O_3 + Cl \rightarrow O_2 + OCl$, $OCl + O \rightarrow O_2 + Cl$

14.76 $CH_3CHO \rightarrow CH_4 + CO$

Time (s)	0	1000	2000	3000	4000
$[CH_3CHO]$(M)	0.250	0.118	0.770	0.572	0.0455
$\ln[CH_3CHO]$	-1.39	-2.14	-2.56	-2.86	-3.09
$1/[CH_3CHO]$	4.00	8.47	13.0	17.5	22.0

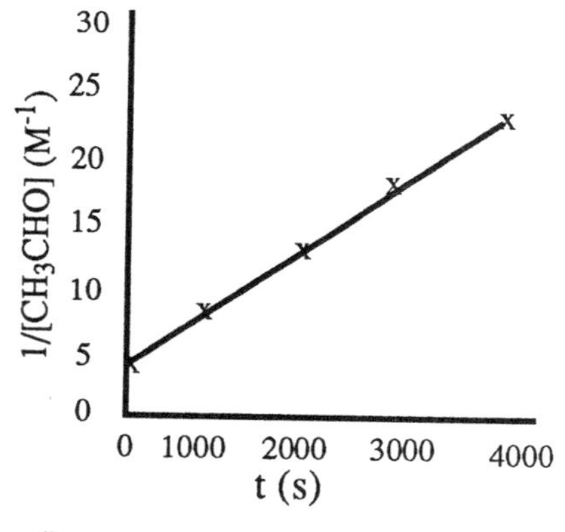

(continued)

(14.76 continued)

a) Rate = $k[CH_3CHO]^2$

b) $k = \dfrac{\Delta x}{\Delta y} = \dfrac{22.0\ M^{-1} - 13.0\ M^{-1}}{4000\ s - 2000\ s} = \dfrac{9.0\ M^{-1}}{2000\ s} = 4.5 \times 10^{-3}\ M^{-1}\ s^{-1}$

c) $\dfrac{1}{[CH_3CHO]} - \dfrac{1}{[CH_3CHO]_0} = kt$

$[CH_3CHO] = (0.25)[CH_3CHO]_0$

$$\frac{1}{0.25[CH_3CHO]_0} - \frac{1}{[CH_3CHO]_0} = (4.5 \times 10^{-3}\ M^{-1}\ s^{-1})t$$

$$\frac{1-0.25}{0.25[CH_3CHO]_0} = (4.5 \times 10^{-3}\ M^{-1}\ s^{-1})t$$

$$\frac{0.75}{0.25[CH_3CHO]_0} = \frac{3}{[CH_3CHO]_0} = (4.5 \times 10^{-3}\ M^{-1}\ s^{-1})t$$

$$t = \frac{3}{(4.5 \times 10^{-3}\ M^{-1}\ s^{-1})[CH_3CHO]_0}$$

Time (t) required for 75% of the acetaldehyde to decompose depends on starting concentration.

Assume starting concentration of experiment for kinetic data:

$[CH_3CHO]_0 = 0.250$ M

$$t = \frac{3}{(4.5 \times 10^{-3}\ M^{-1}\ s^{-1})(0.250\ M} = 2.7 \times 10^3\ s\ \text{(or 44 minutes)}$$

14.77 Rate = k[●●]$^{1/2}$[●]

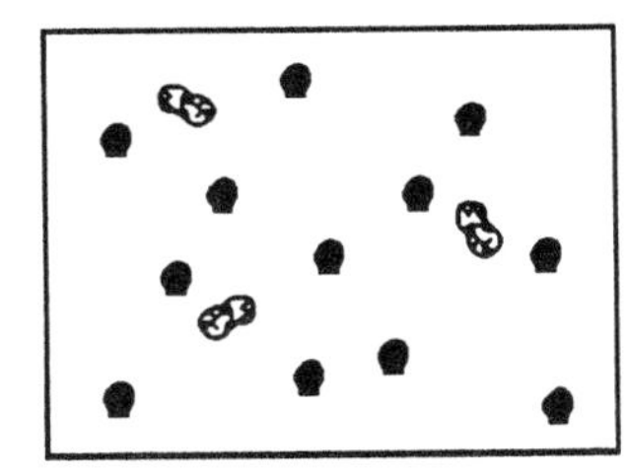

14.78 $Cl_2 \underset{k_{-1}}{\overset{k_1}{\rightleftharpoons}} 2\ Cl$ (fast, reversible) $\qquad Cl + CHCl_3 \xrightarrow{k_2} HCl + CCl_3$ (slow)

$CCl_3 + Cl \xrightarrow{k_2} CCl_4$ (fast)

a) $Cl_2 + CHCl_3 \rightarrow HCl + CCl_4$ $\qquad$ b) intermediates are: Cl and CCl_3

c) $k_1[Cl_2] = k_{-1}[Cl]^2 \qquad [Cl] = \left(\dfrac{k_1}{k_{-1}}\right)^{1/2}[Cl_2]^{1/2}$

$$\text{Rate} = k_2[Cl][CHCl_3] = k_2\left(\frac{k_1}{k_{-1}}\right)^{1/2}[Cl_2]^{1/2}[CHCl_3] = k[Cl_2]^{1/2}[CHCl_3]$$

14.79 The time required for 6 atoms of ^{32}P to decompose to 3 atoms of ^{32}P (i.e., the half-life) will be the same for both flasks. The half-life ($t_{1/2}$) of a first-order process is independent of concentration and equals: $t_{1/2} = \frac{\ln 2}{k} = \frac{0.6931}{k}$

14.80 a)

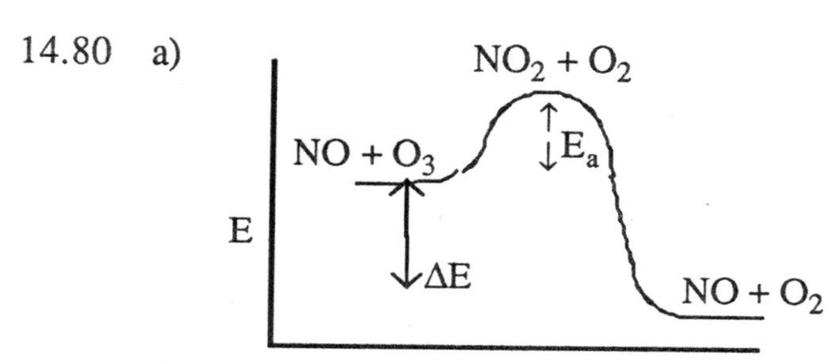

b) NO_2 acts as an intermediate; it is formed in the first step and used in the second step.
NO acts as a catalyst by helping to promote the first step and then being regenerated in the second (last) step.

c) Cl is the more serious threat to the ozone layer because it lowers the activation energy for the decomposition of ozone more than NO does. This is somewhat counteracted by the fact that Cl must be produced from the decomposition of CFC's in the stratosphere while NO is produced by human activities such as driving automobiles.

14.81 Rate = $k[X_2]$
In the first pair of drawings A has 5 X_2's and B has 10 X_2's. The number of Y's is constant. The rate will double with the doubling of the X_2's since the reaction is first order with respect to X_2.
In the second pair of drawings A has 5 X_2's and 8 Y's B has 5 X_2's and 16 Y's. The rate will not change with the doubling of the Y's since the reaction is zero order with respect to Y.

CHAPTER 15: PRINCIPLES OF CHEMICAL EQUILIBRIUM

15.1

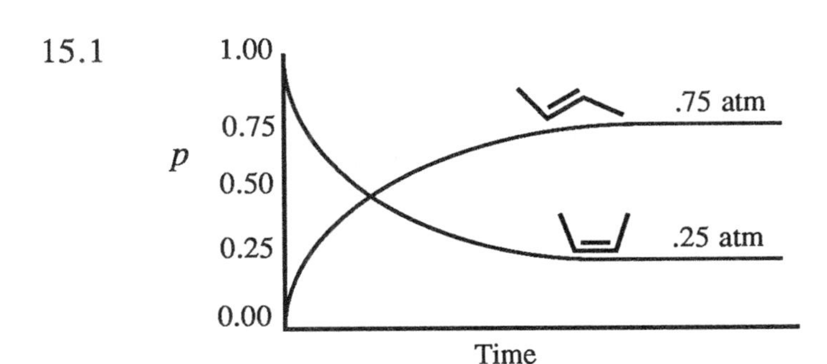

15.2

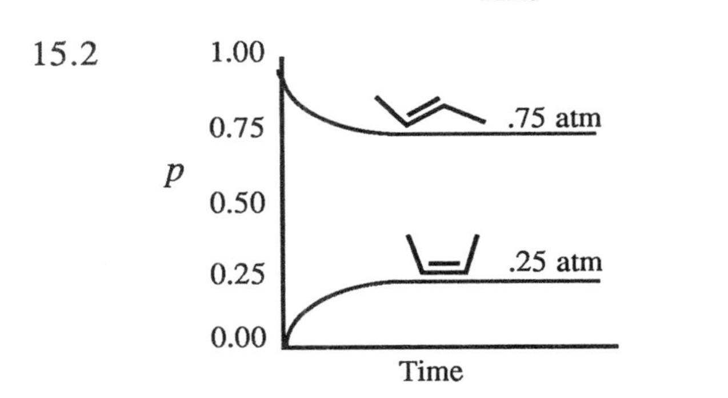

15.3 $ClO^-_{(aq)} + ClO^-_{(aq)} \xrightarrow{k_1} Cl^-_{(aq)} + ClO_2^-{}_{(aq)}$

$ClO^-_{(aq)} + ClO_2^-{}_{(aq)} \xrightarrow{k_2} Cl^-_{(aq)} + ClO_3^-{}_{(aq)}$

a) Every elementary reaction is a molecular rearrangement that goes both in the forward and the reverse direction. We are given elementary reactions in the forward direction, so we need to write elementary reactions in the reverse direction:

$Cl^-_{(aq)} + ClO_2^-{}_{(aq)} \xrightarrow{k_{-1}} 2\ ClO^-_{(aq)}$

$Cl^-_{(aq)} + ClO_3^-{}_{(aq)} \xrightarrow{k_{-2}} ClO^-_{(aq)} + ClO_2^-{}_{(aq)}$

b) $K_{eq} = \dfrac{[Cl^-]^2_{eq}[ClO_3^-]_{eq}}{[ClO^-]^3_{eq}}$

(continued)

(15.3 continued)

c) At equilibrium, each step in the mechanism has a forward rate that equals its reverse rate:

$$2\,ClO^- \underset{k_{-1}}{\overset{k_1}{\rightleftharpoons}} Cl^- + ClO_2^-$$

$$ClO^- + ClO_2^- \underset{k_{-2}}{\overset{k_2}{\rightleftharpoons}} Cl^- + ClO_3^-$$

$$k_1[ClO^-]_{eq}^2 = k_{-1}[Cl^-]_{eq}[ClO_2^-]_{eq} \Rightarrow \frac{k_1}{k_{-1}} = \frac{[Cl^-]_{eq}[ClO_2^-]_{eq}}{[ClO^-]_{eq}^2}$$

$$k_2[ClO^-]_{eq}[ClO_2^-]_{eq} = k_{-2}[Cl^-]_{eq}[ClO_3^-]_{eq} \Rightarrow \frac{k_2}{k_{-2}} = \frac{[Cl^-]_{eq}[ClO_3^-]_{eq}}{[ClO^-]_{eq}[ClO_2^-]_{eq}}$$

$$\frac{k_1 k_2}{k_{-1}k_{-2}} = \frac{[Cl^-]_{eq}^2[ClO_3^-]_{eq}}{[ClO^-]_{eq}^3}$$

$$K_{eq} = \frac{k_{formation}}{k_{decomp.}} = \frac{k_1 k_2}{k_{-1}k_{-2}}$$

15.4 a) $HCOH_{(g)} \xrightarrow{k_{-3}} HCO_{(g)} + H_{(g)}$ $\quad HCO_{(g)} \xrightarrow{k_{-2}} H_{(g)} + CO_{(g)}$

b) $$K_{eq} = \frac{(p_{HCHO})_{eq}}{(p_{H_2})_{eq}(p_{CO})_{eq}}$$

c) $$\frac{k_1}{k_{-1}} = \frac{(p_H)_{eq}^2}{(p_{H_2})_{eq}} \qquad \frac{k_2}{k_{-2}} = \frac{(p_{HCO})_{eq}}{(p_H)_{eq}(p_{CO})_{eq}} \qquad (p_H)_{eq} = \frac{k_{-2}(p_{HCO})_{eq}}{k_2(p_{CO})_{eq}}$$

$$\frac{k_3}{k_{-3}} = \frac{(p_{HCOH})_{eq}}{(p_{HCO})_{eq}(p_H)_{eq}} \qquad (p_H)_{eq} = \frac{k_{-3}}{k_3}\frac{(p_{HCHO})_{eq}}{(p_{HCO})_{eq}}$$

$$\frac{k_1}{k_{-1}} = \frac{(p_H)_{eq}^2}{(p_{H_2})_{eq}} = \frac{1}{(p_{H_2})_{eq}}(p_H)_{eq}(p_H)_{eq}$$

$$= \frac{1}{(p_{H_2})_{eq}}\frac{k_{-2}}{k_2}\frac{(p_{HCO})_{eq}}{(p_{CO})_{eq}}\frac{k_{-3}}{k_3}\frac{(p_{HCHO})_{eq}}{(p_{HCO})_{eq}}$$

$$\frac{k_1}{k_{-1}} = \frac{k_{-2}}{k_2}\frac{k_{-3}}{k_3}\frac{(p_{HCHO})_{eq}}{(p_{H_2})_{eq}(p_{CO})_{eq}} \qquad K_{eq} = \frac{k_1 k_2 k_3}{k_{-1}k_{-2}k_{-3}} = \frac{(p_{HCHO})_{eq}}{(p_{H_2})_{eq}(p_{CO})_{eq}}$$

15.5 Using Figure 15.1 as a guide, your molecular picture should include two hypochlorite ions approaching each other, colliding, and fragmenting to form Cl^- and ClO_2^-. In a second picture another hypochlorite and a ClO_2^- approach each other, collide, and fragment to form chloride and chlorate ions. The reversibility of the reaction can be shown either by drawing molecular pictures of the reaction going in the opposite direction or by showing that the reactions are reversible by using arrows showing the process occurring in both directions. See Figure 15-1.

15.6 Using Figure 15.1 as a guide, your molecular picture should include an H_2 being distorted and then fragmented to form 2 H. In a second picture an H and a CO approach each other, collide, and form HCO. In a third picture an H and an HCO approach each other, collide, and form HCOH. The reversibility of the reactions can be shown by showing each reaction as being reversible as indicated by reaction arrows in both directions.

15.7 ClO^- ions are placed in the reaction vessel, brought to reaction conditions and allowed to react. Analysis of the contents will show the presence of Cl^- and ClO_3^- Cl^- and ClO_3^- ions are placed in the reaction vessel, brought to reaction conditions and allowed to react. Analysis of the contents will show the presence of ClO^- ions.

15.8 H_2 and CO are placed in the reaction vessel, brought to reaction conditions and allowed to react. Analysis of the contents will show the presence of HCHO. HCHO is placed in the reaction vessel, brought to reaction conditions and allowed to react. Analysis of the contents will show the presence of H_2 and CO.

15.9 a) $K_{eq} = \dfrac{\left(p_{PCl_5}\right)_{eq}}{\left(p_{Cl_2}\right)_{eq}\left(p_{PCl_3}\right)_{eq}}$; units = atm^{-1}

b) $K_{eq} = \dfrac{\left(X_{P_4O_{10}}\right)_{eq}}{\left(X_{P_4}\right)_{eq}\left(p_{O_2}\right)_{eq}^5} = \dfrac{1}{\left(p_{O_2}\right)_{eq}^5}$; units = atm^{-5}

c) $K_{eq} = \dfrac{\left(X_{BaO}\right)_{eq}\left(p_{CO}\right)_{eq}^2}{\left(X_{BaCO_3}\right)_{eq}\left(X_C\right)_{eq}} = \left(p_{CO}\right)_{eq}^2$; units = atm^2

d) $K_{eq} = \dfrac{\left(X_{CH_3OH}\right)_{eq}}{\left(p_{CO}\right)_{eq}\left(p_{H_2}\right)_{eq}^2} = \dfrac{1}{\left(p_{CO}\right)_{eq}\left(p_{H_2}\right)_{eq}^2}$; units = atm^{-3}

e) $K_{eq} = \dfrac{\left[PO_4^{3-}\right]_{eq}\left[H_3O^+\right]_{eq}^3}{\left[H_3PO_4\right]_{eq}\left(X_{H_2O}\right)_{eq}^3} = \dfrac{\left[PO_4^{3-}\right]_{eq}\left[H_3O^+\right]_{eq}^3}{\left[H_3PO_4\right]_{eq}}$; units = M^3

•

15.10 a) $$K_{eq} = \frac{(X_{H_2O})^2_{eq}(p_{SO_2})^2_{eq}}{(p_{H_2S})^2_{eq}(p_{O_2})^3_{eq}} = \frac{(p_{SO_2})^2_{eq}}{(p_{H_2S})^2_{eq}(p_{O_2})^3_{eq}}; \text{units:} \frac{(atm)^2}{(atm)^2(atm)^3} = atm^{-3}$$

b) $$K_{eq} = \frac{(X_{Fe})^2_{eq}(p_{CO_2})^3_{eq}}{(p_{CO})^3_{eq}} = \frac{(p_{CO_2})^3_{eq}}{(p_{CO})^3_{eq}}; \text{units:} \frac{(atm)^3}{(atm)^3} = \text{none}$$

c) $$K_{eq} = \frac{\left[Fe(C_2O_4)_3{}^{3-}\right]_{eq}}{\left[Fe^{3+}\right]_{eq}\left[C_2O_4{}^{2-}\right]^3_{eq}}; \text{units:} \frac{M}{M\,M^3} = M^{-3}$$

d) $$K_{eq} = \frac{\left[NH_4{}^+\right]_{eq}\left(X_{H_2O}\right)_{eq}}{\left(p_{NH_3}\right)_{eq}\left[H_3O^+\right]_{eq}} = \frac{\left[NH_4{}^+\right]_{eq}}{\left(p_{NH_3}\right)_{eq}\left[H_3O^+\right]_{eq}}; \text{units:} \frac{M}{(atm)M} = atm^{-1}$$

e) $$K_{eq} = \frac{\left(X_{Sn}\right)_{eq}\left(X_{H_2O}\right)^2_{eq}}{\left(X_{SnO_2}\right)_{eq}\left(p_{H_2}\right)^2_{eq}} = \frac{1}{\left(p_{H_2}\right)^2_{eq}}; \text{units:} \frac{1}{atm^2} = atm^{-2}$$

15.11 a) $$K_{eq} = \frac{\left(X_{P_4}\right)_{eq}\left(p_{O_2}\right)^5_{eq}}{\left[X_{P_4O_{10}}\right]_{eq}} = \left(p_{O_2}\right)^5_{eq}; \text{units} = atm^5$$

b) $$K_{eq} = \frac{\left(X_{BaCO_3}\right)_{eq}\left(X_C\right)_{eq}}{\left(X_{BaO}\right)_{eq}\left(p_{CO}\right)^2_{eq}} = \frac{1}{\left(p_{CO}\right)^2_{eq}}; \text{units} = atm^{-2}$$

c) $$K_{eq} = \frac{\left[H_3PO_4\right]_{eq}\left(X_{H_2O}\right)^3_{eq}}{\left[PO_4^{3-}\right]_{eq}\left[H_3O^+\right]^3_{eq}} = \frac{\left[H_3PO_4\right]_{eq}}{\left[PO_4^{3-}\right]_{eq}\left[H_3O^+\right]^3_{eq}}; \text{units} = M^{-3}$$

15.12 a) $2\ H_2O_{(l)} + 2\ SO_{2(g)} \rightleftarrows 2\ H_2S_{(g)} + 3\ O_{2(g)}$

$$K_{eq} = \frac{\left(p_{H_2S}\right)^2_{eq}\left(p_{O_2}\right)^3_{eq}}{\left(p_{SO_2}\right)^2_{eq}}$$

b) $2\ Fe_{(s)} + 3\ CO_{2(g)} \rightleftarrows Fe_2O_{3(s)} + 3\ CO_{(g)}$

$$K_{eq} = \frac{\left(p_{CO}\right)^3_{eq}}{\left(p_{CO_2}\right)^3_{eq}}$$

(continued)

(15.12 continued)

c) $Fe(C_2O_4)_{3(aq)}^{3-} \rightleftarrows Fe^{3+}_{(aq)} + 3\ C_2O^{2-}_{4(aq)}$

$$K_{eq} = \frac{\left[Fe^{3+}\right]_{eq}\left[C_2O_4{}^{2-}\right]^3_{eq}}{\left[Fe(C_2O_4)_3^{3-}\right]_{eq}}$$

d) $NH^+_{4(aq)} + H_2O_{(l)} \rightleftarrows NH_{3(g)} + H_3O^+_{(aq)}$

$$K_{eq} = \frac{\left(p_{NH_3}\right)_{eq}\left[H_3O^+\right]_{eq}}{\left[NH_4{}^+\right]_{eq}}$$

e) $Sn_{(s)} + 2\ H_2O_{(l)} \rightleftarrows SnO_{2(s)} + 2\ H_{2(g)}$

$$K_{eq} = \left(p_{H_2}\right)^2_{eq}$$

15.13 a) $K_{eq} = \frac{\left(X_{I_2}\right)_{eq}}{\left(p_{I_2}\right)_{eq}} = \frac{1}{\left(p_{I_2}\right)_{eq}} = \frac{1}{\text{vapor pressure}}$

b) $K_{eq} = \frac{\left[I_{2(solution)}\right]_{eq}}{\left(p_{I_2}\right)_{eq}} = K_H$

c) $K_{eq} = \frac{\left(X_{PbI_2}\right)_{eq}}{\left[Pb^{2+}\right]_{eq}\left[I^-\right]^2_{eq}} = \frac{1}{\left[Pb^{2+}\right]_{eq}\left[I^-\right]^2_{eq}} = \frac{1}{K_{sp}}$

d) $K_{eq} = \frac{\left[CN^-\right]_{eq}\left[H_3O^+\right]_{eq}}{\left[HCN\right]_{eq}\left(X_{H_2O}\right)_{eq}} = \frac{\left[CN^-\right]_{eq}\left[H_3O^+\right]_{eq}}{\left[HCN\right]_{eq}} = K_a$

e) $K_{eq} = \frac{\left[Pt^{2+}\right]_{eq}\left[Cl^-\right]^4_{eq}}{\left[PtCl_4{}^{2-}\right]_{eq}} = \frac{1}{K_f}$

15.14 a) $C_2H_5OH_{(l)} \rightleftarrows C_2H_5OH_{(g)}$

$$K_{eq} = (p_{C_2H_5OH})_{eq} \qquad K_{eq} = \text{vapor pressure}$$

b) $Ru^{2+}_{(aq)} + 6\ NH_{3(aq)} \rightleftarrows Ru(NH_3)^{2+}_{6(aq)}$

$$K_{eq} = \frac{[Ru(NH_3)_6^{2+}]_{eq}}{[Ru^{2+}]_{eq}[NH_3]^6_{eq}} \qquad K_{eq} = K_f$$

c) $NH_{3(aq)} \rightleftarrows NH_{3(g)}$

$$K_{eq} = \frac{(p_{NH_3})_{eq}}{[NH_3]_{eq}} \qquad K_{eq} = \frac{1}{K_H}$$

d) $NH^+_{4(aq)} + Cl^-_{(aq)} \rightleftarrows NH_4Cl_{(s)}$

$$K_{eq} = \frac{1}{[NH_4^+]_{eq}[Cl^-]_{eq}} \qquad K_{eq} = \frac{1}{K_{sp}}$$

e) $C_5H_5NH_{(aq)} + H_2O_{(l)} \rightleftarrows C_5H_5NH_2{}^+{}_{(aq)} + OH^-{}_{(aq)}$

$$K_{eq} = \frac{[C_5H_5NH_2^+][OH^-]_{eq}}{[C_5H_5NH]_{eq}} \qquad K_{eq} = K_b$$

15.15 You may wish to draw two pictures. The first picture would be like Figure 15.4 except it should show more molecules of gas escaping and more being dissolved. The second picture would have reduced pressure (fewer molecules) above the liquid. In that second picture you need to show that the number of molecules escaping the liquid is the same as that in the first picture but that the number of molecules being dissolved is less.

15.16 No, the concentration of solid does not change. The pure solid's concentration remains at a mole fraction of 1.00. The concentrations of Na^+ and Cl^- ions do not change either. If there is more surface area of solid NaCl exposed to the solution, more Na^+ and Cl^- ions may pass into solution but this is balanced by more Na^+ and Cl^- ions precipitating out.

15.17 Initial species: $H_{2(g)}$ and $CO_{2(g)}$

Chemical reaction: $H_{2(g)} + CO_{2(g)} \rightleftarrows H_2O_{(g)} + CO_{(g)}$

$$K_{eq} = \frac{(p_{H_2O})_{eq}(p_{CO})_{eq}}{(p_{H_2})_{eq}(p_{CO_2})_{eq}} = \frac{[H_2O]_{eq}[CO]_{eq}}{[H_2]_{eq}[CO_2]_{eq}}$$

$(p_{CO})_{eq} = 0.49$ mol $\Rightarrow (p_{H_2O})_{eq} = (p_{CO})_{eq} = 0.49$ mol

Reaction:	$H_{2(g)}$ +	$CO_{2(g)}$	$\rightleftarrows$	$H_2O_{(g)}$ +	$CO_{(g)}$
Init. (M)	1.0	1.0		0	0
Change (M)	-0.49	-0.49		+0.49	+0.49
Equilibrium (M)	0.51	0.51		0.49	0.49

$$K_{eq} = \frac{(0.49)^2}{(0.51)^2} = 0.92$$

15.18 5 steps

1. *Identify initial chemical species*: HCl and O_2

2. *Identify the chemical reaction*: Given: $4\ HCl_{(g)} + O_{2(g)} \rightleftarrows 2\ Cl_{2(g)} + 2\ H_2O_{(g)}$

3. *Write the equilibrium expression*: $K_{eq} = \dfrac{(p_{Cl_2})_{eq}^2 (p_{H_2O})_{eq}^2}{(p_{HCl})_{eq}^4 (p_{O_2})_{eq}}$

4. *Determine the equilibrium concentrations*:

Reaction:	$4\ HCl_{(g)}$ +	$O_{2(g)}$	$\rightleftarrows$	$2\ Cl_{2(g)}$ +	$2\ H_2O_{(g)}$
Initial (atm)	2.3	1.0		0	0
Change (atm)	-1.86	-0.465		+ 0.93	+0.93
Equilibrium (atm)	0.44	0.54		0.93	0.93

equilibrium conc. = initial conc. + change in conc.

for Cl_2: 0.93 atm = 0 + 0.93 atm

for H_2O: equilibrium conc. = 0 + change in Cl_2 conc. = 0 + 0.93 atm = 0.93 atm

for HCl: equilibrium conc. = 2.3 atm - 2 x change in Cl_2 conc.
= 2.3 atm - 2(0.93 atm) = 2.3 atm - 1.86 atm = 0.4 atm

for O_2: equilibrium conc. = 1.0 atm - (1/2) x change in Cl_2 conc.
= 1.0 atm - 1/2(0.93 atm) = 1.0 atm - 0.465 atm = 0.5 atm

5. *Substitute equilibrium concentrations into equilibrium expression and solve for unknown(s)*:

Unknown is equil. constant. $K_{eq} = \dfrac{[0.93\ atm]^2[0.93\ atm]^2}{[0.44\ atm]^4[0.54\ atm]} = 37 = 4 \times 10^1$ atm

15.19 $C_2H_5CO_2H_{(aq)} + H_2O_{(l)} \rightleftarrows C_2H_5CO_2^-{}_{(aq)} + H_3O^+{}_{(aq)}$

$$K_{eq} = \frac{[C_2H_5CO_2^-]_{eq}[H_3O^+]_{eq}}{[C_2H_5CO_2H]_{eq}} = K_a$$

Reaction:	$C_2H_5CO_2H_{(aq)}$ +	$H_2O_{(l)}$ $\rightleftarrows$	$C_2H_5CO_2^-{}_{(aq)}$ +	$H_3O^+{}_{(aq)}$
Init. (M)	$0.050/.500 = 1.00 \times 10^{-1}$		0	0
Change (M)	-1.15×10^{-3}		$+1.15 \times 10^{-3}$	$+1.15 \times 10^{-3}$
Equilibrium (M)	9.89×10^{-2}		1.15×10^{-3}	1.15×10^{-3}

$$K_a = \frac{(1.15 \times 10^{-3})^2}{(9.89 \times 10^{-2})} = 1.3 \times 10^{-5}\ M$$

15.20 1. *Identify initial chemical species*: HCNO(aq) and $H_2O_{(l)}$

2. *Identify the chemical reaction*: $HCNO_{(aq)} + H_2O_{(l)} \rightleftarrows H_3O^+{}_{(aq)} + CNO^-{}_{(aq)}$

3. *Write the equilibrium expression*: $K_{eq} = K_a = \frac{[H_3O^+]_{eq}[CNO^-]_{eq}}{[HCNO]_{eq}}$

4. *Determine the equilibrium concentrations:*

Reaction:	$HCNO_{(aq)}$	+ $H_2O_{(l)}$ $\rightleftarrows$	$H_3O^+{}_{(aq)}$ +	$CNO^-{}_{(aq)}$
Initial (M)	0.20		0	0
Change (M)	$-\ 6.5 \times 10^{-3}$		$+\ 6.5 \times 10^{-3}$	$+\ 6.5 \times 10^{-3}$
Equilibrium (M)	$0.20 - 6.5 \times 10^{-3}$		$+\ 6.5 \times 10^{-3}$	$+\ 6.5 \times 10^{-3}$

$[CNO^-]_{eq} = [H_3O^+]_{eq}$

5. *Substitute equilibrium concentrations into equilibrium expression and solve*:

$$K_a = \frac{(6.5 \times 10^{-3}\ M)(6.5 \times 10^{-3}\ M)}{(0.20\ M - 6.5 \times 10^{-3}\ M)} = 2.18 \times 10^{-4}\ M$$

$$K_a \cong \frac{(6.5 \times 10^{-3}\ M)(6.5 \times 10^{-3}\ M)}{(0.19\ M)} = 2.2 \times 10^{-4}\ M$$

15.21

Reaction:	$PbF_{2(s)}$	+ $H_2O_{(l)}$ $\rightleftarrows$	$Pb^{2+}{}_{(aq)}$ +	$2\ F^-{}_{(aq)}$
Initial			0	0
Change			$+\ 1.9 \times 10^{-3}$	$+3.8 \times 10^{-3}$
Equilibrium			$+\ 1.9 \times 10^{-3}$	$+3.8 \times 10^{-3}$

$$K_{sp} = K_{eq} = \frac{[Pb^{2+}]_{eq}[F^-]_{eq}^2}{[X_{PbF_2}]_{eq}[X_{H_2O}]} = [Pb^{2+}]_{eq}[F^-]_{eq}^2$$

$$K_{sp} = (1.9 \times 10^{-3} M)(3.8 \times 10^{-3} M)^2 = 2.7 \times 10^{-8}\ M^3$$

15.22 Solubility of $Ca_3(AsO_4)_2$ in H_2O is 0.036 g/L = (0.036 g/L)/398.08 g/mol = 9.0×10^{-5} moles/L

5 steps

1. $Ca_3(AsO_4)_{2(s)}$ in $H_2O_{(l)}$
2. $Ca_3(AsO_4)_{2(s)} \rightleftarrows 3\ Ca^{2+}{}_{(aq)} + 2\ AsO_4{}^{3-}{}_{(aq)}$
3. $K_{sp} = [Ca^{2+}]^3{}_{eq}[AsO_4{}^{3-}]^2{}_{eq}$
4. Reaction:

Reaction:	$Ca_3(AsO_4)_{2(s)}$	$\rightleftarrows$	$3\ Ca^{2+}{}_{(aq)}$	+	$2\ AsO_4{}^{3-}{}_{(aq)}$
Initial	9.0×10^{-5} mol/L		0		0
Change	$- 9.0 \times 10^{-5}$ mol/L		$+ 27 \times 10^{-5}$ M		$+ 18 \times 10^{-5}$ M
Equilibrium	0		2.7×10^{-4} M		1.8×10^{-4} M

5. $K_{sp} = (2.7 \times 10^{-4}\ M)^3(1.8 \times 10^{-4}\ M)^2 = 6.4 \times 10^{-19}\ M^5$

15.23 $H_{2(g)} + Br_{2(g)} \rightleftarrows 2\ HBr_{(g)}$ $\qquad K_{eq} = \dfrac{(p_{HBr})^2_{eq}}{(p_{H_2})_{eq}(p_{Br_2})_{eq}} = 1.6 \times 10^5$

Reaction:	$H_{2(g)}$	$+ Br_{2(g)}$	$\rightleftarrows$	$2\ HBr_{(g)}$
Initial (atm)	0	0		10.0
Change (atm)	+x	+x		-2x
Equilibrium (atm)	+x	+x		10.0 - 2x

$$1.6 \times 10^5 = \frac{(10-2x)^2}{x^2} = \frac{100-40x+4x^2}{x^2}$$

$$0 = 100 - 40x + 4x^2 - 1.6 \times 10^5\ x^2 = -1.6 \times 10^5\ x^2 - 40x + 100$$

$$x = \frac{-b \pm \sqrt{b^2-4ac}}{2a} = \frac{40 \pm \sqrt{1600-4(-1.6 \times 10^5)(100)}}{-2(1.6 \times 10^5)}$$

$$= \frac{40 \pm \sqrt{64001600}}{-3.2 \times 10^5} = \frac{40 \pm 8000.10}{-3.2 \times 10^5} = 2.49 \times 10^{-2} \text{ or } -2.51 \times 10^{-2}$$

$(p_{H_2})_{eq} = (p_{Br_2})_{eq} = 2.5 \times 10^{-2}$ atm $\qquad (p_{HBr})_{eq} = 10.0 - 2(2.5 \times 10^{-2}) = 10$ atm

15.24 5 steps

1. $CO_{(g)}$ and $FeO_{(s)}$
2. $FeO_{(s)} + CO_{(g)} \rightleftarrows Fe_{(s)} + CO_{2(g)}$
3. $K_{eq} = 0.403 = \dfrac{(p_{CO_2})_{eq}}{(p_{CO})_{eq}}$
4. Reaction:

Reaction:	$FeO_{(s)}$	$+ CO_{(g)}$	$\rightleftarrows$	$Fe_{(s)}$	+	$CO_{2(g)}$
Initial (atm)	excess	5.0		0		0
Change (atm)		- x				+ x
Equilibrium (atm)		5.0 - x				+ x

Note: Amounts of $FeO_{(s)}$ and $Fe_{(s)}$ are unnecessary to the calculation as long as there was excess $FeO_{(s)}$ at the beginning.

(continued)

(15.24 continued)

5. $K_{eq} = 0.403 = x/5.0 \text{ atm} - x$ $(0.403)(5.0 \text{ atm} - x) = x$

$(5.0 \text{ atm})(0.403) = x + 0.403\ x$ $2.015 \text{ atm} = 1.403\ x$ $x = 1.4 \text{ atm}$

$(p_{CO_2})_{eq} = x = 1.4 \text{ atm}$ $(p_{CO})_{eq} = 5.0 \text{ atm} - x = 5.0 \text{ atm} - 1.4 \text{ atm} = 3.6 \text{ atm}$

15.25 $$K_{eq} = \frac{(p_{COCl_2})_{eq}}{(p_{CO})_{eq}(p_{Cl_2})_{eq}} = 1.5 \times 10^8 \text{ atm}^{-1}$$

Reaction:	$CO_{(g)}$	$+\ Cl_{2(g)}$	$\rightleftarrows$	$COCl_{2(g)}$
Init. (atm)	0	0		0.250 atm
Change (atm)	+x	+x		-x
Equil. (atm)	x	x		0.250 - x

$$1.5 \times 10^8 = \frac{0.250 - x}{x^2} \qquad 0 = 1.5 \times 10^8\ x^2 + x - 0.250$$

$$x = \frac{-b \pm \sqrt{b^2 - 4ac}}{2a} = \frac{-1 \pm \sqrt{1 - 4(1.5 \times 10^8)(-0.250)}}{2(1.5 \times 10^8)}$$

$$= \frac{-1 \pm 1.22 \times 10^4}{3.0 \times 10^8} = 4.1 \times 10^{-5}$$

$(p_{CO})_{eq} = (p_{Cl_2})_{eq} = 4.1 \times 10^{-5}$ atm

$(p_{COCl_2})_{eq} = 0.250 - 4.1 \times 10^{-5} = 2.50 \times 10^{-1}$ atm

15.26 1. $Cl_{2(g)}$

2. $Cl_{2(g)} \underset{}{\overset{1200\text{ K}}{\rightleftarrows}} 2\ Cl_{(g)}$

3. $$K_{eq} = \frac{(p_{Cl})^2_{eq}}{(p_{Cl_2})_{eq}}$$

$$p_{Cl_2} = nRT/V = \left[\left(\frac{5.0 \text{ g}}{70.90 \text{ g/mol}}\right)\left(0.0821 \frac{\text{L atm}}{\text{mol K}}\right)(1200 \text{ K}) \Big/ 3.0 \text{ L}\right] = 2.3 \text{ atm}$$

4. Reaction:	$Cl_{2(g)}$	$\rightleftarrows$	$2\ Cl_{(g)}$
Initial (atm)	2.3		0
Change (atm)	- x		2 x
Equilibrium (atm)	2.3 - x		2 x

Using the gas law: PV + nRT

5. $$K_{eq} = 2.5 \times 10^{-5} \text{ atm} = \frac{(2x)^2}{2.3 \text{ atm} - x}$$

Assume 2.3 atm - x $\cong$ 2.3 atm, then $K_{eq} = 2.5 \times 10^{-5} \text{ atm} \cong (2x)^2/2.3$ atm

$(2.3 \text{ atm})(2.5 \times 10^{-5} \text{ atm}) = (2\ x)^2 = 4\ x^2$

(continued)

(15.26 continued)

$$x^2 = \frac{(2.3 \text{ atm})(2.5 \times 10^{-5} \text{ atm})}{4} = 1.4 \times 10^{-5} \text{ atm}^2 \qquad x = 3.8 \times 10^{-3} \text{ atm}$$

$(p_{Cl})_{eq} = 2x = 2(3.8 \times 10^{-3} \text{ atm}) = 7.6 \times 10^{-3}$ atm

Approximation: $2.3 - x \cong 2.3$

$2.3 - 0.0038 \cong 2.3$

$$\frac{3.8 \times 10^{-3}}{2.3} \times 100\% = 0.17\%\text{; approximation acceptable}$$

15.27 $PbCl_{2(s)} + H_2O_{(l)} \rightleftarrows Pb^{2+}_{(aq)} + 2\ Cl^-_{(aq)}$

$K_{sp} = [Pb^{2+}]_{eq}[Cl^-]^2_{eq} = 2 \times 10^{-5}\ M^3$

$Q = [Pb^{2+}][Cl^-]^2$

$$[Pb^{2+}] = \frac{0.50 \text{ g} / 278.1 \text{ g mol}^{-1} \times 1 \text{ mol } Pb^{2+} / \text{mol } PbCl_2}{0.300 \text{ L}} = 6.0 \times 10^{-3} \text{ M}$$

$[Cl^-] = 2[Pb^{2+}] = 1.2 \times 10^{-2}$ M $Q = (6.0 \times 10^{-3} \text{ M})(1.2 \times 10^{-2} \text{ M})^2 = 8.6 \times 10^{-7}\ M^3$

$Q < K_{sp}$ so all of the solid dissolves.

15.28 1. $Na^+_{(aq)}$, $Cl^-_{(aq)}$, $Pb^{2+}_{(aq)}$, $NO^{3-}_{(aq)}$

2. $2\ Cl^-_{(aq)} + Pb^{2+}_{(aq)} \rightleftarrows PbCl_{2(s)}$ $\qquad PbCl_{2(s)} \rightleftarrows Pb^{2+}_{(aq)} + 2\ Cl^-_{(aq)}$

3. $Q = [Pb^{2+}][Cl^-]^2$

200 mL + 300 mL = 500 mL

4. Initial conc. (M) $\quad \left[Pb^{2+}\right] = \dfrac{(300 \text{ mL})(4.00 \times 10^{-2} \text{ M})}{(500 \text{ mL})} = 2.4 \times 10^{-2}$ M

$$\left[Cl^-\right] = \frac{(200 \text{ mL})(2.50 \times 10^{-2} \text{ M})}{(500 \text{ mL})} = 1.0 \times 10^{-2} \text{ M}$$

$Q = (2.4 \times 10^{-2} \text{ M})(1.0 \times 10^{-2} \text{ M})^2 = 2.4 \times 10^{-6}\ M^3$

$2.4 \times 10^{-6}\ M^3 < 2 \times 10^{-5}\ M^3$; $\therefore Q < K_{sp}$

No, solid $PbCl_2$ will not form; $PbCl_{2(s)} \rightleftarrows Pb^{2+}_{(aq)} + 2\ Cl^-_{(aq)}$ may go to the right if more $PbCl_{2(s)}$ were available, *i.e.*, even more $PbCl_2$ would dissove in this solution..

15.29 a) $2\ SO_{2(g)} + O_{2(g)} \rightleftarrows 2\ SO_{3(g)}$

$\Delta G° = -RT \ln K_{eq}$ $\qquad \ln K_{eq} = -\Delta G°/RT$ $\qquad R = 8.314$ J/K $\qquad T = 298$ K

$\Delta G^\circ_{rxn} = \Sigma(\text{coeff})\,\Delta G^\circ_f(\text{products}) - \Sigma(\text{coeff})\,\Delta G^\circ_f(\text{reactants})$

$= 2\ \Delta G^\circ_f(SO_3) - 2\ \Delta G^\circ_f(SO_2) - \Delta G^\circ_f(O_2)$

$= 2\ (-371.06 \text{ kJ mol}^{-1}) - 2(-300.194 \text{ kJ mol}^{-1}) - 0 = -141.73$ kJ/mol

$$\ln K_{eq} = \frac{141.73 \text{ kJ / mol}}{(8.314 \text{ J / mol K})(298 \text{ K})} 1000 \text{ J / kJ} = 57.21$$

$K_{eq} = e^{57.21} = 7.0 \times 10^{24}\ \text{atm}^{-1}$

(continued)

(15.29 continued)

b) $2\ CO_{(g)} + O_{2(g)} \rightleftarrows 2\ CO_{2(g)}$

$\Delta G^\circ_{rxn} = 2\ \Delta G^\circ_f(CO_2) - \Delta G^\circ_f(O_2) - 2\ \Delta G^\circ_f(CO)$

$= 2(-394.359) - 0 - 2(-137.168) = -514.38$ kJ/mol

$$\ln K_{eq} = \frac{-\Delta G}{RT} = \frac{(514.38)10^3}{(8.314)(298)} = 207.62 \qquad K_{eq} = 1.5 \times 10^{90}\ \text{atm}^{-1}$$

c) $BaCO_{3(s)} + C_{(s)} \rightleftarrows BaO_{(s)} + 2\ CO_{(g)}$

$\Delta G^\circ_{rxn} = \Delta G^\circ_f(BaO) + 2\ \Delta G^\circ_f(CO) - \Delta G^\circ_f(BaCO_3) - \Delta G^\circ_f(C)$

$= -525.1 + 2(-137.168) + 1137.6 - 0 = 338.16$ kJ/mol

$$\ln K_{eq} = \frac{-(338.16)(1000)}{(8.314)(298)} = -136.49 \qquad K_{eq} = 5.3 \times 10^{-60}\ \text{atm}^2$$

15.30 a) $CO_{2(g)} + H_2O_{(l)} \rightleftarrows CO_{2(g)} + H_{2(g)}$

$\Delta G^\circ = (1)(0.0\ \text{kJ/mol}) + (1)(-394.359\ \text{kJ/mol}) - [(1)(-137.168\ \text{kJ/mol}) + (1)(-237.129\ \text{kJ/mol})] = -20.062$ kJ/mol

$$\ln K_{eq} = -\frac{\Delta G^\circ}{RT} = -\frac{(-20.062\ \text{kJ/mol})(10^3\ \text{J/kJ})}{(8.314\ \text{J/mol K})(298\ \text{K})} = 8.10$$

$K_{eq} = e^{8.10} = 3.3 \times 10^3$ atm

b) $CH_{4(g)} + H_2O_{(l)} \rightleftarrows CO_{(g)} + 3\ H_{2(g)}$

$\Delta G^\circ = (3)(0.0\ \text{kJ/mol}) + (1)(-137.168\ \text{kJ/mol}) - [(1)(-50.72\ \text{kJ/mol}) + (1)(-237.129\ \text{kJ/mol})] = 150.68$ kJ/mol

$$\ln K_{eq} = -\frac{\Delta G^\circ}{RT} = -\frac{(150.68\ \text{kJ/mol})(10^3\ \text{J/kJ})}{(8.314\ \text{J/mol K})(298\ \text{K})} = -60.8$$

$K_{eq} = e^{-60.8} = 3.9 \times 10^{-27}\ \text{atm}^3$

c) $SnO_{2(s)} + 2\ H_{2(g)} \rightleftarrows Sn_{(s)} + 2\ H_2O_{(l)}$

$\Delta G^\circ = (2)(-237.129\ \text{kJ/mol}) + (1)(0.0\ \text{kJ/mol}) - [(1)(-519.6\ \text{kJ/mol}) + (2)(0.0\ \text{kJ/mol})] = 45.3$ kJ/mol (or 45.4 kJ/mol if $Sn_{(s)\ gray}$ is used)

$$\ln K_{eq} = -\frac{\Delta G^\circ}{RT} = -\frac{(45.3\ \text{kJ/mol})(10^3\ \text{J/kJ})}{(8.314\ \text{J/mol K})(298\ \text{K})} = -18.3$$

$K_{eq} = e^{-18.3} = 1.1 \times 10^{-8}\ \text{atm}^{-2}$

d) $4\ NH_{3(g)} + 5\ O_{2(g)} \rightleftarrows 4\ NO_{(g)} + 6\ H_2O_{(l)}$

$\Delta G^\circ = (4)(86.55\ \text{kJ/mol}) + (6)(-237.129\ \text{kJ/mol}) - [(4)(-16.45\ \text{kJ/mol}) + (5)(0.0\ \text{kJ/mol})] = -1011$ kJ/mol

$$\ln K_{eq} - \frac{(-1011\ \text{kJ/mol})(10^3\ \text{J/kJ})}{(8.314\ \text{J/mol K})(298\ \text{K})} = 408 \qquad K_{eq} = e^{408} = 1.5 \times 10^{177}\ \text{atm}^{-5}$$

e) $3\ Fe_{(s)} + 4\ H_2O_{(l)} \rightleftarrows Fe_3O_{4(s)} + 4\ H_{2(g)}$

$\Delta G^\circ = (1)(-1015.4\ \text{kJ/mol}) + (4)(0.0\ \text{kJ/mol}) - [(3)(0.0\ \text{kJ/mol}) + (4)(-237.129\ \text{kJ/mol})] = -66.88$ kJ/mol

$$\ln K_{eq} - \frac{(-66.88\ \text{kJ/mol})(10^3\ \text{J/kJ})}{(8.314\ \text{J/mol K})(298\ \text{K})} = 27.00 \qquad K_{eq} = e^{27.00} = 5.3 \times 10^{11}\ \text{atm}^4$$

15.31 a) $CO_{(g)} + H_2O_{(g)} \rightleftarrows CO_{2(g)} + H_{2(g)}$

$\Delta G° = \Delta G°(CO_2) + \Delta G°(H_2) - \Delta G°(CO) - \Delta G°(H_2O)$
$= -394.359 + 0 + 137.168 + 228.72 = -28.47$ kJ/mol

$$\ln K_{eq} = \frac{-\Delta G°}{RT} = \frac{(28.47)(1000)}{(8.314)(298)} = 11.49 \qquad K_{eq} = 9.8 \times 10^4$$

b) $CH_{4(g)} + H_2O_{(g)} \rightleftarrows CO_{(g)} + 3\ H_{2(g)}$

$\Delta G° = -137.168 + 3(0) - 228.72 + 50.72 = -315.17$ kJ/mol

$$\ln K_{eq} = \frac{-(-315.17)(1000)}{(8.314)(298)} = 127.21 \qquad K_{eq} = 1.8 \times 10^{55}\ \text{atm}^2$$

c) $SnO_{2(s)} + 2\ H_{2(g)} \rightleftarrows Sn_{(s)} + 2\ H_2O_{(g)}$

$\Delta G° = 0 + 2(-228.72) + 519.6 - 2(0) = 62.16$ kJ/mol

$$\ln K_{eq} = \frac{-(62.16)(1000)}{(8.314)(298)} = -25.09 \qquad K_{eq} = 1.3 \times 10^{-11}$$

d) $4\ NH_{3(g)} + 5\ O_{2(g)} \rightleftarrows 4\ NO_{(g)} + 6\ H_2O_{(g)}$

$\Delta G° = 4(86.55) + 6(-228.72) - 4(-16.45) - 5(0) = -960.32$ kJ/mol

$$\ln K_{eq} = \frac{-(-960.32)(1000)}{(8.314)(298)} = 387.61 \qquad K_{eq} = 2.2 \times 10^{168}\ \text{atm}$$

e) $3\ Fe_{(s)} + 4\ H_2O_{(g)} \rightleftarrows Fe_3O_{4(s)} + 2\ H_{2(g)}$

$\Delta G° = -1015.4 + 4(0) - 3(0) - 4(-228.72) = -100.52$ kJ/mol

$$\ln K_{eq} = \frac{-(-100.52)(1000)}{(8.314)(298)} = 40.57 \qquad K_{eq} = 4.2 \times 10^{17}\ \text{atm}^{-2}$$

15.32 a) $2\ SO_{2(g)} + O_{2(g)} \rightleftarrows 2\ SO_{3(g)}$

$\Delta H \approx \Delta H°_{rxn} = (2)(-395.72\ \text{kJ/mol}) - [(2)(-296.830\ \text{kJ/mol}) + (1)(0.0\ \text{kJ/mol})]$
$= -197.78$ kJ/mol

$\Delta S \approx \Delta S°_{rxn} = (2)(256.76\ \text{J/mol K}) - [(2)(248.22\ \text{J/mol K}) + (1)(205.138\ \text{J/mol K}]$
$= -188.06$ J/mol K

$$\ln K_{eq} = -\frac{\Delta H°}{RT} + \frac{\Delta S°}{R} = -\frac{(-197.78\ \text{kJ/mol})(10^3\ \text{J/kJ})}{(8.314\ \text{J/mol K})(500\ \text{K})} + \frac{(-188.06\ \text{J/mol K})}{(8.314\ \text{J/mol K})}$$

$$= 47.6 - 22.6 = 25.0$$

$K_{eq} = e^{25.0} = 7 \times 10^{10}\ \text{atm}^{-1}$

b) $2\ CO_{(g)} + O_{2(g)} \rightleftarrows 2\ CO_{2(g)}$

$\Delta H \approx \Delta H°_{rxn} = (2)(-393.509\ \text{kJ/mol}) - [(2)(-110.525\ \text{kJ/mol}) + (1)(0.0\ \text{kJ/mol})]$
$= -565.968$ kJ/mol

$\Delta S°_{rxn} = (2)(213.74\ \text{J/mol K}) - [(2)(197.674\ \text{J/mol K}) + (1)(205.138\ \text{J/mol K}]$
$= -173.01$ J/mol K

(continued)

(15.32 continued)

$$\ln K_{eq} = \frac{(-565.968\ \text{kJ/mol})(10^3\ \text{J/kJ})}{(8.314\ \text{J/mol K})(500\ \text{K})} + \frac{(-173.01\ \text{J/mol K})}{(8.314\ \text{J/mol K})}$$

$$= 136.1 - 20.8 = 115.3$$

$K_{eq} = e^{115.3} = 1.2 \times 10^{50}\ \text{atm}^{-1}$

c) $BaCO_{3(s)} + C_{(s)} \rightleftarrows BaO_{(s)} + 2\ CO_{(g)}$ Assume $C_{(s)}$ (graphite)

ΔH°_{rxn} = (2)(-110.525 kJ/mol) + (1)(-553.5 kJ/mol) - [(1)(-1216.3 kJ/mol) + (1)(0.0 kJ/mol)] = 441.75 kJ/mol

ΔS°_{rxn} = (2)(197.674 J/mol K) + (1)(70.42 J/mol K) - [(1)(112.1 J/mol K) + (1)(5.740 J/mol K] = 347.93 J/mol K

$$\ln K_{eq} = -\frac{(441.75\ \text{kJ/mol})(10^3\ \text{J/kJ})}{(8.314\ \text{J/mol K})(500\ \text{K})} + \frac{(347.93\ \text{J/mol K})}{(8.314\ \text{J/mol K})}$$

$$= -106.3 + 41.8 = -64.5$$

$K_{eq} = e^{-64.5} = 9.7 \times 10^{-29}\ \text{atm}^2$

15.33 a) $CO_{(g)} + H_2O_{(g)} \rightleftarrows CO_{2(g)} + H_{2(g)}$; T = 700 K

$$\ln K_{eq} = \frac{-\Delta H^\circ}{RT} + \frac{\Delta S^\circ}{R}$$

ΔH° = -393.509 + 0 - (-110.525) - (-241.818) = -41.166 kJ/mol

ΔS° = 213.74 + 130.684 - 197.674 - 188.825 = -42.075 J/mol K

$$\ln K_{eq} = \frac{-(41.166\ \text{kJ / mol})(1000\ \text{J / kJ})}{(8.314\ \text{J / K})(700\ \text{K})} + \frac{(-42.075\ \text{J / mol K})}{(8.314\ \text{J / K})} = 2.013$$

$K_{eq} = e^{2.013} = 7.5$

b) $CH_{4(g)} + H_2O_{(g)} \rightleftarrows CO_{(g)} + 3\ H_{2(g)}$

$\Delta H \approx \Delta H^\circ$= -110.525 + 3(0) - (-74.81) - (-241.818) = 206.103 kJ/mol

$\Delta S \approx \Delta S^\circ$ = 197.674 + 3(130.684) - 186.264 - 188.825 = 214.637 J/mol K

$$\ln K_{eq} = \frac{-\Delta H^\circ}{RT} + \frac{\Delta S^\circ}{R} = \frac{-(206.103)(1000)}{(8.314)(700\ \text{K})} + \frac{214.637}{8.314} = -9.598$$

$K_{eq} = 6.8 \times 10^{-5}\ \text{atm}^2$

c) $SnO_{2(s)} + 2\ H_{2(g)} \rightleftarrows Sn_{(s)} + 2\ H_2O_{(g)}$

$\Delta H \approx \Delta H^\circ$= 0 + 2(-241.818) - (-580.7) - 2(0) = 97.064 kJ/mol

$\Delta S \approx \Delta S^\circ$ = 51.55 + 2(188.825) - 52.3 - 2(130.684) = 115.532 J/mol K

(continued)

(15.33 continued)

$$\ln K_{eq} = \frac{-\Delta H^\circ}{RT} + \frac{\Delta S^\circ}{R} = -2.782 \ ; \quad K_{eq} = 6.2 \times 10^{-2}$$

d) $4\ NH_{3(g)} + 5\ O_{2(g)} \rightleftarrows 4\ NO_{(g)} + 6\ H_2O_{(g)}$

$\Delta H \approx \Delta H^\circ = 4(90.25) + 6(-241.818) - 4(-46.11) - 5(0) = -905.468$ kJ/mol

$\Delta S \approx \Delta S^\circ = 4(210.761) + 6(188.825) - 4(192.45) - 5(205.138) = 180.504$ J/mol K

$\ln K_{eq} = 177.29$; $K_{eq} = 9.96 \times 10^{76}\ atm^1$

e) $3\ Fe_{(s)} + 4\ H_2O_{(g)} \rightleftarrows Fe_3O_{4(s)} + 4\ H_{2(g)}$

$\Delta H \approx \Delta H^\circ = -1118.4 + 4(0) - 3(0) - 4(-241.818) = -151.128$ kJ/mol

$\Delta S \approx \Delta S^\circ = 146.4 + 4(130.684) - 3(27.28) - 4(188.825) = 168.004$ J/mol K

$\ln K_{eq} = 5.76$; $K_{eq} = 3.2 \times 10^2$

15.34 $CO_{(g)} + H_2O_{(g)} \rightleftarrows CO_{2(g)} + H_{2(g)}$

$\Delta G^\circ_{rxn} = (1)(-394.359$ kJ/mol$) + (1)(0$ kJ/mol$) - [(1)(-137.168$ kJ/mol$)$
$+ (1)(-228.72$ kJ/mol$)] = -28.47$ kJ/mol

$\Delta H^\circ_{rxn} = (1)(-393.509$ kJ/mol$) + (1)(0$ kJ/mol$) - [(1)(-110.525$ kJ/mol$)$
$+ (1)(-241.818$ kJ/mol$)] = -41.166$ kJ/mol

$\Delta S^\circ_{rxn} = (1)(213.74$ J/mol K$) + (1)(130.684$ J/mol K$) - [(1)(197.674$ J/mol K$)$
$+ (1)(188.825$ J/mol K$)] = -42.075$ J/mol K

at 298 K:

$$\ln K_{eq} = -\frac{\Delta G^\circ}{RT} = -\frac{(-28.47\ \text{kJ/mol})(10^3\ \text{J/kJ})}{(8.314\ \text{J/mol K})(298\ \text{K})} = 11.5$$

$K_{eq} = e^{11.5} = 9.9 \times 10^4$

$$\text{or}\quad \ln K_{eq} = -\frac{\Delta H^\circ}{RT} + \frac{\Delta S^\circ}{R} = -\frac{(-41.166\ \text{kJ/mol})(10^3\ \text{J/kJ})}{(8.314\ \text{J/mol K})(298\ \text{K})} + \frac{-42.075\ \text{J/mol K}}{8.314\ \text{J/mol K}}$$

$$= 16.6 - 5.1 = 11.5$$

$K_{eq} = e^{11.5} = 9.9 \times 10^4$

At 1200 K (Assuming $\Delta H \approx \Delta H^\circ$ and $\Delta S \approx \Delta S^\circ$)

$$\ln K_{eq} = -\frac{(-41.166\ \text{kJ/mol})(10^3\ \text{J/kJ})}{(8.314\ \text{J/mol K})(1200\ \text{K})} + \frac{-42.075\ \text{J/mol K}}{8.314\ \text{J/mol K}}$$

$$= 4.126 - 5.061 = -0.935$$

$K_{eq} = e^{-0.9355} = 0.393$

15.35 a) $2\ SO_{2(g)} + O_{2(g)} \rightleftarrows 2\ SO_{3(g)}$

$$Q = \frac{\left(p_{SO_3}\right)^2}{\left(p_{SO_2}\right)^2\left(p_{O_2}\right)}$$, so injecting $CO_{(g)}$ into the system has **no effect** on the equilibrium position.

b) $2\ CO_{(g)} + O_{2(g)} \rightleftarrows 2\ CO_{2(g)}$

$$Q = \frac{\left(p_{CO_2}\right)^2}{\left(p_{CO}\right)^2\left(p_{O_2}\right)}$$, so adding $CO_{(g)}$ decreases the value of Q which should cause the reaction to form products from reactants until equilibrium is reestablished. Therefore, **more $CO_{2(g)}$ is produced.**

c) $BaCO_{3(s)} + C_{(s)}\ \ BaO_{(s)} + 2\ CO_{(g)}$

$Q = (p_{CO})^2$, so adding $CO_{(g)}$ makes $Q > K_{eq}$; thus, reaction products will be consumed and reactants will be produced until K_{eq} is reestablished; that is, **$BaCO_{3(s)}$ and $C_{(s)}$ will be produced.**

15.36 a) $CO_{(g)} + H_2O_{(g)} \rightleftarrows CO_{2(g)} + H_{2(g)}$
Additional $H_2O_{(g)}$ into the system results in $Q < K_{eq}$; therefore, the reaction proceeds to the right and $CO_{(g)}$ and $H_2O_{(g)}$ react to form more $CO_{2(g)}$ and $H_{2(g)}$.

b) $CH_{4(g)} + H_2O_{(g)} \rightleftarrows CO_{(g)} + 3\ H_{2(g)}$
Additional $H_2O_{(g)}$ into the system results in $Q < K_{eq}$; therefore, the reaction proceeds to the right and $CH_{4(g)}$ and $H_2O_{(g)}$ react to form more $CO_{(g)}$ and $H_{2(g)}$.

c) $SnO_{2(s)} + 2\ H_{2(g)} \rightleftarrows Sn_{(s)} + 2\ H_2O_{(g)}$
Additional $H_2O_{(g)}$ into the system results in $Q > K_{eq}$; therefore, the reaction proceeds to the left and $Sn_{(s)}$ and $H_2O_{(g)}$ react to form more $SnO_{2(s)}$ and $H_{2(g)}$.

d) $4\ NH_{3(g)} + 5\ O_{2(g)} \rightleftarrows 4\ NO_{(g)} + 6\ H_2O_{(g)}$
Additional $H_2O_{(g)}$ into the system results in $Q > K_{eq}$; therefore, the reaction proceeds to the left and $NO_{(g)}$ and $H_2O_{(g)}$ react to form more $NH_{3(g)}$ and $O_{2(g)}$.

e) $3\ Fe_{(s)} + 4\ H_2O_{(g)} \rightleftarrows Fe_3O_{4(s)} + 4\ H_{2(g)}$
Additional $H_2O_{(g)}$ into the system results in $Q < K_{eq}$; therefore, the reaction proceeds to the right and $Fe_{(s)}$ and $H_2O_{(g)}$ react to form more $Fe_3O_{4(s)}$ and $H_{2(g)}$.

15.37 $PbCl_{2(s)} \rightleftarrows Pb^{2+}{}_{(aq)} + 2\ Cl^-{}_{(aq)}$

a) $Q = [Pb^{2+}][Cl^-]^2$ so adding more $PbCl_{2(s)}$ will not change Q. The system remains at equilibrium and there is no effect on the amount of dissolved $PbCl_{2(s)}$.

b) Adding more H_2O dilutes the solution which lowers $[Pb^{2+}]$ and $[Cl^-]$. Thus, $Q < K_{eq}$ so more $PbCl_{2(s)}$ will dissolve until K_{eq} is reestablished.

c) Adding solid NaCl increases $[Cl^-]$. Thus, $Q > K_{eq}$ so the reaction proceeds to the left and some $PbCl_{2(s)}$ will precipitate.

d) Adding solid KNO_3 has no effect on Q so there is no effect on the amount of dissolved $PbCl_{2(s)}$.

15.38 $2\,SO_{2(g)} + O_{2(g)} \rightleftarrows 2\,SO_{3(g)}$

a) The reaction produces heat (is exothermic). When heat is added by raising the temperature, the reaction proceeds to the left (an endothermic process that consumes heat) and the amount of $SO_{3(g)}$ in the system decreases.

b) Addition of more $O_{2(g)}$ results in $Q < K_{eq}$; therefore, the reaction proceeds to the right and more $SO_{3(g)}$ is formed.

c) The addition of Ar gas increases the total pressure of the system but does not change the partial pressure of each reacting gas. (Ar does not react in this system.) Therefore, Q does not change and $Q = K_{eq}$. The amount of $SO_{3(g)}$ in the system does not change.

15.39 $SO_{2(g)} + Cl_{2(g)} \rightleftarrows SO_2Cl_{2(g)} + \text{heat}$ $\quad Q = \dfrac{(p_{SO_2Cl_2})}{(p_{SO_2})(p_{Cl_2})}$

Changes that would drive the equilibrium to the left:

1. Heating the system (adding heat).
2. Adding $SO_2Cl_{2(g)}$ to the system.
3. Removing $Cl_{2(g)}$ from the system.
4. Removing $SO_{2(g)}$ from the system.
5. Decreasing the pressure of the system by increasing the volume.

15.40 $PCl_{5(g)} + \text{heat} \rightleftarrows PCl_{3(g)} + Cl_{2(g)}$

Changes that would drive the reaction to the left:

1. Cooling the reaction.
2. Addition of $PCl_{3(g)}$ to the system.
3. Addition of $Cl_{2(g)}$ to the system.
4. Increasing the pressure of the system by decreasing the volume
5. Removal of $PCl_{5(g)}$ from the system.

15.41 a) 15 [symbol] $\rightarrow$ 3 [symbol] + 12 [symbol] + 12 [symbol]

3 [symbol] $\rightleftarrows$ 12 [symbol] + 12 [symbol]

b) $K_{eq} = \dfrac{(p\ [symbol])_{eq}\,(p\ [symbol])_{eq}}{(p\ [symbol])_{eq}} = \dfrac{(12)(12)}{3} = 48$

15.42 $HClO_{(aq)} + H_2O_{(l)} \rightleftarrows H_3O^+{}_{(aq)} + ClO^-{}_{(aq)}$

$$K_{eq} = K_a = \frac{[H_3O^+]_{eq}[ClO^-]_{eq}}{[HClO]_{eq}}$$

15.43 Reaction: $H_{2(g)} + F_{2(g)} \rightleftarrows 2\ HF_{(g)}$ $\quad K_{eq} = \dfrac{(p_{HF})_{eq}^2}{(p_{H_2})_{eq}(p_{F_2})_{eq}} = 115$

	$H_{2(g)}$	$F_{2(g)}$	$HF_{(g)}$
Init. (atm)	3.00	3.00	0
Change (atm)	-x	-x	+2x
Equil. (atm)	3.00 - x	3.00 - x	2x

$$115 = \frac{(2x)^2}{(3.00 - x)^2} = \frac{4x^2}{(9.00 - 6.00x + x^2)}$$

$4x^2 = 115(9.00 - 6.00x + x^2) = 1035 - 690x + 115\ x^2$

$0 = 111x^2 - 690x + 1035$

$$x = \frac{-b \pm \sqrt{b^2 - 4ac}}{2a} = \frac{690 \pm \sqrt{(690)^2 - 4(111)(1035)}}{2(111)}$$

$$= \frac{690 \pm \sqrt{16560}}{222} = 3.688 \text{ or } 2.528 \text{ atm}$$

x = 2.53 atm because we would get negative partial pressures for $H_{2(g)}$ and $F_{2(g)}$ if we used x = 3.69 atm.

⇒Equilibrium partial pressures:

$(p_{H_2})_{eq} = 3.00 - 2.53 = 0.47$ atm $\qquad (p_{F_2})_{eq} = (p_{H_2})_{eq} = 0.47$ atm

$(p_{HF})_{eq} = 2(2.53) = 5.06$ atm

15.44 $4\ NH_{3(g)} + 5\ O_{2(g)} \rightleftarrows 4\ NO_{(g)} + 6\ H_2O_{(l)}$

$$K_{eq} = \frac{(p_{NO})_{eq}^4}{(p_{NH_3})_{eq}^4 (p_{O_2})_{eq}^5}$$

15.45 $2\ NO_{2(g)} \rightleftarrows N_2O_{4(g)}$

at 298 K: $\ln K_{eq} = \dfrac{-\Delta G°}{RT}$

$\Delta G° = \Delta G°(N_2O_4) - 2\ \Delta G°(NO_2)$

= 97.89 - 2(51.31) = -4.73 kJ/mol

$\ln K_{eq} = \dfrac{-(-4.73)(1000)}{(8.314)(298)} = +1.91$ $\qquad K_{eq} = 6.75\ atm^{-1}$

At 500 K: $\ln K_{eq} = \dfrac{-\Delta H°}{RT} + \dfrac{\Delta S°}{R}$ (Assuming $\Delta H \approx \Delta H°$ and $\Delta S \approx \Delta S°$)

$\Delta H° = \Delta H°(N_2O_4) - 2\ \Delta H°(NO_2) = 9.16 - 2(33.18) = -57.20$ kJ/mol

$\Delta S° = \Delta S°(N_2O_4) - 2\ \Delta S°(NO_2) = 304.29 - 2(240.06) = -175.83$ J/mol K

$\ln K_{eq} = \dfrac{-(-57.20)(1000)}{(8.314)(500)} + \dfrac{-175.83}{8.314} = -7.39$

$K_{eq} = 6.18 \times 10^{-4}$

15.46 $2\,NO_{2(g)} \rightleftarrows N_2O_{4(g)}$

a) Decreasing the partial pressure of $NO_{2(g)}$ means $Q > K_{eq}$ and the reaction proceeds to the left until $Q = K_{eq}$.

b) The solution to Problem 15.45 shows that the equilibrium is shifted to the left as the temperature is increased from 298 K to 500 K. Therefore, if the temperature is reduced (cut in half), the equilibrium would shift to the right.

c) Addition of $Ar_{(g)}$ does not change the equilibrium. It is a non reactant and does not change the partial pressures of $NO_{2(g)}$ or $N_2O_{4(g)}$.

15.47 $Ca^{2+}{}_{(aq)} + 3\,H_2O_{(l)} + CO_{2(g)} \rightleftarrows CaCO_{3(s)} + 2\,H_3O^{+}{}_{(aq)}$

$$K_{eq} = \frac{[H_3O^+]^2_{eq}}{[Ca^{2+}]_{eq}(p_{CO_2})_{eq}}$$

Concentration units: $Ca^{2+}{}_{(aq)}$ = M (moles per liter) $H_2O_{(l)}$ = X (mole fraction)

$CO_{2(g)}$ = atm (partial pressure) $CaCO_{3(s)}$ = X (mole fraction)

$H_3O^{+}{}_{(aq)}$ = M (moles per liter)

$$K_{eq} = \frac{M^2}{M\ atm} = M\ atm^{-1}$$

15.48 5 steps

1. Initial Species: $Na^{+}{}_{(aq)}$, $F^{-}{}_{(aq)}$, $BaF_{2(s)}$
2. Chemical reaction: $BaF_{2(s)} \rightleftarrows Ba^{2+}{}_{(aq)} + 2\,F^{-}{}_{(aq)}$
3. Equilibrium expression: $K_{sp} = [Ba^{2+}]_{eq}\,[F^-]^2_{eq}$
4. Equilibrium Concentrations:

Reaction:	$BaF_{2(s)}$	$\rightleftarrows$	$Ba^{2+}{}_{(aq)}$	+	$2\,F^{-}{}_{(aq)}$
Initial (M)			0		1.00×10^{-1}
Change (M)			+ x		+ 2x
Equilibrium (M)			x		$2x + 1.00 \times 10^{-1}$

5. Substitution and solution: From Appendix G, pK_{sp} for $BaF_2 = 6.00$,

$\therefore K_{sp} = 1.0 \times 10^{-6}\ M^3$

$K_{sp} = 1.0 \times 10^{-6} = [Ba^{2+}]_{eq}\,[F^-]^2_{eq} = (x)(2x + 1.00 \times 10^{-1})^2\ M^3$

Assume $1.00 \times 10^{-1} + 2x \cong 1.00 \times 10^{-1}$

$1.0 \times 10^{-6} = (x)(1.00 \times 10^{-1})^2$

$x = 1.00 \times 10^{-4}$ M $(2x + 1.00 \times 10^{-1} = 0.1002 \cong 0.100)$

grams of BaF_2 that dissolve:

$(1.00 \times 10^{-4}\ moles/L)(500\ mL)(1\ L/10^3\ mL)(175.3\ g\ BaF_2/mole)$

$= 8.76 \times 10^{-3}\ g\ BaF_2$

15.49 $K_{eq} = 25 = \frac{[\;][\;]}{[\;][\;]}$

The molecular picture requested will contain 10 of each product molecule and two of each reactant molecule when the system reaches equilibrium.

15.50 1. $CO_{2(g)}$, $C_{(s)}$

2. $CO_{2(g)} + C_{(s)} \rightleftarrows 2\ CO_{(g)}$

3. $K_{eq} = \frac{(p_{CO})^2_{eq}}{(p_{CO_2})_{eq}}$

4. Reaction:

	$CO_{2(g)}$	+	$C_{(s)}$	$\rightleftarrows$	$2\ CO_{(g)}$
Initial (atm)	0.464				0
Change (atm)	-x				+ 2x
Equilibrium (atm)	0.464 - x				2x

5. Total pressure = (0.464 atm - x) + (2x) = 0.746 atm

x = 0.746 atm - 0.464 atm = 0.282 atm

$(p_{CO_2})_{eq}$ = 0.464 atm - 0.282 atm = 0.182 atm

$(p_{CO})_{eq}$ = 2(0.282 atm) = 0.564 atm

$$K_{eq} = \frac{(0.564\ \text{atm})^2}{(0.182\ \text{atm})} = 1.75\ \text{atm}$$

15.51 Reaction:

	$N_{2(g)}$	+	$3\ H_{2(g)}$	$\rightleftarrows$	$2\ NH_{3(g)}$
Initial (atm)	5.0		3.0		0
Change (atm)	- x		-3 x		+ 2x
Equilibrium (atm)	5.0- x		3.0-3x		2x

$$K_{eq} = \frac{(p_{NH_3})^2_{eq}}{(p_{N_2})_{eq}(p_{H_2})^3_{eq}} = 2.81 \times 10^{-5}\ \text{atm}^{-2}$$

T = 472°C = 745 K

$$2.81 \times 10^{-5}\ \text{atm}^{-2} = \frac{(2x)^2}{(5.0 - x)(3.0 - 3x)^3}$$

K_{eq} is very small compared to the initial conc. of N_2 and H_2; therefore $5 - x \approx 5$ and $3 - 3x \approx 3$.

$$2.81 \times 10^{-5}\ \text{atm}^{-2} = \frac{4x^2}{5.0\ (3.0)^3} = \frac{4x^2}{135};\quad x = \sqrt{\frac{(2.81 \times 10^{-5})(135)}{4}} = 3.08 \times 10^{-2}$$

$(p_{NH_3})_{eq} = 2(3.08 \times 10^{-2}) = 6.16 \times 10^{-2}$ atm

$(p_{N_2})_{eq} = 5.0 - 3.08 \times 10^{-2} = 5.0$ atm

$(p_{H_2})_{eq} = 3.0 - 3(3.08 \times 10^{-2}) = 2.9$ atm

15.52 1. $C_{(s)}$, $CO_{2(g)}$, $CO_{(g)}$

2. $CO_{2(g)} + C_{(s)} \rightleftarrows 2\, CO_{(g)}$

3. $K_{eq} = \dfrac{(p_{CO})^2_{eq}}{(p_{CO_2})_{eq}} = 167.5 \text{ atm}$

4. $P = \dfrac{nRT}{V} = \dfrac{(0.500 \text{ mol})(0.0821 \text{ L atm mol}^{-1} \text{ K}^{-1})(1273 \text{ K})}{1 \text{ L}} = 52.3 \text{ atm}$

Reaction:	$CO_{2(g)} + C_{(s)} \rightleftarrows$	$2\, CO_{(g)}$
Initial (atm)	52.3	52.3
Change (atm)	- x	+ 2x
Equilibrium (atm)	52.3 - x	52.3 + 2x

5. $K_{eq} = 167.5 = \dfrac{(52.3 + 2x)^2}{(52.3 - x)}$

$(167.5)(52.3 - x) = (52.3 + 2x)^2$

$8760.25 - 167.5\, x = 2735.29 + 209.2\, x + 4x^2$

$0 = 4x^2 + 376.7\, x - 6024.96$

$$x = \frac{-b \pm \sqrt{b^2 - 4\, ac}}{2a} = \frac{-376.7 \pm \sqrt{(376.7)^2 - 4(4)(-6024.96)}}{(2)(4)} = \frac{-376.7 \pm 488.2}{8}$$

$= 13.9$ atm

Equilibrium total pressure = (52.3 atm - x) + (52.3 atm + 2x) = 104.6 atm + x
= 104.6 atm + 13.9 atm = 118.5 atm

15.53

Reaction:	$Sn_{(s)}$ +	$2\, H_{2(g)}$ $\rightleftarrows$	$SnH_{4(g)}$
Initial (atm)		2.00×10^2	0
Change (atm)		-2 x	+x
Equilibrium (atm)		$2.00 \times 10^2 - 2\, x$	x

$K_{eq} = \dfrac{(p_{SnH_4})_{eq}}{(p_{H_2})^2_{eq}} = 1.07 \times 10^{-33} \text{ atm}^{-1}$

$1.07 \times 10^{-33} = \dfrac{x}{(2 \times 10^2 - 2x)^2}$ $x = (1.07 \times 10^{-33})(2 \times 10^2)^2 = 4.28 \times 10^{-29}$ atm

$(p_{SnH_4})_{eq} = 4.28 \times 10^{-29}$ atm

$PV = nRT$

$n = \dfrac{PV}{RT} = \dfrac{(4.28 \times 10^{-29} \text{ atm})(10 \text{ L})}{(0.0821 \text{ L atm / K mol})(298 \text{ K})} = 1.75 \times 10^{-29} \text{ mol}$

# of molecules = nN_A = (1.75×10^{-29} mole)(6.022×10^{23} molecules/mole)
= 1.05×10^{-5} molecules of SnH_4

15.54 1. $CCl_{4(g)}$ (Assume initial pressure is 1.00 atm; 3 significant figures.)

2. $CCl_{4(g)} \rightleftarrows C_{(s)} + 2\ Cl_{2(g)}$

3. $K_{eq} = \dfrac{\left(p_{Cl_2}\right)^2_{eq}}{\left(p_{CCl_4}\right)_{eq}}$

4. Reaction:

	$CCl_{4(g)}$	$\rightleftarrows$	$C_{(s)} + 2\ Cl_{2(g)}$
Initial (atm)	1.00		0
Change (atm)	- x		+ 2x
Equilibrium (atm)	1.00 - x		2x

5. 1.35 atm = (1.00 atm - x) + 2x 0.35 atm = x

$(p_{CCl_4})_{eq}$ = 1.00 atm - 0.35 atm = 0.65 atm $(p_{Cl_2})_{eq}$ = 2(0.35 atm) = 0.70 atm

$$K_{eq} = \frac{(0.70\ \text{atm})^2}{(0.65\ \text{atm})} = 0.75\ \text{atm}$$

15.55

Reaction:	$CCl_{4(g)}$	$\rightleftarrows$	$2\ Cl_{2(g)}$	+	$C_{(g)}$
Initial	0.325		0.35		
Change	- x		+ 2x		
Equilibrium	0.325 - x		0.35 + 2x		

$$0.75\ \text{atm} = \frac{(0.35\ +\ 2x)^2}{(0.325 - x)}$$ $$0.24375 - 0.75\,x = 0.1225 + 1.40 + 4x^2$$

$$4x^2 + 2.15\,x - 0.12125 = 0$$ $$x = \frac{-2.15 \pm \sqrt{4.6225\ +\ 1.94}}{8}$$ $$x = 0.0515$$

$(p_{Cl_2})_{eq}$ = [0.35 + 2(0.0515)] atm = 0.45 atm

$(p_{CCl_4})_{eq}$ = [0.325 - 0.0515] atm = 0.27 atm

15.56 First step: $H_2SO_4 \underset{k_{-1}}{\overset{k_1}{\rightleftarrows}} SO_3 + H_2O$

$$k_1[H_2SO_4]_{eq} = k_{-1}[SO_3]_{eq}$$ $$\frac{k_1}{k_{-1}} = \frac{[SO_3]_{eq}[H_2O]_{eq}}{[H_2SO_4]_{eq}}$$

Second step: $SO_3 + C_6H_6 \underset{k_{-2}}{\overset{k_2}{\rightleftarrows}} C_6H_6SO_3$

$$k_2[SO_3]_{eq}\,[C_6H_6]_{eq} = k_{-2}[C_6H_6SO_3]_{eq}$$ $$\frac{k_2}{k_{-2}} = \frac{[C_6H_6SO_3]_{eq}}{[SO_3]_{eq}[C_6H_6]_{eq}}$$

(continued)

(15.56 continued)

Third step: $C_6H_6SO_3 + H_2O \underset{k_{-3}}{\overset{k_3}{\rightleftharpoons}} C_6H_6SO_3^- + H_3O^+$

$k_3[H_2O]_{eq}[C_6H_6SO_3]_{eq} = k_{-3}[C_6H_6SO_3^-]_{eq}[H_3O^+]_{eq}$

$$\frac{k_3}{k_{-3}} = \frac{[C_6H_5SO_3^-]_{eq}[H_3O^+]_{eq}}{[C_6H_6SO_3]_{eq}}$$

$$\left(\frac{k_1}{k_{-1}}\right)\left(\frac{k_2}{k_{-2}}\right)\left(\frac{k_3}{k_{-3}}\right) = \frac{[SO_3]_{eq}[H_2O]_{eq}[C_6H_6SO_3]_{eq}[C_6H_5SO_3^-]_{eq}[H_3O^+]_{eq}}{[H_2SO_4]_{eq}[SO_3]_{eq}[C_6H_6]_{eq}[H_2O]_{eq}[C_6H_6SO_3]_{eq}}$$

$$= \frac{[C_6H_5SO_3^-]_{eq}[H_3O^+]_{eq}}{[H_2SO_4]_{eq}[C_6H_6]_{eq}} = K_{eq}$$

$$K_{eq} = \left(\frac{k_1}{k_{-1}}\right)\left(\frac{k_2}{k_{-2}}\right)\left(\frac{k_3}{k_{-3}}\right)$$

15.57 a) $BeO_{(s)} + H_2O_{(l)} \rightleftharpoons Be^{2+}{}_{(aq)} + 2\ OH^-{}_{(aq)}$

$K_{eq} = [Be^{2+}]_{eq}[OH^-]^2_{eq}\ M^3$

b) $CO_{2(g)} + 2\ H_2O_{(l)} \rightleftharpoons HCO_3^-{}_{(aq)} + H_3O^+{}_{(aq)}$

$$K_{eq} = \frac{[HCO_3^-]_{eq}[H_3O]^+_{eq}}{(p_{CO_2})_{eq}} M^2\ atm^{-1}$$

c) $NH_{3(aq)} + CH_3CO_2H_{(aq)} \rightleftharpoons NH_4^+{}_{(aq)} + CH_3CO_2^-{}_{(aq)}$

$$K_{eq} = \frac{[NH_4^+]_{eq}[CH_3CO_2^-]_{eq}}{[NH_3]_{eq}[CH_3CO_2H]_{eq}}$$

d) $3\ H_2S_{(g)} + 6\ H_2O_{(l)} + 2\ Fe^{3+}{}_{(aq)} \rightleftharpoons Fe_2S_{3(s)} + 6\ H_3O^+{}_{(aq)}$

$$K_{eq} = \frac{[H_3O^+]^6_{eq}}{(p_{H_2S})^3_{eq}[Fe^{3+}]^2} M^4\ atm^{-3}$$

15.58 a) 1. $HI_{(g)}$

2. $2\ HI_{(g)} \rightleftarrows H_{2(g)} + I_{2(g)}$

3. $$K_{eq} = \frac{(p_{H_2})_{eq}(p_{I_2})_{eq}}{(p_{HI})^2_{eq}}$$

4.

Reaction:	$2\ HI_{(g)}$	$\rightleftarrows$	$H_{2(g)}$	+	$I_{2(g)}$
Initial (atm)	40.9 (1 mol)		0		0
Change (atm)	- 2x		+ x		+ x
Equil. (atm)	40.9 - 2x = 26.01		x = 7.44		x = 7.44

$$(p_{HI})_{init} = \frac{nRT}{V} = \frac{(1.00\ mol)(0.0821\ L\ atm\ mol^{-1}\ K^{-1})(498\ K)}{1.00\ L} = 40.9\ atm$$

$$(p_{I_2})_{eq} = (p_{H_2})_{eq} = \frac{(0.182\ mol)(0.0821\ L\ atm\ mol^{-1}\ K^{-1})(498\ K)}{1.00\ L} = 7.44\ atm$$

5. $$K_{eq} = \frac{(7.44\ atm)(7.44\ atm)}{(26.0\ atm)^2} = 0.0818 \quad or \quad \frac{(0.182\ mol)(0.182\ mol)}{(0.636\ mol)^2} = 0.0819$$

b) $2\ HI_{(g)} \rightleftarrows H_{2(g)} + I_{2(g)}$

$\Delta H \approx \Delta H^\circ_{rxn} = (1)(0.0\ kJ/mol) + (1)(62.438\ kJ/mol) - [(2)(26.48\ kJ/mol)]$
$= +9.48\ kJ/mol$

$\Delta S \approx \Delta S^\circ_{rxn} = (1)(130.684\ J/mol\ K) + (1)(260.69\ J/mol\ K) - [(2)(206.594\ J/mol\ K)]$
$= -21.814\ J/mol\ K$

$$\ln K_{eq} = -\frac{\Delta H}{RT} + \frac{\Delta S}{R} = -\frac{(9.48\ kJ/mol)(10^3\ J/kJ)}{(8.314\ J/mol\ K)(600 + 273)K} + \frac{(-21.814\ J/mol\ K)}{(8.314\ J/mol\ K)}$$

$= -1.31 - 2.624 = -3.93$

$K_{eq} = e^{-3.93} = 0.0196$

15.59 Reaction: $MgF_{2(s)} + H_2O_{(l)} \rightleftarrows Mg^{2+}_{(aq)} + 2\ F^-_{(aq)}$

$K_{eq} = K_{sp} = [Mg^{2+}]_{eq}[F^-]^2_{eq}$

$K_{sp} = (1.14 \times 10^{-3}\ M)[2(1.14 \times 10^{-3}\ M)]^2 = 5.93 \times 10^{-9}\ M^3$

15.60 $K_{eq} = \frac{[●○]^2}{[●●][○○]} = 4$

	●●	+ ○○	$\rightleftarrows$	2 ●○
Initial	6	12		0
Change	- x	- x		+ 2x
Equil.	6 - x	12 - x		+ 2x

$$K_{eq} = \frac{(2x)^2}{(6 - x)(12 - x)} \qquad x = 4$$

The requested molecular drawing needs to show 2 ●●, 8 ○○ and 8 ●○.

15.61 $CO_{2(g)} + 2\,OH^-_{(aq)} \rightleftarrows CO_3^{2-}{}_{(aq)} + H_2O_{(l)}$

a) $K_{eq} = \dfrac{[CO_3^{2-}]_{eq}}{(p_{CO_2})_{eq}[OH^-]^2_{eq}}$

b) $Q = \dfrac{[CO_3^{2-}]}{(p_{CO_2})[OH^-]^2}$, so adding Na_2CO_3 makes $Q > K_{eq}$.

When this happens, reaction products are consumed and reactants are produced until K_{eq} is restored. Thus, the pressure of CO_2 in this system will **increase** if $Na_2CO_{3(s)}$ is dissolved in the solution.

c) Bubbling $HCl_{(g)}$ through the solution would cause the concentration of OH^- to decrease, $HCl + OH^- \rightarrow H_2O + Cl^-$; the reaction would shift to the left and p_{CO_2} would increase.

15.62 a) $[PCl_5]_i = \dfrac{2.00 \text{ g } PCl_5}{(208.22 \text{ g } PCl_5 / \text{mol})(3.00 \text{ L})} = 3.20 \times 10^{-3}$ M

Reaction:	$PCl_{5(g)}$	$\rightleftarrows$	$PCl_{3(g)}$ +	$Cl_{2(g)}$
Initial (M)	3.20×10^{-3}		0	0
Change (M)	- x		+x	+x
Equilibrium (M)	3.20×10^{-3} - x		x	x

$$K_{eq} = 2.24 \times 10^{-2} \text{ M} = \frac{[Cl_2]_{eq}[PCl_3]_{eq}}{[PCl_5]_{eq}} = \frac{x^2}{3.20 \times 10^{-3} \text{ M} - x}$$

$(2.24 \times 10^{-2} \text{ M})(3.20 \times 10^{-3} \text{ M} - x) = x^2$
$0 = x^2 + 2.24 \times 10^{-2} \text{ M } x - 7.168 \times 10^{-5} \text{ M}^2$

$$x = \frac{-2.24 \times 10^{-2} \pm \sqrt{(2.24 \times 10^{-2})^2 - 4(1)(-7.168 \times 10^{-5})}}{2(1)}$$

$$= \frac{-2.24 \times 10^{-2} \pm 2.808 \times 10^{-2}}{2} = 2.84 \times 10^{-3} \text{ M}$$

$(p_{Cl_2})_{eq} = (2.84 \times 10^{-3} \text{ mol/L})(0.0821 \text{ L atm mol}^{-1} \text{ K}^{-1})(500 \text{ K}) = 0.117$ atm

Note: The K_{eq} given is the value for concentration of the gases in mol/L (M), not in atm.

b) $K_{eq} = 33.3 \text{ M} = \dfrac{x^2}{3.20 \times 10^{-3} \text{ M} - x}$

$33.3 \text{ M } (3.20 \times 10^{-3} \text{ M} - x) = x^2$
$0 = x^2 + 33.3 \text{ x M} - 0.10656 \text{ M}^2$

$$x = \frac{-33.3 \pm \sqrt{(33.3)^2 - (4)(1)(-0.10656)}}{2(1)} = \frac{-33.3 \pm 33.306}{2} = 3.20 \times 10^{-3}$$

(continued)

(15.62 continued)

Reaction:	$PCl_{5(g)}$	$\rightleftarrows$	$PCl_{3(g)}$ +	$Cl_{2(g)}$
Initial (M)	3.20×10^{-3}		0	0
Change (M)	-3.20×10^{-3}		$+3.20 \times 10^{-3}$	$+3.20 \times 10^{-3}$
Conc. at completion (M)	0		3.20×10^{-3}	3.20×10^{-3}
Change (M)	+x		-x	-x
Equilibrium (M)	x		3.20×10^{-3} -x	3.20×10^{-3} -x

$$K_{eq} = 33.3\ M = \frac{(3.20 \times 10^{-3}\ M - x)^2}{x}$$

$$33.3x\ M = (3.20 \times 10^{-3}\ M - x)^2 \approx (3.20 \times 10^{-3}\ M)^2$$

i.e., $3.20 \times 10^{-3} - x \approx 3.20 \times 10^{-3}$

$$x = \frac{(3.20 \times 10^{-3}\ M)^2}{33.3\ M} = 3.08 \times 10^{-7}\ M$$

$$[Cl_2] = 3.20 \times 10^{-3}\ M - x = 3.20 \times 10^{-3}\ M - 3.08 \times 10^{-7} \approx 3.20 \times 10^{-3}\ M$$

$$(p_{Cl_2})_{eq} = (3.20 \times 10^{-3}\ mol/L)(0.0821\ L\ atm\ mol^{-1}\ K^{-1})(760\ K) = 0.200\ atm$$

c) $$\ln\left(\frac{k_{T_2}}{k_{T_1}}\right) = \frac{\Delta H°}{R}\left(\frac{1}{T_1} - \frac{1}{T_2}\right)$$

$$\Delta H° = \frac{R \ln\left(\frac{k_{T_2}}{k_{T_1}}\right)}{\left(\frac{1}{T_1} - \frac{1}{T_2}\right)} = \frac{(8.314\ J/mol\ K) \ln\left(\frac{33.3\ M}{2.24 \times 10^{-2}\ M}\right)}{\left(\frac{1}{500\ K} - \frac{1}{760\ K}\right)}$$

$$= \frac{(8.314\ J/mol\ K)(7.304)}{(0.00200\ K^{-1} - 0.00132 K^{-1})} = 89302\ J/mol = 89.3\ kJ/mol$$

15.63 $Hg_{(g)} + Hg_2Cl_{2(s)} \rightleftarrows Hg_2Cl_{2(s)}$

a) $$\ln K_{eq} = \frac{-\Delta G°}{RT}$$

$\Delta G° = \Delta G°(Hg_2Cl_{2(s)}) - \Delta G°(Hg) - \Delta G°(HgCl_2) = -210.745 - 31.82 - (-178.6)$
$= -63.97$ kJ/mol

$$\ln K_{eq} = \frac{-(-63.97)(1000)}{(8.314)(298)} = 25.82;\ K_{eq} = 1.63 \times 10^{11}\ atm^{-1}$$

b) $\Delta H° = -265.22 - 61.32 - (-224.3) = -102.24$ kJ/mol
$\Delta S° = 192.5 - 174.96 - 146.0 = -128.46$ J/mol K
(Assuming $\Delta H \approx \Delta H°$ and $\Delta S \approx \Delta S°$)
$\Delta G = \Delta H - T\Delta S = 0$ at equilibrium $\Rightarrow T\Delta S = \Delta H$

$$T = \frac{\Delta H°}{\Delta S°} = \frac{(-102.24)(1000)}{-128.46} = 796\ K$$

c) $$\ln K_{eq} = \frac{-\Delta H°}{RT} + \frac{\Delta S°}{R} \qquad \ln K_{eq} = \frac{-(-102.24)(1000)}{(8.314)(1000\ K)} + \frac{-128.46}{8.314} = -3.154$$

$K_{eq} = 4.269 \times 10^{-2}\ atm^{-1}$

15.64 $C_{(s)} + 2\,H_2O_{(g)} \rightleftarrows CO_{2(g)} + 2\,H_{2(g)}$

Equilibrium (atm) 280 atm x 2x

$$K_{eq} = 0.38\text{ atm} = \frac{(p_{CO_2})_{eq}(p_{H_2})_{eq}^2}{(p_{H_2O})_{eq}^2} = \frac{x(2x)^2}{(280\text{ atm})^2}$$

$4x^3 = (280\text{ atm})^2\,(0.38\text{ atm}) = 29792\text{ atm}^3$

$x^3 = 7448\text{ atm}^3$ $x = 19.5\text{ atm}$

$(p_{CO_2})_{eq} = x = 19.5\text{ atm}$ $(p_{H_2})_{eq} = 2x = 2(19.5\text{ atm}) = 39.0\text{ atm}$

15.65 $C_{(s)} + 2\,H_2O_{(g)} \rightleftarrows CO_{2(g)} + 2\,H_{2(g)}$

$$K_{eq} = \frac{(p_{CO_2})_{eq}(p_{H_2})_{eq}^2}{(p_{H_2O})_{eq}^2} = 0.38\text{ atm}$$

The research will fail because catalysts do not affect K_{eq} (c.f. p. 745).

15.66 a) $CO + H_2 \rightarrow HCHO$

$\Delta H \approx \Delta H^\circ_{rxn} = (1)(-108.57\text{ kJ/mol}) - [(1)(-110.525\text{ kJ/mol}) + (1)(0.0\text{ kJ/mol})]$
$= 1.955\text{ kJ/mol}$

$\Delta S \approx \Delta S^\circ_{rxn} = (1)(218.77\text{ J/mol K}) - [(1)(197.674\text{ J/mol K}) + (1)(130.684\text{ J/mol K})]$
$= -109.588\text{ J/mol K}$

$$\ln K_{eq} = -\frac{(1.955\text{ kJ/mol})(10^3\text{ J/kJ})}{(8.314\text{ J/mol K})(500\text{ K})} + \frac{-109.588\text{ J/mol K}}{(8.314\text{ J/mol K})}$$

$$= -0.470 - 13.18 = -13.65$$

$K_{eq} = e^{-13.65} = 1.2 \times 10^{-6}\text{ atm}^{-1}$

b) $2\,CO + 3\,H_2 \rightarrow H_2O + CH_3CHO$

$\Delta H \approx \Delta H^\circ_{rxn} = (1)(-241.818\text{ kJ/mol}) + (1)(-166.19\text{ kJ/mol}) - [(2)(-110.525\text{ kJ/mol}) + (3)(0.0\text{ kJ/mol})] = -186.958\text{ kJ/mol}$

$\Delta S \approx \Delta S^\circ_{rxn} = (1)(188.825\text{ J/mol K}) + (1)(250.3\text{ J/mol K}) - [(2)(197.674\text{ J/mol K}) + (3)(130.684\text{ J/mol K})] = -348.275\text{ J/mol K}$

$$\ln K_{eq} = -\frac{(-186.958\text{ kJ/mol})(10^3\text{ J/kJ})}{(8.314\text{ J/mol K})(500\text{ K})} + \frac{(-348.275\text{ J/mol K})}{(8.314\text{ J/mol K})}$$

$$= 44.97 - 41.89 = 3.08$$

$K_{eq} = e^{3.08} = 21.8\text{ atm}^{-3}$

(continued)

(15.66 continued)

c) $3\ CO + 6\ H_2 \rightarrow 3\ H_2O + C_3H_6$

$\Delta H \approx \Delta H^\circ_{rxn}$ = (3)(-241.818 kJ/mol) + (1)(20.41 kJ/mol) - [(3)(-110.525 kJ/mol) + (6)(0.0 kJ/mol)] = -373.469 kJ/mol

$\Delta S \approx \Delta S^\circ_{rxn}$ = (3)(188.825 J/mol K) + (1)(226.9 J/mol K) -[(3)(197.674 J/mol K) + (6)(130.684 J/mol K)] = -583.751 J/mol K

$$\ln K_{eq} = -\frac{(-373.469 \text{ kJ / mol})(10^3 \text{ J / kJ})}{(8.314 \text{ J / mol K})(500 \text{ K})} + \frac{(-583.751 \text{ J / mol K})}{(8.314 \text{ J / mol K})}$$

$$= 89.84 - 70.21 = 19.63$$

$K_{eq} = e^{19.63} = 3.4 \times 10^8$ atm^{-5}

15.67 The Haber synthesis is used to produce ammonia that in turn is used either directly or indirectly as much needed fertilizer. In the United States 14.5 million tons of fertilizer are produced via the Haber method. Without these fertilizers, food production would be greatly reduced.

15.68 $CO_{(g)} + H_2O_{(g)} \rightleftarrows CO_{2(g)} + H_{2(g)}$

if $K_{eq} = 1$ $\quad \ln K_{eq} = 0 \therefore \frac{\Delta S}{R} = \frac{\Delta H}{RT} \quad \Delta S = \frac{\Delta H}{T} \quad T = \frac{\Delta H}{\Delta S}$

$\Delta H \approx \Delta H^\circ_{rxn}$ = (1)(-393.509 kJ/mol) + (1)(0.0 kJ/mol) - [(1)(-110.525 kJ/mol) + (1)(-241.818 kJ/mol)] = -41.166 kJ/mol

$\Delta S \approx \Delta S^\circ_{rxn}$ = (1)(213.74 J/mol K) + (1)(130.684 J/mol K) -[(1)(197.674 J/mol K) + (1)(188.825 J/mol K)] = -42.075 J/mol K

$$T = \frac{\Delta H}{\Delta S} = \frac{(-41.166 \text{ kJ / mol})(10^3 \text{ J / kJ})}{(-42.075 \text{ J / mol K})} = 978.4 \text{ K (or 705.2°C)}$$

15.69 a) Haber synthesis:

$$CH_{4(g)} + H_2O_{(g)} \underset{800\,°C}{\overset{\text{Ni cat.}}{\rightleftarrows}} CO_{(g)} + 3\ H_{2(g)}$$

$$CO_{(g)} + H_2O_{(g)} \underset{250\,°C}{\overset{Fe_2O_3/Cr_2O_3}{\rightleftarrows}} CO_{2(g)} + H_{2(g)}$$

$$N_{2(g)} + 3\ H_{2(g)} \underset{450\,°C/270\text{ atm}}{\overset{\text{Fe powder doped with } K_2O/Al_2O_3}{\rightleftarrows}} 2\ NH_{3(g)}$$

b) Ostwald process:

$$4\ NH_{3(g)} + 5\ O_{2(g)} \xrightarrow[1200\text{ K}]{\text{Pt gauze}} 4\ NO_{(g)} + 6\ H_2O_{(g)}$$

$$2\ NO_{(g)} + O_{2(g)} \rightleftarrows 2\ NO_{2(g)}$$

$$3\ NO_{2(g)} + H_2O_{(l)} \rightarrow 2\ HNO_{3(aq)} + NO_{(g)}$$

(continued)

(15.69 continued)

c) Contact process:

$S_{(s)} + O_{2(g)} \rightarrow SO_{2(g)}$

$2\ SO_{2(g)} + O_{2(g)} \xrightleftharpoons[\text{high T}]{V_2O_5} 2\ SO_{3(g)}$

$SO_{3(g)} + H_2SO_{4(l)} \rightarrow H_2S_2O_{7(l)}$

$H_2S_2O_{7(l)} + H_2O_{(l)} \rightarrow 2\ H_2SO_{4(l)}$

15.70 at 25°C $\dfrac{0.030\text{ g } Sr(IO_3)_2}{(437.43\text{ g } Sr(IO_3)_2\text{ / mol})(0.100\text{ L})} = 6.86 \times 10^{-4}\text{ M}$

Reaction:	$Sr(IO_3)_{2(s)}$ $\rightleftarrows$	$Sr^{2+}_{(aq)}$ +	$2\ IO^-_{3(aq)}$
Initial (M)		0	0
Change (M)		$+6.86 \times 10^{-4}$	$+1.37 \times 10^{-3}$
Equilibrium (M)		6.86×10^{-4}	1.37×10^{-3}

$K_{sp} = [Sr^{2+}]_{eq}[IO_3^-]^2_{eq} = [6.86 \times 10^{-4}\text{ M}][1.37 \times 10^{-3}\text{ M}]^2 = 1.29 \times 10^{-9}\text{ M}^3$

$\Delta G^\circ_{298} = -\ RT \ln K_c = -(8.314\text{ J/mol K})(298\text{ K}) \ln (1.29 \times 10^{-9}) = 50712\text{ J/mol}$ or 50.7 kJ/mol

at 100°C $\dfrac{0.80\text{ g } Sr(IO_3)_2}{(437.43\text{ g } Sr(IO_3)_2\text{ / mol})(0.100\text{ L})} = 1.83 \times 10^{-2}\text{ M}$

$[Sr^{2+}]_{eq} = 1.83 \times 10^{-2}\text{ M}$ $\quad [IO_3^-]_{eq} = 3.66 \times 10^{-2}\text{ M}$

$K_{sp} = (1.83 \times 10^{-2}\text{ M})(3.66 \times 10^{-2}\text{ M})^2 = 2.45 \times 10^{-5}\text{ M}^3$

$\Delta G^\circ_{373} = -\ (8.314\text{ J/mol K})(373\text{ K}) \ln (2.45 \times 10^{-5}) = 3.29 \times 10^4\text{ J/mol}$ or 32.9 kJ/mol

15.71 $Br_{2(g)} + I_{2(g)} \rightarrow 2\ IBr_{(g)}$ $\quad K_{eq} = \dfrac{(p_{IBr})^2_{eq}}{(p_{Br_2})_{eq}(p_{I_2})_{eq}} = 322$

$[p_{Br_2}]_{eq} = 0.512\text{ atm}$ $\quad [p_{I_2}]_{eq} = 0.327\text{ atm}$

$(p_{IBr})_{eq} = \sqrt{322(p_{Br_2})_{eq}(p_{I_2})_{eq}} = \sqrt{322(0.512)(0.327)} = 7.34\text{ atm}$

15.72 $K_{eq} = \dfrac{[Ca^{2+}]_{eq}[HCO_3^-]^2_{eq}}{(p_{CO_2})_{eq}} = 1.56 \times 10^{-8}\text{ M}^3\text{ atm}^{-1}$

if $[Ca^{2+}]_{eq} = x$, then $[HCO_3^-]_{eq} = 2x$

$K_{eq} = \dfrac{(x)(2x)^2}{(3.2 \times 10^{-4}\text{ atm})} = 1.56 \times 10^{-8}\text{ M}^3\text{ atm}^{-1}$

$4x^3 = (1.56 \times 10^{-8}\text{ M}^3\text{ atm}^{-1})(3.2 \times 10^{-4}\text{ atm}) = 4.992 \times 10^{-12}\text{ M}^3$

$x^3 = 1.25 \times 10^{-12}\text{ M}^3$ $\quad x = 1.1 \times 10^{-4}\text{ M} = [Ca^{2+}]_{eq}$

15.73 $2\,SO_{2(g)} + O_{2(g)} \rightleftarrows 2\,SO_{3(g)}$

$$K_{eq} = \frac{\left(p_{SO_3}\right)_{eq}^2}{\left(p_{SO_2}\right)_{eq}^2\left(p_{O_2}\right)_{eq}} = \frac{(5)^2}{(5)^2(4)} = 0.25\ \text{atm}^{-1}$$

New conditions: 3 SO_3, 5 SO_2, 4 O_2

$$Q = \frac{\left(p_{SO_3}\right)^2}{\left(p_{SO_2}\right)^2\left(p_{O_2}\right)} = \frac{(3)^2}{(5)^2(4)} = 9.0 \times 10^{-2}\ \text{atm}^{-1}$$

$\Rightarrow Q < K_{eq}$ so the reaction consumes reactants and forms products until K_{eq} is reestablished.

15.74 Equilibrium at 1100 K: 4 O_2, 5 SO_3, 5 SO_2

Equilibrium at 1300 K: 5 O_2, 3 SO_3, 7 SO_2

Upon increase in temperature the equilibrium: $2\,SO_2 + O_2 \rightleftarrows 2\,SO_3$ shifts to the left. Upon addition of heat, the reaction would shift in the endothermic direction (to the left). Therefore, the reaction is exothermic (to the right).

15.75 $$Q = \frac{\left(p_{SO_3}\right)^2}{\left(p_{SO_2}\right)^2\left(p_{O_2}\right)} = \frac{(9)^2}{(9)^2(4)} = 0.25$$

$\Rightarrow Q = K_{eq}$ so there is no change in the position of the equilibrium.

15.76 $$\frac{1.75\ \text{g Butyric acid}}{(88.0\ \text{g Butyric acid / mol})(0.125\ \text{L})} = 0.159\ \text{M}$$

Reaction:

	$CH_3CH_2CH_2COOH_{(aq)} + H_2O_{(l)}$	$\rightleftarrows$	$CH_3CH_2CH_2COO^-_{(aq)}$	+ $H_3O^+_{(aq)}$
Initial (M)	0.159		0	0
Change (M)	-x		+x	+x
Equilibrium (M)	0.159 - x		x	x

$$K_a = 1.52 \times 10^{-5}\ \text{M} = \frac{[CH_3CH_2CH_2COO^-]_{eq}[H_3O^+]_{eq}}{[CH_3CH_2CH_2COOH]_{eq}}$$

$$= \frac{x^2}{0.159\ \text{M} - x} \approx \frac{x^2}{0.159\ \text{M}}$$

$x = 1.55 \times 10^{-3}\ \text{M} = [H_3O^+]_{eq}$

15.77 $2\ AcOH \rightleftarrows (AcOH)_2$

$$K_{eq} = \frac{\left(p_{(AcOH)_2}\right)_{eq}}{\left(p_{AcOH}\right)_{eq}^2} = 3.72\ atm^{-1}$$

$(p_{AcOH})_{eq} + (p_{(AcOH)_2})_{eq} = 0.75$ atm

$(p_{AcOH})_{eq} = 0.75$ atm - $(p_{(AcOH)_2})_{eq}$

$$\left(p_{AcOH}\right)_{eq} = \sqrt{\frac{\left(p_{(AcOH)_2}\right)_{eq}}{3.72}} = 0.75 - \left(p_{(AcOH)_2}\right)_{eq}$$

$x/3.72 = (0.75 - x)^2 = 0.563 - 1.50x + x^2$

$x = 3.72(x^2 - 1.50x + 0.563) = 3.72x^2 - 5.58x + 2.093$

$0 = 3.72x^2 - 6.58x + 2.093$

$$x = \frac{6.58 \pm \sqrt{(6.58)^2 - 4(3.72)(2.093)}}{2(3.72)} = \frac{6.58 \pm 3.487}{7.44} = 1.35 \text{ or } 0.416$$

a) $(p_{dimer})_{eq} = 0.416$ atm

b) K_{eq} is lower at 200°C

15.78 The solubility (K_{sp}) of an exothermic reaction decreases with increasing temperature. The solubility of sodium sulfate decreases with increasing temperature. Therefore, the dissolution of sodium sulfate is an exothermic reaction. The temperature of the water in the thermos flask will increase as sodium sulfate is dissolved or $\ln K_{eq} = -\Delta H/RT + \Delta S/R$. If a reaction is exothermic, $\Delta H°$ is negative and the first term is positive. As T increases, this term decreases and K_{eq} decreases. This is true for sodium sulfate dissolution. Therefore, sodium sulfate dissolution is exothermic. The water in the thermos flask will increase in temperature.

15.79

Reaction:	$2\ Ef_{(g)}$ +	$3\ N_{2(g)}$	$\rightleftarrows$	$2\ EfN_{3(g)}$
Initial (atm)	0.75	1.00		0
Change (atm)	-2x	-3x		+2x
Equilibrium (atm)	0.75 - 2x	1.00 - 3x		2x

$(0.75 - 2x) + (1.00 - 3x) + 2x = 0.85$ $\qquad 1.75 - 3x = 0.85$

$0.90 = 3x$ $\qquad x = 0.30$ atm

$$K_{eq} = \frac{\left(p_{EfN_3}\right)_{eq}^2}{\left(p_{Ef}\right)_{eq}^2\left(p_{N_2}\right)_{eq}^3} = \frac{(2x)^2}{(0.75 - 2x)^2(1.00 - 3x)^3} = \frac{(0.60)^2}{(0.15)^2(0.10)^3}$$

$= 1.6 \times 10^4\ atm^{-3}$

15.80 $[Ag^+]_i = \frac{(250 \text{ mL})(0.200 \text{ M})}{(600 \text{ mL})} = 0.0833 \text{ M}$

$[CO_3^{2-}]_i = \frac{(350 \text{ mL})(0.300 \text{ M})}{(600 \text{ mL})} = 0.175 \text{ M}$

Reaction:	$Ag_2CO_{3(s)}$	$\rightleftarrows$	$2\ Ag^+_{(aq)}$	+	$CO_3^{2-}{}_{(aq)}$
Initial (M)			0.0833		0.175
Change (M)	+0.0252 moles (solid)		-0.0833		-0.042
Completion (M)	0.0252 moles (solid)		0		0.133
Change (M)	-0.600 y moles		+2y		+y
Equilibrium (M)	0.0252 moles - 0.600y		2y		y + 0.133

$K_{sp} = [Ag^+]^2_{eq}[CO_3^{2-}]_{eq} = (2y)^2(y + 0.133 \text{ M}) \cong 4y^2(0.133 \text{ M}) = 8.2 \times 10^{-12} \text{ M}^3$

$y = 3.9 \times 10^{-6} \text{ M}$

$[Ag^+]_{eq} = 2y = 2(3.9 \times 10^{-6} \text{ M}) = 7.8 \times 10^{-6} \text{ M}$

$$Ag_2CO_{3(s)} = \left[\frac{0.0833 \text{ mol } Ag^+}{L} - \frac{7.8 \times 10^{-6} \text{ mol } Ag^+}{L}\right] \times 0.600 \text{ L}$$

$$\times \frac{1 \text{ mol } Ag_2CO_3}{2 \text{ mol } Ag^+} \times \frac{275.73 \text{ g } Ag_2CO_3}{\text{mol}} = 6.9 \text{ g } Ag_2CO_{3(s)}$$

CHAPTER 16: AQUEOUS EQUILIBRIA

16.1 a) H_2O, CH_3CO_2H
b) NH_4^+, Cl^-, H_2O
c) K^+, Cl^-, H_2O
d) Na^+, $CH_3CO_2^-$, H_2O
e) Na^+, OH^-, H_2O

16.2 a) HClO and H_2O
b) Ca^{2+}, Br^- and H_2O
c) K^+, ClO^- and H_2O
d) H_3O^+, NO_3^- and H_2O
e) HCN and H_2O

16.3 a) $CH_3CO_2H_{(l)} + H_2O_{(l)} \rightleftarrows CH_3CO_2^-{}_{(aq)} + H_3O^+{}_{(aq)}$ $K_{eq} = K_a = 1.8 \times 10^{-5}$ M
b) $NH_4^+ + H_2O_{(l)} \rightleftarrows NH_{3(aq)} + H_3O^+$ $K_{eq} = K_a = 5.6 \times 10^{-10}$ M
c) $KCl_{(s)} + H_2O_{(l)} \rightleftarrows K^+{}_{(aq)} + Cl^-{}_{(aq)}$ $H_2O + H_2O \rightleftarrows H_3O^+ + OH^-$
$K_w = 1 \times 10^{-14}$
d) $CH_3CO_2^- + H_2O_{(l)} \rightleftarrows CH_3CO_2H_{(aq)} + OH^-{}_{(aq)}$ $K_{eq} = K_w/K_a = 5.6 \times 10^{-10}$ M
e) $OH^- + H_2O_{(l)} \rightleftarrows H_2O_{(l)} + OH^-{}_{(aq)}$ $K_{eq} = 1$
Ranking of K_{eq}'s: (e) > (a) > (b) = (d) > (c)

16.4 a) $HClO + H_2O \rightleftarrows H_3O^+ + ClO^-$ $K_a = 3.0 \times 10^{-8}$
$H_2O + H_2O \rightleftarrows H_3O^+ + OH^-$ $K_w = 1 \times 10^{-14}$
b) $H_2O + H_2O \rightleftarrows H_3O^+ + OH^-$ $K_w = 1 \times 10^{-14}$
c) $ClO^- + H_2O \rightleftarrows HClO + OH^-$ $K_b = 3.3 \times 10^{-7}$
$H_2O + H_2O \rightleftarrows H_3O^+ + OH^-$ $K_w = 1 \times 10^{-14}$
d) $H_2O + H_2O \rightleftarrows H_3O^+ + OH^-$ $K_w = 1 \times 10^{-14}$
e) $HCN + H_2O \rightleftarrows H_3O^+ + CN^-$ $K_a = 4.9 \times 10^{-10}$
$H_2O + H_2O \rightleftarrows H_3O^+ + OH^-$ $K_w = 1 \times 10^{-14}$
Ranking of K_{eq}'s: (c) > (a) > (e) > (b) = (d)

16.5 a) $AgCl_{(s)} \rightleftarrows Ag^+_{(aq)} + Cl^-_{(aq)}$ $K_{sp} = [Ag^+]_{eq}[Cl^-]_{eq}$

b) $BaSO_{4(s)} \rightleftarrows Ba^{2+}_{(aq)} + SO_4{}^{2-}_{(aq)}$ $K_{sp} = [Ba^{2+}]_{eq}[SO_4{}^{2-}]_{eq}$

c) $Fe(OH)_{2(s)} \rightleftarrows Fe^{2+}_{(aq)} + 2\ OH^-_{(aq)}$ $K_{sp} = [Fe^{2+}]_{eq}[OH^-]^2_{eq}$

d) $Ca_3(PO_4)_2 \rightleftarrows 3\ Ca^{2+}_{(aq)} + 2\ PO_4{}^{3-}_{(aq)}$ $K_{sp} = [Ca^{2+}]^3_{eq}[PO_4{}^{3-}]^2_{eq}$

16.6 a) $AgCl_{(s)} \rightleftarrows Ag^+_{(aq)} + Cl^-_{(aq)}$

(solid) x x

$K_{sp} = [Ag^+]_{eq}[Cl^-]_{eq} = 1.8 \times 10^{-10}\ M^2 = x^2$

$x = 1.3 \times 10^{-5}\ M = [Ag^+]_{eq} = [Cl^-]_{eq}$ = amount of AgCl that dissolves per liter

Mass = $(1.3 \times 10^{-5}$ mol/L)(0.500 L)(143.4 AgCl/mol) = 9.3×10^{-4} g AgCl

b) $BaSO_{4(s)} \rightleftarrows Ba^{2+}_{(aq)} + SO_4{}^{2-}_{(aq)}$

(solid) x x

$K_{sp} = [Ba^{2+}]_{eq}[SO_4{}^{2-}]_{eq} = 1.1 \times 10^{-10}\ M^2 = x^2$

$x = 1.0 \times 10^{-5}\ M = [Ba^{2+}]_{eq} = [SO_4{}^{2-}]_{eq}$ = amount of $BaSO_4$ that dissolves per liter

Mass = $(1.0 \times 10^{-5}$ mol/L)(0.500 L)(233.37 $BaSO_4$/mol) = 1.2×10^{-3} g $BaSO_4$

c) $Fe(OH)_{2(s)} \rightleftarrows Fe^{2+}_{(aq)} + 2\ OH^-_{(aq)}$

(solid) x 2x

$K_{sp} = [Fe^{2+}]_{eq}[OH^-]^2_{eq} = 7.9 \times 10^{-16}\ M^3 = x(2x)^2 = 4x^3$

$x = 5.8 \times 10^{-6}\ M = [Fe^{2+}]_{eq} = 1/2\ [OH^-]_{eq}$ = amount of $Fe(OH)_2$ that dissolves per liter

Mass = $(5.8 \times 10^{-6}$ mol/L)(0.500 L)(89.87 g $Fe(OH)_2$/mol) = 2.6×10^{-4} g $Fe(OH)_2$

d) $Ca_3(PO_4)_{2(s)} \rightleftarrows 3\ Ca^{2+}_{(aq)} + 2\ PO_4{}^{3-}_{(aq)}$

(solid) 3x 2x

$K_{sp} = [Ca^{2+}]^3_{eq}[PO_4{}^{3-}]^2_{eq} = 2.0 \times 10^{-29}\ M^5 = (3x)^3(2x)^2 = 108\ x^5$

$x = 7.1 \times 10^{-7}\ M = 1/3[Ca^{2+}]_{eq} = 1/2[PO_4{}^{3-}]_{eq}$ = amount of $Ca_3(PO_4)_2$ that dissolves per liter

Mass = $(7.1 \times 10^{-7}$ mol/L)(0.500 L)(310.20 g $Ca_3(PO_4)_2$/mol)
= 1.1×10^{-4} g $Ca_3(PO_4)_2$

16.7 $CaC_2O_{4(s)} \rightleftarrows Ca^{2+}_{(aq)} + C_2O_4{}^{2-}_{(aq)}$ $K_{sp} = [Ca^{2+}]_{eq}[C_2O_4{}^{2-}]_{eq}$

$MM_{CaC_2O_4} = 128.10$ g/mol

$$[CaC_2O_4] = \left(\frac{6.1\text{ mg}}{1.0\text{ L}}\right)\left(\frac{1\text{ g}}{1000\text{ mg}}\right)\left(\frac{1\text{ mol}}{128.10\text{ g}}\right) = 4.762 \times 10^{-5}\text{ M}$$

$= [Ca^{2+}]_{eq} = [C_2O_4{}^{2-}]_{eq}$

$K_{sp} = (4.762 \times 10^{-5}\ M)^2 = 2.3 \times 10^{-9}\ M^2$

16.8 $Na_2SO_{4(s)} \rightleftarrows 2\ Na^+_{(aq)} + SO_4^{2-}{}_{(aq)}$

(solid) 1.34 M 0.669 M

$$\left(\frac{9.5\text{ g }Na_2SO_4}{0.100\text{ L}}\right)\left(\frac{1\text{ mol }Na_2SO_4}{142.05\text{ g }Na_2SO_4}\right) = 0.669\text{ mol }Na_2SO_4\text{ / L}$$

$[SO_4^{2-}]_{eq} = 0.669$ M $[Na^+]_{eq} = 1.34$ M

$K_{sp} = [Na^+]^2_{eq}[SO_4^{2-}]_{eq} = (1.34\text{ M})^2(0.669\text{ M}) = 1.2\text{ M}^3$

16.9 $Cu(NO_3)_{2(aq)} \rightleftarrows Cu^{2+}{}_{(aq)} + 2\ NO_3^-{}_{(aq)}$

$KOH_{(aq)} \rightleftarrows K^+{}_{(aq)} + OH^-{}_{(aq)}$ Spectator ions: K^+, NO_3^-

	$Cu^{2+}{}_{(aq)}$ +	$OH^-{}_{(aq)}$ $\rightleftarrows$	$Cu(OH)_2$
Init. (M)	0.15 M	0.20 M	
Change (M)	-0.10	-0.20	
Compl. (M)	0.05 M	0	
Change to Equil. (M)	+y	+2y	
Equil. (M)	0.05 + y	2y	

$K_{eq} = 1/K_{sp} = 10^{19.66} = 4.6 \times 10^{19}$

$K_{sp} = [Cu^{2+}]_{eq}[OH^-]^2_{eq} = 2.2 \times 10^{-20}\text{ M}^3$

$= (0.05 + y)(2y)^2 \approx (0.05)(2y)^2 = 0.20y^2$

$y = \sqrt{2.2 \times 10^{-20}/0.20} = 3.3 \times 10^{-10}$ M

$[Cu^{2+}]_{eq} = 0.05$ M $[OH^-]_{eq} = 2y = 6.6 \times 10^{-10}$ M $[K^+] = 0.20$ M

$[NO_3^-] = 0.30$ M

16.10 a) Reaction:

a) Reaction:	$Ca_3(PO_4)_2$ $\rightleftarrows$	$3\ Ca^{2+}{}_{(aq)}$ +	$2\ PO_4^{3-}{}_{(aq)}$
Init. (M)	0	4.00×10^{-2}	2.2×10^{-3}
Change (M)	+3.30 moles (solid)	-3.3×10^{-3}	-2.2×10^{-3}
Compl. (M)	3.30 moles (solid)	3.7×10^{-2}	0
Change to Equil. (M)	-y(3000) moles	+3y	+2y
Equil. (M)	3.30 moles - y(3000) moles	3.7×10^{-2} +3y	2y

$K_{sp} = [Ca^{2+}]^3_{eq}[PO_4^{3-}]^2_{eq} = 2.0 \times 10^{-29}\text{ M}^5 = (3.7 \times 10^{-2}\text{ M} + 3y)^3(2y)^2$

$\cong (3.7 \times 10^{-2}\text{ M})^3(2y)^2$

$$4y^2 = \frac{2.0 \times 10^{-29}\text{ M}^5}{(3.7 \times 10^{-2}\text{ M})^3} = 3.9 \times 10^{-25}\text{ M}^2 \qquad y = 3.1 \times 10^{-13}\text{ M}$$

$[PO_4^{3-}]_{eq} = 2y = 2(3.1 \times 10^{-13}\text{ M}) = 6.2 \times 10^{-13}$ M

b) 3.3 moles - y(3000)moles $\cong$ 3.3 moles of $Ca_3(PO_4)_2$ precipitates.

Mass = (3.30 mol $Ca_3(PO_4)_2$)(310.20 g $Ca_3(PO_4)_2$/mol $Ca_3(PO_4)_2$)

= 1.0×10^3 g $Ca_3(PO_4)_2$ precipitated or 1.0 kg $Ca_3(PO_4)_2$.

16.11 pH = - log [H_3O^+]

a) pH = - log (4.0) = -0.60

b) pH = - log (3.75 x 10^{-6}) = 5.43

c) pH = - log (0.0048) = 2.32

d) pH = - log (7.45 x 10^{-12}) = 11.13

16.12 [H_3O^+] = 10^{-pH}

a) pH = 0.66 [H_3O^+] = $10^{-0.66}$ = 0.22 M

b) pH = 7.85 [H_3O^+] = $10^{-7.85}$ = 1.4 x 10^{-8} M

c) pH = 3.68 [H_3O^+] = $10^{-3.68}$ = 2.1 x 10^{-4} M

d) pH = 14.33 [H_3O^+] = $10^{-14.33}$ = 4.7 x 10^{-15} M

16.13 pH + pOH = 14.00 pH = 14.00 - pOH

pH = 14.00 - (-log [OH^-]) = 14.00 + log [OH^-]

a) pH = 14.00 + log (2.0) = 14.30

b) pH = 14.00 + log (3.75 x 10^{-6}) = 8.57

c) pH = 14.00 + log (0.0048) = 11.68

d) pH = 14.00 + log (7.45 x 10^{-12}) = 2.87

16.14 a) pH = 0.66 [H_3O^+] = $10^{-0.66}$ = 0.22 M

$$[OH^-] = \frac{K_w}{[H_3O^+]} = \frac{1.0 \times 10^{-14}\ M^2}{0.22\ M} = 4.5 \times 10^{-14}\ M$$

or pOH = 14.00 - pH pOH = 14.00 - 0.66 = 13.34

[OH^-] = $10^{-13.34}$ = 4.6 x 10^{-14} M

b) pH = 7.85 [H_3O^+] = $10^{-7.85}$ = 1.4 x 10^{-8} M

$$[OH^-] = \frac{1.0 \times 10^{-14}\ M^2}{1.4 \times 10^{-8}\ M} = 7.1 \times 10^{-7}\ M$$

or pOH = 14.00 - 7.85 = 6.15 [OH^-] = $10^{-6.15}$ = 7.1 x 10^{-7} M

c) pH = 3.68 [H_3O^+] = $10^{-3.68}$ = 2.1 x 10^{-4} M

$$[OH^-] = \frac{1.0 \times 10^{-14}\ M^2}{2.1 \times 10^{-4}\ M} = 4.8 \times 10^{-11}\ M$$

or pOH = 14.00 - 3.68 = 10.32 [OH^-] = $10^{-10.32}$ = 4.8 x 10^{-11} M

d) pH = 14.33 [H_3O^+] = $10^{-14.33}$ = 4.7 x 10^{-15} M

$$[OH^-] = \frac{1.0 \times 10^{-14}\ M^2}{4.7 \times 10^{-15}\ M} = 2.1\ M$$

or pOH = 14.00 - 14.33 = -0.33 [OH^-] = $10^{-0.33}$ = 2.1 M

16.15 a) CH_3CO_2H = weak acid

b) NH_4Cl = salt

c) KCl = salt

d) $NaCH_3CO_2$ = salt

e) NaOH = strong base

16.16 a) $HClO$ weak acid
b) $CaBr_2$ salt
c) $KClO$ salt
d) HNO_3 strong acid
e) HCN weak acid

16.17 a) NH_3 = weak base
b) $HClO_4$ = strong acid
c) $HClO$ = weak acid
d) $Ba(OH)_2$ = strong base

16.18 a) $HONO_2 \equiv HNO_3$ strong acid
b) $HOCH_3$ not an acid in aqueous solution
c) HOH weak acid (and weak base)
d) $HOCl$ weak acid
e) NH_2OH weak base

16.19

weak acids	conjugate base	weak bases	**conj. acids**
$HClO$ (or $HOCl$)	ClO^- (or OCl^-)	NH_3	**NH_4^+**
H_2O	OH^-	H_2O	**H_3O^+**
		NH_2OH	**NH_3OH^+**

16.20 a)

	$NH_{3(aq)}$ + $H_2O_{(l)}$	$\rightleftarrows$ $NH_4^+{}_{(aq)}$ +	$OH^-{}_{(aq)}$
Initial (M)	2.5×10^{-2}	0	0
Change (M)	-x	+x	+x
Equil. (M)	2.5×10^{-2} - x	x	x

$$K_b = 1.8 \times 10^{-5}\ M = \frac{[NH_4^+]_{eq}[OH^-]_{eq}}{[NH_3]_{eq}} = \frac{(x)(x)}{(2.5 \times 10^{-2}\ M - x)} \cong \frac{x^2}{2.5 \times 10^{-2}}$$

$x = 6.7 \times 10^{-4}$ M Note: 2.5×10^{-2} M - 6.7×10^{-4} M = 2.4×10^{-2} M

$[OH^-]_{eq} = x = 6.7 \times 10^{-4}$ M pOH = - log(6.7×10^{-14}) = 3.17

pH = 14.00 - 3.17 = 10.83

b)

	$HClO_{4(aq)}$ + $H_2O_{(l)}$	$\rightarrow$ $H_3O^+{}_{(aq)}$ +	$ClO^-{}_{(aq)}$
Initial (M)	2.5×10^{-2}	0	0
Change (M)	-2.5×10^{-2}	$+2.5 \times 10^{-2}$	$+2.5 \times 10^{-2}$
Equil. (M)	≈ 0	2.5×10^{-2}	2.5×10^{-2}

Note: $HClO_4$ is a strong acid and completely ionizes.

pH = - log $[H_3O^+]$ = - log $[2.5 \times 10^{-2}]$ = 1.60

c)

	$HClO_{(aq)}$ + $H_2O_{(l)}$	$\rightleftarrows$ $H_3O^+{}_{(aq)}$ +	$ClO^-{}_{(aq)}$
Initial (M)	2.5×10^{-2}	0	0
Change (M)	-x	+x	+x
Equil. (M)	2.5×10^{-2} -x	x	x

(continued)

(16.20 continued)

$$K_a = 3.0 \times 10^{-8}\ M = \frac{[H_3O^+]_{eq}[ClO^-]_{eq}}{[HClO]_{eq}} = \frac{(x)(x)}{(2.5 \times 10^{-2}\ M - x)} \approx \frac{x^2}{2.5 \times 10^{-2}}$$

$x = 2.7 \times 10^{-5}$ M   Note: 2.5×10^{-2} M - 2.7×10^{-5} M = 2.5×10^{-2} M

$[H_3O^+]_{eq} = x = 2.7 \times 10^{-5}$ M   pH = - log (2.7×10^{-5}) = 4.57

d)	$Ba(OH)_{2(aq)}$	$\rightleftarrows$	$Ba^{2+}{}_{(aq)}$	+	$2\ OH^-{}_{(aq)}$
Initial (M)	2.5×10^{-2}		0		0
Change (M)	-2.5×10^{-2}		$+2.5 \times 10^{-2}$		$+5.0 \times 10^{-2}$
Equil. (M)	≈ 0		2.5×10^{-2}		5.0×10^{-2}

Note: $Ba(OH)_2$ is a strong base and completely dissociates.

$[OH^-]_{eq} = 5.0 \times 10^{-2}$ M   pOH = - log(5.0×10^{-2}) = 1.30

pH = 14.00 - 1.30 = 12.70

16.21 a) $NaOH \rightleftarrows Na^+ + OH^-$ $[NaOH] = [Na^+] = [OH^-] = 1.5$ M

pH = 14.00 - pOH = 14.00 - (- log $[OH^-]$)

= 14.00 + log $[OH^-]$ = 14.00 + log (1.5) = 14.18

b)	$C_5H_5N + H_2O$	$\rightleftarrows$	$C_5H_5NH^+$	+	OH^-
Init. (M)	1.5 M		0		0
Change (M)	-x		+x		+x
Equil. (M)	1.5-x		x		x

$$pK_b = 8.72 \quad K_b = 10^{-8.72} = 1.91 \times 10^{-9} = \frac{[C_5H_5NH^+]_{eq}[OH^-]_{eq}}{[C_5H_5N]_{eq}}$$

$$1.91 \times 10^{-9} = \frac{x^2}{1.5 - x} \approx \frac{x^2}{1.5} \qquad x^2 = 1.5(1.91 \times 10^{-9}) = 2.86 \times 10^{-9}$$

$x = 5.35 \times 10^{-5} = [OH^-]$

pH = 14.00 - pOH = 14.00 - (- log $[OH^-]$) = 14.00 + log (5.35×10^{-5}) = 9.73

c)	$NH_2OH + H_2O$	$\rightleftarrows$	NH_3OH^+	+	OH^-
Init. (M)	1.5 M		0		0
Change (M)	-x		+x		+x
Equil. (M)	1.5-x		x		x

$$pK_b = 7.96 \quad K_b = 10^{-7.96} = 1.10 \times 10^{-8} = \frac{[NH_3OH^+]_{eq}[OH^-]_{eq}}{[NH_2OH]_{eq}}$$

$$1.10 \times 10^{-8} = \frac{x^2}{1.5 - x} \qquad x^2 = 1.5(1.10 \times 10^{-8}) = 1.65 \times 10^{-8}$$

$x = 1.28 \times 10^{-4}$ M = $[OH^-]$ pH = 14.00 + log (1.28×10^{-4}) = 10.11

(continued)

(16.21 continued)

d) $HCO_2H + H_2O_{(l)} \rightleftarrows HCO_2^-{}_{(aq)} + H_3O^+{}_{(aq)}$

	HCO_2H	HCO_2^-	H_3O^+
Init. (M)	1.5 M	0	0
Change (M)	-x	+x	+x
Equil. (M)	1.5-x	x	x

$$K_a = \frac{[HCO_2^-]_{eq}[H_3O^+]_{eq}}{[HCO_2H]_{eq}} = 1.82 \times 10^{-4}$$

$$K_a = \frac{(x)(x)}{1.5-x} = \frac{x^2}{1.5} = 1.82 \times 10^{-4};\ x = 1.65 \times 10^{-2} M$$

$pH = -\log 1.65 \times 10^{-2} = 1.78$

16.22 a) Major species: HNO_2 and H_2O Minor species: H_3O^+ and NO_2^-

b) Reaction: $HNO_{2(aq)} + H_2O_{(l)} \rightleftarrows H_3O^+{}_{(aq)}$ and $NO_2^-{}_{(aq)}$

	HNO_2	H_3O^+	NO_2^-
Init. (M)	0.150	0	0
Change (M)	-x	+x	+x
Equil. (M)	0.150-x	x	x

$$K_a = 4.5 \times 10^{-4}\ M = \frac{[H_3O^+]_{eq}[NO_2^-]_{eq}}{[HNO_2]_{eq}} = \frac{(x)(x)}{(0.150\ M - x)}$$

$4.5 \times 10^{-4}\ M\ (0.150\ M - x) = x^2$ $\qquad 0 = x^2 + 4.5 \times 10^{-4}\ M - 6.75 \times 10^{-5}\ M^2$

$$x = \frac{-4.5 \times 10^{-4} \pm \sqrt{(4.5 \times 10^{-4})^2 - (4)(1)(-6.75 \times 10^{-5})}}{2(1)}$$

$$= \frac{4.5 \times 10^{-4} \pm 1.64 \times 10^{-2}}{2} = 8.0 \times 10^{-3}\ M$$

($x = 8.3 \times 10\text{-}3$ M if $0.150\ M - x \cong 0.150$ M)

$[H_3O^+]_{eq} = [NO_2^-]_{eq} = 8.0 \times 10^{-3}$ M $[HNO_2]_{eq} = 0.150\ M - 8.0 \times 10^{-3}\ M = 0.142$ M

c) $pH = -\log [H_3O^+] = -\log (8.0 \times 10^{-3}) = 2.10$

d) Your molecular drawing should look like the molecular drawings in Figure 16-5 and the answer to Exercise 16.3.2 but with the following species:

$HNO_2 + H_2O \rightleftarrows H_3O^+ + NO_2^-$

16.23 $N(CH_3)_{3(l)} + H_2O_{(l)} \rightleftarrows HN(CH_3)_3^+{}_{(aq)} + OH^-{}_{(aq)}$; $K_b = \dfrac{[HN(CH_3)_3^+]_{eq}[OH^-]_{eq}}{[N(CH_3)_3]_{eq}}$

$pK_b = 4.19$ $\qquad K_b = 10^{-4.19} = 6.46 \times 10^{-5}$

a) Major species: H_2O and $N(CH_3)_3$ Minor species: $HN(CH_3)_3^+$ and OH^-

(continued)

(16.23 continued)

b)

	$NMe_3 + H_2O \rightleftarrows$	$^{+}HNMe_3$	+ OH^-
Init. (M)	0.350 M	0	0
Change (M)	-x	+x	+x
Equil. (M)	0.350-x	x	x

$$K_b = 6.46 \times 10^{-5} = \frac{(x)(x)}{(0.350 - x)} \approx \frac{x^2}{0.350}$$

$$x = \sqrt{(0.350)(6.46 \times 10^{-5})} = 4.75 \times 10^{-3}\ M$$

$[N(CH_3)_3]_{eq} = 0.345\ M \quad [^{+}HN(CH_3)_3]_{eq} = [OH^-]_{eq} = 4.75 \times 10^{-3}\ M$

c) pH = 14.00 - pOH = 14.00 + log [OH] = 11.68

d) Your molecular drawing should look like the molecular drawings in Figure 16-5, the answer to Exercise 16.3.2 or the answer to Problem 16.22 but with the following species:

$$NMe_3 \quad + \quad H_2O \quad \rightleftarrows \quad ^{+}HNMe_3 \quad + \quad OH^-$$

16.24 a) Major species: Na^+, SO_3^{2-} and H_2O

b)

	$SO_3^{2-}{}_{(aq)} + H_2O_{(l)} \rightleftarrows$	$HSO_3^-{}_{(aq)}$ +	$OH^-{}_{(aq)}$
c) Init. (M)	0.45	0	0
Change (M)	-x	+x	+x
Equil. (M)	0.45-x	x	x

$$K_b = \frac{[HSO_3^-]_{eq}[OH^-]_{eq}}{[SO_3^{2-}]_{eq}} \qquad pK_b = 14.00 - 7.20 = 6.80$$

$$K_b = 1.6 \times 10^{-7}\ M = \frac{(x)(x)}{(0.45\ M - x)} \cong \frac{x^2}{0.45\ M}$$

$x = 2.7 \times 10^{-4}\ M = [OH^-]_{eq} \quad pOH = -\log(2.7 \times 10^{-4}) = 3.57$

pH = 14.00 - 3.57 = 10.43

16.25 $NH_4NO_{3(aq)} \rightleftarrows NH_4^+{}_{(aq)} + NO_3^-{}_{(aq)}$

$NH_4^+{}_{(aq)} + H_2O \rightleftarrows NH_{3(aq)} + H_3O^+{}_{(aq)}$

a) Major species: $NH_4^+{}_{(aq)}$, $NO_3^-{}_{(aq)}$ and H_2O

b) $NH_4^+{}_{(aq)} + H_2O_{(l)} \rightleftarrows NH_{3(aq)} + H_3O^+{}_{(aq)}$

(continued)

(16.25 continued)

c)	NH_4^+ +	H_2O $\rightleftarrows$	NH_3 +	H_3O^+
Init. (M)	0.0100 M		0	0
Change (M)	-x		+x	+x
Equil. (M)	0.0100 M-x		x	x

$pK_a = 9.25$ $\quad K_a = 10^{-9.25} = 5.6 \times 10^{-10}$ M

$$K_a = \frac{[NH_3]_{eq}[H_3O^+]_{eq}}{[NH_4^+]} = \frac{(x)(x)}{(0.0100\text{ M} - x)} \approx \frac{x^2}{0.0100} = 5.6 \times 10^{-10}$$

$$x = \sqrt{(0.0100)(5.6 \times 10^{-10})} = 2.37 \times 10^{-6}\text{ M} = [H_3O^+]$$

$pH = -\log [H_3O^+] = -\log (2.37 \times 10^{-6}) = 5.63$

16.26 $HPO_4^{2-}{}_{(aq)} + H_3O^+{}_{(aq)} \rightleftarrows H_2PO_4^-{}_{(aq)} + H_2O_{(l)}$

Reverse of: $H_2PO_4^-{}_{(aq)} + H_2O_{(l)} \rightleftarrows HPO_4^{2-}{}_{(aq)} + H_3O^+{}_{(aq)}$

$$K_{a_2} = \frac{[HPO_4^{2-}]_{eq}[H_3O^+]_{eq}}{[H_2PO_4^-]_{eq}} = 6.2 \times 10^{-8}\text{ M}$$

$$K_{eq} = \frac{[H_2PO_4^-]_{eq}}{[HPO_4^{2-}]_{eq}[H_3O^+]_{eq}} = \frac{1}{K_{a_2}} = \frac{1}{6.2 \times 10^{-8}\text{ M}} = 1.6 \times 10^7\text{ M}^{-1}$$

16.27 $HPO_4^{2-}{}_{(aq)} + OH^-{}_{(aq)} \rightleftarrows PO_4^{3-}{}_{(aq)} + H_2O_{(l)}$

$HPO_4^{2-}{}_{(aq)} + H_2O_{(l)} \rightleftarrows PO_4^{3-}{}_{(aq)} + H_3O^+{}_{(aq)}$ $\qquad K_{a_3} = 4.8 \times 10^{-13}$ M

$$K_{a_3} = \frac{[PO_4^{3-}]_{eq}[H_3O^+]_{eq}}{[HPO_4^{2-}]_{eq}}$$

$H_3O^+{}_{(aq)} + OH^-{}_{(aq)} \rightleftarrows 2\,H_2O_{(l)}$ $\qquad \dfrac{1}{K_w} = \dfrac{1}{[H_3O^+]_{eq}[OH^-]_{eq}}$

$$\frac{K_{a_3}}{K_w} = \frac{[PO_4^{3-}]_{eq}[H_3O^+]_{eq}}{[HPO_4^{2-}]_{eq}} \times \frac{1}{[H_3O^+]_{eq}[OH^-]_{eq}} = \frac{[PO_4^{3-}]_{eq}}{[HPO_4^{2-}]_{eq}[OH^-]_{eq}} = K_{eq}$$

$$K_{eq} = \frac{K_{a_3}}{K_w} = \frac{4.8 \times 10^{-13}\text{ M}}{1.0 \times 10^{-14}\text{ M}^2} = 48\text{ M}^{-1}$$

16.28 a) No buffer properties; resulting solution will have H_3O^+, Cl^-, H_2O and CH_3CO_2H as major species.

b) Yes, will have buffer properties. Resulting solution will have CH_3CO_2H and $CH_3CO_2^-$ among major species.

c) No buffer properties. Resulting solution will have Na^+, OH^-, $CH_3CO_2^-$ and H_2O as major species.

d) Yes, will have buffer properties. Resulting solution will have $CH_3CO_2^-$ and CH_3CO_2H among major species.

16.29 (b) mmoles H_3O^+ = mmoles HCl = (50 mL)(0.25 M HCl) = 12.5 mmoles H_3O^+
mmoles $CH_3CO_2^-$ = mmoles $NaCH_3CO_2$= (100 mL)(0.25 M $NaCH_3CO_2$) = 25 mmoles $CH_3CO_2^-$

Reaction:	$CH_3CO_2^-{}_{(aq)}$ +	$H_3O^+{}_{(aq)}$ $\rightleftarrows$	$CH_3CO_2H_{(aq)} + H_2O_{(l)}$
Initial, mmoles	25	12.5	0
Change, mmoles	-12.5	-12.5	+12.5
Completion, mmoles	12.5	~ 0	12.5

$$pH = pK_a + \log\frac{[CH_3CO_2^-]}{[CH_3CO_2H]} = 4.75 + \log\left(\frac{12.5}{12.5}\right) = 4.75$$

Note: Remember that in the equation you can substitute amounts of acid and base in moles or mmoles.

d) mmol CH_3CO_2H = (100 mL)(0.25 M CH_3CO_2H) = 25 mmol CH_3CO_2H
mmol OH^- = mmol NaOH= (50 mL)(0.25 M NaOH) = 12.5 mmol OH^-

Reaction:	$CH_3CO_2H_{(aq)}$ +	$OH^-{}_{(aq)}$ $\rightleftarrows$	$CH_3CO_2^-{}_{(aq)} + H_2O_{(l)}$
Initial, mmol	25	12.5	
Change, mmol	-12.5	-12.5	+12.5
Completion, mmol	12.5	~ 0	12.5

$$pH = pK_a + \log\frac{[CH_3CO_2^-]}{[CH_3CO_2H]} = 4.75 + \log\left(\frac{12.5}{12.5}\right) = 4.75$$

16.30 Assuming addition of 5.0 mmol of a strong, monoprotic acid:

a) Not buffered

b) mmol $CH_3CO_2^-$ = (100 L)(0.25 M $NaCH_3CO_2$) = 25 mmol $CH_3CO_2^-$
mmol of H_3O^+ = (50 mL)(0.25 M HCl) = 12.5 mmol H_3O^+

Reaction:	$CH_3CO_2^-{}_{(aq)}$ +	$H_3O^+{}_{(aq)}$ $\rightleftarrows$	$CH_3CO_2H_{(aq)} + H_2O_{(l)}$
Initial, mmol	25	12.5	0
Change, mmol	-12.5	-12.5	+12.5
Amts in Buffer, mmol	12.5	small	12.5

$$pH = pK_a + \log\frac{[CH_3CO_2^-]}{[CH_3CO_2H]} = pK_a + \log\left(\frac{mol/V}{mol/V}\right) = pK_a + \log\left(\frac{mol}{mol}\right)$$

Note: In the log term the number of mmol (or mol) of each species can be substituted since the volumes are the same for both species and cancel.

(continued)

(16.30 continued)

$$pH = 4.75 + \log\left(\frac{12.5}{12.5}\right) = 4.75 + 0 = 4.75 \text{ (of buffer solution)}$$

Upon additiion of 5.0 mmol of strong, monoprotic acid:

Reaction:	$CH_3CO_2^-$ +	H_3O^+ $\rightleftarrows$	$CH_3CO_2H + H_2O$
Initial, mmol	12.5	small	12.5
Amt on add. HCl, mmol		+5.0	
Change, mmol	-5.0	-5.0	+5.0
Amt after HCl reacts, mmol	7.5	small	17.5

$$pH = 4.75 + \log\left(\frac{7.5}{17.5}\right) = 4.75 + \log(0.428) = 4.75 - 0.37 = 4.38$$

Change in pH = 4.38 - 4.75 = -0.37

c) not buffered

d) mmol CH_3CO_2H = (100 mL)(0.25 M CH_3CO_2H) = 25 mmol CH_3CO_2H

mmol OH^- = mmol NaOH= (50 mL)(0.25 M NaOH) = 12.5 mmol OH^-

Reaction:	$CH_3CO_2H_{(aq)}$ +	$OH^-_{(aq)}$ $\rightleftarrows$	$CH_3CO_2^-{}_{(aq)} + H_2O_{(l)}$
Initial, mmol	25	12.5	0
Change, mmol	-12.5	-12.5	+12.5
Amt in Buffer, mmol	12.5	small	12.5

$$pH = pK_a + \log\frac{[CH_3CO_2^-]}{[CH_3CO_2H]} = 4.75 + \log\left(\frac{12.5}{12.5}\right) = 4.75 + 0 = 4.75$$

Upon addition of 5.0 mmol of strong, monoprotic acid:

Reaction:	$CH_3CO_2^-{}_{(aq)}$ +	$H_3O^+{}_{(aq)}$ $\rightleftarrows$	$CH_3CO_2H_{(aq)} + H_2O_{(l)}$
Initial, mmol	12.5		12.5
Amt on add. HCl, mmol		+5.0	
Change, mmol	-5.0	-5.0	+5.0
Amt after HCl reacts, mmol	7.5	small	17.5

$$pH = pK_a + \log\frac{[CH_3CO_2^-]}{[CH_3CO_2H]} = 4.75 + \log\left(\frac{7.5}{17.5}\right) = 4.75 + -0.37 = 4.38$$

Change in pH = 4.38 - 4.75 = -0.37

16.31 b) Addition of base would change pH 0.1 unit from 4.75 to 4.85.

$$4.85 = 4.75 + \log\frac{[CH_3CO_2^-]}{[CH_3CO_2H]} \quad \log\frac{[CH_3CO_2^-]}{[CH_3CO_2H]} = 0.10 \quad \frac{[CH_3CO_2^-]}{[CH_3CO_2H]} = 1.26$$

$$\text{or} \quad \frac{\text{mmoles } CH_3CO_2^-/\text{mL}}{\text{mmoles } CH_3CO_2H/\text{mL}} = \frac{\text{mmoles } CH_3CO_2^-}{\text{mmoles } CH_3CO_2H} = 1.26$$

mmoles $CH_3CO_2^-$ = 1.26(mmoles CH_3CO_2H)

(continued)

(16.31 continued)

Reaction:	$CH_3CO_2H_{(aq)}$ +	$OH^-_{(aq)}$ $\rightleftarrows$	$CH_3CO_2^-{}_{(aq)} + H_2O_{(l)}$	
Initial, mmol	12.5		12.5	
Change, mmol	-x	+x	+x	(from Prob. 16.29 (b))
Compl., mmol	12.5-x	x	12.5+x	

12.5 + x = 1.26(12.5 mmol - x)
x + 1.26x = 1.26(12.5 mmol) - 12.5 mmol = 1.26(12.5 mmol) - 1(12.5 mmol)
= 0.26(12.5 mmol)

2.26x = 0.26(12.5 mmol) $\quad x = \dfrac{0.26(12.5\text{ mmol})}{2.26} = 1.4$ mmoles = OH^-

moles of base = 1.4×10^{-3} moles

d) The calculation is exactly the same as for part (b) because we have a buffer containing the same species at the same concentration.

16.32 pK_a (formic acid) = 3.74 $\quad pH = pK_a + \log([A^-]/[HA])$
$4.07 = 3.74 + \log([A^-]/[HA])$

$$\log\left(\frac{[A^-]}{[HA]}\right) = 0.33 \qquad \left(\frac{[A^-]}{[HA]}\right) = 10^{0.33} = 2.14$$

The molecular picture that this question calls for needs to show a ratio of A^- (formate ions) to HA (formic acid molecules) that is 21 to 10.

16.33 0.50 M NaH_2PO_4 $\quad$ 0.20 M Na_2HPO_4

$H_2PO_4^-{}_{(aq)} + H_2O_{(l)} \rightleftarrows HPO_4^{2-}{}_{(aq)} + H_3O^+{}_{(aq)} \qquad pK_{a_2} = 7.21$

$$pH = pK_a + \log\left\{\frac{[A^-]_{initial}}{[HA]_{initial}}\right\}; \; [A^-] = [HPO_4^{2-}] = 0.20\text{ M};$$

$[HA] = [H_2PO_4^-] = 0.50$ M $\qquad pH = 7.21 + \log\left(\dfrac{0.20}{0.50}\right) = 6.81$

16.34 mmol $H_2PO_4^-$ = (150 mL)(0.50 M NaH_2PO_4) = 75 mmol $H_2PO_4^-$
mmol HPO_4^{2-} = (150 mL)(0.20 M Na_2HPO_4) = 30 mmol HPO_4^{2-}
(0.040 g NaOH) / (0.040 g / mmol) = 1.0 mmol NaOH = 1.0 mmol OH^-

$$pH = pKa + \log\frac{[HPO_4^{2-}]}{[H_2PO_4^-]} = 7.21 + \log\left(\frac{30/V}{75/V}\right) = 7.21 + \log\left(\frac{30}{75}\right)$$
$$= 7.21 - 0.40 = 6.81$$

Reaction:	$H_2PO_4^-{}_{(aq)}$ +	$OH^-_{(aq)}$ $\rightleftarrows$	$HPO_4^{2-} + H_2O_{(l)}$
Initial, mmol	75		30
Amt on add. NaOH, mmol		1.0	
Change, mmol	-1.0	-1.0	+1.0
Amts after NaOH reacts, mmol	74	small	31

$pH = 7.21 + \log\left(\dfrac{31}{74}\right) = 7.21 - 0.38 = 6.83 \qquad$ Change in pH = 6.83 - 6.81 = +0.02

16.35 $[H_2PO_4^-]_{init.} = 0.50$ M $HPO_4^{2-} + H_3O^+ \rightleftarrows H_2PO_4^- + H_2O_{(l)}$

$[HPO_4^{2-}]_{init.} = 0.20$ M volume of buffer = 250 mL

pH range = 6.81 ± 0.2 = 6.61

$$6.61 = 7.21 + \log\left(\frac{0.20 - x}{0.50 + x}\right);\ \log\left(\frac{0.20 - x}{0.50 + x}\right) = 6.61 - 7.21 = -0.60$$

$$\left(\frac{0.20 - x}{0.50 + x}\right) = 10^{-0.60} = 0.251$$

0.20 - x = 0.251(0.50 + x) = 0.126 + 0.251x

1.251x = 0.20 - 0.126 = 0.074 x = 0.059 M H_3O^+

moles H_3O^+ = (0.059 mol/L)(250 mL)(1 L/1000 mL) = 1.49 x 10^{-2} moles

16.36

pH	conjugate acid-base pair	substance added to HCl for buffer preparation
3.50	HCO_2H—HCO_2^-	HCO_2Na
6.85	$H_2PO_4^-$—HPO_4^{2-}	Na_2HPO_4
11.00	HCO_3^-—CO_3^{2-}	Na_2CO_3
12.60	HPO_4^{2-}—PO_4^{3-}	Na_3PO_4

16.37 $HCO_3^-{}_{(aq)} + OH^-{}_{(aq)} \rightleftarrows CO_3^{2-}{}_{(aq)} + H_2O_{(l)}$ $K_{a_2} = 5.6 \times 10^{-11}$ M $pK_a = 10.25$

$$pH = pK_a + \log\frac{[Base]_{eq}}{[Acid]_{eq}} \qquad \log\frac{[Base]_{eq}}{[Acid]_{eq}} = pH - pK_a = 10.60 - 10.25 = 0.35$$

$$\frac{[Base]_{eq}}{[Acid]_{eq}} = 10^{0.35} = 2.23 \qquad K_{a_2} = \frac{[CO_3^{2-}]_{eq}}{[HCO_3^-]_{eq}[OH^-]_{eq}}$$

OH^- is a strong base so every OH^- added takes a proton off HCO_3^- to form CO_3^{2-};

∴ $[OH^-] = [CO_3^{2-}]$ = base.

$[Base]_{eq} = 2.23\ [Acid]_{eq}$ $[Acid]_{eq} = [Acid]_{init} - [Base]_{eq}$

or $(\text{mmoles base})_{eq} = 2.23\ \{(\text{mmoles acid})_{init} - (\text{mmoles base})_{eq}\}$

x = $(\text{mmoles base})_{eq}$ x = 2.23 {(0.200 M)(250 mL) - x}

3.23 x = 111.5 mmoles x = 34.5 mmoles base

$$\frac{34.5 \text{ mmoles base}}{1.0 \text{ M NaOH}} = 34.5 \text{ mL of } 1.0 \text{ M NaOH}$$

16.38 a) less than 7 $HClO_{(aq)} + H_2O_{(l)} \rightleftarrows H_3O^+{}_{(aq)} + ClO^-{}_{(aq)}$

b) equal to 7 $H_2O_{(l)} + H_2O_{(l)} \rightleftarrows H_3O^+{}_{(aq)} + OH^-{}_{(aq)}$

c) less than 7 $HNO_{2(aq)} + H_2O_{(l)} \rightleftarrows H_3O^+{}_{(aq)} + NO_2^-{}_{(aq)}$

d) greater than 7 $NH_{3(aq)} + H_2O_{(l)} \rightleftarrows NH_4^+{}_{(aq)} + OH^-{}_{(aq)}$

16.39 a) $HA_{(aq)} + H_2O_{(l)} \rightleftarrows A^-_{(aq)} + H_3O^+_{(aq)}$

	$HA_{(aq)}$	$A^-_{(aq)}$	$H_3O^+_{(aq)}$
Initial	10^{-2} M		
Change	-x	+x	+x
Equil.	10^{-2} -x	x	x

$$K_a = \frac{[A^-]_{eq}[H_3O^+]_{eq}}{[HA]_{eq}} = 3.0 \times 10^{-4}\ M \qquad K_a = \frac{(x)(x)}{(10^{-2} - x)} = 3.0 \times 10^{-4}\ M$$

$x^2 = 3.0 \times 10^{-4}(10^{-2} - x) = 3.0 \times 10^{-6} - 3.0 \times 10^{-4}\ x$

$0 = x^2 + 3.0 \times 10^{-4}\ x - 3.0 \times 10^{-6}$

$$= \frac{-3.0 \times 10^{-4} \pm \sqrt{(3.0 \times 10^{-4}) - 4(1)(-3.0 \times 10^{-6})}}{2} \qquad x = 1.6 \times 10^{-3} \text{ or } -1.89 \times 10^{-3}$$

$[H_3O^+] = 1.6 \times 10^{-3}$ M $\qquad$ pH = - log$[H_3O^+]$ = - log(1.6×10^{-3}) = 2.80

b) At the stoichiometric point the amount of added OH^- = amt. of HA originally present; thus, aspirin is no longer a major species in solution and pH is determined by proton transfer from H_2O to $A^-_{(aq)}$:

$$K_{eq} = \frac{K_w}{K_a} = \frac{1 \times 10^{-14}\ M^2}{3.0 \times 10^{-4}\ M} = 3.3 \times 10^{-11}\ M$$

$A^-_{(aq)} + H_2O_{(l)} \rightleftarrows HA_{(aq)} + OH^-_{(aq)}$

	$A^-_{(aq)}$	$HA_{(aq)}$	$OH^-_{(aq)}$
Initial	10^{-2} M		
Change	-x	+x	+x
Equil.	10^{-2} -x	x	x

$$3.3 \times 10^{-11} \approx \frac{x^2}{10^{-2}} \Rightarrow x = 5.77 \times 10^{-7} = [OH^-]; \quad pOH = -\log[OH^-]$$

pH = 14.00 + log(5.77×10^{-7})= 7.76

c) pH = pK_a of weak acid at midpoint, so pH = - log K_a = -log(3.0×10^{-4}) = 3.52

16.40 $K_{a_1}(H_3PO_4) = 10^{-2.12} = 7.5 \times 10^{-3}$

At stoichiometric point: moles OH^- = moles of H_3PO_4 originally present.
mol OH^- = (0.250 L)(0.025 M) = 0.00625 mol

$H_3PO_{4(aq)} + OH^-_{(aq)} \rightleftarrows H_2PO_4^-{}_{(aq)} + H_2O_{(l)}$

At first stoichiometric point, reaction goes to completion.
The initial amounts: $H_2PO_4^-$ = 0.00625 mol, OH^- = 0, H_3PO_4 = 0

Reaction: $H_2PO_4^-{}_{(aq)} + H_2O_{(l)} \rightleftarrows H_3PO_{4(aq)} + OH^-_{(aq)}$ $\quad K_{eq} = \frac{K_w}{K_a} = 1.3 \times 10^{-12}$ M

Reaction: $H_2PO_4^-{}_{(aq)} + H_2O_{(l)} \rightleftarrows H_3O^+_{(aq)} + H_2PO_4^{2-}{}_{(aq)}$ $\quad K_{a_2} = 6.2 \times 10^{-8}$ M

The second equilibrium has the larger K_{eq}, so it generates the larger changes from initial concentrations.

(continued)

(16.40 continued)

	Reaction: $H_2PO_4^-{}_{(aq)}$ + $H_2O_{(l)}$ ⇄	$H_3O^+{}_{(aq)}$ +	$H_2PO_4^{2-}{}_{(aq)}$
Init. (M)	$\frac{0.00625 \text{ mol}}{0.256 \text{ L}} = 0.0244$	0	0
Change (M)	-x	+x	+x
Equil. (M)	0.0244-x	x	x

$$K_{a_2} = \frac{[H_3O^+]_{eq}[HPO_4^{2-}]_{eq}}{[H_2PO_4^-]_{eq}} = \frac{(x)(x)}{0.0244 \text{ M} - x} \cong \frac{x^2}{0.0244 \text{ M}} = 6.2 \times 10^{-8} \text{ M}$$

$x = 3.9 \times 10^{-5}$ M = $[H_3O^+]$ pH = 4.41

Bromocresol green would be the best indicator.

16.41 At the midpoint of a titration, the pH of the solution = pK_a of the weak acid.
$pH_{midpoint} = 3.88 = pK_a$

16.42 formic acid $pK_a = 3.74$; acetic acid $pK_a = 4.74$; benzoic acid $pK_a = 4.20$
An acetic acid-acetate buffer solution would be the best choice, but no source of acetate or method of preparation is available. Benzoic acid-benzoate buffer solution would be second choice except the solubility of benzoic acid is less than 0.1M at room temperature.
If we assume that the sodium benzoate is available as a pure solid, we could prepare the buffer.

$$4.80 = 4.20 + \log\left(\frac{[\text{benzoate}]}{[\text{benzoic acid}]}\right) \qquad \log\left(\frac{[\text{benzoate}]}{[\text{benzoic acid}]}\right) = 0.60$$

$$\left(\frac{[\text{benzoate}]}{[\text{benzoic acid}]}\right) = 3.98 \cong 4 \qquad [\text{benzoate}] = 4[\text{benzoic acid}]$$

total molarity = 0.35 M = [benzoate] + [benzoic acid] = 5[benzoic acid]
[benzoic acid] = 0.07 M [benzoate] = 0.28 M
(0.07 M)(1 L) = (0.1 M)(V) V = 0.7 L
(0.28 M)(1 L)(144 g $NaC_6H_5CO_2$/mol) = 40 g $NaC_6H_5CO_2$
Take 0.7 L of 0.1 M benzoic acid (the solution may have to be at an elevated temperature); add and dissolve 40 g of $NaC_6H_5CO_2$; and dilute to 1 L. (Cool to room temperature if necessary).

16.43 $Zn(OH)_{2(s)} \rightleftarrows Zn^{2+}{}_{(aq)} + 2\ OH^-{}_{(aq)}$ $K_{sp} = [Zn^{2+}]_{eq}[OH^-]^2_{eq} = 1.2 \times 10^{-17}\ M^3$

	Zn^{2+}	OH^-
Initial (M)	0	0
Change (M)	+x	+2x
Equil. (M)	x	2x

$1.2 \times 10^{-17} = x(2x)^2 = 4x^3$

$$x = \sqrt[3]{(1.2 \times 10^{-17})/4} = 1.44 \times 10^{-6} \text{ M} = [Zn^{2+}]_{eq} = [Zn(OH)_2]_{dissolved}$$

$$(1.44 \times 10^{-6} \text{ M})(99.40 \text{ g/mole})(1.0 \text{ L}) = 1.43 \times 10^{-4} \text{ g } Zn(OH)_{2(s)}$$

16.44 a) only weak acid present:

	$C_8H_{12}N_2O_3(aq) + H_2O(l) \rightleftarrows H_3O^+(aq) + C_8H_{11}N_2O_3^-(aq)$		
Init. (M)	0.0120 M		
Change (M)	-x	+x	+x
Equil. (M)	0.0120 M-x	x	x

$$K_a = 1 \times 10^{-8}\ M = \frac{(x)(x)}{0.0120\ M - x} \cong \frac{x^2}{0.0120\ M}$$

$x = 1.1 \times 10^{-5}\ M = [H_3O^+]eq$ $\quad$ $pH = -\log(1.1 \times 10^{-5}\ M) = 4.96$

b) 5.0 mL of strong base (0.200 M); mmol base = (5.0 mL)(0.200 M) = 1.0 mmol
mmol acid = (200 mL)(0.120 M) = 2.40 mmol acid (originally)

	$C_8H_{12}N_2O_3(aq)$	+ $OH^-(aq)$	$\rightleftarrows H_2O(l)$ +	$C_8H_{11}N_2O_3^-(aq)$
Init.	2.40 mmol			0
Add. Base		1.0 mmol		
Change	-1.0 mmol	-1.0 mmol		+1.0 mmol
Equil.	1.4 mmol	small		1.0 mmol

A buffer solution:

$$pH = pK_a + \log\frac{[C_8H_{11}N_2O_3^-]}{[C_8H_{12}N_2O_3]} = pK_a + \log\left(\frac{\text{mmol } C_8H_{11}N_2O_3^-}{\text{mmol } C_8H_{12}N_2O_3}\right)$$

= 8.0 + log (1.0 mmol/1.4 mmol) = 8.0 - 0.1 = 7.9

c) 12.00 mL of st. base (0.200 M); mmol base = (12.0 mL)(0.200 M) = 2.40 mmol

	$C_8H_{12}N_2O_3(aq)$	+ $OH^-(aq)$	$\rightleftarrows H_2O(l)$ +	$C_8H_{11}N_2O_3^-(aq)$
Init.	2.40 mmol			0
Add. Base		2.40 mmol		
Change	-2.40 mmol	-2.40 mmol		+2.40 mmol
Compl.	0	0		2.40 mmol

Only 2.40 mmol $C_8H_{11}N_2O_3^-$ present initially

	$C_8H_{11}N_2O_3^-(aq)$	$+ H_2O(l) \rightleftarrows$	$C_8H_{12}N_2O_3(aq)$	+ $OH^-(aq)$
Init. (M)	2.40 mmol/212 mL		0	0
Change (M)	-x		+x	+x
Equil. (M)	(2.40 mmol/212 mL)-x		x	x

$$K_b = \frac{K_w}{K_a} = \frac{1.0 \times 10^{-14}\ M^2}{1.0 \times 10^{-8}\ M} = 1 \times 10^{-6} = \frac{[C_8H_{12}N_2O_3]_{eq}[OH^-]_{eq}}{[C_8H_{11}N_2O_3^-]_{eq}}$$

$$= \frac{(x)(x)}{\frac{2.40}{212}M - x} \cong \frac{x^2}{\frac{2.40}{212}M}$$

$x^2 = 1.1 \times 10^{-8}\ M^2$ $\quad$ $x = 1.1 \times 10^{-4}\ M = [OH^-]_{eq}$

pOH = 4.0 $\quad$ pH = 14.0 - 4.0 = 10.0

d) Addition of 13.0 mL of 0.200M base results in 1.0 mL of excess 0.200M base. (It takes 12.0 mL of the base to reach the stoichiometric point.) The major species (besides H_2O and $C_8H_{11}N_2O_3^-$) is OH^- from this excess base.

Concentration of OH^- = (1.0 mL)(0.200 M)/(213 mL) = 9.4×10^{-4} M

$[OH^-] = 9.4 \times 10^{-4}$ $\quad$ pOH = 3.03 $\quad$ pH = 14.00 - 3.03 = 10.97

16.45 a) $H_2SO_{4(aq)} + H_2O_{(l)} \rightleftarrows HSO_4^-{}_{(aq)} + H_3O^+{}_{(aq)}$

$HSO_4^-{}_{(aq)} + H_2O_{(l)} \rightleftarrows SO_4^{2-}{}_{(aq)} + H_3O^+{}_{(aq)}$

Major Species: $H_2O_{(l)}$, $H_3O^+{}_{(aq)}$, $HSO_4^-{}_{(aq)}$

b) $Na_2SO_{4(s)} + H_2O_{(l)} \rightleftarrows 2\,Na^+{}_{(aq)} + SO_4^{2-}{}_{(aq)}$

$SO_4^{2-}{}_{(aq)} + H_2O_{(l)} \rightleftarrows HSO_4^-{}_{(aq)} + OH^-{}_{(aq)}$

Major Species: $H_2O_{(l)}$, $Na^+{}_{(aq)}$, $SO_4^{2-}{}_{(aq)}$

c) $CO_{2(g)} + H_2O_{(l)} \rightleftarrows H_2CO_{3(aq)}$

$H_2CO_{3(aq)} + H_2O_{(l)} \rightleftarrows HCO_3^-{}_{(aq)} + H_3O^+{}_{(aq)}$

Major species: $CO_{2(aq)} = H_2CO_{3(aq)}$

d) $NH_4Cl_{(s)} + H_2O_{(l)} \rightleftarrows NH_4^+{}_{(aq)} + Cl^-$

$NH_4^+{}_{(aq)} + H_2O_{(l)} \rightleftarrows NH_{3(aq)} + H_3O^+{}_{(aq)}$

Major species: $H_2O_{(l)}$, $NH_4^+{}_{(aq)}$, $Cl^-{}_{(aq)}$

16.46 a) The molecular picture that you will need to draw for this problem will need to show four water molecules, two oxalic acid molecules and four hydrogen oxalate ions ($HC_2O_4^-$).

b) The molecular picture that you will need to draw for this problem will need to show eight water molecules, four hydrogen oxalate ions ($HC_2O_4^-$) and two oxalate ions ($C_2O_4^{2-}$).

c) The molecular picture that you will need to draw for this problem will need to show two oxalic acid molecules, four ammonium ions (NH_4^+) and four hydrogen oxalate ions ($HC_2O_4^-$).

16.47 A $H_2PO_4^-$/ HPO_4^{2-} buffer is required.

$NaH_2PO_{4(s)}$/1.00 M NaOH $NaH_2PO_{4(s)}/Na_2HPO_{4(s)}/H_2O$

$Na_2HPO_{4(s)}$/1.00 M HCl

16.48 $K_b = 1.7 \times 10^{-6}$ M $pK_b = 5.77$

	$N_2H_{4(aq)}$ + $H_2O_{(l)}$ $\rightleftarrows$	$N_2H_5^+{}_{(aq)}$ +	$OH^-{}_{(aq)}$
Init. (M)	2.00×10^{-1}	0	0
Change (M)	-x	+x	+x
Equil. (M)	2.00×10^{-1}-x	x	x

$$K_b = \frac{[N_2H_5^+]_{eq}[OH^-]_{eq}}{[N_2H_4]_{eq}} = 1.7 \times 10^{-6}\ M = \frac{(x)(x)}{2.00 \times 10^{-1}\ M - x} \cong \frac{x^2}{2.00 \times 10^{-1}\ M}$$

$x = 5.8 \times 10^{-4} = [OH^-]_{eq}$ pOH = 3.23   pH = 14.00 - 3.23 = 10.77

$$H_2\ddot{N}-\ddot{N}H_2 + H-\ddot{\underset{..}{O}}-H \rightleftarrows H_2\ddot{N}-\overset{+}{N}H_3 + {}^-:\ddot{\underset{..}{O}}-H$$

16.49 $H_2\ddot{N}\text{–(chain)–}\ddot{N}H_2 + H\text{–}\ddot{O}\text{–}H \rightleftarrows H_2\ddot{N}\text{–(chain)–}NH_3^+ + {}^-:\ddot{O}\text{–}H$

$H_2\ddot{N}\text{–(chain)–}NH_3^+ + H\text{–}\ddot{O}\text{–}H \rightleftarrows H_3N^+\text{–(chain)–}NH_3^+ + {}^-:\ddot{O}\text{–}H$

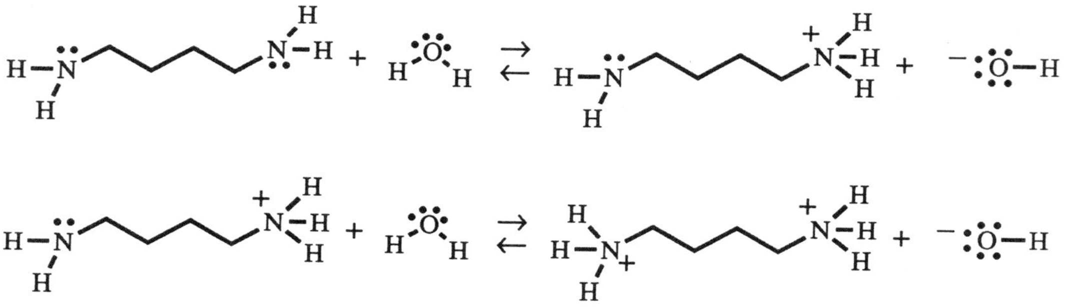

16.50 Calculations show that there will be four ammonium ions in this solution for every ammonia molecule in solution. Therefore, the molecular picture that you are to draw needs to show that ratio of species. Remember to use the symbols indicated in the problem.

16.51 $HA_{(aq)} + H_2O_{(l)} \rightleftarrows A^-_{(aq)} + H_3O^+_{(aq)}$ $\qquad K_a = \dfrac{[A^-]_{eq}[H_3O^+]_{eq}}{[HA]_{eq}}$

0.060	0	0
-x	+x	+x
0.060-x	x	x

$x = [H_3O^+]_{eq} = 10^{-pH} = 10^{-2.71} = 1.95 \times 10^{-3}$ M

$[HA]_{eq} = 0.060 - 1.95 \times 10^{-3} = 5.81 \times 10^{-2}$ M

$$K_a = \frac{(1.95 \times 10^{-3})^2}{5.81 \times 10^{-2}} = 6.54 \times 10^{-5}$$

$pK_a = -\log K_a = -\log(6.54 \times 10^{-5}) = 4.18$; benzoic acid

16.52 $H_3C\text{–}C(=\ddot{O}:)\text{–}\ddot{O}\text{–}H + H\text{–}\ddot{N}(H)\text{–}H \rightleftarrows H\text{–}N^+(H)(H)\text{–}H + H_3C\text{–}C(=\ddot{O}:)\text{–}:\ddot{O}:^-$

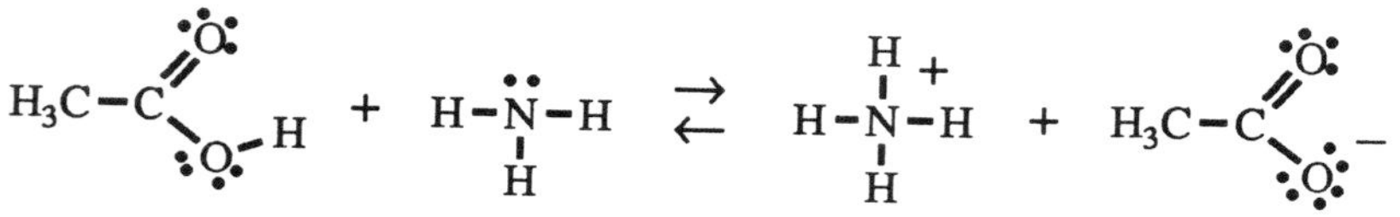

16.53 a) $CH_3CO_2H_{(aq)} + H_2O_{(l)} \rightleftarrows CH_3CO_2^-{}_{(aq)} + H_3O^+_{(aq)}$

1.00 M	0	0
-x	+x	+x
1.00-x	x	x

$$K_a = 1.8 \times 10^{-5}\ M = \frac{[CH_3CO_2^-]_{eq}[H_3O^+]_{eq}}{[CH_3CO_2H]_{eq}}$$

$1.8 \times 10^{-5} = x^2/1.00\text{-}x \approx x^2/1$; $x = 4.2 \times 10^{-3}$ M $= [H_3O^+]_{eq}$

$pH = -\log [H_3O^+]_{eq} = -\log(4.2 \times 10^{-3}) = 2.38$; acidic

(continued)

(16.53 continued)

b) $NH_{3(aq)} + H_2O_{(l)} \rightleftarrows NH_4^+{}_{(aq)} + OH^-{}_{(aq)}$ $K_b = \dfrac{[NH_4^+]_{eq}[OH^-]_{eq}}{[NH_3]_{eq}} = 1.8 \times 10^{-5}$

pOH = 2.38 pH = 14.00 - pOH = 14.00 - 2.38 = 11.62; basic

c) $NH_4Cl_{(s)} \rightleftarrows NH_4^+{}_{(aq)} + Cl^-$

$NH_4^+{}_{(aq)} + H_2O_{(l)} \rightleftarrows NH_3^+{}_{(aq)} + H_3O^+{}_{(aq)}$

$$K_a = \frac{[NH_3]_{eq}[H_3O^+]_{eq}}{[NH_4^+]_{eq}} = \frac{K_w}{K_b} = \frac{10^{-14}}{1.8 \times 10^{-5}} = 5.6 \times 10^{-10}$$

$$K_a = 5.6 \times 10^{-10} = \frac{x^2}{1.00 - x} \cong x^2;\ x = [H_3O^+] = \sqrt{5.6 \times 10^{-10}} = 2.4 \times 10^{-5}$$

pH = - log(2.4 x 10^{-5}) = 4.62; acidic

d) $NaCH_3CO_{2(s)} \rightleftarrows Na^-{}_{(aq)} + CH_3CO_2^-{}_{(aq)}$

	$CH_3CO_{2(aq)} + H_2O_{(l)} \rightleftarrows$	$CH_3CO_2H_{(aq)}$	$+ OH^-{}_{(aq)}$
Init.	1.00 M	0	0
Change	-x	+x	+x
Equil.	1.00-x	x	x

$$K_b = \frac{[CH_3CO_2H]_{eq}[OH^-]_{eq}}{[CH_3CO_2^-]_{eq}} = \frac{K_w}{K_a} = \frac{10^{-14}}{1.8 \times 10^{-5}} = 5.6 \times 10^{-10}$$

$x = [OH^-]_{eq} = \sqrt{5.6 \times 10^{-10}} = 2.4 \times 10^{-5}$

pOH = 4.62; pH = 14.00 - pOH = 14.00 - 4.62 = 9.38; basic

e) $NH_4CH_3CO_2 \rightleftarrows NH_4^+{}_{(aq)} + CH_3CO_2^-{}_{(aq)}$

$NH_4^+{}_{(aq)} + H_2O_{(l)} \rightleftarrows NH_3^+{}_{(aq)} + H_3O^+{}_{(aq)}$ $K_a = 5.6 \times 10^{-10}$

$CH_3CO_{2(aq)} + H_2O_{(l)} \rightleftarrows CH_3CO_2H_{(aq)} + OH^-{}_{(aq)}$ $K_b = 5.6 \times 10^{-10}$; neutral

16.54 The major species are NH_4^+, H_2O and the spectator ion is Cl^-

Reaction:	$NH_4^+{}_{(aq)} + H_2O_{(l)} \rightleftarrows$	$H_3O^+{}_{(aq)}$ +	$NH_{3(aq)}$
Initial (M)	0.025	0	0
Change (M)	-x	+x	+x
Equil. (M)	0.025-x	x	x

$$K_a = \frac{K_w}{K_a} = \frac{1.0 \times 10^{-14}\ M^2}{1.8 \times 10^{-5}\ M} = \frac{[H_3O^+]_{eq}[NH_3]_{eq}}{[NH_4^+]_{eq}} = \frac{(x)(x)}{0.025\ M - x} \cong \frac{x^2}{0.025\ M}$$

$x = 3.7 \times 10^{-6}\ M = [H_3O^+]_{eq}$ pH = - log(3.7 x 10^{-6}) = 5.43

16.55 a) $H_2CO_{3(aq)} + H_2O_{(l)} \rightleftarrows HCO_3^-{}_{(aq)} + H_3O^+{}_{(aq)}$

$HCO_3^-{}_{(aq)} + H_2O_{(l)} \rightleftarrows CO_3^{2-}{}_{(aq)} + H_3O^+{}_{(aq)}$ acidic

b) $KHCO_{3(s)} \rightleftarrows K^+{}_{(aq)} + HCO_3^-{}_{(aq)}$

acidic: $HCO_3^-{}_{(aq)} + H_2O_{(l)} \rightleftarrows CO_3^{2-}{}_{(aq)} + H_3O^+{}_{(aq)}$

basic: $HCO_3^-{}_{(aq)} + H_2O_{(l)} \rightleftarrows H_2CO_{3(s)} + OH^-{}_{(aq)}$ both

c) $NH_{3(aq)} + H_2O_{(l)} \rightleftarrows NH_4^+{}_{(aq)} + OH^-{}_{(aq)}$ basic

d) $NaCl_{(s)} \rightleftarrows Na^+{}_{(aq)} + Cl^-{}_{(aq)}$ neither

16.56 $[K^+]_{eq} = 0.400$ M, spectator ion

Reaction:	$CO_3^{2-}{}_{(aq)} + H_2O_{(l)} \rightleftarrows$	$HCO_3^-{}_{(aq)}$	$+ OH^-{}_{(aq)}$
Initial (M)	0.200		
Change (M)	-x	+x	+x
Equil. (M)	0.200-x	x	x

$$K_b = \frac{K_w}{K_a} = \frac{1.0 \times 10^{-14}\ M^2}{5.6 \times 10^{-11}\ M} = \frac{[HCO_3^-]_{eq}[OH^-]_{eq}}{[CO_3^{2-}]_{eq}} = \frac{(x)(x)}{0.200\ M - x} \cong \frac{x^2}{0.200\ M}$$

$x = 6.0 \times 10^{-3}\ M = [OH^-]_{eq} = [HCO_3^-]_{eq}$

$[CO_3^{2-}] = 0.200\ M - 6.0 \times 10^{-3}\ M = 0.194$ M (Note: This small difference from the initial concentration of CO_3^{2-} does not affect OH^- or HCO_3^- concentrations.)

Reaction:	$HCO_3^-{}_{(aq)} + H_2O_{(l)} \rightleftarrows$	$H_2CO_{3(aq)}$	$+ OH^-{}_{(aq)}$
Initial (M)	6.0×10^{-3}	0	6.0×10^{-3}
Change (M)	-z	+z	+z
Equil. (M)	6.0×10^{-3}-z	z	6.0×10^{-3}+z

$$K_b = \frac{K_w}{K_a} = \frac{1.0 \times 10^{-14}\ M^2}{4.3 \times 10^{-7}\ M} = \frac{[H_2CO_3]_{eq}[OH^-]_{eq}}{[HCO_3^-]_{eq}} = \frac{z(6.0 \times 10^{-3}\ M + z)}{(6.0 \times 10^{-3}\ M - z)}$$

$$\cong \frac{z(6.0 \times 10^{-3}\ M)}{(6.0 \times 10^{-3}\ M)}$$

$z = 2.3 \times 10^{-8}\ M = [H_2CO_3]_{eq}$ $\qquad [H_3O^+] = \frac{1.0 \times 10^{-14}\ M^2}{6.0 \times 10^{-3}\ M} = 1.7 \times 10^{-12}$

In summary:

$[K^+]_{eq} = 0.400$ M $\qquad [OH^-]_{eq} = 6.0 \times 10^{-3}$ M

$[H_3O^+]_{eq} = 1.7 \times 10^{-12}$ M $\qquad [CO_3^{2-}]_{eq} = 0.194$ M

$[HCO_3^-]_{eq} = 6.0 \times 10^{-3}$ M $\qquad [H_2CO_3]_{eq} = 2.3 \times 10^{-8}$ M

16.57 a) $NH_4NO_3 \rightleftarrows NH_4^+ + NO_3^-$ (NO_3^-; spectator ion)

$NH_4^+(aq) + H_2O(l) \rightleftarrows NH_3(aq) + H_3O^+(aq)$

Major species: $H_2O(l)$, $NH_4^+(aq)$

b) $KH_2PO_4 \rightleftarrows K^+ + H_2PO_4^-$

$H_2PO_4^-(aq) + H_2O(l) \rightleftarrows HPO_4^{2-}(aq) + H_3O^+(aq)$

Major species: $H_2PO_4^-(aq)$, $H_2O(l)$

c) $Na_2O + H_2O(l) \rightleftarrows 2\ OH^-(aq)^+ + 2\ Na^+$

Major species: $OH^-(aq)$, $H_2O(l)$

d) $C_6H_5CO_2H(aq) + H_2O(l) \rightleftarrows C_6H_5CO_2^-(aq) + H_3O^+(aq)$

Major species: $C_6H_5CO_2H(aq)$, $H_2O(l)$

16.58 $[NH_4^+]_i = \left(\frac{35.0\text{ g }NH_4Cl}{1\text{ L}}\right)\left(\frac{1\text{ mol}}{53.5\text{ g }NH_4Cl}\right) = 0.654\text{ M}$ $[NH_3]_i = 1.00\text{ M}$

$pK_a = 14.00 - 4.75 = 9.25$

$$pH = pK_a + \log\frac{[NH_3]}{[NH_4^+]} = 9.25 + \log\left(\frac{1.00\text{ M}}{0.654\text{ M}}\right) = 9.25 + 0.18 = 9.43$$

16.59 $pH = pK_a + \log\frac{[A^-]_{init}}{[HA]_{init}} = pK_a + \log\frac{[NH_3]}{[NH_4^+]}$

$[NH_3] = 1.00\text{ M}$ $[NH_4^+] = \frac{35.0\text{ g}}{53.5\text{ g / mol}} \Big/ 1\text{ L} = 0.65\text{ M}$

$$pH = pK_a + \log\frac{1}{0.65};\ pK_a = \frac{pK_w}{pK_b} = 14.00 - 4.75 = 9.25$$

$pH = 9.25 + 0.18 = 9.43$

$NH_3(aq) + H_3O^+(aq) \rightleftarrows NH_4^+(aq) + H_2O(l)$

NH_3	NH_4^+
1.00 M	0.65 M
-x	+x
1.00-x	0.65+x

$$pH = 9.25 + \log\frac{1.00 - x}{0.65 + x} = 9.43 - 0.05$$

$$\log\frac{1.00 - x}{0.65 + x} = 9.43 - 0.05 - 9.25 = 0.13$$

$\frac{1.00 - x}{0.65 + x} = 10^{0.13} = 1.35$ $1.00 - x = 1.35(0.65 + x) = 0.88 + 1.35x$

$0.12 = 2.35x$; $x = 0.05$ moles $H_3O^+(aq)$

16.60 pH = 3.00 $[H_3O^+] = [HCl] = 1.0 \times 10^{-3}$ M

(5.0 mL HCl)(1.0×10^{-3} M) = 5.0×10^{-3} mmoles H_3O^+

(0.654 M)(250 mL) = 163.5 mmoles NH_4^+

(1.00 M)(250 mL) = 250 mmoles NH_3

Reaction:	$NH_{3(aq)}$ +	$H_3O^+_{(aq)} \rightleftarrows$	$H_2O_{(l)}$ + $NH_4^+{}_{(aq)}$
Initial	250 mmol	0.005 mmol	163.5 mmol

The amount of HCl added is so small that there is no change to be calculated within the significant figures given.

16.61 $HPO_4^{2-}{}_{(aq)} + H_2O_{(l)} \rightleftarrows PO_4^{3-}{}_{(aq)} + H_3O^+{}_{(aq)}$

$$K_{a_3} = \frac{[PO_4^{3-}]_{eq}[H_3O^+]_{eq}}{[HPO_4^{2-}]_{eq}} = 4.8 \times 10^{-13} \text{ M}$$

$HPO_4^{2-}{}_{(aq)} + H_2O_{(l)} \rightleftarrows H_2PO_4^-{}_{(aq)} + OH^-{}_{(aq)}$

$$K_b = \frac{K_w}{K_{a_2}} = \frac{1 \times 10^{-14}}{6.2 \times 10^{-8}} = 1.6 \times 10^{-7} \text{ M}$$

$K_b > K_{a_3}$ ∴ second reaction dominates

$HPO_4^{2-}{}_{(aq)} + H_2O_{(l)} \rightleftarrows$	$H_2PO_4^-{}_{(aq)}$ +	$OH^-{}_{(aq)}$
0.250 M	0	0
-x	+x	+x
0.250 M - x	x	x

$$K_b = 1.6 \times 10^{-7} = \frac{[H_2PO_4^-]_{eq}[OH^-]_{eq}}{[HPO_4^{2-}]_{eq}} = \frac{x^2}{(0.250 - x)} \approx \frac{x^2}{0.250}$$

$x = [OH^-] = 2.0 \times 10^{-4}$

pH = 14.00 - pOH = 14.00 - 3.70 = 10.30

16.62 a) $K_{sp} = [Zn^{2+}][OH^-]^2 = 1.8 \times 10^{-14}$ M^3 *(From problem 16.43 and Appendix G $K_{sp}(Zn(OH)_2) = 1.2 \times 10^{-17}$. Values calculated using this value are shown below in parentheses.)*

$K_{sp} = [Fe^{3+}][OH^-]^3 = 1.1 \times 10^{-36}$ M^4

$[OH^-]_{max}$ if $[Zn^{2+}]$ = 0.300 M;

$$[OH^-] = \sqrt{\frac{K_{sp}}{[Zn^{2+}]}} = \sqrt{\frac{1.8 \times 10^{-14} \text{ M}^3}{0.300 \text{ M}}} = 2.4 \times 10^{-7} \text{ M } (= 6.3 \times 10^{-9})$$

$[Fe^{3+}]$ if $[OH^-] = 2.4 \times 10^{-7}$ M;

$$[Fe^{3+}] = \frac{K_{sp}}{[OH^-]^3} = \frac{1.1 \times 10^{-36} \text{ M}^4}{(2.4 \times 10^{-7} \text{ M})^3} = 8.0 \times 10^{-17} \text{ M } (= 4.4 \times 10^{-12})$$

(continued)

(16.62 continued)

Yes, Fe^{3+} can be separated from Zn^{2+} by addition of NaOH. Ignoring volume change, the concentration of Fe^{3+} will decrease to about 8.0×10^{-17} M *(4.4×10^{-12})* before the Zn^{2+} starts to precipitate.

b) Ignoring volume change, $[OH^-] = 2.4 \times 10^{-7}$ M; pOH = 6.62; pH = 7.38
($[OH^-] = 6.3 \times 10^{-9}$ M; pOH = 8.20; pH = 5.80)

c) The Fe^{3+} concentration at this pH would be 8.0×10^{-17} M *(4.4×10^{-12})*.

16.63 pH = 8.5
a) methyl orange = yellow
b) phenol red = red
c) bromocresol green = blue
d) thymol blue = yellow to green

16.64 a) $HCO_3^-(aq) + H_2O(l) \rightleftarrows H_3O^+(aq) + CO_3^{2-}(aq)$ $K_{a_2} = 5.6 \times 10^{-11}$ M

$HCO_3^-(aq) + H_2O(l) \rightleftarrows H_2CO_3(aq) + OH^-(aq)$

$$K_b = \frac{K_w}{K_a} = \frac{1.0 \times 10^{-14}\ M^2}{4.3 \times 10^{-7}\ M} = 2.3 \times 10^{-8}\ M$$

The second equilibrium above determines the pH of the solution because its K_{eq} is larger.

b) Reaction:	$HCO_3^-(aq)$ +	$H_2O(l) \rightleftarrows$	$H_2CO_3(aq)$ +	$OH^-(aq)$
Init. (M)	0.0228 mol/0.150 L		0	0
Change (M)	-x		+x	+x
Equil. (M)	0.152-x		x	x

$$K_b = 2.3 \times 10^{-8}\ M = \frac{[H_2CO_3]_{eq}[OH^-]_{eq}}{[H_2CO_3^-]_{eq}} = \frac{(x)(x)}{(0.152 - x)} \cong \frac{x^2}{0.152}$$

$x = 5.9 \times 10^{-5} = [OH^-]$

pOH = 4.23 pH = 14.00 - 4.23 = 9.77

16.65 $PhO^-(aq) + H_2O \rightleftarrows PhOH(aq) + OH^-(aq)$ $K_b = \frac{[PhOH]_{eq}[OH^-]_{eq}}{[PhO^-]_{eq}}$

	$PhO^-(aq)$	$PhOH(aq)$	$OH^-(aq)$
Init. (M)	0.010 M	0	0
Change (M)	-x	+x	+x
Equil. (M)	0.010-x	+x	$x = [OH^-] = 1.00 \times 10^{-3}$

pH = 11 = 14 - pOH; pOH = 3.00

$$K_b = \frac{x^2}{(0.010 - x)} = \frac{(1.00 \times 10^{-3})^2}{(0.010 - 1.00 \times 10^{-3})} = 1.11 \times 10^{-4}\ M$$

$$K_a = \frac{K_w}{K_b} = \frac{1.00 \times 10^{-14}\ M^2}{1.11 \times 10^{-4}\ M} = 9.00 \times 10^{-11}\ M$$

$pK_a = -\log K_a = 10.05$

16.66 For acetate - acetic acid buffer:

$$5.2 = 4.75 + \log\frac{[\text{acetate}]}{[\text{acetic acid}]} \qquad \log\frac{[\text{acetate}]}{[\text{acetic acid}]} = 0.45 \qquad \frac{[\text{acetate}]}{[\text{acetic acid}]} = 2.82$$

[acetate] = 2.82[acetic acid] = 2.82(0.50 M) = 1.41 M

(1.41 mol/L)(1.0 L)(82 g/mol) = 116 g

Take a clean 1.0 L volumetric flask and fill about half full with distilled water. Add 0.50 mol of glacial acetic acid {(60 g/mol)(0.50 mol) = 30 g} to the flask and mix. Then add 116 g of sodium acetate and mix until the sodium acetate dissolves. Add distilled water to the 1.0 L mark and mix thoroughly.

16.67

Reaction:	$H_3BO_{3(aq)} + H_2O_{(l)} \rightleftarrows$	$H_3O^+_{(aq)}$ +	$H_2BO_3^-{}_{(aq)}$
Initial (M)	0.050	0	0
Change (M)	-x	+x	+x
Equil. (M)	0.050-x	x	x

$$K_a = 7.3 \times 10^{-10}\ M = \frac{(x)(x)}{0.050\ M - x} \cong \frac{x^2}{0.050\ M}$$

$x = 6.0 \times 10^{-6} = [H_3O^+]_{eq}$ pH = 5.22

16.68 a) $pH = 4.75 + \log\frac{[\text{acetate}]}{[\text{acetic acid}]} = 4.75 + \log\left(\frac{7}{5}\right) = 4.75 + 0.15 = 4.90$

b) The molecular drawing required needs to show eight acetate ions and four acetic acid molecules since the added hydroxide ion will have reacted with one acetic acid to produce a water molecule and an acetate ion. Your drawing needs to use symbols similar to those used in the problem.

c) Since the HSO_4^- is a stronger acid than acetic acid, the one added ion of HSO_4^- will react with an acetate ion to produce a molecule of acetic acid and one ion of SO_4^{2-}. The molecular drawing requested needs to show a final solution that contains six acetate ions, six acetic acid molecules and one SO_4^{2-} ion. Your drawing needs to use symbols similar to those used in the problem.

16.69 $KCN \rightleftarrows K^+ + CN^-$

	$CN^-{}_{(aq)}$ + $H_2O_{(l)} \rightleftarrows$	$HCN_{(aq)}$ +	$OH^-{}_{(aq)}$
Init. (M)	2.00×10^{-2} M	0	0
Change (M)	-x	+x	+x
Equil. (M)	2.00×10^{-2}-x	x	x

$$K_b = \frac{[HCN]_{eq}[OH^-]_{eq}}{[CN^-]_{eq}} = 10^{-4.69} = 2.04 \times 10^{-5}$$

$x^2 = (2.04 \times 10^{-5})(2.00 \times 10^{-2} - x) = 4.08 \times 10^{-7} - 2.04 \times 10^{-5}x$

$0 = x^2 + 2.04 \times 10^{-5}x - 4.08 \times 10^{-7}$

$$x = \frac{-2.04 \times 10^{-5} \pm \sqrt{(2.04 \times 10^{-5})^2 - 4(1)(-4.08 \times 10^{-7})}}{2(1)}$$

$= 6.29 \times 10^{-4}\ M = [OH^-]_{eq}$ pOH = 3.20

$pH = 14.00 - pOH = 14.00 - (-\log[OH^-]_{eq}) = 14.00 + \log[OH^-]_{eq} = 10.80$

16.70 a) Spectator ions Na^+ and H_2O are major species. HCO_3^- also is a major species.

b) $HCO_3^-(aq) + H_2O(l) \rightleftarrows H_3O^+(aq) + CO_3^{2-}(aq)$ $\quad K_{a2} = 5.6 \times 10^{-11}$ M

$HCO_3^-(aq) + H_2O(l) \rightleftarrows H_2CO_3(aq) + OH^-(aq)$ $\quad K_b = 2.3 \times 10^{-8}$ M

c) The second equilibrium determines the pH of the solution because its K_{eq} is larger.

d)

Reaction:	$HCO_3^-(aq)$ + $H_2O(l)$ $\rightleftarrows$	$H_2CO_3(aq)$ +	$OH^-(aq)$
Initial (M)	0.200	0	0
Change (M)	-x	+x	+x
Equil. (M)	0.200-x	x	x

$$K_b = \frac{[H_2CO_3]_{eq}[OH^-]_{eq}}{[HCO_3^-]_{eq}} = 2.3 \times 10^{-8}\text{ M} = \frac{(x)(x)}{0.200\text{ M} - x} \cong \frac{x^2}{0.200\text{ M}}$$

$x = 6.8 \times 10^{-5}$ M = $[OH^-]_{eq} = [H_2CO_3]_{eq}$

$[HCO_3^-] = 0.200$ $\quad [Na^+] = 0.200$ M $\quad [H_2O] = 55.6$ M

$$[H_3O^+] = \frac{1.0 \times 10^{-14}\text{ M}^2}{6.8 \times 10^{-5}\text{ M}} = 1.5 \times 10^{-10}\text{ M}$$

e) pH = - log(1.5×10^{-10}) = 9.82, basic solution

16.71 $Na_2SO_3(aq) \rightleftarrows 2\,Na^+(aq) + SO_3^{2-}(aq)$

$SO_3^{2-}(aq) + H_2O(l) \rightleftarrows HSO_3^-(aq) + OH^-(aq)$

$CH_3CO_2H(aq) + H_2O(l) \rightleftarrows CH_3CO_2^-(aq) + H_3O^+(aq)$

a) Major species in each solution: $SO_3^{2-}(aq)$, $Na^+(aq)$, $H_2O(l)$

: $CH_3CO_2H(aq)$, $H_2O(l)$

b) $CH_3CO_2H(aq) + SO_3^{2-}(aq) \rightleftarrows HSO_3^-(aq) + CH_3CO_2^-(aq)$

c) acid: CH_3CO_2H

base: SO_3^{2-}

conjugate acid: HSO_3^-

conjugate base: $CH_3CO_2^-$

16.72 $$pH = pK_a + \log\frac{[\text{acetate}]}{[\text{acetic acid}]} \qquad 5.00 = 4.75 + \log\frac{[\text{acetate}]}{(0.100\text{ M})}$$

$$\log\frac{[\text{acetate}]}{(0.100\text{ M})} = 0.25 \qquad \frac{[\text{acetate}]}{(0.100\text{ M})} = 1.78 \qquad [\text{acetate}] = 0.178\text{ M}$$

$$(750\text{ mL})\left(\frac{1\text{ L}}{10^3\text{ mL}}\right)\left(\frac{0.178\text{ mol}}{\text{L}}\right)\left(\frac{82.0\text{ g sodium acetate}}{1\text{ mol}}\right)$$

= 10.9 g sodium acetate

16.73 White solid would be $Zn(OH)_2$. pH = 8.00 $[H_3O^+] = 10^{-8}$ $[OH^-] = 10^{-6}$

$Zn(OH)_{2(s)} \rightleftarrows Zn^{2+}_{(aq)} + 2\ OH^-_{(aq)}$

$pK_{sp} = 16.92$ $K_{sp} = 1.20 \times 10^{-17}$

$K_{sp} = [Zn^{2+}]_{eq}[OH^-]^2_{eq} = 1.20 \times 10^{-17}$ M

$$[Zn^{2+}]_{eq} = \frac{K_{sp}}{[OH^-]^2_{eq}} = \frac{1.20 \times 10^{-17}\ M^3}{(10^{-6}\ M)^2} = 1.20 \times 10^{-5}\ M$$

total mmoles of Zn^{2+} = (100 mL)(0.100 M) = 10.0 mmoles

mmoles of Zn^{2+} in solution = (400 mL)(1.20×10^{-5} M) = 4.80×10^{-3} mmoles

mmoles of Zn^{2+} precipitated = 10.0 mmoles - 4.80×10^{-3} mmoles = 9.9952 $\cong$ 10.0 mmoles

$$\text{mass precipitated} = (10.0\ \text{mmoles}\ Zn^{2+})\left(\frac{1\ \text{mmole}\ Zn(OH)_2}{1\ \text{mmole}\ Zn^{2+}}\right)\left(\frac{99.40 \times 10^{-3}\ g\ Zn(OH)_2}{\text{mmole}\ Zn(OH)_2}\right) = 0.994\ g\ Zn(OH)_2$$

16.74 $pH = pK_a = 2.36$ at the midpoint of the titration

$K_a = 10^{-2.36} = 4.4 \times 10^{-3}$ M

16.75 $$K_{eq} = \frac{[HClO]_{eq}[OH^-]_{eq}}{[ClO^-]_{eq}} = \frac{K_w}{K_a} = \frac{1.00 \times 10^{-14}\ M^2}{3.02 \times 10^{-8}\ M} = 3.31 \times 10^{-7}\ M$$

$NaClO \rightarrow Na^+_{(aq)} + ClO^-_{(aq)}$

	$ClO^-_{(aq)} + H_2O_{(l)} \rightleftarrows$	$HClO_{(aq)}$	$+ OH^-_{(aq)}$
Init.	1.00×10^{-2}	0	0
Change	-x	+x	+x
Equil.	1.00×10^{-2}-x	x	x

$$K_{eq} = \frac{x^2}{1.00 \times 10^{-2} - x} \approx \frac{x^2}{1.00 \times 10^{-2}} = 3.31 \times 10^{-7}$$

$x = \sqrt{(1.00 \times 10^{-2})(3.31 \times 10^{-7})} = 5.75 \times 10^{-5}$

$x = [OH^-]_{eq} = 5.75 \times 10^{-5}$ M

pOH = 4.24 pH = 14.00 - pOH = 14.00 - 4.24 = 9.76

16.76 a) KH_2PO_4 and K_2HPO_4 will be used to prepare the buffer solution.

b) $7.25 = 7.21 + \log\frac{[HPO_4^{2-}]}{[H_2PO_4^-]}$ $0.04 = \log\frac{[HPO_4^{2-}]}{(0.085\ M)}$

$\frac{[HPO_4^{2-}]}{0.085\ M} = 1.1$ $[HPO_4^{2-}] = (1.1)(0.085\ M) = 0.094$ M

KH_2PO_4: (0.085 mol/L)(1.5 L)(136 g KH_2PO_4/mol) = 17 g KH_2PO_4

K_2HPO_4: (0.094 mol/L)(1.5 L)(174 g K_2HPO_4/mol) = 25 g K_2HPO_4

16.77 $H_2PO_4^-{}_{(aq)} + H_2O_{(l)} \rightleftarrows HPO_4^{2-}{}_{(aq)} + H_3O^+{}_{(aq)}$

$K_a = 6.2 \times 10^{-8}$ M

$pK_a = 7.21$

$7.25 = 7.21 + \log\frac{[HPO_4^{2-}]}{[H_2PO_4^-]}$ $\quad\quad 0.04 = \log\frac{[HPO_4^{2-}]}{(0.085\ M)}$

$\frac{[HPO_4^{2-}]}{0.085\ M} = 1.1$ $\quad\quad [HPO_4^{2-}] = (1.1)(0.085\ M) = 0.094\ M$

Enzyme loses activity at pH = 7.1 $\quad 7.1 = 7.21 + \log\frac{[HPO_4^{2-}]}{[H_2PO_4^-]}$

$\log\frac{[HPO_4^{2-}]}{[H_2PO_4^-]} = -0.11$ $\quad\quad \frac{[HPO_4^{2-}]}{[H_2PO_4^-]} = 10^{-0.11} = 0.78$

$[HPO_4^{2-}]_{eq} = 0.78[H_2PO_4^-]_{eq}$

molarity of added $H_3O^+ = x\ M_{H_3O^+}$

$[HPO_4^{2-}]_{eq} = HPO_4^{2-}]_{init} - x\ M_{H_3O^+}$

$[H_2PO_4^-]_{eq} = H_2PO_4^-]_{init} + x\ M_{H_3O^+}$

$[HPO_4^{2-}]_{eq} = 0.094\ M - x\ M_{H_3O^+}$

$[H_2PO_4^-]_{eq} = 0.085\ M + x\ M_{H_3O^+}$

$0.094\ M - x\ M_{H_3O^+} = (0.78)(0.085\ M + x\ M_{H_3O^+})$

$0.094\ M - 0.066\ M = 1.78 \times M_{H_3O^+}$ $\quad\quad x\ M_{H_3O^+} = \frac{0.028\ M}{1.78} = 0.016\ M$

$$\text{moles } H_3O^+ = (0.016)\left(\frac{250\ mL}{1000\ mL/L}\right) = 4.0 \times 10^{-3} \text{ moles } H_3O^+$$

Note: This calculation is for the buffer with a pH of 7.25.

16.78 Unknown acid, HA, concentration: $\left(\frac{8.45\ g\ HA}{0.750\ L}\right)\left(\frac{1\ mol}{74.1\ g\ HA}\right) = 0.152\ M$

$[H_3O^+]_{eq} = 10^{-pH} = 10^{-2.846} = 1.426 \times 10^{-3}$

Reaction:	$HA_{(aq)}$ + $H_2O_{(l)}$ ⇄	$H_3O^+{}_{(aq)}$ +	$A^-{}_{(aq)}$
Initial (M)	0.152	0	0
Change (M)	-1.426×10^{-3}	$+1.426 \times 10^{-3}$	$+1.426 \times 10^{-3}$
Equil. (M)	0.151	1.426×10^{-3}	1.426×10^{-3}

$$K_a = \frac{[H_3O^+]_{eq}[A^-]_{eq}}{[HA]_{eq}} = \frac{(1.426 \times 10^{-3}\ M)^2}{0.151\ M} = 1.347 \times 10^{-5}\ M$$

$pK_a = 4.871$

16.79 $NH_{3(aq)}$ + $NH_4^+{}_{(aq)}$

M: 0.300 M 0.300 M

Vol.: 0.360 L 0.640 L

$$[NH_4^+]_{init.} = \frac{(0.640\ L)(0.300\ M)}{(0.640\ L + 0.360\ L)} = 0.192\ M$$

$$[NH_3]_{init.} = \frac{(0.360)(0.300)}{(0.360 + 0.640)} = 0.108\ M$$

a) $$pH = pK_a + \log\left\{\frac{[A^-]_{init.}}{[HA]_{init.}}\right\} = (pK_w - pK_b) + \log\frac{[NH_3]_{init.}}{[NH_4^+]_{init.}}$$

$$= (14.00 - 4.75) + \log\frac{0.108}{0.192} = 9.25 - 0.25 = 9.00$$

b) $NH_{3(aq)} + H_3O^+{}_{(aq)} \rightarrow NH_4^+{}_{(aq)} + H_2O_{(l)}$

c) 0.005 moles of H_3O^+ is added to 0.108 moles NH_3 and 0.192 mole NH_4^+, so a small amount of NH_3 gets protonated.

$[NH_3]$ = (0.108 mol - 0.005 mol)/1 L = 0.103 M

$[NH_4^+]$ = (0.192 + 0.005)/1 L = 0.197 M

pH = 9.25 + log(0.103/0.197) = 8.97

d) As long as the amount of added OH^- is less than the amount of NH_4^+ initially present, the balanced chemical equation is as follows:

$$NH_4^+{}_{(aq)} + OH^-{}_{(aq)} \rightarrow NH_{3(aq)} + H_2O_{(l)}$$

e) $$\frac{14.0\ g\ KOH}{56.11\ g/mol} = 0.250\ mol\ OH^-$$

Therefore, all of the NH_4^+ is converted to NH_3, with 0.058 M OH^- left over. The excess OH^- will determine the pH of the solution.

pH = 14.00 - (- log$[OH^-]$) = 14.00 - 1.24 = 12.76

16.80 Reaction: $C_{17}H_{19}NO_{3(aq)} + H_2O_{(l)} \rightleftarrows C_{17}H_{20}NO_3^+{}_{(aq)} + OH^-{}_{(aq)}$

Reaction:	$C_{17}H_{19}NO_{3(aq)}$	$C_{17}H_{20}NO_3^+{}_{(aq)}$	$OH^-{}_{(aq)}$
Initial (M)	0.015	0	0
Change (M)	-x	+x	+x
Equil. (M)	0.015-x	x	x

$$K_b = 10^{-pK_b} = 7.9 \times 10^{-7} = \frac{[C_{17}H_{20}NO_3^+]_{eq}[OH^-]_{eq}}{[C_{17}H_{19}NO_3]_{eq}} = \frac{(x)(x)}{0.015\ M - x} \cong \frac{x^2}{0.015\ M}$$

$x = 1 \times 10^{-4}\ M = [OH^-]_{eq}$

pOH = 4

pH = 14 - 4 = 10

16.81 $MgCO_{3(s)} \rightleftarrows Mg^{2+}{}_{(aq)} + CO_3{}^{2-}{}_{(aq)}$ $\qquad$ $pK_{sp} = 5.00$; $K_{sp} = 1 \times 10^{-5}\ M^2$

$CO_3{}^{2-}{}_{(aq)} + H_2O_{(l)} \rightleftarrows HCO_3{}^{-}{}_{(aq)} + OH^{-}{}_{(aq)}$

a) Major species: $Mg^{2+}{}_{(aq)}$, $CO_3{}^{2-}$, $H_2O_{(l)}$

b) $K_{sp} = [Mg^{2+}]_{eq}[CO_3{}^{2-}]_{eq} = 1.00 \times 10^{-5}\ M^2$

$[Mg^{2+}]_{eq} = \sqrt{1.00 \times 10^{-5}\ M^2} = 3.16 \times 10^{-3}\ M$

c) $$K_b = \frac{[HCO_3{}^-]_{eq}[OH^-]_{eq}}{[CO_3{}^{2-}]_{eq}} = \frac{K_w}{K_a} = \frac{1.00 \times 10^{-14}\ M^2}{5.6 \times 10^{-11}\ M} = 1.79 \times 10^{-4}\ M$$

	$CO_3{}^{2-}{}_{(aq)} + H_2O_{(l)} \rightleftarrows$	$HCO_3{}^{-}{}_{(aq)}$	$+ OH^{-}{}_{(aq)}$
Init. (M)	3.16×10^{-3} M	0	0
Change	-x	+x	+x
Equil. (M)	3.16×10^{-3}-x	x	x

$$K_b = \frac{x^2}{3.16 \times 10^{-3} - x} = 1.79 \times 10^{-4};$$

$x^2 = 1.79 \times 10^{-4}(3.16 \times 10^{-3} - x) = 5.65 \times 10^{-7} - 1.79 \times 10^{-4}x$

$0 = x^2 + 1.79 \times 10^{-4} - 5.65 \times 10^{-7}$

$$x = \frac{-1.79 \times 10^{-4} \pm \sqrt{(1.79 \times 10^{-4})^2 - 4(1)(-5.65 \times 10^{-7})}}{2(1)} = 6.67 \times 10^{-4}\ M = [OH^-]_{eq}$$

pH =14.00 - pOH = 14.00 - 3.18 = 10.82 = 10.8

d) $CaCO_{3(s)} \rightleftarrows Ca^{2+}{}_{(aq)} + CO_3{}^{2-}{}_{(aq)}$

$pK_{sp} = 8.55$ $\qquad$ $K_{sp} = [Ca^{2+}]_{eq}[CO_3{}^{2-}]_{eq} = 2.82 \times 10^{-9}\ M$

$[CO_3{}^{2-}]_{eq} = 3.16 \times 10^{-3} - 6.67 \times 10^{-4}\ M = 2.50 \times 10^{-3}\ M$

Note: This equilibrium results in a slight adjustment to $[Mg^{2+}]$ in part (b) to ~3.5 x 10^{-3} M.

$[Ca^{2+}]_{eq} = 2.82 \times 10^{-9}\ M^2/2.50 \times 10^{-3}\ M = 1.13 \times 10^{-6}\ M$

16.82 a) 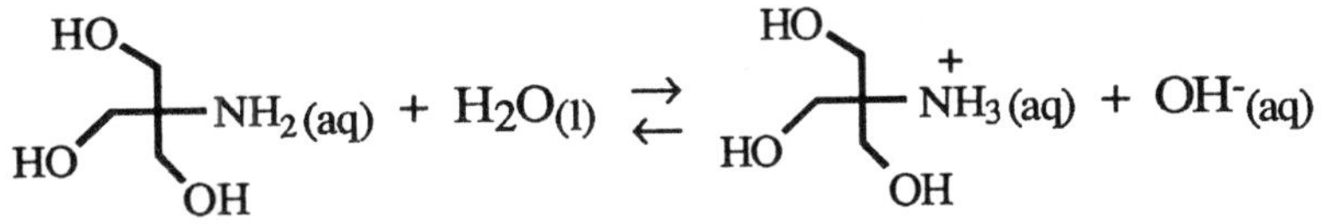 HO, HO, OH — $NH_{2(aq)}$ + $H_2O_{(l)} \rightleftarrows$ HO, HO, OH — $\overset{+}{N}H_{3(aq)}$ + $OH^{-}{}_{(aq)}$

or $(HOCH_2)_3CNH_{2(aq)} + H_2O_{(l)} \rightleftarrows (HOCH_2)_3CNH_3{}^{+}{}_{(aq)} + OH^{-}{}_{(aq)}$

or $TRIS_{(aq)} + H_2O_{(l)} \rightleftarrows TRISH^{+}{}_{(aq)} + OH^{-}{}_{(aq)}$

b) $H_3O^{+}{}_{(aq)} + Cl^{-}{}_{(aq)} + OH^{-}{}_{(aq)} \rightleftarrows 2\ H_2O_{(l)} + Cl^{-}{}_{(aq)}$

followed by: $TRIS_{(aq)} + H_2O_{(l)} \rightleftarrows TRISH^{+}{}_{(aq)} + OH^{-}{}_{(aq)}$

c) $TRISH^{+}{}_{(aq)} + OH^{-}{}_{(aq)} \rightleftarrows TRIS_{(aq)} + H_2O_{(l)}$

d) $TRIS_{(aq)} + H_3O^{+}{}_{(aq)} \rightleftarrows TRISH^{+}{}_{(aq)} + H_2O_{(l)}$

16.83 [TRIS] = 0.30 M $\qquad$ [TRISH+] = 0.60 M

$[H_3O^+]_{init.}$ = 12 M $\qquad$ Total vol. = 1.0 L

$TRIS_{(aq)} + H_3O^+_{(aq)} \rightleftarrows TRISH^+_{(aq)} + H_2O_{(l)}$

a) $pH = pKa + \log\left\{\frac{[A^-]_{init}}{[HA]_{init.}}\right\}$

$$= (14.00 - pK_b) + \log\frac{[TRIS]}{[TRISH^+]} = (14.0 - 5.7) + \log\frac{0.30}{0.60} = 8.0$$

b) $\frac{(5.0\ mL\ H_3O^+)}{1000\ mL/L} 12\ M\ H_3O^+ = 6.0 \times 10^{-2}$ moles $H_3O^+_{(aq)}$

$$[TRIS] = \frac{0.30\ mol - 0.06\ mol}{1.005\ L} = 0.239\ M \quad [TRISH^+] = \frac{0.60 + 0.06}{1.005} = 0.657\ M$$

$$pH = pK_a + \log\frac{0.239}{0.657} = 8.3 - 0.439 = 7.86 = 7.9$$

16.84 Moles of HCl = moles H_3O^+ = (0.035 L)(12 mol/L) = 0.42 moles

Reaction:	$TRIS_{(aq)}$ +	$H_3O^+_{(aq)} \rightleftarrows$	$TRISH^+_{(aq)} + H_2O_{(l)}$
Initial	0.30 moles		0.60 moles
Add. of HCl		+0.42 moles	
Change	-0.30 moles	-0.30 moles	+0.30 moles
Equil.	0	0.12 moles	.090 moles

for buffer: $pH = pK_a + \log\frac{[TRIS]}{[TRISH^+]} = 8.3 + \log\left(\frac{0.30}{0.60}\right) = 8.3 + (-0.3) = 8.0$

new solution, excess acid: $[H_3O^+] = \frac{0.12\ moles}{1.035\ L} \cong 0.12\ M$

pH = 0.9 $\quad$ The buffer capacity has been exceeded. The pH depends on the excess acid only.

16.85 From Problem 16.84: $\frac{35\ mL\ H_3O^+}{1000\ mL/L} 12\ M\ H_3O^+ = 0.42$ moles H_3O^+

$TRISH^+_{(aq)} + OH^-_{(aq)} \rightleftarrows TRIS_{(aq)} + H_2O_{(l)}$

Adding 0.42 moles of $OH^-_{(aq)}$ will restore the buffer to its original pH. Adding 0.63 moles of TRIS would also restore the buffer to its original pH.

16.86 $K_{sp} = [Fe^{2+}]_{eq}[OH^-]^2_{eq} = 10^{-15.10} = 7.9 \times 10^{-16}\ M^3$

$K_{sp} = [Ca^{2+}]_{eq}[OH^-]^2_{eq} = 10^{-5.26} = 5.5 \times 10^{-6}\ M^3$

$K_{sp} = [Mg^{2+}]_{eq}[OH^-]^2_{eq} = 10^{-10.74} = 1.8 \times 10^{-11}\ M^3$

a) $Fe(OH)_2$ precipitates first.

(continued)

(16.86 continued)

b) $[OH^-]_{eq} = \sqrt{\dfrac{K_{sp}}{[Fe^{2+}]_{eq}}} = \sqrt{\dfrac{7.9 \times 10^{-16}\ M^3}{0.10\ M}} = 8.9 \times 10^{-8}\ M$

pOH = 7.05   pH = 14.00 - 7.05 = 6.95

c) $[OH^-]_{eq} = \sqrt{\dfrac{K_{sp}}{[Mg^{2+}]_{eq}}} = \sqrt{\dfrac{1.8 \times 10^{-11}\ M^3}{0.10\ M}} = 1.3 \times 10^{-5}\ M$

pOH = 4.87   pH = 14.00 - 4.87 = 9.13

d) $[Fe^{2+}]_{eq} = \dfrac{K_{sp}}{[OH^-]^2_{eq}} = \dfrac{7.9 \times 10^{-16}\ M^3}{(1.3 \times 10^{-5}\ M)^2} = 4.7 \times 10^{-6}\ M$

16.87 Choose an indicator whose color changes as close to the stoichiometric point as possible: $pH_{stoic} = pK_{indicator} \pm 1$

$CH_3CO_2H_{(aq)} + OH^-_{(aq)} \rightleftarrows CH_3CO_2^-{}_{(aq)} + H_2O_{(l)}$

At the stoichiometric point all CH_3CO_2H has been converted to $CH_3CO_2^-$; the pH is determined by the acid-base equilibrium of $CH_3CO_2^-{}_{(aq)}$:

	$CH_3CO_2^-{}_{(aq)} + H_2O_{(l)}$	$\rightleftarrows$ $CH_3CO_2H_{(aq)}$	$+ OH^-{}_{(aq)}$
Init.	0.3 M	0	0
Change	-x	+x	+x
Equil.	0.3-x	x	x

$K_{eq} = \dfrac{K_w}{K_a} = \dfrac{1.00 \times 10^{-14}\ M^2}{1.8 \times 10^{-5}\ M} = 5.6 \times 10^{-10}\ M$

$5.6 \times 10^{-10}\ M = \dfrac{x^2}{0.3 - x} \approx \dfrac{x^2}{0.3}$; $x^2 = (0.3)(5.6 \times 10^{-10})$ assuming minimal volume change during titration.

$x = 1.3 \times 10^{-5}\ M = [OH^-]_{eq}$ pOH = 4.9   pH = 14 - pOH = 14 - 4.9 = 9.1

Possible indicators: thymol blue ($pK_{In} = 8.9$) phenolphthalein ($pK_{In} = 9.4$; best choice), or thymolphthalein ($pK_{In} = 10.0$). The pH at the stoichiometric point would be 9.0 if it is assumed an equal volume of base is added during titration.

16.88 $Mg(OH)_{2(s)} \rightleftarrows Mg^{2+}{}_{(aq)} + 2\ OH^-{}_{(aq)}$

As the pH increases, the $[OH^-]$ increases. The equilibrium shifts to the left and more $Mg(OH)_2$ precipitates. As the pH decreases, the $[OH^-]$ decreases (and $[H^+]$ increases). The equilibrium shifts to the right and more $Mg(OH)_2$ dissolves.

16.89 Some salts that would be more soluble in acidic solution than in pure water are: $Ca(OH)_2$, $Cu(OH)_2$, $Fe(OH)_2$, $Fe(OH)_3$, $Ni(OH)_2$.

Salts that would be less soluble in acidic solution than in pure water include: NH_4Cl, NH_4NO_3, NH_4ClO_4, H_2NH_3Cl, CH_3NH_3Cl.

Salts whose solubility does not depend on pH include: KCl, $KClO_4$, $NaClO_4$, NaCl, KNO_3, $NaNO_3$.

16.90 a) Use $pK_a = 10.25$ for HCO_3^- - CO_3^{2-} buffer system from Table 16-5.

$$pH = pKa + \log\frac{[CO_3^{2-}]}{[HCO_3^-]} \qquad 10.00 = 10.25 + \log\frac{[CO_3^{2-}]}{(0.25\ M)}$$

$$\log\frac{[CO_3^{2-}]}{(0.25\ M)} = -0.25$$

$$\frac{[CO_3^{2-}]}{(0.25\ M)} = 10^{-0.25} = 0.56 \qquad [CO_3^{2-}] = (0.25\ M)(0.56) = 0.14\ M$$

b) amount of Na_2CO_3:

$$(500\ mL)\left(\frac{1\ L}{10^3\ mL}\right)\left(\frac{0.14\ mol}{1\ L}\right)\left(\frac{106.00\ g\ Na_2CO_3}{mol}\right) = 7.4\ g\ Na_2CO_3$$

amount of $NaHCO_3$:

$$(500\ mL)\left(\frac{1\ L}{10^3\ mL}\right)\left(\frac{0.25\ mol}{1\ L}\right)\left(\frac{84.00\ g\ NaHCO_3}{mol}\right) = 11\ g\ NaHCO_3$$

Note: The buffer could also be prepared from 0.195 moles of Na_2CO_3 (21 g Na_2CO_3) and 0.125 moles of HCl (20.8 mL of 6.0 M HCl) by dissolving and diluting to 500 mL. The buffer could also be prepared from 0.195 moles of $NaHCO_3$ (16 g $NaHCO_3$) and 0.070 moles of NaOH (70 mL of 1.0 M NaOH) by dissolving and diluting to 500 mL.

16.91 a) At the stoichiometric point, the amount of added OH^- = NH_4^+ originally present.

$$\text{mol } OH^- = \frac{(0.375\ g\ NH_4Cl)}{53.5\ g/mol} = 7.01 \times 10^{-3}\ mol$$

$$\text{vol. of titrant} = \frac{7.01 \times 10^{-3}\ mol\ OH^-}{0.08775\ M\ OH^-} = 79.9\ mL$$

vol. final soln. = 25 + 79.9 = 104.9 mL

$$NH_4^+ + OH^- \rightarrow NH_3 + H_20 \qquad [NH_3] = \frac{7.01 \times 10^{-3}\ mol}{105\ mL/1000\ mL/L} = 6.68 \times 10^{-2} M$$

$$K_b = \frac{[NH_4^+]_{eq}[OH^-]_{eq}}{[NH_3]_{eq}} = 10^{-4.75} = 1.75 \times 10^{-5} M$$

	$NH_{3(aq)}$ + $H_2O_{(l)}$ $\rightleftarrows$	$NH_4^+{}_{(aq)}$ +	$OH^-{}_{(aq)}$
Init.	6.68×10^{-2}	0	0
Change	-x	+x	+x
Equil.	6.68×10^{-2}-x	x	x

$$K_b = \frac{x^2}{6.68 \times 10^{-2} - x} \approx \frac{x^2}{6.68 \times 10^{-2}} = 1.78 \times 10^{-5};$$

$$x = \sqrt{(6.68 \times 10^{-2})(1.78 \times 10^{-5})} = 1.09 \times 10^{-3} M$$

$pOH = -\log 1.09 \times 10^{-3} = 2.96 \qquad pH = 14.00 - 2.96 = 11.04 = 11.0$

(continued)

(16.91 continued)

b) $NH_{3(aq)} + H_3O^+_{(aq)} \rightarrow NH_4^+ + H_2O_{(l)}$

35 mL   x mL

0.15 M   0.537 M

mol added H_3O^+ = mol NH_3 =

$$\left(\frac{0.15 \text{ mol}}{\text{L}}\right)35 \text{ mL}\left(\frac{1 \text{ L}}{1000 \text{ mL}}\right) = 5.25 \times 10^{-3} \text{ moles}$$

$$\text{vol } H_3O^+ \text{ added} = \frac{5.25 \times 10^{-3} \text{ mol}}{0.537 \text{ mol / L}}\left(\frac{1000 \text{ mL}}{\text{L}}\right) = 9.78 \text{ mL}$$

Total vol. final soln = 35 + 9.78 = 44.8 mL

$[NH_4^+]_{init.} = (0.15)35/44.8 = 0.12$ M

$$K_a = \frac{K_w}{K_b} = \frac{10^{-14} \text{ M}^2}{1.78 \times 10^{-5} \text{M}} = 5.62 \times 10^{-10} \text{M}$$

	$NH_4^+{}_{(aq)} + H_2O_{(l)} \rightleftarrows$	$NH_{3(aq)}$	$+ H_3O^+{}_{(aq)}$
Init.	0.12	0	0
Change	-x	+x	+x
Equil.	0.12-x	x	x

$$K_a = \frac{x^2}{0.12 - x} \approx \frac{x^2}{0.12} = 5.62 \times 10^{-10};\ x^2 = (0.12)(5.62 \times 10^{-10})$$

$$x = [H_3O^+]_{eq} = \sqrt{(5.62 \times 10^{-10})(0.12)} = \sqrt{6.59 \times 10^{-11}} = 8.21 \times 10^{-6} \text{ M}$$

$pH = -\log[H_3O^+]_{eq} = 5.09$

c) $H_3O^+{}_{(aq)} + OH^-{}_{(aq)} \rightarrow 2\ H_2O_{(l)}$ pH = 7.0

16.92 Use the $H_2PO_4^-$ - HPO_4^{2-} buffer system.

$$pH = pK_a + \log\frac{[HPO_4^{2-}]}{[H_2PO_4^-]} \qquad 7.50 = 7.21 + \log\frac{[HPO_4^{2-}]}{[H_2PO_4^-]}$$

$$\log\frac{[HPO_4^{2-}]}{[H_2PO_4^-]} = 0.29 \qquad \frac{[HPO_4^{2-}]}{[H_2PO_4^-]} = 1.95 \qquad [HPO_4^{2-}] = 1.95\,[H_2PO_4^-]$$

0.125 mol of $H_2PO_4^-$ is needed to neutralize 0.125 mol of OH- ions.
Then (0.125 mol)(1.95) = 0.244 mol of $H_2PO_4^-$ would be needed for the buffer.
Amount of KH_2PO_4: (0.125 mol)(136.1 g/mol) = 17.0 g KH_2PO_4
Amount of K_2HPO_4: (0.244 mol)(174.2 g/mol) = 42.5 g K_2HPO_4
Dissolve 17.0 g KH_2PO_4 and 42.5 g K_2HPO_4 in about a liter of distilled water. Then dilute the solution to 1.5 L and mix thoroughly. Note: This buffer could also be prepared from 0.369 moles of KH_2PO_4 (50.2 g KH_2PO_4) and 0.244 moles of NaOH (2.44 g NaOH) by dissolving and diluting to 1.5 L. Or, this buffer could also be prepared from 0.369 moles of K_2HPO_4 (64.3 g K_2HPO_4) and 0.125 mol of HCl (0.125 mol ÷ 12.0 mol/L = 0.0104 L = 10.4 mL) by dissolving and diluting to 1.5 L.

(continued)

(16.92 continued)

Also note: This problem does not define the buffer capacity nor the pH range to be maintained. If 0.125 moles of H^+ ion were added to the buffer, the pH would be:

$$pH = 7.21 + \log\frac{[HPO_4^{2-}]}{[H_2PO_4^-]} = 7.21 + \log\left(\frac{0.119 \text{ mol}}{0.250 \text{ mol}}\right) = 7.21 + (-0.32) = 6.89$$

If 0.125 moles of OH- ion were added to the buffer, all of the $H_2PO_4^-$ would be converted to HPO_4^{2-}. As a result, 0.369 moles of HPO_4^{2-} would be present.

Reaction:	$HPO_4^{2-}{}_{(aq)} + H_2O \rightleftarrows$	$H_2PO_4^-{}_{(aq)}$	+ $OH^-{}_{(aq)}$
Init. (M)	0.369/1.5	0	0
Change (M)	-x	+x	+x
Equil.(M)	(0.369/1.5)-x	x	x

$$K_b = \frac{K_w M}{K_a} = \frac{1.0 \times 10^{-14} \text{ M}^2}{6.2 \times 10^{-8} \text{ M}} = 1.6 \times 10^{-7} \text{ M} = \frac{(x)(x)}{0.246 \text{ M} - x} \approx \frac{x^2}{0.246 \text{ M}}$$

$x = 2.0 \times 10^{-4}$ M = $[OH^-]$ pOH = 3.7 pH = 14.0 - 3.7 = 10.3

To keep the solution buffered near a pH of 7.50 if 0.125 moles of OH^- ions were added, more than the calculated amounts of KH_2PO_4 and K_2HPO_4 would have to be used (in the ratio of 1 to 1.95).

16.93 $Al(OH)_3 \rightleftarrows Al^{3+}{}_{(aq)} + 3\,OH^-{}_{(aq)}$ $K_{sp} = [Al^{3+}]_{eq}[OH^-]^3_{eq} = 1.8 \times 10^{-33}$ M^4

a) $K_{sp} = x(3x)^3$ $27x^4 = 1.8 \times 10^{-33}$ $x = \sqrt[4]{\frac{1.8 \times 10^{-33}}{27}} = 2.86 \times 10^{-9}$ M

$[Al^{3+}]_{eq} = x = 2.9 \times 10^{-9}$ M

b) $$[OH^-] = \frac{(1200 \text{ L})(0.250 \text{ M})}{1200 \text{ L} + 1300 \text{ L}} = 0.120 \text{ M}$$

$$[Al^{3+}] = \frac{(1300 \text{ L})(0.223 \text{ M})}{1200 \text{ L} + 1300 \text{ L}} = 0.116 \text{ M}$$

$1/K_{sp} = 1/1.8 \times 10^{-33}$ M^4 = 5.56×10^{32} M^{-4}

	$Al^{3+}{}_{(aq)}$ +	$3\,OH^-{}_{(aq)} \rightleftarrows Al(OH)_{3(s)}$
Init.	0.116	0.120
Change	-0.040	-0.120
Complete	0.076	0
Change to Equil.	y	3y
Equil.	0.076+y	3y

$K_{sp} = 1.8 \times 10^{-33}$ M^4 = $(0.076 + y)(3y)^3$ $y = 9.6 \times 10^{-12}$

Amount $Al(OH)_3$ that ppts = (0.040 M)(2500 L) = 100 moles

Mass of $Al(OH)_3$ = (100 moles)(78.0 g/mol = 7800 g = 7.80 kg

Residual conc. of Al^{3+} = 0.076 M = 7.6×10^{-2} M

16.94 a) $[H_3PO_4]_i = (14.7\ M)\left(\frac{35\ mL}{1000\ mL}\right) = 0.515\ M$

$$[OH^-]_i = \left(\frac{46.8\ g\ KOH}{1\ L}\right)\left(\frac{1\ mol}{56.1\ g\ KOH}\right) = 0.834\ M$$

Reaction:	$H_3PO_{4(aq)}$	+ $OH^-_{(aq)}$	$\rightleftarrows$ $H_2PO_4^-{}_{(aq)}$ + $H_2O_{(l)}$
Initial (M)	0.515	0.834	0
Change (M)	-0.515	-0.515	+0.515
Completion (M)	~0	0.319	0.515

Reaction:	$H_2PO_4^-{}_{(aq)}$	+ $OH^-_{(aq)}$	$\rightleftarrows$ $HPO_4^{2-}{}_{(aq)}$ + $H_2O_{(l)}$
Initial (M)	0.515	0.319	0
Change (M)	-0.319	-0.319	+0.319
Completion (M)	0.196	0	0.319

$$pH = pK_a + \log\frac{[HPO_4^{2-}]}{[H_2PO_4^-]} = 7.21 + \log\left(\frac{0.319}{0.196}\right) = 7.21 + 0.21 = 7.42$$

b) Reaction:	$HPO_4^{2-}{}_{(aq)}$	+ $H_3O^+{}_{(aq)}$	$\rightleftarrows$ $H_2PO_4^-{}_{(aq)}$ + $H_2O_{(l)}$
Initial (moles)	0.319	0.105	0.196
Change (moles)	-0.105	-0.105	+0.105
Completion (moles)	0.214	0	0.301

$$pH = 7.21 + \log\left(\frac{0.214}{0.301}\right) = 7.21 + (-0.15) = 7.06$$

c) $[H_3PO_4]_i = 0.515\ M$ (Same as for part (a))

$$[OH^-]_i = \left(\frac{46.8\ g\ NaOH}{1\ L}\right)\left(\frac{1\ mol}{40.0\ g\ NaOH}\right) = 1.17\ M$$

Reaction:	$H_3PO_{4(aq)}$	+ $2\ OH^-_{(aq)}$	$\rightleftarrows$ $HPO_4^{2-}{}_{(aq)}$ + $2\ H_2O_{(l)}$
Initial (M)	0.515	1.17	
Change (M)	-0.515	-1.03	+0.515
Completion (M)	0	0.14	0.515

Reaction:	$HPO_4^{2-}{}_{(aq)}$	+ $OH^-_{(aq)}$	$\rightleftarrows$ $PO_4^{3-}{}_{(aq)}$ + $H_2O_{(l)}$
Initial (M)	0.515	0.14	0
Change (M)	-0.14	-0.14	+0.14
Completion (M)	0.375	0	0.14

$$pH = pK_a + \log\frac{[PO_4^{3-}}{HPO_4^{2-}} = 12.31 + \log\left(\frac{0.14}{0.375}\right) = 12.31 + (-0.43) = 11.88$$

This is a buffer solution, but it is the HPO_4^{2-}— PO_4^{3-} system not the $H_2PO_4^-$ — HPO_4^{2-} system and it buffers at a pH range of 11-13 instead of 6-8.

16.95 a) $A = HCO_2H$, $B = HCO_2H + HCO_2^-$, $C = HCO_2^-$, $D = OH^- + HCO_2^-$

b) At the stoichiometric point, mol OH^- = mol HCO_2H originally present:
mol OH^- = (0.125 L HCO_2H)(0.135 M HCO_2H) = 1.688×10^{-2} moles
or mol OH^- = {29.8 mL/(1000 mL/L)}(0.567 M) = 1.690×10^{-2} moles

$HCO_2H_{(aq)} + OH^-_{(aq)} \rightarrow HCO_2^-{}_{(aq)} + H_2O_{(l)}$

Goes to completion to $[HCO_2^-]_{init} = [OH^-]$

$$[HCO_2^-]_{init.} = \frac{1.69 \times 10^{-2} \text{ moles}}{0.125 \text{ L} + 0.0298 \text{ L}} = 0.109 \text{ M}$$

$$K_{eq} = \frac{K_w}{K_a} = \frac{1 \times 10^{-14} \text{ M}^2}{10^{-3.75}} = 5.62 \times 10^{-11} \text{ M}$$

	$HCO_2^-{}_{(aq)}$ + $H_2O_{(l)}$	$\rightleftarrows$	$HCO_2H_{(aq)}$	+ $OH^-_{(aq)}$
Initial (M)	0.109 M		0	0
Change (M)	-x		+x	+x
Completion (M)	0.109-x		x	x

$$K_{eq} = \frac{x^2}{0.109 - x} \approx \frac{x^2}{0.109} = 5.62 \times 10^{-11} \text{ M}$$

$$x = \sqrt{(0.109)(5.62 \times 10^{-11})} = \sqrt{6.14 \times 10^{-12}} = 2.48 \times 10^{-6} \text{ M}$$

$x = [OH^-]_{eq}$ pOH = - log 2.48×10^{-6} = 5.61

$pH_{stoich.\ pt.}$ = 14.00 - 5.61 = 8.39

16.96 $$[C_{20}H_{24}N_2O_2]_i = \left(\frac{1.00 \text{ g}}{1.90 \times 10^2 \text{ L}}\right)\left(\frac{1 \text{ mol}}{324 \text{ g}}\right) = 1.62 \times 10^{-5} \text{ M}$$

a) Reaction: $C_{20}H_{24}N_2O_{2(aq)} + H_2O_{(l)} \rightleftarrows C_{20}H_{25}N_2O_2^+{}_{(aq)} + OH^-_{(aq)}$

	$C_{20}H_{24}N_2O_2$	$C_{20}H_{25}N_2O_2^+$	OH^-
Initial (M)	1.62×10^{-5}	0	0
Change (M)	-x	+x	+x
Equil. (M)	1.62×10^{-5}-x	x	x

$pK_b = 5.1$ $$K_b = 7.9 \times 10^{-6} \text{ M} = \frac{[C_{20}H_{25}N_2O_2^+]_{eq}[OH^-]_{eq}}{[C_{20}H_{24}N_2O_2]_{eq}}$$

Note: concentration and K_b are too similar for approximation.

$$7.9 \times 10^{-6} \text{ M} = \frac{x^2}{1.623 \times 10^{-5} \text{ M} - x} \qquad 0 = x^2 + 7.9 \times 10^{-6} x - 1.28 \times 10^{-10}$$

$$x = \frac{-7.9 \times 10^{-6} \pm \sqrt{(7.9 \times 10^{-6})^2 - (4)(1)(-1.28 \times 10^{-10})}}{2(1)}$$

$$= \frac{-7.9 \times 10^{-6} \pm 2.4 \times 10^{-5}}{2} = 8.0 \times 10^{-6} = [OH^-]$$

pOH = 5.10 pH = 14.00 - 5.10 = 8.90

(continued)

(16.96 continued)

b) $(100.0\ mL)(L/10^3\ mL)(1.62 \times 10^{-5}\ M) = 1.62 \times 10^{-6}$ moles $C_{20}H_{24}N_2O_2$

Volume of HCl needed = $(1.62 \times 10^{-6}\ moles)(1\ L/0.0100\ mol) = 1.62 \times 10^{-4}\ L$ or 0.162 mL

Reaction:	$C_{20}H_{25}N_2O_2^+{}_{(aq)}$ + $H_2O_{(l)}$ $\rightleftarrows$	$C_{20}H_{24}N_2O_{2(aq)}$ +	$H_3O^+{}_{(aq)}$
Initial (M)	1.62×10^{-5}	0	0
Change (M)	-x	+x	+x
Equil. (M)	1.62×10^{-5}-x	x	x

$$K_a = \frac{K_w}{K_b} = \frac{1.0 \times 10^{-14}\ M^2}{7.9 \times 10^{-6}\ M} = 1.27 \times 10^{-9}\ M = \frac{(x)(x)}{1.62 \times 10^{-5}\ M - x} \cong \frac{x^2}{1.62 \times 10^{-5}\ M}$$

$x = 1.4 \times 10^{-7}\ M = [H_3O^+]_{eq}$ pH = 6.85

16.97 $Cd(OH)_{2(s)} + H_2O_{(l)} \rightleftarrows Cd^{2+}{}_{(aq)} + 2\ OH^-{}_{(aq)}$ $pK_{sp} = 13.60$

$K_{sp} = 2.51 \times 10^{-14}\ M$

$K_{sp} = [Cd^{2+}]_{eq}[OH^-]^2_{eq} = x(2x)^2 = 4x^3 = 2.51 \times 10^{-14}$

$$x = \sqrt[3]{\frac{2.51 \times 10^{-14}}{4}} = \sqrt[3]{6.28 \times 10^{-15}} = 1.84 \times 10^{-5}\ M$$

$[OH^-]_{eq} = 2x = 2(1.84 \times 10^{-5}\ M) = 3.69 \times 10^{-5}\ M$; pOH = 4.43

pH = 14.00 - pOH = 14.00 - 4.43 = 9.57

16.98 The stoichiometric point is the point at which stoichiometric equivalent amounts of reactants have been combined. For example, for an acid-base titration the stoichiometric point is reached when the number of moles of OH^- ion added equals the number of moles of H_3O^+ ion present. The midpoint of a titration is reached when half of the volume required to reach the stoichiometric point has been added.

16.99 a) $NaOH_{(aq)} \rightleftarrows Na^+{}_{(aq)} + OH^-{}_{(aq)}$ completion

$C_6H_5CO_2H_{(s)} + H_2O_{(l)} \rightleftarrows C_6H_5CO_2^-{}_{(aq)} + H_3O^+{}_{(aq)}$ small extent

Sum: $NaOH_{(aq)} + C_6H_5CO_2H_{(s)} + H_2O_{(l)} \rightleftarrows$

$Na^+{}_{(aq)} + OH^-{}_{(aq)} + C_6H_5CO_2^-{}_{(aq)} + H_3O^+{}_{(aq)}$

Net: $OH^-{}_{(aq)} + C_6H_5CO_2H_{(s)} \rightarrow C_6H_5CO_2^-{}_{(aq)} + H_2O_{(l)}$ completion

b) $(CH_3)_3N_{(aq)} + H_2O_{(l)} \rightleftarrows (CH_3)_3NH^+{}_{(aq)} + OH^-{}_{(aq)}$ small extent

$HNO_{3(aq)} + H_2O_{(l)} \rightleftarrows H_3O^+{}_{(aq)} + NO_3^-{}_{(aq)}$ completion

Sum:

$Me_3N_{(aq)} + HNO_{3(aq)} + 2\ H_2O_{(l)} \rightleftarrows Me_3NH^+{}_{(aq)} + NO_3^-{}_{(aq)} + H_3O^+{}_{(aq)} + OH^-{}_{(aq)}$

Net: $(CH_3)_3N_{(aq)} + H_3O^+{}_{(aq)} \rightarrow (CH_3)_3NH^+{}_{(aq)} + H_2O_{(l)}$ completion

(continued)

(16.99 continued)

c) $Na_2SO_{4(aq)} \rightleftarrows 2\ Na^+_{(aq)} + SO_4^{2-}{}_{(aq)}$ completion

$CH_3CO_2H_{(aq)} + H_2O_{(l)} \rightleftarrows CH_3CO_2^-{}_{(aq)} + H_3O^+_{(aq)}$ small extent

Sum: $Na_2SO_{4(aq)} + CH_3CO_2H_{(aq)} + H_2O_{(l)}$

$\rightleftarrows 2\ Na^+_{(aq)} + CH_3CO_2^-{}_{(aq)} + SO_4^{2-}{}_{(aq)} + H_3O^+_{(aq)}$

Net: $SO_4^{2-}{}_{(aq)} + CH_3CO_2H_{(aq)} \rightleftarrows CH_3CO_2^-{}_{(aq)} + HSO_4^-{}_{(aq)}$ small extent

HSO_4^- $K_a = 1.1 \times 10^{-2}$; CH_3CO_2H $K_a = 1.8 \times 10^{-5}$. HSO_4^- is a stronger acid than CH_3CO_2H; ∴ reaction proceeds to only a small extent.

d) $NH_4Cl_{(aq)} \rightleftarrows NH_4^+{}_{(aq)} + Cl^-{}_{(aq)}$ completion

$NH_4^+ + H_2O_{(l)} \rightleftarrows NH_{3(aq)} + H_3O^+_{(aq)}$

$Ca(OH)_{2(aq)} \rightarrow Ca^{2+}{}_{(aq)} + 2\ OH^-{}_{(aq)}$

$OH^-{}_{(aq)} + H_3O^+_{(aq)} \rightarrow 2\ H_2O_{(l)}$ completion

Net: $NH_4^+{}_{(aq)} + OH^-{}_{(aq)} \rightarrow NH_{3(aq)} + H_2O_{(l)}$ completion

e) $K_2HPO_{4(aq)} \rightleftarrows 2\ K^+_{(aq)} + HPO_4^{2-}{}_{(aq)}$ completion

$HPO_4^{2-}{}_{(aq)} + NH_{3(aq)} \rightleftarrows NH_4^+{}_{(aq)} + PO_4^{3-}{}_{(aq)}$ small extent

$K_{a,\ HPO_4^{2-}} = 4.8 \times 10^{-13}$ $K_{a,\ NH_4^+} = 5.6 \times 10^{-10}$

So, NH_4^+ is a stronger acid than HPO_4^{2-}.

16.100 a)

Reaction:	$NO_2^-{}_{(aq)}$ + $H_2O_{(l)}$ $\rightleftarrows$	$HNO_{2(aq)}$ +	$OH^-{}_{(aq)}$
Initial (M)	0.150		
Change (M)	-x	+x	+x
Equil. (M)	0.150-x	x	x

$$K_b = \frac{K_w}{K_a} = \frac{1.0 \times 10^{-14}\ M^2}{7.1 \times 10^{-4}\ M} = 1.4 \times 10^{-11}\ M = \frac{(x)(x)}{0.150\ M - x} \cong \frac{x^2}{0.150\ M}$$

$x = 1.4 \times 10^{-6}\ M = [OH^-]_{eq}$

pOH = 5.85

pH = 14.00 - 5.85 = 8.15

b) At the halfway point, the solution is a buffer solution containing equal amounts of HNO_2 and NO_2^-.

$$pH = pK_a + \log\frac{[NO_2^-]}{[HNO_2]} = 3.15 + \log(1) = 3.15$$

(continued)

(16.100 continued)

c) moles of NO_2^- $= (200\ mL)\left(\frac{L}{10^3\ mL}\right)\left(\frac{0.150\ mol}{L}\right) = 0.0300$ moles

moles of NO_2^- = moles of HCl = 0.0300 moles

volume of 6 M HCl used to reach stoichiometric point

= (0.0300 moles)(1 L/ 6 moles) = 0.005 L = 5 mL

Reaction:	$HNO_{2(aq)}$	+	$H_2O_{(l)}$	$\rightleftarrows$	$H_3O^+{}_{(aq)}$	+	$NO_2^-{}_{(aq)}$
Initial (M)	0.0300 mol/0.205 L				0		0
Change (M)	-x				+x		+x
Equil. (M)	0.146-x				x		x

$$K_a = 7.1 \times 10^{-4}\ M = \frac{(x)(x)}{0.146\ M - x} \qquad x^2 + 7.1 \times 10^{-4}\ M\ x - 1.04 \times 10^{-4} = 0$$

$$x = \frac{-7.1 \times 10^{-4} \pm \sqrt{(7.1 \times 10^{-4})^2 - (4)(1)(-1.04 \times 10^{-4})}}{2}$$

$$= \frac{-7.1 \times 10^{-4} \pm 2.04 \times 10^{-2}}{2} = 9.8 \times 10^{-3}\ M = [H_3O^+]_{eq}$$

pH = - log(9.8×10^{-3}) = 2.01

d) Thymol blue (pH range of 1.2 - 2.8) is a suitable indicator for this titration.

16.101 $pH = pK_a + \log\frac{[CH_3CO_2^-]}{[CH_3CO_2H]}$ $\qquad 4.26 = 4.75 + \log\frac{[CH_3CO_2^-]}{[CH_3CO_2H]}$

$\frac{[CH_3CO_2^-]}{[CH_3CO_2H]} = 0.32$ or about one $CH_3CO_2^-$ for every three CH_3CO_2H

Using the symbols requested, draw a molecular picture that shows the one to three ratio in this buffer solution.

16.102 a) major species: H_2O, Cl^- and NH_4^+

equilibrium: $NH_4^+{}_{(aq)} + H_2O_{(l)} \rightleftarrows H_3O^+{}_{(aq)} + NH_{3(aq)}$

equilibrium constant: K_a

b) major species: H_2O, (Ca^{2+} and SO_4^{2-} ions are present at concentrations of 3×10^{-3} M)

equilibrium: $H_2O_{(l)} + H_2O_{(l)} \rightleftarrows H_3O^+{}_{(aq)} + OH^-{}_{(aq)}$

equilibrium constant: K_w

c) major species: H_2O, CH_3CO_2H, $CH_3CO_2^-$ and Na^+

equilibrium: $CH_3CO_2H_{(aq)} + H_2O_{(l)} \rightleftarrows CH_3CO_2^-{}_{(aq)} + H_3O^+{}_{(aq)}$

equilibrium constant: K_a

16.103

	$C_{20}H_{25}N_3O$ + H_2O $\rightleftarrows$	$C_{20}H_{26}N_3O^+(aq)$	+ $OH^-(aq)$
Init.	0.55 M	0	0
Change	-x	+x	+x
Equil.	0.55-x	x	x

$$K_b = \frac{[LSDH^+][OH^-]}{[LSD]} = 10^{-6.12} = 7.59 \times 10^{-7} \text{ M}$$

$$K_b = \frac{x^2}{0.55 - x} \cong \frac{x^2}{0.55} = 7.59 \times 10^{-7}$$

$$x = [OH^-]_{eq} = \sqrt{(0.55)(7.59 \times 10^{-7})} = 6.46 \times 10^{-4}$$

$pOH = -\log[OH^-]_{eq} = 3.19$

$pH = 14.00 - pOH = 14.00 - 3.19 = 10.81$

16.104 a) major species: H_2O, H_3O^+, Br^-, Ca^{2+}, OH^-

$$H_3O^+(aq) + OH^-(aq) \rightleftarrows 2\ H_2O(l)$$

b) major species: H_2O, Na^+, HSO_4^-, OH^-

$$OH^-(aq) + HSO_4^-(aq) \rightleftarrows H_2O(l) + SO_4^{2-}(aq)$$

c) major species: H_2O, NH_4^+, I^-, Pb^{2+}, NO_3^-

$$2\ I^-(aq) + Pb^{2+}(aq) \rightleftarrows PbI_2(s)$$

16.105 Glycine = H_2NCH_2COH

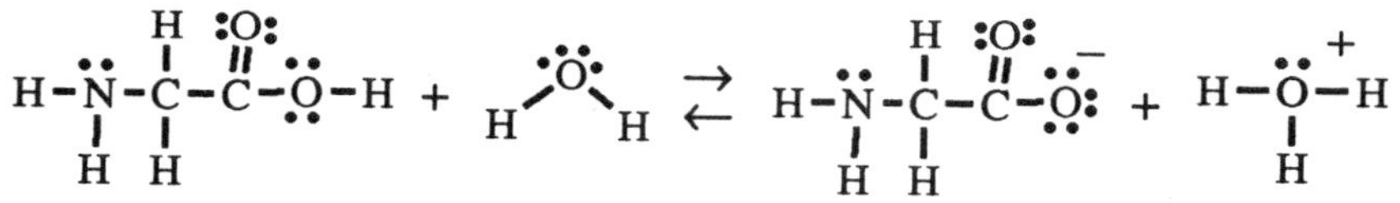

$$H_2N\text{-}CH_2\text{-}C(=O)\text{-}OH + H_2O \rightleftarrows H_2N\text{-}CH_2\text{-}C(=O)\text{-}O^- + H_3O^+$$

and

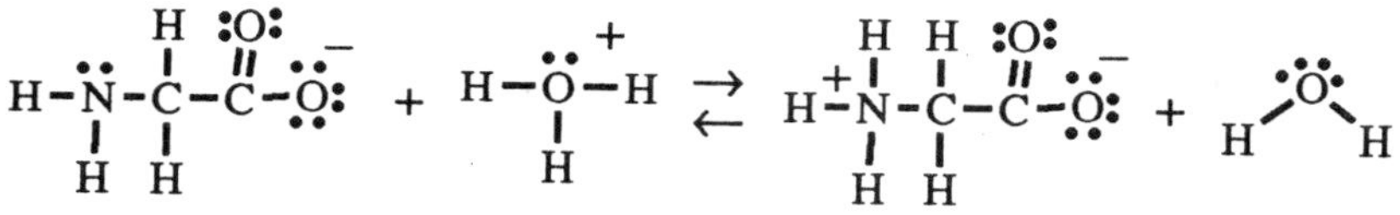

$$H_2N\text{-}CH_2\text{-}C(=O)\text{-}O^- + H_3O^+ \rightleftarrows H_3N^+\text{-}CH_2\text{-}C(=O)\text{-}O^- + H_2O$$

net reaction:

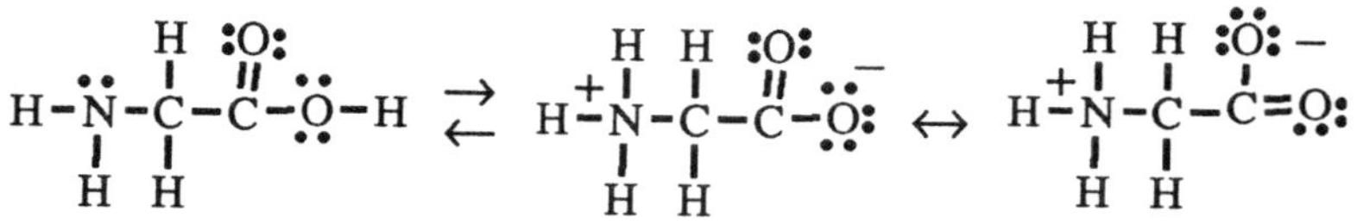

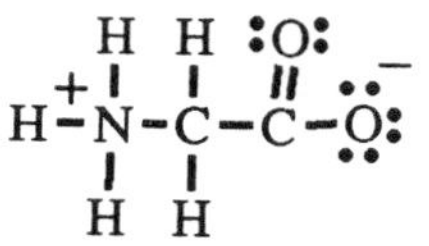

$$H_2N\text{-}CH_2\text{-}C(=O)\text{-}OH \rightleftarrows H_3N^+\text{-}CH_2\text{-}C(=O)\text{-}O^- \leftrightarrow H_3N^+\text{-}CH_2\text{-}C(\text{-}O^-)=O$$

$$H_3N^+\text{-}CH_2\text{-}C(=O)\text{-}O^- = \text{zwitterion.}$$

CHAPTER 17: ELECTRON TRANSFER REACTIONS: REDOX AND ELECTROCHEMISTRY

17.1 a) The oxidation numbers of the atoms in $Fe(OH)_3$ are: Fe = +3, O = -2, H = +1.
b) The oxidation numbers of the atoms in NH_3 are: N = -3 and H = +1.
c) The oxidation numbers of the atoms in PCl_5 are: P = +5 and Cl = -1.
d) The oxidation numbers of the atoms in K_2CO_3 are: K = +1, C = +4, O = -2.
e) The oxidation number of the atoms in P_4 is: P = 0.

17.2 a) The oxidation numbers of the atoms in $(CH_3)_2O$ are: C = -2, H = +1, O = -2.
b) The oxidation numbers of the atoms in $Al(OH)_4^-$ are: Al = +3, O = -2, H = +1.
c) The oxidation numbers of the atoms in XeF_4 are: Xe = +4 and F = -1.
d) The oxidation numbers of the atoms in F_2O (or OF_2) are: O = +2 and F = -1.
e) The oxidation numbers of the atoms in $KMnO_4$ are: K = +1, Mn = +7, O = -2.

17.3 (b) is a redox reaction in which Fe^{2+} is oxidized to Fe^{3+} and H_2O_2 is reduced to OH^-.

17.4 (a) is a redox reaction. O is reduced from 0 in O_2 to -2 in CH_3COOH. C is oxidized from -2 in CH_3OH to 0 in CH_3COOH.
(d) is a redox reaction. Fe is reduced from +3 in $Fe(C_2O_4)_3^{3-}$ to +2 in FeC_2O_4. C is oxidized from +3 in $C_2O_4^{2-}$ to +4 in CO_2.

17.5 a) Cl_2 b) HCl c) $HClO$ d) $HClO_2$ e) $HClO_3$ f) $HClO_4$

17.6 a) S_8 b) FeS_2 c) H_2S d) SCl_2 e) SO_2 or H_2SO_3
f) SO_3 or H_2SO_4

17.7 a) Oxidation $Na \rightarrow Na^+$ Reduction $H_2O \rightarrow H_2$

b) Oxidation $Au \rightarrow AuCl_4^-$ Reduction $HNO_3 \rightarrow NO$

c) Oxidation $C_2O_4^{2-} \rightarrow CO_2$ Reduction $MnO_4^- \rightarrow Mn^{2+}$

17.8 a) Oxidation $Co^{2+} \rightarrow Co^{3+}$ Reduction $H_2O_2 \rightarrow OH^-$
b) Oxidation $C_6H_{12}O_6 \rightarrow CO_2$ Reduction $O_2 \rightarrow H_2O$
c) Oxidation $PbSO_4 \rightarrow PbO_2 + SO_4^{2-}$ Reduction $PbSO_4 \rightarrow Pb + SO_4^{2-}$

17.9 a) Step 1: $Cu^+ \rightarrow CuO$ (acid solution)

Step 2: $\mathbf{H_2O} + Cu^+ \rightarrow CuO$

Step 3: $H_2O + Cu^+ \rightarrow CuO + \mathbf{2\ H^+_{(aq)}}$

Step 4: $H_2O + Cu^+ \rightarrow CuO + 2\ H^+_{(aq)} + \mathbf{e^-}$

b) Steps 1 & 2: $S \rightarrow H_2S$ (acid solution)

Step 3: $\mathbf{2\ H^+_{(aq)}} + S \rightarrow H_2S$

Step 4: $\mathbf{2\ e^-} + 2\ H^+_{(aq)} + S \rightarrow H_2S$

c) $AgCl \rightarrow Ag$ (basic solution)

Steps 1, 2 & 3: $AgCl \rightarrow Ag + \mathbf{Cl^-}$

Step 4: $\mathbf{e^-} + AgCl \rightarrow Ag + Cl^-$

d) Step 1: $I^- \rightarrow IO_3^-$ (basic solution)

Step 2: $\mathbf{3\ H_2O} + I^- \rightarrow IO_3^-$

Step 3: $3\ H_2O + I^- \rightarrow IO_3^- + \mathbf{6\ H^+_{(aq)}}$

Step 3 a: $3\ H_2O + I^- \rightarrow IO_3^- + 6\ H^+_{(aq)}$

$6\ OH^- + 6\ H^+_{(aq)} \rightarrow 6\ H_2O$

$\mathbf{6\ OH^-} + 3\ H_2O + I^- \rightarrow IO_3^- + \mathbf{6\ H_2O}$

Cancel H_2O on both sides: $6\ OH^- + I^- \rightarrow IO_3^- + \mathbf{3\ H_2O}$

Step 4: $6\ OH^- + I^- \rightarrow IO_3^- + 3\ H_2O + \mathbf{6\ e^-}$

e) Step 1: $IO_3^- \rightarrow IO^-$ (basic solution)

Step 2: $IO_3^- \rightarrow IO^- + \mathbf{2\ H_2O}$

Step 3: $\mathbf{4\ H^+_{(aq)}} + IO_3^- \rightarrow IO^- + 2\ H_2O$

Step 3 a: $4\ H^+_{(aq)} + IO_3^- \rightarrow IO^- + 2\ H_2O$

$4\ H_2O \rightarrow 4\ H^+_{(aq)} + 4\ OH^-$

$\mathbf{4\ H_2O} + IO_3^- \rightarrow IO^- + 2\ H_2O + \mathbf{4\ OH^-}$

Cancel H_2O on both sides: $\mathbf{2\ H_2O} + IO_3^- \rightarrow IO^- + 4\ OH^-$

Step 4: $\mathbf{4\ e^-} + 2\ H_2O + IO_3^- \rightarrow IO^- + 4\ OH^-$

f) Step 1: $H_2CO \rightarrow CO_2$ (acid solution)

Step 2: $\mathbf{H_2O} + H_2CO \rightarrow CO_2$

Step 3: $H_2O + H_2CO \rightarrow CO_2 + \mathbf{4\ H^+_{(aq)}}$

Step 4: $H_2O + H_2CO \rightarrow CO_2 + 4\ H^+_{(aq)} + \mathbf{4\ e^-}$

17.10 a) Step 1: $SbH_3 \rightarrow Sb$ (acid solution)

Steps 2 & 3: $SbH_3 \rightarrow Sb + \mathbf{3\ H^+_{(aq)}}$

Step 4: $SbH_3 \rightarrow Sb + 3\ H^+_{(aq)} + \mathbf{3\ e^-}$

(continued)

(17.10 continued)

b) Step 1: $AsO_2^- \rightarrow As$ (basic solution)

Step 2: $AsO_2^- \rightarrow As + \mathbf{2\ H_2O}$

Step 3: $\mathbf{4\ H^+_{(aq)}} + AsO_2^- \rightarrow As + 2\ H_2O$

Step 3 a: $4\ H^+_{(aq)} + AsO_2^- \rightarrow As + 2\ H_2O$

$4\ H_2O \rightarrow 4\ H^+_{(aq)} + 4\ OH^-$

$\mathbf{4\ H_2O} + AsO_2^- \rightarrow As + 2\ H_2O + \mathbf{4\ OH^-}$

Cancel H_2O on both sides: $\mathbf{2\ H_2O} + AsO_2^- \rightarrow As + 4\ OH^-$

Step 4: $2\ H_2O + AsO_2^- + \mathbf{3\ e^-} \rightarrow As + 4\ OH^-$

c) $BrO_3^- \rightarrow Br_2$ (acid solution)

Step 1: $\mathbf{2}\ BrO_3^- \rightarrow Br_2$

Step 2: $2\ BrO_3^- \rightarrow Br_2 + \mathbf{6\ H_2O}$

Step 3: $\mathbf{12\ H^+_{(aq)}} + 2\ BrO_3^- \rightarrow Br_2 + 6\ H_2O$

Step 4: $12\ H^+_{(aq)} + 2\ BrO_3^- + \mathbf{10\ e^-} \rightarrow Br_2 + 6\ H_2O$

d) Step 1: $Cl^- \rightarrow ClO_2^-$ (basic solution)

Step 2: $\mathbf{2\ H_2O} + Cl^- \rightarrow ClO_2^-$

Step 3: $2\ H_2O + Cl^- \rightarrow ClO_2^- + \mathbf{4\ H^+_{(aq)}}$

Step 3 a: $2\ H_2O + Cl^- \rightarrow ClO_2^- + 4\ H^+_{(aq)}$

$4\ H^+_{(aq)} + 4\ OH^- \rightarrow 4\ H_2O$

$2\ H_2O + Cl^- + \mathbf{4\ OH^-} \rightarrow ClO_2^- + \mathbf{4\ H_2O}$

Cancel H_2O on both sides: $Cl^- + 4\ OH^- \rightarrow ClO_2^- + \mathbf{2\ H_2O}$

Step 4: $Cl^- + 4\ OH^- \rightarrow ClO_2^- + 2\ H_2O + \mathbf{4\ e^-}$

e) Step 1: $Sb_2O_5 \rightarrow Sb_2O_3$ (acid solution)

Step 2: $Sb_2O_5 \rightarrow Sb_2O_3 + \mathbf{2\ H_2O}$

Step 3: $\mathbf{4\ H^+_{(aq)}} + Sb_2O_5 \rightarrow Sb_2O_3 + 2\ H_2O$

Step 4: $\mathbf{4\ e^-} + 4\ H^+_{(aq)} + Sb_2O_5 \rightarrow Sb_2O_3 + 2\ H_2O$

f) Step 1 and 2: $H_2O_2 \rightarrow O_2$ (basic solution)

Step 3: $H_2O_2 \rightarrow O_2 + \mathbf{2\ H^+_{(aq)}}$

Step 3 a: $H_2O_2 \rightarrow O_2 + 2\ H^+_{(aq)}$

$2\ H^+_{(aq)} + 2\ OH^- \rightarrow 2\ H_2O$

$\mathbf{2\ OH^-} + H_2O_2 \rightarrow O_2 + \mathbf{2\ H_2O}$

Step 4: $2\ OH^- + H_2O_2 \rightarrow O_2 + 2\ H_2O + \mathbf{2\ e^-}$

17.11 a) Oxidation: $H_2O + Cu^+ \rightarrow CuO + 2\ H^+_{(aq)} + e^-$

Reduction: $2\ e^- + 2\ H^+_{(aq)} + S \rightarrow H_2S$

Multiply the oxidation half-reaction by 2:

$2\ H_2O + 2\ Cu^+ \rightarrow 2\ CuO + 4\ H^+_{(aq)} + 2\ e^-$

Combine the half reactions:

$2\ H_2O + 2\ Cu^+ + 2\ e^- + 2\ H^+_{(aq)} + S \rightarrow 2\ CuO + 4\ H^+_{(aq)} + 2\ e^- + H_2S$

Cancel duplicate species:

$2\ H_2O + 2\ Cu^+ + S \rightarrow 2\ CuO + 2\ H^+_{(aq)} + H_2S$

b) Oxidation: $6\ OH^- + I^- \rightarrow IO_3^- + 3\ H_2O + 6\ e^-$

Reduction: $e^- + AgCl \rightarrow Ag + Cl^-$

Multiply the reduction half-reaction by 6:

$6\ e^- + 6\ AgCl \rightarrow 6\ Ag + 6\ Cl^-$

Combine the half reactions:

$6\ OH^- + I^- + 6\ e^- + 6\ AgCl \rightarrow IO_3^- + 3\ H_2O + 6\ e^- + 6\ Ag + 6\ Cl^-$

Cancel duplicate species:

$6\ OH^- + I^- + 6\ AgCl \rightarrow IO_3^- + 3\ H_2O + 6\ Ag + 6\ Cl^-$

c) Oxidation: $6\ OH^- + I^- \rightarrow IO_3^- + 3\ H_2O + 6\ e^-$

Reduction: $4\ e^- + 2\ H_2O + IO_3^- \rightarrow IO^- + 4\ OH^-$

Multiply the oxidation half-reaction by 2:

$12\ OH^- + 2\ I^- \rightarrow 2\ IO_3^- + 6\ H_2O + 12\ e^-$

Multiply the reduction half-reaction by 3:

$12\ e^- + 6\ H_2O + 3\ IO_3^- \rightarrow 3\ IO^- + 12\ OH^-$

Combine the half-reactions:

$12\ OH^- + 2\ I^- + 12\ e^- + 6\ H_2O + 3\ IO_3^- \rightarrow$

$2\ IO_3^- + 6\ H_2O + 12\ e^- + 3\ IO^- + 12\ OH^-$

Cancel duplicate species: $2\ I^- + IO_3^- \rightarrow 3\ IO^-$

d) Oxidation: $H_2CO + H_2O \rightarrow CO_2 + 4\ H^+_{(aq)} + 4\ e^-$

Reduction: $2\ e^- + 2\ H^+_{(aq)} + S \rightarrow H_2S$

Multiply the reduction half-reaction by 2:

$4\ e^- + 4\ H^+_{(aq)} + 2\ S \rightarrow 2\ H_2S$

Combine the half-reactions:

$H_2CO + H_2O + 4\ e^- + 4\ H^+_{(aq)} + 2\ S \rightarrow CO_2 + 4\ H^+_{(aq)} + 4\ e^- + 2\ H_2S$

Cancel duplicate species:

$H_2CO + H_2O + 2\ S \rightarrow CO_2 + 2\ H_2S$

17.12 a) Reduction: $12\ H^+_{(aq)} + 2\ BrO_3^- + 10\ e^- \rightarrow Br_2 + 6\ H_2O$

Oxidation: $SbH_3 \rightarrow Sb + 3\ H^+_{(aq)} + 3\ e^-$

Multiply the reduction half-reaction by 3:

$36\ H^+_{(aq)} + 6\ BrO_3^- + 30\ e^- \rightarrow 3\ Br_2 + 18\ H_2O$

(continued)

(17.12 continued)

Multiply the oxidation half-reaction by 10: $10\ SbH_3 \rightarrow 10\ Sb + 30\ H^+_{(aq)} + 30\ e^-$

Combine the half-reactions: $10\ SbH_3 + 36\ H^+_{(aq)} + 6\ BrO_3^- + 30\ e^- \rightarrow$

$10\ Sb + 30\ H^+_{(aq)} + 30\ e^- + 3\ Br_2 + 18\ H_2O$

Cancel duplicate species:

$10\ SbH_3 + 6\ H^+_{(aq)} + 6\ BrO_3^- \rightarrow 10\ Sb + 3\ Br_2 + 18\ H_2O$

b) Reduction: $4\ e^- + 4\ H^+_{(aq)} + Sb_2O_5 \rightarrow Sb_2O_3 + 2\ H_2O$

Oxidation: $SbH_3 \rightarrow Sb + 3\ H^+_{(aq)} + 3\ e^-$

Multiply the reduction half-reaction by 3:

$12\ e^- + 12\ H^+_{(aq)} + 3\ Sb_2O_5 \rightarrow 3\ Sb_2O_3 + 6\ H_2O$

Multiply the oxidation half-reaction by 4: $4\ SbH_3 \rightarrow 4\ Sb + 12\ H^+_{(aq)} + 12\ e^-$

Combine the half-reactions: $12\ e^- + 12\ H^+_{(aq)} + 3\ Sb_2O_5 + 4\ SbH_3$

$\rightarrow 3\ Sb_2O_3 + 6\ H_2O + 4\ Sb + 12\ H^+_{(aq)} + 12\ e^-$

Cancel duplicate species: $3\ Sb_2O_5 + 4\ SbH_3 \rightarrow 3\ Sb_2O_3 + 6\ H_2O + 4\ Sb$

c) Reduction: $2\ H_2O + AsO_2^- + 3\ e^- \rightarrow As + 4\ OH^-$

Oxidation: $Cl^- + 4\ OH^- \rightarrow ClO_2^- + 2\ H_2O + 4\ e^-$

Multiply the red. half-reaction by 4: $8\ H_2O + 4\ AsO_2^- + 12\ e^- \rightarrow 4\ As + 16\ OH^-$

Multiply the ox. half-reaction by 3: $3\ Cl^- + 12\ OH^- \rightarrow 3\ ClO_2^- + 6\ H_2O + 12\ e^-$

Combine the half-reactions: $8\ H_2O + 4\ AsO_2^- + 12\ e^- + 3\ Cl^- + 12\ OH^-$

$\rightarrow 4\ As + 16\ OH^- + 3\ ClO_2^- + 6\ H_2O + 12\ e^-$

Cancel duplicate species: $2\ H_2O + 4\ AsO_2^- + 3\ Cl^- \rightarrow 4\ As + 4\ OH^- + 3\ ClO_2^-$

d) Reduction: $2\ H_2O + AsO_2^- + 3\ e^- \rightarrow As + 4\ OH^-$

Oxidation: $2\ OH^- + H_2O_2 \rightarrow O_2 + 2\ H_2O + 2\ e^-$

Multiply the reduction half-reaction by 2: $4\ H_2O + 2\ AsO_2^- + 6\ e^- \rightarrow 2\ As + 8\ OH^-$

Multiply the oxidation half-reaction by 3: $6\ OH^- + 3\ H_2O_2 \rightarrow 3\ O_2 + 6\ H_2O + 6\ e^-$

Combine the half-reactions: $4\ H_2O + 2\ AsO_2^- + 6\ e^- + 6\ OH^- + 3\ H_2O_2$

$\rightarrow 2\ As + 8\ OH^- + 3\ O_2 + 6\ H_2O + 6\ e^-$

Cancel duplicate species: $2\ AsO_2^- + 3\ H_2O_2 \rightarrow 2\ As + 2\ OH^- + 3\ O_2 + 2\ H_2O$

17.13 a) $PbO + Co(NH_3)_6^{3+} \rightarrow PbO_2 + Co(NH_3)_6^{2+}$ (basic)

I and II. Oxidation:

Step 1: $PbO \rightarrow PbO_2$

Step 2: $\mathbf{H_2O} + PbO \rightarrow PbO_2$

Step 3: $H_2O + PbO \rightarrow PbO_2 + \mathbf{2\ H^+_{(aq)}}$

Step 3 a: $H_2O + PbO \rightarrow PbO_2 + 2\ H^+_{(aq)}$

$2\ H^+_{(aq)} + 2\ OH^- \rightarrow 2\ H_2O$

$\mathbf{2\ OH^-} + H_2O + PbO \rightarrow PbO_2 + \mathbf{2\ H_2O}$

(continued)

(17.13 continued)

Cancel duplicate species: $2\ OH^- + PbO \rightarrow PbO_2 + \mathbf{H_2O}$

Step 4: $2\ OH^- + PbO \rightarrow PbO_2 + H_2O + \mathbf{2\ e^-}$

Reduction:

Steps 1 to 3: $Co(NH_3)_6^{3+} \rightarrow Co(NH_3)_6^{2+}$

Step 4: $\mathbf{e^-} + Co(NH_3)_6^{3+} \rightarrow Co(NH_3)_6^{2+}$

III. Multiply the red. half-reaction by 2: $2\ e^- + 2\ Co(NH_3)_6^{3+} \rightarrow 2\ Co(NH_3)_6^{2+}$

IV. Combine the half-reactions: $2\ OH^- + PbO + 2\ e^- + 2\ Co(NH_3)_6^{3+} \rightarrow$
$PbO_2 + H_2O + 2\ e^- + Co(NH_3)_6^{2+}$

Cancel duplicate species:

$2\ OH^- + PbO + 2\ Co(NH_3)_6^{3+} \rightarrow PbO_2 + H_2O + Co(NH_3)_6^{2+}$

b) $O_2 + As \rightarrow HAsO_2 + H_2O$ (acidic)

I and II. Oxidation

Step 1: $As \rightarrow HAsO_2$

Step 2: $\mathbf{2\ H_2O} + As \rightarrow HAsO_2$

Step 3: $2\ H_2O + As \rightarrow HAsO_2 + \mathbf{3\ H^+_{(aq)}}$

Step 4: $2\ H_2O + As \rightarrow HAsO_2 + 3\ H^+_{(aq)} + \mathbf{3\ e^-}$

Reduction:

Step 1: $O_2 \rightarrow H_2O$

Step 2: $O_2 \rightarrow \mathbf{2}\ H_2O$

Step 3: $\mathbf{4\ H^+_{(aq)}} + O_2 \rightarrow 2\ H_2O$

Step 4: $\mathbf{4\ e^-} + 4\ H^+_{(aq)} + O_2 \rightarrow 2\ H_2O$

III. Multiply the ox. half-rxn. by 4: $8\ H_2O + 4\ As \rightarrow 4\ HAsO_2 + 12\ H^+_{(aq)} + 12\ e^-$

Multiply the reduction half-reaction by 3: $12\ e^- + 12\ H^+_{(aq)} + 3\ O_2 \rightarrow 6\ H_2O$

IV. Combine the half-reactions: $8\ H_2O + 4\ As + 12\ e^- + 12\ H^+_{(aq)} + 3\ O_2 \rightarrow$
$4\ HAsO_2 + 12\ H^+_{(aq)} + 12\ e^- + 6\ H_2O$

Cancel duplicate species: $2\ H_2O + 4\ As + 3\ O_2 \rightarrow 4\ HAsO_2$

c) $Br^- + MnO_4^- \rightarrow MnO_2 + BrO_3^-$ (basic)

I and II. Oxidation:

Step 1: $Br^- \rightarrow BrO_3^-$

Step 2: $\mathbf{3\ H_2O} + Br^- \rightarrow BrO_3^-$

Step 3: $3\ H_2O + Br^- \rightarrow BrO_3^- + \mathbf{6\ H^+_{(aq)}}$

Step 3 a: $3H_2O + Br^- \rightarrow BrO_3^- + 6\ H^+_{(aq)}$

$6\ H^+_{(aq)} + 6\ OH^- \rightarrow 6\ H_2O$

$\mathbf{6\ OH^-} + 3\ H_2O + Br^- \rightarrow BrO_3^- + \mathbf{6\ H_2O}$

Cancel duplicate species: $6\ OH^- + Br^- \rightarrow BrO_3^- + \mathbf{3}\ H_2O$

(continued)

(17.13 continued)

Step 4: $6\ OH^- + Br^- \rightarrow BrO_3^- + 3\ H_2O +$ **$6\ e^-$**

Reduction:

Step 1: $MnO_4^- \rightarrow MnO_2$

Step 2: $MnO_4^- \rightarrow MnO_2 + \mathbf{2\ H_2O}$

Step 3: $\mathbf{4\ H^+_{(aq)}} + MnO_4^- \rightarrow MnO_2 + 2\ H_2O$

Step 3 a: $4\ H^+_{(aq)} + MnO_4^- \rightarrow MnO_2 + 2\ H_2O$

$4\ H_2O \rightarrow 4\ H^+_{(aq)} + 4\ OH^-$

$\mathbf{4\ H_2O} + MnO_4^- \rightarrow MnO_2 + 2\ H_2O + \mathbf{4\ OH^-}$

Cancel duplicate species: $\mathbf{2}\ H_2O + MnO_4^- \rightarrow MnO_2 + 4\ OH^-$

Step 4: $\mathbf{3\ e^-} + 2\ H_2O + MnO_4^- \rightarrow MnO_2 + 4\ OH^-$

III. Multiply the red. half-rxn. by 2: $6\ e^- + 4\ H_2O + 2\ MnO_4^- \rightarrow 2\ MnO_2 + 8\ OH^-$

IV. Combine the half-reactions: $6\ OH^- + Br^- + 6\ e^- + 4\ H_2O + 2\ MnO_4^- \rightarrow$
$BrO_3^- + 3\ H_2O + 6\ e^- + 2\ MnO_2 + 8\ OH^-$

Cancel duplicate species: $Br^- + H_2O + 2\ MnO_4^- \rightarrow BrO_3^- + 2\ MnO_2 + 2\ OH^-$

d) $NO_2 \rightarrow NO_3^- + NO$ (acidic)

I and II. Oxidation

Step 1: $NO_2 \rightarrow NO_3^-$

Step 2: $\mathbf{H_2O} + NO_2 \rightarrow NO_3^-$

Step 3: $H_2O + NO_2 \rightarrow NO_3^- + \mathbf{2\ H^+_{(aq)}}$

Step 4: $H_2O + NO_2 \rightarrow NO_3^- + 2\ H^+_{(aq)} + \mathbf{e^-}$

Reduction:

Step 1: $NO_2 \rightarrow NO$

Step 2: $NO_2 \rightarrow NO + \mathbf{H_2O}$

Step 3: $\mathbf{2\ H^+_{(aq)}} + NO_2 \rightarrow NO + H_2O$

Step 4: $\mathbf{2\ e^-} + 2\ H^+_{(aq)} + NO_2 \rightarrow NO + H_2O$

III. Multiply the ox. half-rxn. by 2: $2\ H_2O + 2\ NO_2 \rightarrow 2\ NO_3^- + 4\ H^+_{(aq)} + 2\ e^-$

IV. Combine the half-reactions: $2\ H_2O + 2\ NO_2 + 2\ e^- + 2\ H^+_{(aq)} + NO_2 \rightarrow$
$2\ NO_3^- + 4\ H^+_{(aq)} + 2\ e^- + NO + H_2O$

Cancel duplicate species: $H_2O + 3\ NO_2 \rightarrow 2\ NO_3^- + 2\ H^+_{(aq)} + NO$

e) $ClO_4^- + Cl^- \rightarrow ClO^- + Cl_2$ (acidic)

I and II. Oxidation: $Cl^- \rightarrow Cl_2$

Step 1: $\mathbf{2}\ Cl^- \rightarrow Cl_2$

(continued)

(17.13 continued)

Steps 2, 3, 4: $2\ Cl^- \rightarrow Cl_2 + \mathbf{2\ e^-}$

Reduction:

Step 1: $ClO_4^- \rightarrow ClO^-$

Step 2: $ClO_4^- \rightarrow ClO^- + \mathbf{3\ H_2O}$

Step 3: $\mathbf{6\ H^+_{(aq)}} + ClO_4^- \rightarrow ClO^- + 3\ H_2O$

Step 4: $\mathbf{6\ e^-} + 6\ H^+_{(aq)} + ClO_4^- \rightarrow ClO^- + 3\ H_2O$

III. Multiply the oxidation half-reaction by 3: $6\ Cl^- \rightarrow 3\ Cl_2 + 6\ e^-$

IV. Combine the half-reactions:

$$6\ Cl^- + 6\ e^- + 6\ H^+_{(aq)} + ClO_4^- \rightarrow 3\ Cl_2 + ClO^- + 3\ H_2O + 6\ e^-$$

Cancel duplicate series:

$$6\ Cl^- + 6\ H^+_{(aq)} + ClO_4^- \rightarrow 3\ Cl_2 + 6\ e^- + ClO^- + 3\ H_2O$$

Note: The oxidation and reduction half-reactions chosen could also be:

Oxidation: $Cl^- \rightarrow ClO^-$

Reduction: $ClO_4^- \rightarrow Cl_2$

The balanced reaction would then be:

$7\ Cl^- + 2\ H^+_{(aq)} + 2\ ClO_4^- \rightarrow 7\ ClO^- + Cl_2 + H_2O$

f) $AlH_4^- + H_2CO \rightarrow Al^{3+} + CH_3OH$ (basic)

I and II. Oxidation

Steps 1 & 2: $AlH_4^- \rightarrow Al^{3+}$

Step 3: $AlH_4^- \rightarrow Al^{3+} + \mathbf{4\ H^+_{(aq)}}$

Step 3 a: $AlH_4^- \rightarrow Al^{3+} + 4\ H^+_{(aq)}$

$4\ H^+_{(aq)} + 4\ OH^- \rightarrow 4\ H_2O$

$\mathbf{4\ OH^-} + AlH_4^- \rightarrow Al^{3+} + \mathbf{4\ H_2O}$

Step 4: $4\ OH^- + AlH_4^- \rightarrow Al^{3+} + 4\ H_2O + \mathbf{8\ e^-}$

Reduction:

Steps 1 & 2: $H_2CO \rightarrow CH_3OH$

Step 3: $\mathbf{2\ H^+_{(aq)}} + H_2CO \rightarrow CH_3OH$

Step 3 a: $2\ H^+_{(aq)} + H_2CO \rightarrow CH_3OH$

$2\ H_2O \rightarrow 2\ H^+_{(aq)} + 2\ OH^-$

$\mathbf{2\ H_2O} + H_2CO \rightarrow CH_3OH + \mathbf{2\ OH^-}$

Step 4: $\mathbf{2\ e^-} + 2\ H_2O + H_2CO \rightarrow CH_3OH + 2\ OH^-$

III. Multiply the red. half-rxn. by 4: $8\ e^- + 8\ H_2O + 4\ H_2CO \rightarrow 4\ CH_3OH + 8\ OH^-$

IV. Combine the half-reactions: $4\ OH^- + AlH_4^- + 8\ e^- + 8\ H_2O + 4\ H_2CO \rightarrow$
$Al^{3+} + 4\ H_2O + 8\ e^- + 4\ CH_3OH + 8\ OH^-$

Cancel duplicate species: $AlH_4^- + 4\ H_2O + 4\ H_2CO \rightarrow Al^{3+} + 4\ CH_3OH + 4\ OH^-$

17.14 a) $H_5IO_6 + Cr \rightarrow IO_3^- + Cr^{3+}$ (acidic)

I and II. Oxidation:

Steps 1-3: $Cr \rightarrow + Cr^{3+}$

Step 4: $Cr \rightarrow + Cr^{3+} + \mathbf{3\ e^-}$

Reduction:

Step 1: $H_5IO_6 \rightarrow IO_3^-$

Step 2: $H_5IO_6 \rightarrow IO_3^- + \mathbf{3\ H_2O}$

Step 3: $\mathbf{H^+_{(aq)}} + H_5IO_6 \rightarrow IO_3^- + 3\ H_2O$

Step 4: $H^+_{(aq)} + H_5IO_6 + \mathbf{2\ e^-} \rightarrow IO_3^- + 3\ H_2O$

III. Multiply the oxidation half-reaction by 2: $2\ Cr \rightarrow 2\ Cr^{3+} + 6\ e^-$

Multiply the red. half-reaction by 3: $3\ H^+_{(aq)} + 3\ H_5IO_6 + 6\ e^- \rightarrow 3\ IO_3^- + 9\ H_2O$

IV. Combine the half-reactions:

$$2\ Cr + 3\ H^+_{(aq)} + 3\ H_5IO_6 + 6\ e^- \rightarrow 2\ Cr^{3+} + 6\ e^- + 3\ IO_3^- + 9\ H_2O$$

Cancel duplicate species: $2\ Cr + 3\ H^+_{(aq)} + 3\ H_5IO_6 \rightarrow 2\ Cr^{3+} + 3\ IO_3^- + 9\ H_2O$

b) $Se + Cr(OH)_3 \rightarrow Cr + SeO_3^{2-}$ (basic)

I and II. Oxidation:

Step 1: $Se \rightarrow SeO_3^{2-}$

Step 2: $\mathbf{3\ H_2O} + Se \rightarrow SeO_3^{2-}$

Step 3: $3\ H_2O + Se \rightarrow SeO_3^{2-} + \mathbf{6\ H^+_{(aq)}}$

Step 3 a: $3\ H_2O + Se \rightarrow SeO_3^{2-} + 6\ H^+_{(aq)}$

$6\ H^+_{(aq)} + 6\ OH^- \rightarrow 6\ H_2O$

$\mathbf{6\ OH^-} + 3\ H_2O + Se \rightarrow SeO_3^{2-} + \mathbf{6\ H_2O}$

Cancel duplicate species: $6\ OH^- + Se \rightarrow SeO_3^{2-} + \mathbf{3}\ H_2O$

Step 4: $6\ OH^- + Se \rightarrow SeO_3^{2-} + 3\ H_2O + \mathbf{4\ e^-}$

Reduction:

Step 1: $Cr(OH)_3 \rightarrow Cr$

Step 2: $Cr(OH)_3 \rightarrow Cr + \mathbf{3\ H_2O}$

Step 3: $\mathbf{3\ H^+_{(aq)}} + Cr(OH)_3 \rightarrow Cr + 3\ H_2O$

Step 3 a: $3\ H^+_{(aq)} + Cr(OH)_3 \rightarrow Cr + 3\ H_2O$

$3\ H_2O \rightarrow 3\ H^+_{(aq)} + 3\ OH^-$

$\mathbf{3\ H_2O} + Cr(OH)_3 \rightarrow \mathbf{3\ OH^-} + Cr + 3\ H_2O$

Cancel duplicate species: $Cr(OH)_3 \rightarrow 3\ OH^- + Cr$

Step 4: $\mathbf{3\ e^-} + Cr(OH)_3 \rightarrow 3\ OH^- + Cr$

III. Multiply the ox. half-rxn. by 3: $18\ OH^- + 3\ Se \rightarrow 3\ SeO_3^{2-} + 9\ H_2O + 12\ e^-$

Multiply the reduction half-reaction by 4: $12\ e^- + 4\ Cr(OH)_3 \rightarrow 12\ OH^- + 4\ Cr$

(continued)

(17.14 continued)

IV. Combine the half-reactions: $18\ OH^- + 3\ Se + 12\ e^- + 4\ Cr(OH)_3 \rightarrow$
$3\ SeO_3^{2-} + 9\ H_2O + 12\ e^- + 12\ OH^- + 4\ Cr$

Cancel duplicate species: $6\ OH^- + 3\ Se + 4\ Cr(OH)_3 \rightarrow 3\ SeO_3^{2-} + 9\ H_2O + 4\ Cr$

c) $HClO + Co \rightarrow Cl_2 + Co^{2+}$ (acidic)

I and II. Oxidation:

Steps 1-3: $Co \rightarrow Co^{2+}$

Step 4: $Co \rightarrow Co^{2+} + \mathbf{2\ e^-}$

Reduction: $HClO \rightarrow Cl_2$

Step 1: $\mathbf{2}\ HClO \rightarrow Cl_2$

Step 2: $2\ HClO \rightarrow Cl_2 + \mathbf{2\ H_2O}$

Step 3: $\mathbf{2\ H^+_{(aq)}} + 2\ HClO \rightarrow Cl_2 + 2\ H_2O$

Step 4: $\mathbf{2\ e^-} + 2\ H^+_{(aq)} + 2\ HClO \rightarrow Cl_2 + 2\ H_2O$

III. The oxidation half-reaction produces 2 e^- and the reduction half-reaction consumes 2 e^-.

IV. Combine the half-reactions: $2\ e^- + 2\ H^+_{(aq)} + 2\ HClO + Co \rightarrow$
$Cl_2 + 2\ H_2O + Co^{2+} + 2\ e^-$

Cancel duplicate species: $2\ H^+_{(aq)} + 2\ HClO + Co \rightarrow Cl_2 + 2\ H_2O + Co^{2+}$

d) $CH_3CHO + Cu^{2+} \rightarrow CH_3COO^- + Cu_2O$ (basic)

I and II. Oxidation:

Step 1: $CH_3CHO \rightarrow CH_3COO^-$

Step 2: $\mathbf{H_2O} + CH_3CHO \rightarrow CH_3COO^-$

Step 3: $H_2O + CH_3CHO \rightarrow CH_3COO^- + \mathbf{3\ H^+_{(aq)}}$

Step 3 a: $H_2O + CH_3CHO \rightarrow CH_3COO^- + 3\ H^+_{(aq)}$

$3\ H^+_{(aq)} + 3\ OH^- \rightarrow 3\ H_2O$

$\mathbf{3\ OH^-} + H_2O + CH_3CHO \rightarrow CH_3COO^- + \mathbf{3\ H_2O}$

Cancel duplicate species: $3\ OH^- + CH_3CHO \rightarrow CH_3COO^- + \mathbf{2}\ H_2O$

Step 4: $3\ OH^- + CH_3CHO \rightarrow CH_3COO^- + 2\ H_2O + \mathbf{2\ e^-}$

Reduction: $Cu^{2+} \rightarrow Cu_2O$

Step 1: $\mathbf{2}\ Cu^{2+} \rightarrow Cu_2O$

Step 2: $2\ Cu^{2+} + \mathbf{H_2O} \rightarrow Cu_2O$

Step 3: $2\ Cu^{2+} + H_2O \rightarrow Cu_2O + \mathbf{2\ H^+_{(aq)}}$

Step 3 a: $2\ Cu^{2+} + H_2O \rightarrow Cu_2O + 2\ H^+_{(aq)}$

$2\ H^+_{(aq)} + 2\ OH^- \rightarrow 2\ H_2O$

$2\ Cu^{2+} + H_2O + \mathbf{2\ OH^-} \rightarrow Cu_2O + \mathbf{2\ H_2O}$

Cancel duplicate species: $2\ Cu^{2+} + 2\ OH^- \rightarrow Cu_2O + H_2O$

(continued)

(17.14 continued)

Step 4: $\mathbf{2\ e^-} + 2\ Cu^{2+} + 2\ OH^- \rightarrow Cu_2O + H_2O$

III. The oxidation half-reaction produces 2 e^- and the reduction half-reaction consumes 2 e^-.

IV. Combine the half-reactions: $3\ OH^- + CH_3CHO + 2\ e^- + 2\ Cu^{2+} + 2\ OH^- \rightarrow$
$CH_3COO^- + 2\ H_2O + 2\ e^- + Cu_2O + H_2O$

Combining and canceling duplicate species:

$$5\ OH^- + CH_3CHO + 2\ Cu^{2+} \rightarrow CH_3COO^- + 3\ H_2O + Cu_2O$$

e) $NO_3^- + H_2O_2 \rightarrow NO + O_2$ (acidic)

I and II. Oxidation:

Steps 1 & 2: $H_2O_2 \rightarrow O_2$

Step 3: $H_2O_2 \rightarrow O_2 + \mathbf{2\ H^+_{(aq)}}$

Step 4: $H_2O_2 \rightarrow O_2 + 2\ H^+_{(aq)} + \mathbf{2\ e^-}$

Reduction: $NO_3^- \rightarrow NO$

Steps 1 & 2: $NO_3^- \rightarrow NO + \mathbf{2\ H_2O}$

Step 3: $\mathbf{4\ H^+_{(aq)}} + NO_3^- \rightarrow NO + 2\ H_2O$

Step 4: $4\ H^+_{(aq)} + \mathbf{3\ e^-} + NO_3^- \rightarrow NO + 2\ H_2O$

III. Multiply the ox. half-reaction by 3: $3\ H_2O_2 \rightarrow 3\ O_2 + 6\ H^+_{(aq)} + 3\ e^-$

Multiply the red. half-reaction by 2: $6\ e^- + 8\ H^+_{(aq)} + 2\ NO_3^- \rightarrow 2\ NO + 4\ H_2O$

IV. Combine the half-reactions: $3\ H_2O_2 + 6\ e^- + 8\ H^+_{(aq)} + 2\ NO_3^- \rightarrow$
$3\ O_2 + 6\ H^+_{(aq)} + 6\ e^- + 2\ NO + 4\ H_2O$

Cancel duplicate species: $2\ H^+_{(aq)} + 3\ H_2O_2 + 2\ NO_3^- \rightarrow 3\ O_2 + 4\ H_2O + 2\ NO$

f) $BrO_3^- + Fe^{2+} \rightarrow Br^- + Fe^{3+}$ (acidic)

I and II. Oxidation:

Steps 1-4: $Fe^{2+} \rightarrow Fe^{3+} + \mathbf{e^-}$

Reduction:

Steps 1 & 2: $BrO_3^- \rightarrow Br^- + \mathbf{3\ H_2O}$

Step 3: $\mathbf{6\ H^+_{(aq)}} + BrO_3^- \rightarrow Br^- + 3\ H_2O$

Step 4: $\mathbf{6\ e^-} + 6\ H^+_{(aq)} + BrO_3^- \rightarrow Br^- + 3\ H_2O$

III. Multiply the oxidation half-reaction by 6: $6\ Fe^{2+} \rightarrow 6\ Fe^{3+} + 6\ e^-$

IV. Combine the half-reactions: $6\ e^- + 6\ H^+_{(aq)} + BrO_3^- + 6\ Fe^{2+} \rightarrow$
$Br^- + 3\ H_2O + 6\ Fe^{3+} + 6\ e^-$

Cancel duplicate species: $6\ H^+_{(aq)} + BrO_3^- + 6\ Fe^{2+} \rightarrow Br^- + 3\ H_2O + 6\ Fe^{3+}$

17.15 a) $O_2 + Cu \rightarrow 2\ CuO$

$\Delta G^\circ_{rxn} = 2\ \text{mol}\left(\Delta G^\circ_{f\ CuO}\right) = 2\ \text{mol}\ (-129.7\ \text{kJ/mol}) = -259.4\ \text{kJ}$; spontaneous

(continued)

(17.15 continued)

b) $O_2 + 2\ Hg \rightarrow 2\ HgO$

$\Delta G^\circ_{rxn} = 2\ mol\ (\Delta G^\circ_{f\ HgO}) = 2\ mol\ (-58.539\ kJ/mol) = -117.078\ kJ$; spontaneous

c) $CuS + O_2 \rightarrow Cu + SO_2$ $\quad \Delta G^\circ_{rxn} = 1\ mol(\Delta G^\circ_{f\ SO_2}) - 1\ mol(\Delta G^\circ_{f\ CuS})$

$= 1\ mol(-300.2\ kJ/mol) - 1\ mol(-53.6\ kJ/mol) = -246.6\ kJ$; spontaneous

d) $FeS + O_2 \rightarrow Fe + SO_2$ $\quad \Delta G^\circ_{rxn} = 1\ mol(\Delta G^\circ_{f\ SO_2}) - 1\ mol(\Delta G^\circ_{f\ FeS})$

$= 1\ mol(-300.2\ kJ/mol) - 1\ mol(-100.4\ kJ/mol) = -199.8\ kJ$; spontaneous

Note: value for $\Delta G^\circ_{f\ FeS}$ is from CRC HANDBOOK OF CHEMISTRY AND PHYSICS.

17.16 a) $H_2O + CO \rightarrow CO_2 + H_2$

$\Delta G^\circ_{rxn} = 1\ mol\ (\Delta G^\circ_{f\ CO_2}) - [1\ mol(\Delta G^\circ_{f\ H_2O}) + 1\ mol(\Delta G^\circ_{f\ CO})]$

$= 1\ mol(-394.359\ kJ/mol) - [1\ mol(-237.129\ kJ/mol) + 1\ mol(-137.168\ kJ/mol)]$

$= -20.062\ kJ$; spontaneous

for $H_2O_{(g)}$

$\Delta G^\circ_{rxn} = 1\ mol(-394.359\ kJ/mol) - [1\ mol(-228.72\ kJ/mol) + 1\ mol(-137.168\ kJ/mol)]$

$= -28.47\ kJ$; spontaneous

b) $2Al + 3\ MgO \rightarrow 3\ Mg + Al_2O_3$

$\Delta G^\circ_{rxn} = 1\ mol\ (\Delta G^\circ_{f\ Al_2O_3}) - 3\ mol(\Delta G^\circ_{f\ MgO})$

$= 1\ mol(-1582.3\ kJ/mol) - 3\ mol(-569.43\ kJ/mol)$

$= +126.0\ kJ$; not spontaneous

c) $PbS + Cu \rightarrow CuS + Pb$ $\quad \Delta G^\circ_{rxn} = 1\ mol\ (\Delta G^\circ_{fCuS}) - 1\ mol(\Delta G^\circ_{fPbS})$

$= 1\ mol(-53.6\ kJ/mol) - 1\ mol(-98.7\ kJ/mol) = +45.1\ kJ$; not spontaneous

d) $N_2 + 2\ O_2 \rightarrow 2\ NO_2$ $\quad \Delta G^\circ_{rxn} = 2\ mol\ (\Delta G^\circ_{fNO_2}) = 2\ mol(51.31\ kJ/mol)$

$= 102.62\ kJ$; not spontaneous

17.17 Your sketch should illustrate an electrode similar to the electrode shown on the left in the answer given at the end of the chapter for Exercise 17.3.1. Your sketch needs to show solid AgCl and solid Ag in contact with each other, electrons flowing into the solid AgCl from an inert electrode, and AgCl becoming solid Ag and aqueous chloride ions. The aqueous chloride ions move away into the aqueous medium surrounding the electrode and solids.

17.18 Your sketch should be similar to that shown as the left compartment of Figure 17-10 and/or the right portion of Figure 17-19. In your sketch you need to show a molecular view that has a solid lead electrode with electrons flowing out an external circuit, HSO_4^- moving from the aqueous surroundings toward the lead electrode, lead sulfate being formed at the surface of the electrode by the reaction of lead and HSO_4^-, and hydrogen ions moving into the solution also from the reaction of lead and HSO_4^-.

17.19 Draw a sketch that uses two Pt electrodes like the one shown on the right in Figure 17-7 or on the left in the drawing in Figure 17-13. One side would need to be the $H^+_{(aq)}/H_{2(g)}$ shown in Figure 17-13. The other compartment would need to be $Cl_{2(g)}/Cl^-_{(aq)}$ in place of $H^+_{(aq)}/H_{2(g)}$. A porous plate will need to separate the two compartments.

17.20 The reaction being studied can be separated into the following half-reactions:

$$Cu^+_{(aq)} + e^- \rightarrow Cu_{(s)} \qquad Cu^+_{(aq)} \rightarrow e^- + Cu^{2+}_{(aq)}$$

A sketch of a cell that could be used to study the above would involve a copper electrode in contact with a solution of aqueous Cu^+ in one compartment and an inert electrode in contact with an aqueous solution of Cu^+ and Cu^{2+} in the other compartment. An external circuit and the usual porous plate separator would also need to be shown in your sketch of this system.

17.21 $Pb + PbO_2 + 2\ HSO_4^- + 2\ H^+ \rightarrow 2\ PbSO_4 + 2\ H_2O$

$$15\text{ s x}\frac{5.9\text{ C}}{\text{s}}\text{x}\frac{1\text{ mol e}^-}{96,485\text{ C}}\text{x}\frac{1\text{ mole Pb}}{2\text{ mol e}^-}\text{x}\frac{207.2\text{ g Pb}}{\text{mole Pb}} = 0.095\text{ g Pb}$$

$$15\text{ s x}\frac{5.9\text{ C}}{\text{s}}\text{x}\frac{1\text{ mol e}^-}{96,485\text{ C}}\text{x}\frac{1\text{ mole PbO}_2}{2\text{ mol e}^-}\text{x}\frac{239.2\text{ g PbO}_2}{\text{mole PbO}_2} = 0.11\text{ g PbO}_2$$

17.22 $2\ Al + Fe_2O_3 \rightarrow Al_2O_3 + 2\ Fe$

$$15\text{ s x}\frac{5.9\text{ C}}{\text{s}}\text{x}\frac{1\text{ mol e}^-}{96,485\text{ C}}\text{x}\frac{1\text{ mol Al}}{3\text{ mol e}^-}\text{x}\frac{26.98\text{ g Al}}{1\text{ mol Al}} = 0.0082\text{ g Al}$$

$$15\text{ s x}\frac{5.9\text{ C}}{\text{s}}\text{x}\frac{1\text{ mol e}^-}{96,485\text{ C}}\text{x}\frac{1\text{ mol Fe}_2\text{O}_3}{6\text{ mol e}^-}\text{x}\frac{159.7\text{ g Fe}_2\text{O}_3}{1\text{ mol Fe}_2\text{O}_3} = 0.024\text{ g Fe}_2\text{O}_3$$

17.23 1.750 amp - 1.350 amp = 0.400 amp

$$0.850\text{ g PbSO}_4\text{ x}\frac{\text{mol PbSO}_4}{303.3\text{ g PbSO}_4}\text{x}\frac{2\text{ mol e}^-}{\text{mol PbSO}_4}\text{x}\frac{96,485\text{ C}}{\text{mol e}^-}$$

$$\text{x}\frac{\text{s}}{0.400\text{ C}}\text{x}\frac{\text{min}}{60\text{ s}} = 22.5\text{ min}$$

17.24 $$\frac{1.00\text{ g HgO}}{216.6\text{ g/mol}} = 4.617\text{ x }10^{-3}\text{ mol HgO} \qquad \frac{1.00\text{ g Zn}}{65.38\text{ g/mol}} = 1.530\text{ x }10^{-2}\text{ mol Zn}$$

$$(4.617\text{ x }10^{-3}\text{ mol HgO})\left(\frac{2\text{ mol e}^-}{1\text{ mol HgO}}\right)\left(\frac{96,485\text{ C}}{\text{mol e}^-}\right)\left(\frac{1\text{ A}\bullet\text{s}}{1\text{ C}}\right) = 890.9\text{ A}\bullet\text{s}$$

$$\left(\frac{890.9\text{ A}\bullet\text{s}}{0.20\text{ x }10^{-3}\text{ A}}\right)\left(\frac{1\text{ min}}{60\text{ s}}\right)\left(\frac{1\text{ hr}}{60\text{ min}}\right) = 1.24\text{ x }10^3\text{ hr}$$

17.25 a) $Cr_2O_7^{2-} + C_2H_4O \rightarrow CH_3CO_2H + Cr^{3+}$ (acidic)

I and II. Oxidation:

Step 1: $C_2H_4O \rightarrow CH_3CO_2H$

Step 2: $\mathbf{H_2O} + C_2H_4O \rightarrow CH_3CO_2H$

Step 3: $H_2O + C_2H_4O \rightarrow CH_3CO_2H + \mathbf{2\ H^+_{(aq)}}$

Step 4: $H_2O + C_2H_4O \rightarrow CH_3CO_2H + 2\ H^+_{(aq)} + \mathbf{2\ e^-}$

Reduction: $Cr_2O_7^{2-} \rightarrow Cr^{3+}$

Step 1: $Cr_2O_7^{2-} \rightarrow 2\ Cr^{3+}$

Step 2: $Cr_2O_7^{2-} \rightarrow 2Cr^{3+} + \mathbf{7\ H_2O}$

Step 3: $\mathbf{14\ H^+_{(aq)}} + Cr_2O_7^{2-} \rightarrow 2\ Cr^{3+} + 7\ H_2O$

Step 4: $\mathbf{6\ e^-} + 14\ H^+_{(aq)} + Cr_2O_7^{2-} \rightarrow 2\ Cr^{3+} + 7\ H_2O$

III. Multiply the oxidation half-reaction by 3:

$3\ H_2O + 3\ C_2H_4O \rightarrow 3\ CH_3CO_2H + 6\ H^+_{(aq)} + 6\ e^-$

IV. Combine the half-reactions: $3\ H_2O + 3\ C_2H_4O + 6\ e^- + 14\ H^+_{(aq)} + Cr_2O_7^{2-} \rightarrow 3\ CH_3CO_2H + 6\ H^+_{(aq)} + 6\ e^- + 2\ Cr^{3+} + 7\ H_2O$

Cancel duplicate species:

$3\ C_2H_4O + 8\ H^+_{(aq)} + Cr_2O_7^{2-} \rightarrow 3\ CH_3CO_2H + 2\ Cr^{3+} + 4\ H_2O$

b) $1.00\text{ g } C_2H_4O\left(\frac{1\text{ mol } C_2H_4O}{44.0\text{ g } C_2H_4O}\right)\left(\frac{2\text{ mol } e^-}{1\text{ mol } C_2H_4O}\right) = 0.0455\text{ mol } e^-$

17.26 $MnO_4^- + C_2H_4O \rightarrow CH_3CO_2H + MnO_2$ (weakly acidic)

I and II. Oxidation:

Step 1: $C_2H_4O \rightarrow CH_3CO_2H$

Step 2: $\mathbf{H_2O} + C_2H_4O \rightarrow CH_3CO_2H$

Step 3: $H_2O + C_2H_4O \rightarrow CH_3CO_2H + \mathbf{2\ H^+_{(aq)}}$

Step 4: $H_2O + C_2H_4O \rightarrow CH_3CO_2H + 2\ H^+_{(aq)} + \mathbf{2\ e^-}$

Reduction:

Step 1: $MnO_4^- \rightarrow MnO_2$

Step 2: $MnO_4^- \rightarrow MnO_2 + \mathbf{2\ H_2O}$

Step 3: $\mathbf{4\ H^+_{(aq)}} + MnO_4^- \rightarrow MnO_2 + 2\ H_2O$

Step 4: $\mathbf{3\ e^-} + 4\ H^+_{(aq)} + MnO_4^- \rightarrow MnO_2 + 2\ H_2O$

III. Multiply the oxidation half-reaction by 3:

$3\ H_2O + 3\ C_2H_4O \rightarrow 3\ CH_3CO_2H + 6\ H^+_{(aq)} + 6\ e^-$

Multiply the red. half-rxn by 2: $6\ e^- + 8\ H^+_{(aq)} + 2\ MnO_4^- \rightarrow 2\ MnO_2 + 4\ H_2O$

IV. Combine the half-reactions: $3\ H_2O + 3\ C_2H_4O + 6\ e^- + 8\ H^+_{(aq)} + 2\ MnO_4^- \rightarrow 3\ CH_3CO_2H + 6\ H^+_{(aq)} + 6\ e^- + 2\ MnO_2 + 4\ H_2O$

Cancel duplicate species:

$3\ C_2H_4O + 2\ H^+_{(aq)} + 2\ MnO_4^- \rightarrow 3\ CH_3CO_2H + 2\ MnO_2 + H_2O$

17.27 a) $Co(NH_3)_6^{3+} + e^- \rightarrow Co(NH_3)_6^{2+}$ $E° = 0.108$ V

$PbO_2 + H_2O + 2\,e^- \rightarrow PbO + 2\,OH^-$ $E° = 0.247$ V (from CRC)

$E° = 0.108\text{ V} - 0.247\text{ V} = -0.139\text{ V}$

b) $O_2 + 4\,H^+_{(aq)} + 4\,e^- \rightarrow 2\,H_2O$ $E° = 1.229$ V

$HAsO_2 + 3\,H^+_{(aq)} + 3\,e^- \rightarrow As + 2\,H_2O$ $E° = 0.248$ V

$E° = 1.229\text{ V} - 0.248\text{ V} = 0.981\text{ V}$

c) $BrO_3^- + 3\,H_2O + 6\,e^- \rightarrow Br^- + 6\,OH^-$ $E° = 0.61$ V (from CRC)

$MnO_4^- + 2\,H_2O + 3\,e^- \rightarrow MnO_2 + 4\,OH^-$ $E° = 0.595$ V (from CRC)

$E° = 0.595\text{ V} - 0.61\text{ V} = -0.02\text{ V}$

17.28 c) $Co^{2+} + 2\,e^- \rightarrow Co$ $E° = -0.277$ V

$2\,HClO + 2\,H^+_{(aq)} + 2\,e^- \rightarrow Cl_2 + 2\,H_2O$ $E° = 1.611$ V

$E° = 1.611\text{ V} - (-0.277\text{ V}) = +1.888\text{ V}$

e) $O_2 + 2\,H^+_{(aq)} + 2\,e^- \rightarrow H_2O_2$ $E° = +0.695$

$NO_3^- + 4\,H^+_{(aq)} + 3\,e^- \rightarrow NO + 2\,H_2O$ $E° = +0.957$

$E° = 0.957\text{ V} - (+0.695\text{ V}) = 0.262\text{ V}$

17.29 Draw a sketch that includes two Pt electrodes, compartments, etc., like the compartment on the right in Figure 17-7 or the left in Figure 17-13. One side would need to be the $H^+_{(aq)}/H_{2(g)}$ as shown in Figures 17-7 and 17-13. The other compartment would need to be $F_{2(g)}/F^-_{(aq)}$ in place of $H^+_{(aq)}/H_{2(g)}$. An external circuit and a porous plate to connect the compartments are needed.

$F_{2\,(g)} + 2\,e^- \rightarrow 2\,F^-_{(aq)}$ $E^\circ_{F_2/F^-} = ?$

$2\,H^+_{(aq)} + 2\,e^- \rightarrow H_{2(g)}$ $E^\circ_{H^+/H_2} = 0.000$ V

$E^\circ_{cell} = E^\circ_{F_2/F^-} - E^\circ_{H^+/H_2}$ $E^\circ_{cell} = E^\circ_{F_2/F^-}$

The anode would be the standard hydrogen electrode.

17.30 A galvanic cell consisting of a Ru electrode (the wire) immersed in a solution of $RuCl_3$ and the lead electrode immersed in 1M H_2SO_4. The reduction potentials are:

$Ru^{3+} + 3\,e^- \rightarrow Ru$ $E^\circ_{Ru^{3+}/Ru} = ?$

$PbSO_4 + 2\,e^- \rightarrow Pb + SO_4^{2-}$ $E^\circ = -0.3588$ V

$E^\circ_{cell} = E^\circ_{Ru^{3+}/Ru} - (-0.3588\text{ V})$ $E^\circ_{Ru^{3+}/Ru} = E^\circ_{cell} - 0.3588\text{ V}$

Note: $RuCl_3$ is not very soluble. It is unlikely that a standard Ru electrode (1M Ru^{3+}) could be prepared. The actual cell potential would have to be corrected for the real concentration.

17.31 $E^\circ_{cell} = E^\circ_{cathode} - E^\circ_{anode}$

anode: $Ru \rightarrow Ru^{3+} + 3\,e^-$

cathode: $PbSO_4 + 2\,e^- \rightarrow Pb + SO_4^{2-}$ $E° = -0.3588$ V

(continued)

(17.31 continued)

$$E^\circ_{cell} = 0.745\ V$$

$E^\circ_{cell} = E^\circ_{cathode} - E^\circ_{anode}$

$E^\circ_{anode} = E^\circ_{cathode} - E^\circ_{cell} = -0.3588\ V - 0.745\ V = -1.104\ V$

If the Pb electrode were the anode instead of the cathode

$E^\circ_{cell} = E^\circ_{cathode} - E^\circ_{anode}$

$E^\circ_{cathode} = 0.745\ V + (-0.3588\ V) = 0.386\ V$ (more logical value as reduction potentials of different ions of Ru are all positive)

17.32 $Cu^{2+} + 2\ e^- \rightarrow Cu$ $\quad E^\circ = 0.3419\ V$

$Cu^+ + e^- \rightarrow Cu$ $\quad E^\circ = 0.521\ V$

$E^\circ = 0.521\ V - (0.3419\ V) = 0.179\ V$

$Cu \rightarrow Cu^{2+} + 2\ e^-$ is the anode.

Your drawing needs to look very much like the drawing in Figure 17-16 with the following modifications: a) the Cu^{2+} solution in the compartment on the right is replaced by a Cu^+ solution and b) NO_3^{2-} is present in both compartments in place of the SO_4^{2-}. The reaction in the left compartment remains $Cu \rightarrow Cu^{2+} + 2e^-$ while the one on the right becomes $Cu^+ + e^- \rightarrow Cu$. The direction of flow of electrons and anions will be the same as in Figure 17-16.

17.33 The standard potentials for a - c were determined in Problem 17.27.

a) $\Delta G^\circ = -nFE^\circ = -(2\ mol)(9.6485 \times 10^4\ C/mol)(-0.139\ V) = +2.68 \times 10^4\ J$ or $+26.8\ kJ$

b) $\Delta G^\circ = -nFE^\circ = -(12\ mol)(9.6485 \times 10^4\ C/mol)(0.981\ V) = -1.14 \times 10^6\ J$ or $-1.14 \times 10^3\ kJ$

c) $\Delta G^\circ = -nFE^\circ = -(6\ mol)(9.6485 \times 10^4\ C/mol)(-0.02\ V) = 1 \times 10^4\ J$ or 10 kJ

d) $NO_2 + 2\ H^+_{(aq)} + 2\ e^- \rightarrow NO + H_2O$ $\quad E^\circ = 1.03\ V$

$NO_3^- + 2\ H^+_{(aq)} + e^- \rightarrow NO_2 + H_2O$ $\quad E^\circ = 0.81\ V$ (calculated from CRC)

$E^\circ = 1.03\ V - (0.81\ V) = 0.22\ V$

$\Delta G^\circ = -nFE^\circ = -(2\ mol)(9.6485 \times 10^4\ C/mol)(0.22\ V) = -4.2 \times 10^4\ J$ or $-42\ kJ$

e) $Cl_2 + 2\ e^- \rightarrow 2\ Cl^-$ $\quad E^\circ = 1.35827\ V$

$ClO_4^- + 6\ H^+_{(aq)} + 6\ e^- \rightarrow ClO^- + 3\ H_2O$ $\quad E^\circ = 1.349\ V$ (calculated from other E° in App. H)

$E = 1.349\ V - (1.358\ V) = -0.009\ V$

$\Delta G^\circ = -nFE^\circ = -(6\ mol)(9.6485 \times 10^4\ C/mol)(-0.009\ V) = +5 \times 10^3\ J$ or 5 kJ

17.34 The standard potentials were determined in Problem 17.28.

c) $\Delta G^\circ = -nFE^\circ$

$= -(2\ mol)(9.6485 \times 10^4\ C/mol)(1.888\ V) = -3.64 \times 10^5\ J$ or $-364\ kJ$

e) $\Delta G^\circ = -nFE^\circ = -(6\ mol)(9.6485 \times 10^4\ C/mol)(0.262\ V) = -1.52 \times 10^5\ J$ or $-152\ kJ$

17.35 $Ca^{2+} + 2\,e^- \rightarrow Ca$ $\quad E° = -2.868$ V

$Ca(OH)_2 + 2\,e^- \rightarrow Ca + 2\,OH^-$ $\quad E° = -3.02$ V

$Ca(OH)_2 + 2\,e^- \rightarrow Ca + 2\,OH^-$

$Ca \rightarrow Ca^{2+} + 2\,e^-$

$Ca(OH)_2 \rightarrow Ca^{2+} + 2\,OH^-$ $\quad K_{sp} = [Ca^{2+}][OH^-]^2$

$E° = -3.02\text{ V} - (-2.868\text{ V}) = -0.15\text{ V}$

$\Delta G° = -nFE° = -RT \ln K \qquad \ln K = nFE°/RT$

$$\log K = \frac{nE°}{0.05916\text{ V}} = \frac{2(-0.15\text{ V})}{0.05916\text{ V}} = -5.07 \quad K = K_{sp} = 8.5 \times 10^{-6}$$

Table 17-1 contains sufficient data only for the calculation for $Ca(OH)_2$. Other possible compounds are either missing data or are not metal hydroxides.

17.36 $Cu^{2+} + H_2 \rightarrow Cu + 2\,H^+(aq)$

$$E^\circ_{cell} = E^\circ_{Cu^{2+}/Cu} - E^\circ_{H^+_{(aq)}/H_2} - \frac{0.05916\text{ V}}{n}\log Q$$

$$= 0.34\text{ V} - 0\text{ V} - \frac{0.05916\text{ V}}{2}\log\frac{[H^+(aq)]^2}{[Cu^{2+}](P_{H_2})} \qquad (\text{Assuming } P_{H_2} = 1.00\text{ atm})$$

$$= 0.34\text{ V} - \frac{0.05916\text{ V}}{2}\log\frac{(1.00 \times 10^{-3})^2}{(1.00 \times 10^{-3})(1.00\text{ atm})}$$

$$= 0.34\text{ V} + 0.08874\text{ V} = 0.43\text{ V}$$

17.37 $E° = 1.35$ V $\quad 2NiO(OH)_{(s)} + 2\,H_2O_{(l)} + Cd_{(s)} \rightarrow 2\,Ni(OH)_{2(s)} + Cd(OH)_{2(s)}$

$$E = E° - \frac{RT}{nF}\log Q \qquad Q = 1 \qquad E = 1.35\text{ V}$$

The concentration of OH^- does not appear in Q. If one looks at the individual half-cells and their potentials, one finds that the concentration of OH^- affects each half-cell equally. Therefore, upon subtraction of the half-cell potentials, any changes due to changes in the concentration of OH^- will cancel.

17.38 $$E = E^\circ_{cathode} - E^\circ_{anode} - \frac{0.05916\text{ V}}{n}\log Q$$

$Zn^{2+}{(aq)} + 2\,e^- \rightarrow Zn_{(s)}$ $\quad E° = -0.7618$ V

$$E = -\frac{0.05916\text{ V}}{2}\log\frac{[Zn^{2+}]_{dilute}}{[Zn^{2+}]_{conc}}$$

$$0.040\text{ V} = -\frac{0.05916\text{ V}}{2}\log\frac{[Zn^{2+}]_{dilute}}{0.500\text{ M}}$$

$$\log\frac{[Zn^{2+}]_{dilute}}{0.500\text{ M}} = \frac{(0.040\text{ V})2}{-0.05916\text{ V}} = -1.35$$

$$\frac{[Zn^{2+}]_{dilute}}{0.500\text{ M}} = 10^{-1.35} = 0.0444 \qquad [Zn^{2+}]_{dilute} = 0.033\text{ M} = 2.2 \times 10^{-2}\text{ M}$$

17.39 Net reaction for the standard dry cell:

$Zn_{(s)} + 2\ NH_4^+{}_{(aq)} \rightleftarrows Zn^{2+}{}_{(aq)} + 2\ NH_{3(g)} + H_{2(g)}$

$$E = E° - \frac{0.05916\ V}{2} \log \frac{[Zn^{2+}](p_{NH_3})^2(p_{H_2})}{[NH_4^+]^2}$$

As the reaction proceeds, the numerator (products) of the log term increases and the denominator (reactants) decreases, the log value increases, and E decreases as the cell is used.

17.40 Net reaction for the lead storage cell:

$Pb_{(s)} + 2\ HSO_4^-{}_{(aq)} + PbO_{2(s)} + 2\ H^+{}_{(aq)} \rightleftarrows 2\ PbSO_{4(s)} + H_2O_{(l)}$

E° = 2.051 V

$$E = E° - \frac{0.05916\ V}{2} \log \frac{1}{[H^+{}_{(aq)}]^2[HSO_4^-]^2}$$

As the reaction proceeds, the denominator of the log term (reactants) decreases and the numerator (products) increases, the log value increases, and E decreases as the cell is used.

17.41 $Cr^{2+} + 2\ e^- \rightarrow Cr$ E° = -0.913 V

$Fe^{2+} + 2\ e^- \rightarrow Fe$ E° = -0.447 V

Chromium will corrode because the potential for the oxidation of Cr is more positive than the potential for the oxidation of Fe.

17.42 $Zn^{2+} + 2\ e^- \rightarrow Zn$ E° = -0.7618 V

$Fe^{2+} + 2\ e^- \rightarrow Fe$ E° = -0.447 V

The potential for the oxidation of Zn is more positive than the potential for the oxidation of Fe. Therefore, the Zn will oxidize (corrode) preferentially. The steel will not oxidize as long as there is zinc present.

17.43 The presence of $NaHCO_3$ makes the water (solution) slightly basic.

$Ag_2S + 2\ e^- \rightarrow 2\ Ag + S^{2-}$ reduction E° = -0.691 V

$3\ Ag_2S + 6\ e^- \rightarrow 6\ Ag + 3\ S^{2-}$

$Al + 4\ OH^- \rightarrow Al(OH)_4^- + 3\ e^-$ oxidation E° = +2.33 V

$2\ Al + 8\ OH^- \rightarrow 2\ Al(OH)_4^- + 6\ e^-$

$3\ Ag_2S + 2\ Al + 8\ OH^- \rightarrow 6\ Ag + 2\ Al(OH)_4^- + 3\ S^{2-}$ overall

E° = +1.64 V

17.44 Sacrificial anode for Fe

From Table 17-1: Zn and Mg have oxidation potentials greater than the oxidation potential for Fe. Al and Cr also have oxidation potentials greater than the oxidation potential for Fe, but they both form oxides that adhere to the metal surface (passivation) which limits their oxidation. Na, Ca and Li also have greater oxidation potentials than Fe, but they are too reactive and will quickly oxidize if exposed to water or the atmosphere.

From the Appendix: Only Mn might be another reasonable possibility.

Only Mg could serve as a sacrificial anode for Al.

17.45 $2\ Cl^- \rightarrow Cl_2 + 2\ e^-$

$$(200\ \text{min})\left(\frac{60\ \text{s}}{\text{min}}\right)\left(\frac{4.50\ \text{C}}{\text{s}}\right)\left(\frac{\text{mol e}^-}{9.6485 \times 10^4\ \text{C}}\right)\left(\frac{1\ \text{mol Cl}_2}{2\ \text{mol e}^-}\right)\left(\frac{70.9\ \text{g Cl}_2}{\text{mol Cl}_2}\right)$$

$$= 19.8\ \text{g Cl}_2$$

17.46 a) The anode, which supplies electrons, would be attached to the Pb electrode where the reaction would be: $PbSO_{4(s)} + 2\ e^- \rightarrow 2\ Pb_{(s)} + SO_4^{2-}{}_{(aq)}$

b) $n = It/F$ $\quad t = nF/I$

$t = (4.80\ \text{g}\ PbSO_4)(1\ \text{mol}\ PbSO_4/303.37\ PbSO_4)(2\ \text{mol}\ e^-/\text{mol}\ PbSO_4)$
$\times(96485\ \text{C/mol e-})(1\ \text{A s/C}) \div 0.120\ \text{A}$

$= 2.55 \times 10^4\ \text{s} = 425\ \text{min} = 7.08\ \text{hr}$

17.47 $Cu^{2+} + 2\ e^- \rightarrow Cu$

$E° = 0.3419\ V$

$$(0.250\ \text{L})\left(\frac{0.245\ \text{mol CuSO}_4}{\text{L}}\right)\left(\frac{2\ \text{mol e}^-}{\text{mol CuSO}_4}\right)\left(\frac{9.6485 \times 10^4\ \text{C}}{\text{mol e}^-}\right)\left(\frac{\text{s}}{2.45\ \text{C}}\right)\left(\frac{\text{min}}{60\ \text{s}}\right)$$

$$= 80.4\ \text{min}$$

17.48 $Ag^+ + e^- \rightarrow Ag$ $\qquad$ 12.89 g Ag - 10.77 g Ag = 2.12 g Ag deposited

$$(2.12\ \text{g Ag})\left(\frac{1\ \text{mol Ag}}{107.9\ \text{g Ag}}\right)\left(\frac{1\ \text{mol e}^-}{1\ \text{mol Ag}}\right)\left(\frac{9.6485\ \text{C}}{\text{mol e}^-}\right) = 1.90 \times 10^3\ \text{C}$$

$$A = C/s \qquad A = \frac{1.90 \times 10^3\ \text{C}}{(15.0\ \text{min})(60\ \text{s / min})} = 2.11\ \text{A}$$

17.49 $Zn^{2+} + 2\ e^- \rightarrow Zn$

$$(7.55\ \text{g Zn})\left(\frac{1\ \text{mol Zn}}{65.39\ \text{g Zn}}\right)\left(\frac{2\ \text{mol e}^-}{1\ \text{mol Zn}}\right)\left(\frac{9.6485 \times 10^4\ \text{C}}{\text{mol e}^-}\right) = 2.23 \times 10^4\ \text{C}$$

17.50 $(150 \times 10^{-3}\ A)(65\ min)(60\ s/min)(C/A\ s) = 585\ C$

$Cd_{(s)} + 2\ OH^-_{(aq)} \rightleftarrows Cd(OH)_{2(s)} + 2\ e^-$

$NiO_{2(s)} + 2\ H_2O_{(l)} + 2\ e^- \rightleftarrows Ni(OH)_{2(s)} + 2\ OH^-_{(aq)}$

$$(585\ C)\left(\frac{mol\ e^-}{96485\ C}\right)\left(\frac{mol\ Cd}{2\ mol\ e^-}\right)\left(\frac{112.4\ g\ Cd}{mol\ Cd}\right) = 0.34\ g\ Cd$$

$$(585\ C)\left(\frac{mol\ e^-}{96485\ C}\right)\left(\frac{mol\ NiO_2}{2\ mol\ e^-}\right)\left(\frac{90.7\ g\ NiO_2}{mol\ NiO_2}\right) = 0.27\ g\ NiO_2$$

17.51 a) $2\ CuFeS_{2(s)} + 3\ O_{2(g)} \xrightarrow{\Delta} 2\ FeO_{(s)} + 2\ CuS_{(s)} + 2\ SO_{2(g)}$

or $2\ CuFeS_{2(s)} + 4\ O_{2(g)} \xrightarrow{\Delta} 2\ FeO_{(s)} + Cu_2S_{(s)} + 3\ SO_{2(g)}$

b) $Al(O)OH_{(s)} + NaOH_{(aq)} + H_2O_{(l)} \rightarrow Na[Al(OH)_4]_{(aq)}$

c) $Si_{(g)} + O_{2(g)} \rightarrow SiO_{2(s)} + CaO_{(s)} \rightarrow CaSiO_{3(l)}$

d) $TiCl_{4(l)} + 4\ Na_{(s)} \rightarrow Ti_{(s)} + 4\ NaCl_{(s)}$

17.52 a) $2\ NiS_{(s)} + 3\ O_{2(g)} \rightarrow 2\ NiO_{(s)} + 2\ SO_{2(g)}$

b) $3\ Co_3O_{4(s)} + 8\ Al_{(s)} \rightarrow 9\ Co_{(s)} + 4\ Al_2O_{3(s)}$

c) $SnO_{2(s)} + 2\ C_{(s)} \rightarrow Sn_{(s)} + 2\ CO_{(g)}$

17.53 $$(5.60 \times 10^4\ kg\ ore)\left(\frac{10^3\ g}{kg}\right)\left(\frac{2.37\ g\ Cu_2S}{100\ g\ ore}\right)\left(\frac{mol\ Cu_2S}{159.2\ g\ Cu_2S}\right)\left(\frac{2\ mol\ Cu}{mol\ Cu_2S}\right)$$

$$\left(\frac{63.55\ g\ Cu}{mol\ Cu}\right) = 1.06 \times 10^6\ g\ Cu\ (\text{or } 1.06 \times 10^3\ kg\ Cu)$$

$$(5.60 \times 10^4\ kg\ ore)\left(\frac{10^3\ g}{kg}\right)\left(\frac{2.37\ g\ Cu_2S}{100\ g\ ore}\right)\left(\frac{mol\ Cu_2S}{159.2\ g\ Cu_2S}\right)\left(\frac{1\ mol\ SO_2}{1\ mol\ Cu_2S}\right)$$

$= 8.34 \times 10^3\ mol\ SO_2$

$PV = nRT$ $$V_{SO_2} = \frac{nRT}{P} = \frac{(8.34 \times 10^3\ mol)\left(0.0821\frac{L\ atm}{K\ mol}\right)(296.6\ K)}{(755\ torr / 760\ torr\ atm^{-1})}$$

$V_{SO_2} = 2.04 \times 10^5\ L$

17.54 Reaction: $CaCO_{3(s)} \xrightarrow{heat} CaO_{(s)} + CO_{2(g)}$

$SiO_{2(s)} + CaO_{(s)} \xrightarrow{\Delta} CaSiO_{3(l)}$

$$(1 \text{ kg iron ore})\left(\frac{10^3 \text{ g}}{\text{kg}}\right)\left(\frac{9.75 \text{ g } SiO_2}{100 \text{ g iron ore}}\right)\left(\frac{1 \text{ mol } SiO_2}{60.09 \text{ g } SiO_2}\right)\left(\frac{1 \text{ mol CaO}}{1 \text{ mol } SiO_2}\right)$$

$$\text{x}\left(\frac{1 \text{ mol } CaCO_3}{1 \text{ mol Ca}}\right)\left(\frac{100.09 \text{ g } CaCO_3}{1 \text{ mol } CaCO_3}\right)\left(\frac{\text{kg}}{10^3 \text{ g}}\right) = 0.162 \text{ kg } CaCO_3$$

$$(0.162 \text{ kg } CaCO_3)\left(\frac{100 \text{ kg limestone}}{95.5 \text{ kg } CaCO_3}\right) = 0.170 \text{ kg limestone}$$

17.55 $ZnO_{(s)} + C_{(s)} \rightarrow Zn_{(s)} + CO_{(g)}$

$$\Delta G^\circ_{rxn} = 1 \text{ mol}\left(\Delta G^\circ_{f_{CO}}\right) - 1 \text{ mol}\left(\Delta G^\circ_{f_{ZnO}}\right)$$

$= 1$ mol(-137.168 kJ/mol) - 1 mol(-318.30 kJ/mol) = 181.13 kJ

$ZnO_{(s)} + CO_{(g)} \rightarrow Zn_{(s)} + CO_{2(g)}$

$$\Delta G^\circ_{rxn} = 1 \text{ mol}\left(\Delta G^\circ_{f_{CO_2}}\right) - \left[1 \text{ mol}\left(\Delta G^\circ_{f_{CO}}\right) + 1 \text{ mol}\left(\Delta G^\circ_{f_{ZnO}}\right)\right]$$

$$\Delta G^\circ_{rxn} = 1 \text{ mol}(-394.359 \text{ kJ / mol})$$

$$-[1 \text{ mol}(-137.168 \text{ kJ / mol}) + 1 \text{ mol}(-318.30 \text{ kJ / mol})] = 61.11 \text{ kJ}$$

17.56 $Si + O_2 \rightarrow SiO_2$ $\Delta G^\circ = -856.64$ kJ

$4 P + 5 O_2 \rightarrow P_4O_{10}$ $\Delta G^\circ = -2697.7$ kJ (white P) or -2649.3 (red P)

{$P + 5/4O_2 \rightarrow 1/4P_4O_{10}$ $\Delta G^\circ = -674.4$ kJ}

$C + O_2 \rightarrow CO_2$ $\Delta G^\circ = -394.359$ kJ

$S + O_2 \rightarrow SO_2$ $\Delta G^\circ = -300.194$ kJ

$2 Fe + O_2 \rightarrow 2 FeO$ $\Delta G^\circ = -502.8$ kJ (LANGE'S HANDBOOK)

{$Fe + 1/2O_2 \rightarrow FeO$ $\Delta G^\circ = -251.4$ kJ}

$4 Fe + 3 O_2 \rightarrow 2 Fe_2O_3$ $\Delta G^\circ = -1484.4$ kJ

{$Fe + 3/4O_2 \rightarrow 1/2Fe_2O_3$ $\Delta G^\circ = -371.1$ kJ}

The ΔG°'s of reaction are more negative for the oxidation of the impurities (with the exception of sulfur) than for the oxidations of iron when ΔG°'s per mole of reactant are compared. The oxidations of the impurities are more favored than the oxidation of iron except for S which is between the iron oxidations. All reactions are spontaneous.

17.57 $SO_4^{2-}{}_{(aq)} + 4\ H^+{}_{(aq)} + 2\ e^- \rightarrow SO_{2(g)} + 2\ H_2O$

$H_2SO_{3(aq)} + H_2O \rightarrow 2\ e^- + 4\ H^+{}_{(aq)} + SO_4^{2-}{}_{(aq)}$

$H_2SO_{3(aq)} \rightarrow SO_{2(g)} + H_2O$

$E° = 0.20\ V - 0.17\ V = 0.03\ V$ $\Delta G° = -nFE° = -RT \ln K$

$$\ln K = \frac{nFE°}{RT} = \frac{2(9.6485 \times 10^4\ C/mol)(0.03\ V)}{(8.314\ J/K \cdot mol)(298\ K)} = 2.3 \qquad C = J/V$$

$K = 10$ $K = \frac{p_{SO_2}}{[H_2SO_3]}$ $[H_2SO_3] = 1.00\ M$ $p_{SO_2} = 10\ atm$

17.58 $C_2H_5OH + Cr_2O_7^{2-} \rightarrow CH_3CO_2H + Cr^{3+}$ (assume acidic solution)

$C_2H_5OH \rightarrow CH_3CO_2H$

$C_2H_5OH + \mathbf{H_2O} \rightarrow CH_3CO_2H$

$C_2H_5OH + H_2O \rightarrow CH_3CO_2H + \mathbf{4\ H^+{}_{(aq)}}$

$C_2H_5OH + H_2O \rightarrow CH_3CO_2H + 4\ H^+{}_{(aq)} + \mathbf{4\ e^-}$

$Cr_2O_7^{2-} \rightarrow Cr^{3+}$

$Cr_2O_7^{2-} \rightarrow \mathbf{2}\ Cr^{3+}$

$Cr_2O_7^{2-} \rightarrow 2\ Cr^{3+} + \mathbf{7\ H_2O}$

$Cr_2O_7^{2-} + \mathbf{14\ H^+{}_{(aq)}} \rightarrow 2\ Cr^{3+} + 7\ H_2O$

$Cr_2O_7^{2-} + 14\ H^+{}_{(aq)} + \mathbf{6\ e^-} \rightarrow 2\ Cr^{3+} + 7\ H_2O$

Combine half-reactions after multiplying by 3 and 2 respectively:

$3\ C_2H_5OH + 3\ H_2O + 2\ Cr_2O_7^{2-} + 28\ H^+{}_{(aq)} + 12\ e^- \rightarrow$
$3\ CH_3CO_2H + 12\ H^+{}_{(aq)} + 12\ e^- + 4\ Cr^{3+} + 14\ H_2O$

Cancel duplicate species:

$3\ C_2H_5OH + 2\ Cr_2O_7^{2-} + 16\ H^+{}_{(aq)} \rightarrow 3\ CH_3CO_2H + 4\ Cr^{3+} + 11\ H_2O$

$$(50.0\ mL\ solution)\left(\frac{L}{10^3\ mL}\right)\left(\frac{4.5 \times 10^{-4}\ mol\ Cr^{3+}}{L\ solution}\right)\left(\frac{3\ mol\ C_2H_5OH}{4\ mol\ Cr^{3+}}\right)$$

$$\times \left(\frac{46.07\ g\ C_2H_5OH}{1\ mol\ C_2H_5OH}\right)\left(\frac{10^3\ mg}{g}\right) = 0.78\ mg\ C_2H_5OH$$

17.59 $I_2 + 2\ e^- \rightleftarrows 2\ I^-$ $E° = 0.5355\ V$

$2\ IO_3^- + 12\ H^+{}_{(aq)} + 10\ e^- \rightleftarrows I_2 + 6\ H_2O$ $E° = 1.195\ V$

$10\ I^- + 2\ IO_3^- + 12\ H^+{}_{(aq)} \rightleftarrows 6\ I_2 + 6\ H_2O$

Divide through by 2:

$5\ I^-{}_{(aq)} + IO_3^-{}_{(aq)} + 6\ H^+{}_{(aq)} \rightleftarrows 3\ I_{2(s)} + 3\ H_2O_{(l)}$

$E° = 1.195\ V - 0.5355\ V = 0.660\ V$

(continued)

(17.59 continued)

a) pH = 2.00 $\qquad [H^+] = 0.010$ M

$$E = E° - \frac{0.05916\ V}{n}\log Q = E° - \frac{0.05916\ V}{n}\log\frac{1}{[I^-]^5[IO_3^-][H^+]^6}$$

$$= 0.660\ V - \frac{0.05916\ V}{5}\log\frac{1}{(0.100)^5(0.100)(0.010)^6}$$

$$= 0.660\ V - \frac{0.05916\ V}{5}\log 1.0 \times 10^{18}$$

= 0.660 V - 0.213 V = 0.45 V (spontaneous as written)

b) pH = 11.00 $\quad [H^+] = 1.0 \times 10^{-11}$ $\quad$ Using E° = 0.660 V (acidic solution)

$$E = 0.660\ V - \frac{0.05916\ V}{5}\log\frac{1}{(0.100)^5(0.100)(1.0 \times 10^{-11})^6}$$

$$E = 0.660\ V - \frac{0.05916\ V}{5}\log 1.0 \times 10^{72}$$

E = 0.660 V - 85 V = -0.19 V (spontaneous in the reverse direction from written)

c) At equilibrium, E = 0 $\quad E° = \frac{0.05916\ V}{n}\log K$

$$\log K = \frac{nE°}{0.05916\ V} = \frac{5(0.660\ V)}{0.05916\ V} = 55.8$$

$K = 6 \times 10^{55}$ $\qquad K = 1/[I^-]^5[IO_3^-][H^+]^6$

$$[H^+] = \left(\frac{1}{[I^-]^5[IO_3^-]K}\right)^{1/6} \qquad [H^+] = \left(\frac{1}{(0.100)^5(0.100)(6 \times 10^{55})}\right)^{1/6}$$

$[H^+] = 5 \times 10^{-9}$ $\qquad$ pH = 8.3

17.60 Corrosion of iron and aluminum is highly spontaneous but slow during exposure to dry air. Corrosion is faster in contact with water. The oxidation of the metal and the reduction of the oxygen occur at different locations in the water droplet. In the absence of dissolved ions, the electrical circuit is incomplete and the redox reaction is still slow. The ions from the salt complete the circuit and corrosion will be quite rapid.

17.61 $AgCl + e^- \rightarrow Ag + Cl^-$ $\qquad$ E° = 0.22233 V

$E_{cathode} = E°_{cathode} - (0.05916\ V/1)\log[Cl^-] = 0.22233\ V - 0.05916\ V \log(0.500)$

$= 0.22233\ V + 0.0178\ V = 0.240\ V$

$Mg^{2+} + 2\ e^- \rightarrow Mg$ $\qquad$ E° = -2.372 V

net reaction: $\quad 2\ AgCl + Mg \rightarrow 2\ Ag + Mg^{2+} + 2\ Cl^-$

b) $E_{cell} = E_{cathode} - E_{anode} = 0.240\ V - (-2.372\ V) = 2.612\ V$

or $E_{cell} = 0.22233\ V - (-2.372\ V) - \frac{0.05916}{2}\log[Mg^{2+}][Cl^-]^2 = 2.612\ V$

(continued)

(17.61 continued)

c) Your molecular picture should depict an electrochemical cell in which one vessel is like the left compartment in the drawing in Figure 17-5 except that the zinc is replaced with magnesium (both the atoms and the ions) and the SO_4^- is replaced with Cl^-. The other vessel needs to be a drawing of an electrode like the iron electrode shown in the answer given on page 881 for Section Exercise 17.3.1. The Fe would need to be replaced with Ag and the Fe_2O_3 replaced by AgCl. In this vessel the anion would need to be the chloride ion just as it is in the other vessel. The two vessels need to be connected by a porous separator. The chloride ions will flow from the vessel containing the silver electrode to the vessel containing the magnesium electrode.

17.62 $SnCl_2 + K_2Cr_2O_7 \rightarrow Cr^{3+} + Sn^{4+}$

$SnCl_2 \rightarrow Sn^{4+}$ $\qquad$ $K_2Cr_2O_7 \rightarrow Cr^{3+}$

$SnCl_2 \rightarrow Sn^{4+} + \mathbf{2\ Cl^-}$ $\qquad$ $K_2Cr_2O_7 \rightarrow 2\ Cr^{3+} + \mathbf{2\ K^+}$

$SnCl_2 \rightarrow Sn^{4+} + 2\ Cl^- + \mathbf{2\ e^-}$ $\qquad$ $K_2Cr_2O_7 \rightarrow 2\ Cr^{3+} + 2\ K^+ + \mathbf{7\ H_2O}$

$K_2Cr_2O_7 + \mathbf{14\ H^+_{(aq)}} \rightarrow 2\ Cr^{3+} + 2\ K^+ + 7\ H_2O$

$K_2Cr_2O_7 + 14\ H^+_{(aq)} + \mathbf{6\ e^-} \rightarrow 2\ Cr^{3+} + 2\ K^+ + 7\ H_2O$

Multiply oxidation half-reaction by 3 and combine half-reactions:

$$3\ SnCl_2 + K_2Cr_2O_7 + 14\ H^+_{(aq)} + 6\ e^- \rightarrow 3\ Sn^{4+} + 6\ Cl^- + 6\ e^- + 2\ Cr^{3+} + 2\ K^+ + 7\ H_2O$$

Cancel duplicate species:

$$3\ SnCl_2 + K_2Cr_2O_7 + 14\ H^+_{(aq)} \rightarrow 3\ Sn^{4+} + 6\ Cl^- + 2\ Cr^{3+} + 2\ K^+ + 7\ H_2O$$

$$\left(\frac{22.50\ \text{mL}\ K_2Cr_2O_7}{15.00\ \text{mL}\ SnCl_2}\right)\left(\frac{0.100\ \text{mM}\ K_2Cr_2O_7}{1\ \text{mL}\ K_2Cr_2O_7}\right)\left(\frac{3\ \text{mM}\ SnCl_2}{1\ \text{mM}\ K_2Cr_2O_7}\right)$$

$$= \frac{0.450\ \text{mM}\ SnCl_2}{\text{mL}\ SnCl_2} = 0.450\ \text{M}\ SnCl_2$$

17.63 $Fe^{2+} + 2\ e^- \rightarrow Fe$ $\qquad$ $E° = -0.447$ V

$O_2 + 4\ H^+_{(aq)} + 4\ e^- \rightarrow 2\ H_2O$ $\qquad$ $E° = 1.229$ V

$O_2 + 2\ H_2O + 4\ e^- \rightarrow 4\ OH^-$ $\qquad$ $E° = 0.401$ V

pH = 10 $\qquad$ $2\ Fe + O_2 + 2\ H_2O \rightarrow 2\ Fe^{2+} + 4\ OH^-$

$E° = 0.401\ V - (-0.447\ V) = 0.848\ V$

pH = 3 $\qquad$ $2\ Fe + O_2 + 4\ H^+_{(aq)} \rightarrow 2\ Fe^{2+} + 2\ H_2O$

$E° = 1.229\ V - (0.447\ V) = 1.676\ V$

Iron is more likely to rust at pH = 3. Similar results will be obtained if one looks at Fe^{3+} instead of Fe^{2+} or if one calculates adjustments in the values for nonstandard concentrations of H^+ or OH^-.

17.64 $\text{n moles e}^- = \frac{\text{I t}}{\text{F}} = \frac{(27.6\text{ A})(24\text{ hr})(60\text{ min / hr})(60\text{ s / min})}{(96485\text{ C / mol e}^-)(1\text{ A s / C})}$

$= 24.7\text{ mol e}^-$

$Ca^{2+} + 2\,e^- \rightarrow Ca$

$$(24.7\text{ mol e}^-)\left(\frac{\text{mol Ca}}{2\text{ mol e}^-}\right)\left(\frac{40.08\text{ g Ca}}{\text{mol Ca}}\right) = 495\text{ g Ca}$$

17.65 From Table 17-1

$F_2 + 2\,e^- \rightarrow 2\,F^-$	$E° = 2.866$ V	cathode
$Mg^{2+} + 2\,e^- \rightarrow Mg$	$E° = -2.372$ V	anode

$E°_{cell} = 2.866\text{ V} - (-2.372\text{ V}) = 5.238\text{ V}$

$Na^+ + e^- \rightarrow Na$	$E° = -2.714$ V	anode

$E°_{cell} = 2.866\text{ V} - (-2.714\text{ V}) = 5.580\text{ V}$

$Ca^{2+} + 2\,e^- \rightarrow Ca$	$E° = -2.868$ V	anode

$E°_{cell} = 2.866\text{ V} - (-2.868\text{ V}) = 5.734\text{ V}$

$Ca(OH)_2 + 2\,e^- \rightarrow Ca + 2\,OH^-$	$E° = -3.02$ V	anode

$E°_{cell} = 2.866\text{ V} - (-3.02\text{ V}) = 5.89\text{ V}$

$Li^+ + e^- \rightarrow Li$	$E° = -3.0401$ V	anode

$E°_{cell} = 2.866\text{ V} - (-3.0401\text{ V}) = 5.906\text{ V}$

The reaction rate and battery life must be considered as well as E°. It would be difficult to contain F_2 and very dangerous if the container should fail and free F_2.

17.66 $2\,MnO_{2(s)} + Zn^{2+} + 2\,e^- \rightarrow ZnMn_2O_{4(s)}$ 1 C = 1 A s   1A = 1C s^{-1}

$$(4.0\text{ g MnO}_2)\left(\frac{90\text{ g MnO}_2\text{ used}}{100\text{ g MnO}_2}\right)\left(\frac{1\text{ mol MnO}_2}{86.94\text{ g MnO}_2}\right)\left(\frac{2\text{ mol e}^-}{1\text{ mol MnO}_2}\right)\left(\frac{96485\text{ C}}{\text{mol e}^-}\right) \text{x}$$

$$\left(\frac{\text{A s}}{\text{C}}\right)\left(\frac{1}{0.0048\text{ A}}\right) = 1.7 \times 10^6\text{ s} = 2.8 \times 10^4\text{ min} = 4.7 \times 10^2\text{ hr}$$

17.67 $E° = \frac{0.05916\text{ V}}{n}\log K = \frac{0.05916\text{ V}}{2}\log(2.69 \times 10^{12}) = 0.368\text{ V}$

17.68 The better reducing agent will have the more positive (or less negative) potential for its oxidation reaction.

a) Cu is a better reducing agent than Ag.

b) Cr^{2+} is a better reducing agent than Fe^{2+}.

c) H_2 is a better reducing agent than I^- (in acidic solution).

17.69 The better oxidizing agent will have the more positive (or less negative) potential for its reduction reaction.

a) MnO_4^- is a better oxidizing agent than $Cr_2O_7^{2-}$ (in acidic solution).
b) H_2O_2 is a better oxidizing agent than O_2 (in basic solution).
c) Sn^{2+} is a better oxidizing agent than Fe^{2+}.

17.70 $O_2 + 4\ H^+(aq) + 4\ e^- \rightleftarrows 2\ H_2O$ $\quad E° = +1.229\ V$

O_2 will oxidize the following under standard conditions in acidic solution: Cu, Ag, Fe^{2+}, Cr^{2+}, H_2 and I^-. It will oxidize each of the substances in Problem 17.68.

17.71 $H_2O_2(aq) + 2\ H^+(aq) + 2\ e^- \rightarrow 2\ H_2O(l)$ $\quad E° = 1.776\ V$ (from CRC)

$O_2(g) + 2\ H^+(aq) + 2\ e^- \rightarrow H_2O_2(aq)$ $\quad E° = +0.695\ V$

$2\ H_2O_2(aq) \rightarrow 2\ H_2O(l) + O_2(g)$ $\quad E° = 1.776\ V - 0.695\ V = 1.081\ V$

$$\ln K_{eq} = \frac{nFE°}{RT} = \frac{nE°}{0.05916\ V} = \frac{(2)(1.081\ V)}{(0.05916\ V)} = 36.54 \qquad K_{eq} = 7.4 \times 10^{15}\ atm\ M^{-2}$$

Solutions of hydrogen peroxide are unstable. The equilibrium lies far to the right. As the hydrogen peroxide decomposes, the oxygen gas is given off and the reaction continues to completion.

17.72 For Na:

$$(1\ ton\ Na)\left(\frac{2000\ lb}{1\ ton}\right)\left(\frac{454\ g}{lb}\right)\left(\frac{1\ mol\ Na}{22.99\ g\ Na}\right)\left(\frac{1\ mol\ e^-}{1\ mol\ Na}\right)\left(\frac{96485\ C}{1\ mol\ e^-}\right) = 3.8 \times 10^9\ C$$

For Al:

$$(1\ ton\ Al)\left(\frac{2000\ lb}{1\ ton}\right)\left(\frac{454\ g}{lb}\right)\left(\frac{1\ mol\ Al}{26.98\ g\ Na}\right)\left(\frac{3\ mol\ e^-}{1\ mol\ Al}\right)\left(\frac{96485\ C}{1\ mol\ e^-}\right) = 9.7 \times 10^9\ C$$

Aluminum requires about 2.5 times more electricity than sodium to produce a ton of metal.

17.73 $MnO_4^- + H_2SO_3 \rightarrow Mn^{2+} + HSO_4^-$ $\quad$ (acidic)

Oxidation:

$H_2SO_3 \rightarrow HSO_4^-$

$\mathbf{H_2O} + H_2SO_3 \rightarrow HSO_4^-$

$H_2O + H_2SO_3 \rightarrow HSO_4^- + \mathbf{3\ H^+(aq)}$

$H_2O + H_2SO_3 \rightarrow HSO_4^- + 3\ H^+(aq) + \mathbf{2\ e^-}$

Reduction:

$MnO_4^- \rightarrow Mn^{2+}$

$MnO_4^- \rightarrow Mn^{2+} + \mathbf{4\ H_2O}$

$\mathbf{8\ H^+(aq)} + MnO_4^- \rightarrow Mn^{2+} + 4\ H_2O$

$\mathbf{5\ e^-} + 8\ H^+(aq) + MnO_4^- \rightarrow Mn^{2+} + 4\ H_2O$

(continued)

(17.73 continued)

Multiply the oxid. half-reaction by 5 and the red. half-reaction by 2:

$5\ H_2O + 5\ H_2SO_3 \rightarrow 5\ HSO_4^- + 15\ H^+_{(aq)} + 10\ e^-$

$10\ e^- + 16\ H^+_{(aq)} + 2\ MnO_4^- \rightarrow 2\ Mn^{2+} + 8\ H_2O$

Combine the half-reactions:

$5\ H_2O + 5\ H_2SO_3 + 10\ e^- + 16\ H^+_{(aq)} + 2\ MnO_4^- \rightarrow$
$5\ HSO_4^- + 15\ H^+_{(aq)} + 10\ e^- + 2\ Mn^{2+} + 8\ H_2O$

Cancel duplicate species:

$5\ H_2SO_3 + H^+_{(aq)} + 2\ MnO_4^- \rightarrow 5\ HSO_4^- + 2\ Mn^{2+} + 3\ H_2O$

$MnO_4^- + SO_2 \rightarrow Mn^{2+} + HSO_4^-$ (acidic)

Oxidation:

$SO_2 \rightarrow HSO_4^-$

$\mathbf{2\ H_2O} + SO_2 \rightarrow HSO_4^-$

$2\ H_2O + SO_2 \rightarrow HSO_4^- + \mathbf{3\ H^+_{(aq)}}$

$2\ H_2O + SO_2 \rightarrow HSO_4^- + 3\ H^+_{(aq)} + \mathbf{2\ e^-}$

Reduction (same as part a):

$5\ e^- + 8\ H^+_{(aq)} + MnO_4^- \rightarrow Mn^{2+} + 4\ H_2O$

Multiply the oxid. half-reaction by 5 and the red. half-reaction by 2:

$10\ H_2O + 5\ SO_2 \rightarrow 5\ HSO_4^- + 15\ H^+_{(aq)} + 10\ e^-$

$10\ e^- + 16\ H^+_{(aq)} + 2\ MnO_4^- \rightarrow 2\ Mn^{2+} + 8\ H_2O$

Combine the half-reactions:

$10\ H_2O + 5\ SO_2 + 10\ e^- + 16\ H^+_{(aq)} + 2\ MnO_4^- \rightarrow$
$5\ HSO_4^- + 15\ H^+_{(aq)} + 10\ e^- + 2\ Mn^{2+} + 8\ H_2O$

Cancel duplicate species:

$2\ H_2O + 5\ SO_2 + H^+_{(aq)} + 2\ MnO_4^- \rightarrow 5\ HSO_4^- + 2\ Mn^{2+}$

$MnO_4^- + H_2S \rightarrow Mn^{2+} + HSO_4^-$ (acidic)

Oxidation:

$H_2S \rightarrow HSO_4^-$

$\mathbf{4\ H_2O} + H_2S \rightarrow HSO_4^-$

$4\ H_2O + H_2S \rightarrow HSO_4^- + \mathbf{9\ H^+_{(aq)}}$

$4\ H_2O + H_2S \rightarrow HSO_4^- + 9\ H^+_{(aq)} + \mathbf{8\ e^-}$

Reduction (same as part a):

$5\ e^- + 8\ H^+_{(aq)} + MnO_4^- \rightarrow Mn^{2+} + 4\ H_2O$

Multiply the oxid. half-reaction by 5 and the red. half-reaction by 8:

$20\ H_2O + 5\ H_2S \rightarrow 5\ HSO_4^- + 45\ H^+_{(aq)} + 40\ e^-$

$40\ e^- + 64\ H^+_{(aq)} + 8\ MnO_4^- \rightarrow 8\ Mn^{2+} + 32\ H_2O$

(continued)

(17.73 continued)

Combine the half-reactions:

$20\ H_2O + 5\ H_2S + 40\ e^- + 64\ H^+_{(aq)} + 8\ MnO_4^- \rightarrow$
$5\ HSO_4^- + 45\ H^+_{(aq)} + 40\ e^- + 8\ Mn^{2+} + 32\ H_2O$

Cancel duplicate species:

$5\ H_2S + 19\ H^+_{(aq)} + 8\ MnO_4^- \rightarrow 5\ HSO_4^- + 8\ Mn^{2+} + 12\ H_2O$

$MnO_4^- + H_2S_2O_3 \rightarrow Mn^{2+} + HSO_4^-$ (acidic)

Oxidation:

$H_2S_2O_3 \rightarrow HSO_4^-$

$H_2S_2O_3 \rightarrow \mathbf{2}\ HSO_4^-$

$\mathbf{5\ H_2O} + H_2S_2O_3 \rightarrow 2\ HSO_4^-$

$5\ H_2O + H_2S_2O_3 \rightarrow 2\ HSO_4^- + \mathbf{10\ H^+_{(aq)}}$

$5\ H_2O + H_2S_2O_3 \rightarrow 2\ HSO_4^- + 10\ H^+_{(aq)} + \mathbf{8\ e^-}$

Reduction (same as part a):

$5\ e^- + 8\ H^+_{(aq)} + MnO_4^- \rightarrow Mn^{2+} + 4\ H_2O$

Multiply the oxid. half-reaction by 5 and the red. half-reaction by 8:

$25\ H_2O + 5\ H_2S_2O_3 \rightarrow 10\ HSO_4^- + 50\ H^+_{(aq)} + 40\ e^-$

$40\ e^- + 64\ H^+_{(aq)} + 8\ MnO_4^- \rightarrow 8\ Mn^{2+} + 32\ H_2O$

Combine the half-reactions:

$25\ H_2O + 5\ H_2S_2O_3 + 40\ e^- + 64\ H^+_{(aq)} + 8\ MnO_4^- \rightarrow$
$10\ HSO_4^- + 50\ H^+_{(aq)} + 40\ e^- + 8\ Mn^{2+} + 32\ H_2O$

Cancel duplicate species:

$5\ H_2S_2O_3 + 14\ H^+_{(aq)} + 8\ MnO_4^- \rightarrow 10\ HSO_4^- + 8\ Mn^{2+} + 7\ H_2O$

17.74 $Fe^{3+} + e^- \rightleftarrows Fe^{2+}$ $E° = 0.771\ V$

$Cu^{2+} + 2\ e^- \rightleftarrows Cu$ $E° = 0.3419\ V$

$E° = 0.771\ V - 0.3419\ V = 0.429\ V$

$2\ Fe^{3+}_{(aq)} + Cu_{(s)} \rightleftarrows 2\ Fe^{2+}_{(aq)} + Cu^{2+}_{(aq)}$ $E° = 0.429\ V$

$$E = E° - \frac{0.05916\ V}{n} \log \frac{[Fe^{2+}]^2[Cu^{2+}]}{[Fe^{3+}]^2}$$

$$0.00\ V = 0.429\ V - \frac{0.05916\ V}{2} \log \frac{(1.00)^2(1.00)}{[Fe^{3+}]^2}$$

$$(-0.429\ V)\left(\frac{2}{-0.05916\ V}\right) = \log \frac{1.00}{[Fe^{3+}]^2}$$

(continued)

(17.74 continued)

$$14.5 = \log \frac{1.00}{[Fe^{3+}]^2}$$

$$\frac{1.00}{[Fe^{3+}]^2} = 10^{14.5} = 3 \times 10^{14}$$

$$[Fe^{3+}]^2 = \frac{1.00}{3 \times 10^{14}}$$

$[Fe^{3+}] = 6 \times 10^{-8}$ M

17.75 $HgS_{(s)} + O_{2(g)} \xrightarrow{\Delta} Hg_{(l)} + SO_{2(g)}$

$PbS_{(s)} + O_{2(g)} \xrightarrow{\Delta} Pb_{(l)} + SO_{2(g)}$

$ZnO_{(s)} + C_{(s)} \xrightarrow{\Delta} Zn_{(l)} + CO_{(g)}$

$TiCl_{4(l)} + 2\ Mg_{(s)} \xrightarrow{\Delta} Ti_{(s)} + 2\ MgCl_{2(g)}$

$MgCl_{2(l)} \xrightarrow{\text{electrolysis}} Mg_{(s)} + Cl_{2(g)}$

$2\ NaCl_{(l)} \xrightarrow{\text{electrolysis}} 2\ Na_{(l)} + Cl_{2(g)}$

$Cr_2O_{3(s)} + 2\ Al_{(s)} \xrightarrow{\Delta} Al_2O_{3(s)} + 2\ Cr_{(s)}$

17.76 $AgCl_{(s)} + e^- \rightleftarrows Ag_{(s)} + Cl^-_{(aq)}$ $\qquad E° = 0.22233$ V

$Ni^{2+}_{(aq)} + 2\ e^- \rightleftarrows Ni_{(s)}$ $\qquad E° = -0.257$ V

a) $2\ AgCl_{(s)} + Ni_{(s)} \rightleftarrows 2\ Ag_{(s)} + Ni^{2+}_{(aq)} + 2\ Cl^-_{(aq)}$

b) $$E = E^{\circ}_{cathode} - E^{\circ}_{anode} - \frac{0.05916\ V}{n} \log([Ni^{2+}][Cl^-]^2)$$

$$= 0.22233\ V - (-0.257\ V) - \frac{0.05916\ V}{2} \log\{(1.50 \times 10^{-2})(3.00 \times 10^{-2})^2\}$$

$$= 0.22233\ V + 0.257\ V + 0.144\ V = 0.623\ V$$

c) Your sketch of the system and the electron transfer occurring at each electrode should show one vessel like the left compartment in the drawing in Figure 17-5 except that the zinc atoms and ions are replaced with nickel and the SO_4^- is replaced with Cl^-. The other vessel needs to be like the iron electrode shown in the answer given on page 881 for Section Exercise 17.3.1. Replace the Fe with Ag and the Fe_2O_3 with AgCl. The two vessels need to be connected by a porous separator. The chloride ions will flow from the vessel containing the silver electrode to the vessel containing the nickel electrode.

17.77 $Cr(OH)_3 + H_2 \rightarrow Cr + H_2O$ (basic solution)

a) Oxidation:

$2\ OH^- + H_2 \rightarrow 2\ H_2O + 2\ e^-$

Reduction:

$3\ e^- + Cr(OH)_3 \rightarrow Cr + 3\ OH^-$

Multiply the oxid. half-reaction by 3 and the red. half-reaction by 2:

$6\ OH^- + 3\ H_2 \rightarrow 6\ H_2O + 6\ e^-$

$6\ e^- + 2\ Cr(OH)_3 \rightarrow 2\ Cr + 6\ OH^-$

Combine the half-reactions:

$6\ OH^- + 3\ H_2 + 6\ e^- + 2\ Cr(OH)_3 \rightarrow 6\ H_2O + 6\ e^- + 2\ Cr + 6\ OH^-$

Cancel duplicate species:

$3\ H_2 + 2\ Cr(OH)_3 \rightarrow 6\ H_2O + 2\ Cr$

b) $Cr(OH)_3 + 3\ e^- \rightarrow Cr + OH^-$ $E° = -1.48\ V$

$2\ H_2O + 2\ e^- \rightarrow H_2 + 2\ OH^-$ $E° = -0.828\ V$

$E° = -1.48 -(-0.828\ V) = -\ 0.65\ V$

c) $$\Delta G° = -nFE° = -(6\ mol)\left(\frac{9.6485 \times 10^4\ C}{mol}\right)(-0.65\ V)$$

$$= 3.8 \times 10^5\ J = 3.8 \times 10^2\ kJ$$

17.78 a) $MnO_{2(s)} \rightarrow MnO_4^-{}_{(aq)} + Mn^{2+}{}_{(aq)}$ (acid solution)

Oxidation:

$MnO_2 \rightarrow MnO_4^-$

$\mathbf{2\ H_2O} + MnO_2 \rightarrow MnO_4^-$

$2\ H_2O + MnO_2 \rightarrow MnO_4^- + \mathbf{4\ H^+{}_{(aq)}}$

$2\ H_2O + MnO_2 \rightarrow MnO_4^- + 4\ H^+{}_{(aq)} + \mathbf{3\ e^-}$

Reduction:

$MnO_2 \rightarrow Mn^{2+}$

$MnO_2 \rightarrow Mn^{2+} + \mathbf{2\ H_2O}$

$\mathbf{4\ H^+{}_{(aq)}} + MnO_2 \rightarrow Mn^{2+} + 2\ H_2O$

$4\ H^+{}_{(aq)} + MnO_2 + \mathbf{2\ e^-} \rightarrow Mn^{2+} + 2\ H_2O$

Multiply oxidation half-reaction by 2:

$4\ H_2O + 2\ MnO_2 \rightarrow 2\ MnO_4^- + 8\ H^+{}_{(aq)} + 6\ e^-$

Multiply reduction half-reaction by 3:

$12\ H^+{}_{(aq)} + 3\ MnO_2 + 6\ e^- \rightarrow 3\ Mn^{2+} + 6\ H_2O$

(continued)

(17.78 continued)

Combine half-reactions:

$4\ H_2O + 2\ MnO_2 + 12\ H^+_{(aq)} + 3\ MnO_2 + 6\ e^- \rightarrow$
$2\ MnO_4^- + 8\ H^+_{(aq)} + 6\ e^- + 3\ Mn^{2+} + 6\ H_2O$

Combine and/or cancel duplicate species:

$5\ MnO_{2(s)} + 4\ H^+_{(aq)} \rightarrow 2\ MnO_4^-{}_{(aq)} + 3\ Mn^{2+} + 2\ H_2O$

b) $Q = \dfrac{[MnO_4^-]^2[Mn^{2+}]^3}{[H^+_{(aq)}]^2}$

MnO_2 is insol. in H_2O and acid soln. unless it reacts with the acid such as HCl.

c) **n** = 6 electrons

$MnO_2 + 4\ H^+_{(aq)} + 2\ e^- \rightleftarrows Mn^{2+} + 2\ H_2O \qquad E° = 1.23\ V$

$MnO_4^- + 4\ H^+_{(aq)} + 3\ e^- \rightleftarrows MnO_2 + 2\ H_2O \qquad E° = 1.695\ V$

E° = 1.23 V -(1.695) = -0.47 V

$$\Delta G° = -nFE° = -(6\ mol)\left(\frac{96485\ C}{mol}\right)(-0.47\ V) = 2.7 \times 10^5\ C\ V$$

$= 2.7 \times 10^5\ J = 2.7 \times 10^2\ kJ$

$$\ln K_{eq} = \frac{\Delta G°}{-RT} = \frac{2.7 \times 10^5\ J}{-(8.314\ J\ mol^{-1}\ K^{-1})(298\ K)} = -1.1 \times 10^2$$

$\mathbf{K_{eq}} = e^{-1.1 \times 10^2} = 2 \times 10^{-48}$

$$\text{or} \quad \ln K_{eq} = \frac{nFE°}{RT} = \frac{(6\ mol)(96485\ J/V\ mol)(-0.47\ V)}{(8.314\ J/mol\ K)(298\ K)} = -110$$

$\mathbf{K_{eq}} = e^{-110} = 2 \times 10^{-48}$

17.79 a) Your molecular picture should include a drawing like that in Figure 17-16 except the copper electrodes are replaced with zinc electrodes. The Cu^{2+} and SO_4^- in the right compartment are replaced with 1.25 M Zn^{2+} and 2.5 M NO_3^-. The Cu^{2+} and SO_4^- in the left compartment are replaced with 0.250 M Zn^{2+} and 0.500 M Cl^-. The nitrate ions will flow from the right vessel to the left one. Your molecular picture needs to show the spontaneous production of zinc ions and electrons from zinc metal in the left compartment while zinc ions and electrons form zinc metal at the right electrode.

b) Cathode: $2\ e^- + Zn^{2+} \rightarrow Zn$

Anode: $Zn \rightarrow Zn^{2+} + 2\ e^-$

$$E = E° - \frac{0.05916\ V}{n}\log Q \qquad E° = 0$$

(continued)

(17.79 continued)

$$E = -\frac{0.05916\ V}{n}\log Q = -\frac{0.05916\ V}{n}\log \frac{[Zn^{2+}]_{dilute}}{[Zn^{2+}]_{concentrated}}$$

$$E = -\frac{0.05916\ V}{2}\log \frac{(0.250\ M)}{(1.25\ M)} = 0.0207\ V$$

17.80 In moist air the iron of the cans is rapidly oxidized to Fe_2O_3 which flakes off the cans and exposes the iron underneath to further oxidation. This continues until all of the iron has oxidized. The aluminum of cans also rapidly oxidizes to Al_2O_3 but this Al_2O_3 forms a tightly bound layer on the surface of the remaining aluminum and protects it from further oxidation.

17.81 $Tl^+ + e^- \rightarrow Tl$ $\qquad E° = -0.34\ V$

$2\ H^+ + 2\ e^- \rightarrow H_2$ $\qquad E° = 0.00\ V$

$2\ H^+_{(aq)} + 2\ Tl_{(s)} \rightarrow H_{2(g)} + 2\ Tl^+_{(aq)}$

$E° = 0.00\ V - (-0.34\ V) = 0.34\ V$

$$E = E° - \frac{0.05916\ V}{n}\log Q = E° - \frac{0.05916\ V}{n}\log\frac{p_{H_2}[Tl^+]^2}{[H^+]^2}$$

$$= 0.34\ V - \frac{0.05916\ V}{2}\log\frac{(0.90)(0.050)^2}{(0.50)^2} = 0.34\ V - (-0.061\ V) = 0.40\ V$$

17.82 Draw a sketch similar to that in Figure 17-7 except the platinum electrode on the left needs to be replaced with a thallium electrode, the Fe^{3+} is replaced by Tl^+ solution and the flow of electrons is reversed. Your molecular view of the process taking place at the electrodes needs to show the spontaneous production of thallium ions and electrons from thallium metal in the left compartment while hydrogen ion and electrons form hydrogen gas at the right electrode.

CHAPTER 18: THE CHEMISTRY OF LEWIS ACIDS AND BASES

18.1 a) $Ni + 4\ CO \rightarrow [Ni(CO)_4]$ Lewis acid: Ni Lewis base: CO

b) $SbCl_3 + 2\ Cl^- \rightarrow SbCl_5^{2-}$ Lewis acid: $SbCl_3$ Lewis base: Cl^-

c) $(CH_3)_3P + AlBr_3 \rightarrow (CH_3)_3P\text{–}AlBr_3$ Lewis acid: $AlBr_3$ Lewis base: $(CH_3)_3P$

d) $BF_3 + ClF_3 \rightarrow ClF_2^+ + BF_4^-$ Lewis acid: BF_3 Lewis base: ClF_3

e) $I_2 + I^- \rightarrow I_3^-$ Lewis acid: I_2 Lewis base: I^-

18.2 a) p orbital used; changes from sp^2 to sp^3

b) 5 d used

c) 3 p used

d) 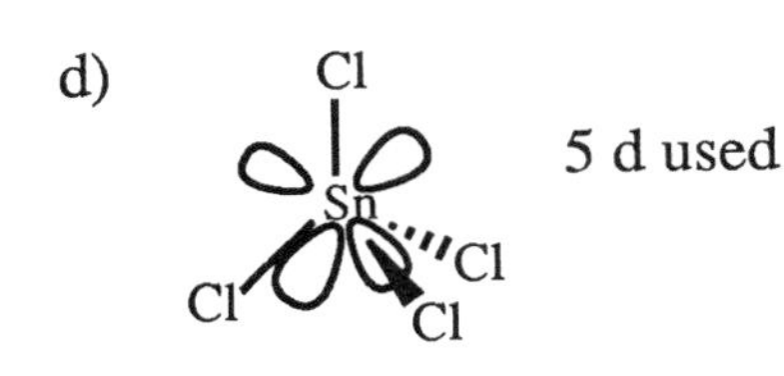

5 d used

18.3 BF_3 and AlF_3 are electron-deficient with an empty orbital that accepts a pair of electrons to form a fourth bond. The central atoms of SiF_4 and PF_5 have empty valence d orbitals that allow them to act as Lewis acids by accepting a pair of electrons. All of the valence orbitals of the C in CF_4 are used to form bonds with F. The valence orbitals of the N in NF_3 are used forming the 3 bonds to F and holding the non-bonding pair of electrons. The S in SF_6 uses sp^3d^2 hybridized orbitals to form 6 bonds to F that occupy the space around the S making it difficult for any further bonds to be formed.

18.4 a) $Cr = [Ar]4s^13d^5$ $Cr^{2+} = [Ar]3d^4$ $Cr^{3+} = [Ar]3d^3$

b) $V^- = [Ar]3d^6$ (or $4s^23d^4$) $V = [Ar]4s^23d^3$

$V^+ = [Ar]3d^4$ (or $4s^13d^3$) $V^{2+} = [Ar]3d^3$

$V^{3+} = [Ar]3d^2$ $V^{4+} = [Ar]3d^1$ $V^{5+} = [Ar]$

(continued)

(18.4 continued)

c) Ti = [Ar]$4s^2 3d^2$ $\quad$ $T1^{2+}$ = [Ar]$3d^2$ $\quad$ Ti^{4+} = [Ar]

Note: For the neutral atoms of transition elements the 3d and 4s orbitals have quite similar energies so that most configurations are of the $4s^2 3d^n$ type. The stabilization of filled and half-filled shells gives $4s^1 3d^5$ for Cr (not $4s^2 3d^4$). When the atoms are ionized, the 3d orbitals become appreciably more stable than the 4s orbitals and the ions all have $3d^n$ configurations. The less common ions are usually unstable and exist only in complexed forms.

18.5 a) Au = [Xe]$6s^1 5d^{10}$ $\quad$ Au^+ = [Xe]$5d^{10}$ $\quad$ Au^{3+} = [Xe]$5d^8$

b) Ni = [Ar]$4s^2 3d^8$ $\quad$ Ni^{2+} = [Ar]$3d^8$ $\quad$ Ni^{3+} = [Ar]$3d^7$

c) Mn^- = [Ar]$4s^2 3d^6$ $\quad$ Mn = [Ar]$4s^2 3d^5$ $\quad$ Mn^+ = [Ar]$4s^1 3d^5$

18.6 a) $[Ru(NH_3)_6]Cl_2$ $\quad$ Ru(II), $4d^6$

b) $[Cr(en)_2I_2]I$ $\quad$ Cr(III), $3d^3$

c) *cis*-$[PdCl_2(P(CH_3)_3)_2]$ $\quad$ Pd(II), $4d^8$

d) *fac*-$[Ir(NH_3)_3Cl_3]$ $\quad$ Ir(III), $5d^6$

e) $[Ni(CO)_4]$ $\quad$ Ni(0), $3d^{10}$ (or $4s^2 3d^8$)

18.7 a) $[Ru(NH_3)_6]Cl_2$ = hexaammineruthenium(II) chloride

b) $[Cr(en)_2I_2]I$ = bis(ethylenediamine)diiodochromium(III) iodide

c) *cis*-$[PdCl_2(P(CH_3)_3)_2]$ = *cis*-dichlorobis(trimethylphosphine)palladium(II)

d) *fac*-$[Ir(NH_3)_3Cl_3]$ = *fac*-triamminetrichloroiridium(III)

e) $[Ni(CO)_4]$ = tetracarbonylnickel(0)

18.8 a) $[Ru(NH_3)_6]^{2+}$, $2Cl^-$

NH3
NH3 NH3
Ru
NH3 NH3
NH3

b)

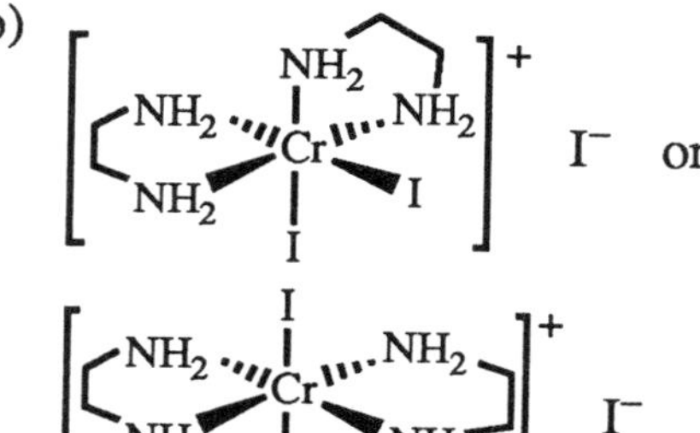

I^- or I^-

c) Cl, Pd, $P(CH_3)_3$, Cl, $P(CH_3)_3$

d) Cl, NH_3, Cl, Ir, NH_3, Cl, NH_3

e)

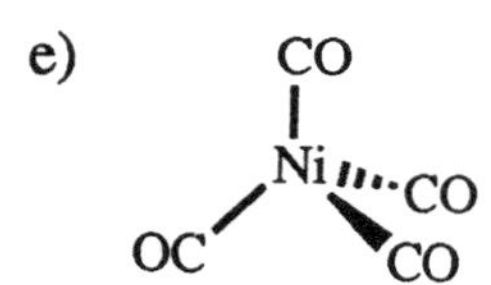

18.9 a) $[Rh(en)_3]Cl_3$ Rh(III), $4d^6$

b) *cis*-$[Mo(CO)_4Br_2]$ Mo(II), $4d^4$

c) $Na_3[IrCl_6]$ Ir(III), $5d^6$

d) *mer*-$[Ir(NH_3)_3Cl_3]$ Ir(III), $5d^6$

e) $[Mn(CO)_5Cl]$ Mn(I), $3d^6$

18.10 a) tris(ethylenediamine)rhodium(III) chloride

b) *cis*-dibromotetracarbonylmolybdenum(II)

c) sodium hexachloroiridate(III)

d) *mer*-triamminetrichloroiridiium(III)

e) pentacarbonylchloromanganese(I)

18.11 a) $[Rh(en)_3]^{3+}$, 3 Cl^-

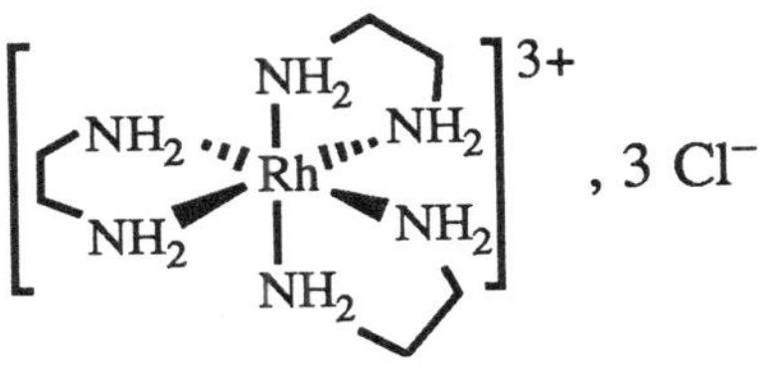

b)

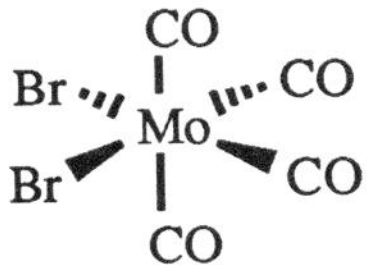

c) 3 Na^+, $[IrCl_6]^{3-}$

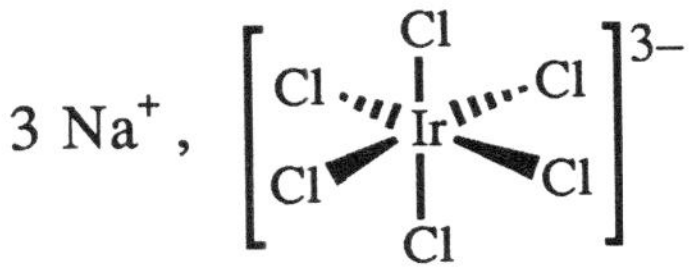

d)

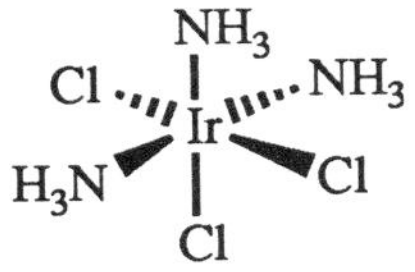

e)

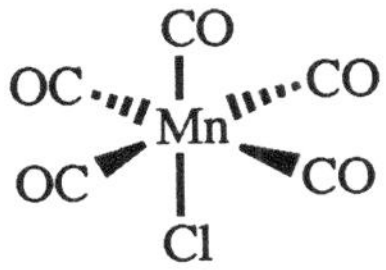

18.12 $[M(CO_3)_3]^{4-}$ or $[M(CO_3)_6]^{10-}$

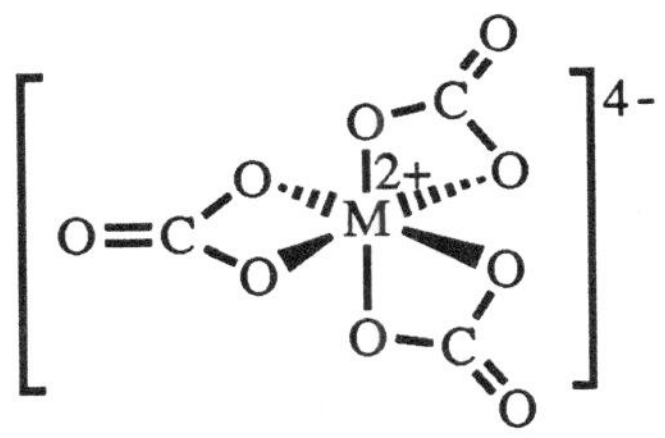

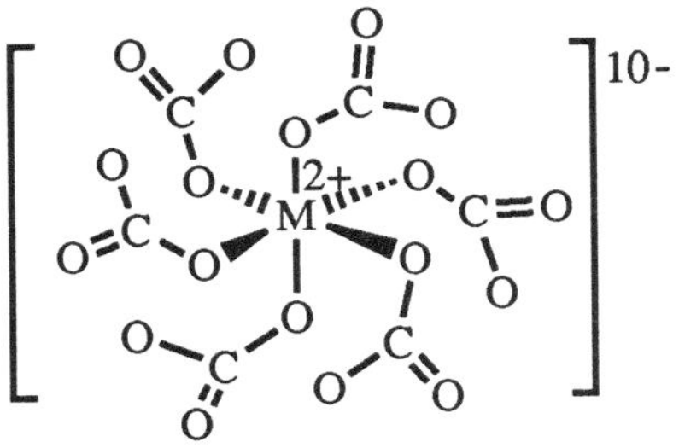

18.13 a) 4 NH_3, Cl^-, NO_2^-, Co^{3+} *cis*-$[Co(NH_3)_4Cl(NO_2)]^+$

b) NH_3, 3 Cl^-, Pt^{2+} $[Pt(NH_3)Cl_3]^-$

c) 2 H_2O, 2 en, Cu^{2+} *trans*-$[Cu(en)_2(H_2O)_2]^{2+}$

d) 4 Cl^-, Fe^{3+} $[FeCl_4]^-$

18.14 a) $K_2[PtCl_4]$

b) $[Cr(NH_3)_5H_2O]I_3$

c) $[Mn(en)_3]Cl_2$

d) $[Co(NH_3)_5I](NO_3)_2$

18.15 $[Ni(CO)_4]$ geometry = tetrahedral color = colorless

$Ni(CO)_4$ is poisonous because it is so volatile. Once inhaled, some of the CO ligands dissociate from the metal and dissolve in the blood stream, replacing O_2 in hemoglobin. Suffocation results.

18.16

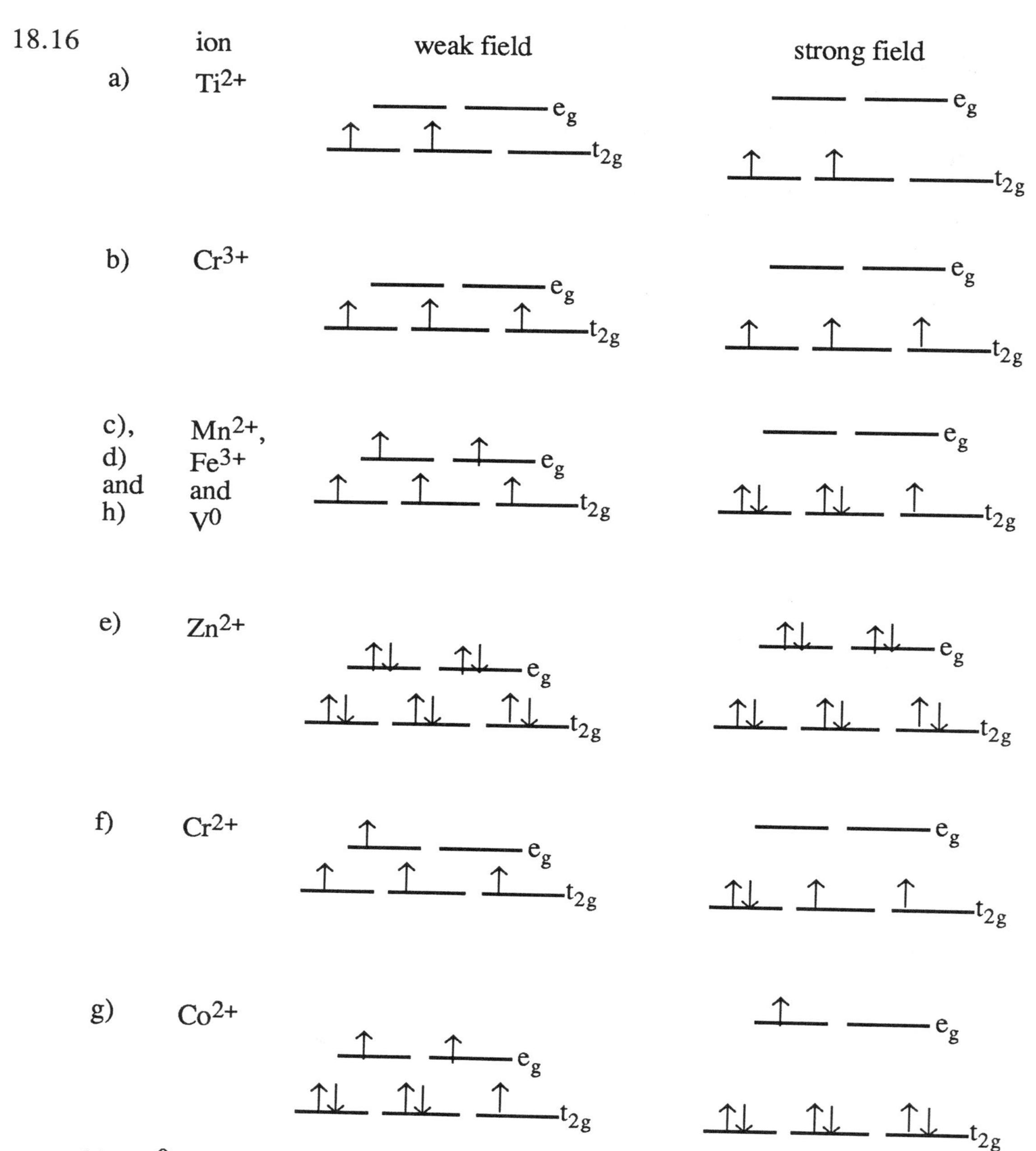

h) V^0 Vanadium has 5 valent electrons ($4s^23d^3$). In the presence of an octahedral field, all 5 electrons are likely to occupy the d orbitals and the energy diagrams for weak and strong fields will be the same as Mn^{2+} and Fe^{3+} (c and d above).

18.17 $E = \frac{hc}{\lambda} = \frac{(6.626 \times 10^{-34}\ Js)(2.998 \times 10^{8}\ m/s)}{(430.5\ nm)(10^{-9}\ m/nm)} \frac{(6.022 \times 10^{23})}{1000\ J/kJ} = 278\ kJ/mol$

$$= \frac{(6.626 \times 10^{-34}\ Js)(2.998 \times 10^{8}\ m/s)(6.022 \times 10^{23})}{(611.5\ nm)(10^{-9}\ m/nm)(1000\ J/kJ)} = 196\ kJ/mol$$

a) $\Delta = 278$ kJ/mol and $\Delta = 196$ kJ/mol

b) green/blue green
Note: The wavelengths around 510 nm are the only ones in the visible region that are transmitted. This area is the wavelength of blue-green to green colored light.

c) tetraaquadichlorochromium(III) ion

d) $[Cr(OH_2)_4Cl_2]^+$ (OH$_2$, H$_2$O, H$_2$O, OH$_2$, Cl, Cl around Cr)
cis
yellow

$[Cr(OH_2)_4Cl_2]^+$ (Cl, H$_2$O, OH$_2$, H$_2$O, OH$_2$, Cl around Cr)
trans
blue-green

e) — — e_g → ↑ — e_g

↑ ↑ ↑ t_{2g} → ↑ ↑ — t_{2g}

18.18 a) $[Ir(NH_3)_6]^{3+}$ The crystal field splitting is very likely to be high for 3 reasons:
1. The ligand (NH_3) is just above the middle of the spectrochemical series.
2. The oxidation state of the metal (Ir) is +3.
3. The metal (Ir) is from row 6 and a 5d metal has a large crystal field.
The Ir complex has 6 ligands and is, therefore, octahedral.
The electron population is: — — e_g

↑↓ ↑↓ ↑↓ t_{2g}

The $[Ir(NH_3)_6]^{3+}$ is diamagnetic.

b) $[Cr(H_2O)_6]^{2+}$ The crystal field splitting is very likely to be low because:
1. The ligand (H_2O) is at the middle of the spectrochemical series and so has little effect.
2. The oxidation state of the metal (Cr) is +2.
3. The metal (Cr) is from row 4 and a 3d metal has a smaller crystal field than 4d or 5d metals.
The Cr complex has 6 ligands and is, therefore, octahedral.
The electron population is:

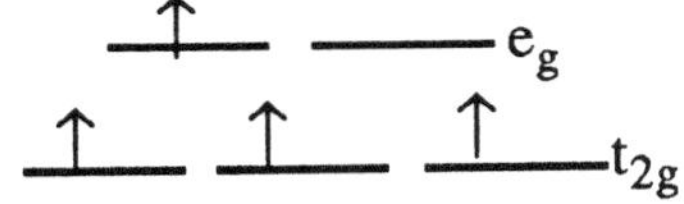

The $[Cr(H_2O)_6]^{2+}$ is paramagnetic and has 4 unpaired electrons.

(continued)

(18.18 continued)

c) $[CoBr_4]^{2-}$ This complex ion must be paramagnetic because Co^{2+} has an odd number of electrons. The Co complex has 4 ligands and is tetrahedral. The crystal field splitting is likely to be low because:

1. The ligand (Br^-) is at the low end of the spectrochemical series.
2. The oxidation state of the metal (Co) is +2.
3. The metal (Co) is from row 4 and a 3d metal.

The electron population is: ↑ ↑ ↑ t

↑↓ ↑↓ e

Note: the electron population would be the same whether the tetrahedral field were weak or strong.

The $[CoBr_4]^{2-}$ is paramagnetic and has 3 unpaired electrons.

d) $[Pd(P(CH_3)_3)_4]$ This complex ion is probably tetrahedral. Pd(0) has 10 valence electrons that likely fill the d orbitals. The compound would be diamagnetic.

e) $[V(en)_3]^{3+}$ The crystal field splitting is likely to be high because:

1. The ligand (en) is in the upper third of the spectrochemical series.
2. The metal (V) has an oxidation state of +3.

This V complex has 3 bidentate ligands and is, therefore, octahedral.

The electron population is: — — e_g

↑ ↑ — t_{2g}

Note: the electron population would be the same whether the octahedral field were weak or strong.

The $[V(en)_3]^{3+}$ is paramagnetic and has 2 unpaired electrons.

18.19 Zr^{2+} complexes are dark purple; Zr^{2+} is d^2

Zr^{4+} complexes are colorless; Zr^{4+} is d^0

Since Zr^{4+} complexes have no d electrons to undergo transition, they are colorless.

18.20 a) $[Ir(NH_3)_3Cl_3]$

mer-$[Ir(NH_3)_3Cl_3]$ (Ir with ligands NH₃, Cl, Cl, Cl, NH₃, NH₃) — *fac*-$[Ir(NH_3)_3Cl_3]$ (Ir with ligands NH₃, Cl, Cl, H₃N, NH₃, Cl)

b) $[PdP(CH_3)_3)_2Cl_2]$

cis-$[PdP(CH_3)_3)_2Cl_2]$ (Pd with Cl, $P(CH_3)_3$, Cl, $P(CH_3)_3$) — *trans*-$[PdP(CH_3)_3)_2Cl_2]$ (Pd with Cl, $P(CH_3)_3$, $(CH_3)_3P$, Cl)

c) $[Cr(CO)_4Br_2]$

cis-$[Cr(CO)_4Br_2]$ (Cr with CO, Br, CO, Br, CO, CO) — *trans*-$[Cr(CO)_4Br_2]$ (Cr with CO, OC, Br, Br, CO, CO)

(continued)

(18.20 continued)

d) $[Cr(en)(NH_3)_2I_2]$

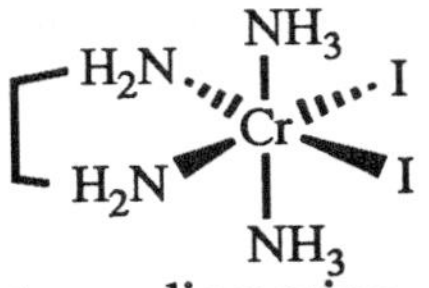

trans-diammine-(ethylenediamine)-*cis*-diiodochromium(II)

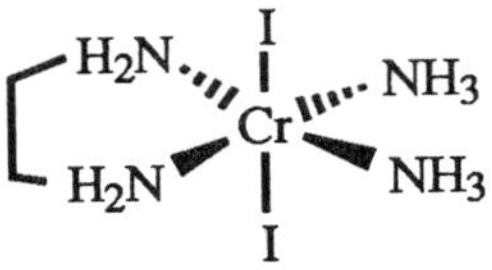

cis-diammine-(ethylenediamine)-*trans*-diiodochromium(II)

and the optical isomers of

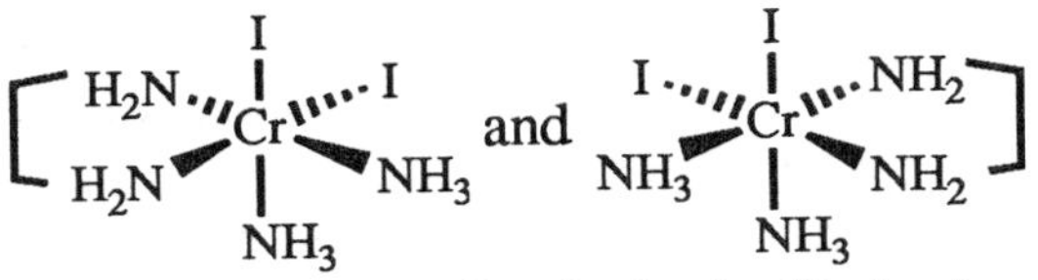

cis-diammine(ethylenediamine)-*cis*-diiodochromium(II)

18.21 $[Cr(H_2O)_6]^{3+}$; $[Cr(NH_3)_6]^{3+}$
Both are Cr^{3+}, d^3, but NH_3 gives complexes with larger splitting than H_2O. According to Table 18-2, the larger Δ should be associated with the orange complex. Therefore, $[Cr(NH_3)_6]^{3+}$ is orange; $[Cr(H_2O)_6]^{3+}$ is violet.

18.22 a) $[Mo(CO)_6]$ has six ligands, therefore, is octahedral. Molybdenum is a fifth row element and CO is the ligand of the spectrochemical series with the highest energy level splitting. This complex has a strong octahedral field. Molybdenum has 6 valence electrons. The configuration of this complex is $(t_{2g})^6$. There are zero unpaired electrons.

b) $[Cr(H_2O)_6]^{2+}$ has six ligands, therefore, is octahedral. Chromium is a 3d transition metal with an oxidation state of +2 (less than +3) and H_2O is a ligand in the middle of the spectrochemical series. This complex ion has a weak octahedral field. Chromium(II) is d^4. The configuration of this complex ion is $(t_{2g})^3 (e_g)^1$. There are 4 unpaired electrons.

c) $[Co(NH_3)_6]^{3+}$ has six ligands, therefore, is octahedral. Cobalt is a 3d transition metal with an oxidation state of +3 and NH_3 is a ligand in the upper half of the spectrochemical series. This complex ion has a strong octahedral field. Cobalt(III) is d^6. The configuration of this complex ion is $(t_{2g})^6$. There are zero unpaired electrons.

d) $[CoBr_4]^{2-}$ (Tetrahedral) Cobalt is a 3d transition metal with an oxidation state of +2 (less than +3) and Br^- is a ligand of the spectrochemical series with a low energy level splitting. This complex ion has a weak tetrahedral field. Cobalt(II) is d^7. The configuration of this complex ion is $(e)^4 (t)^3$. There are 3 unpaired electrons.

e) $[FeCl_4]^-$ (Tetrahedral) Iron is a 3d transition metal with an oxidation state of +3 and Cl^- is a ligand of the spectrochemical series with a low energy level splitting. This complex ion probably has a weak tetrahedral field. Iron(III) is d^5. The configuration of this complex ion is $(e)^2 (t)^3$. There are 5 unpaired electrons.

18.23 a) $BF_3 > BCl_3 > AlCl_3$ Boron is above aluminum in the periodic table; therefore, BF_3 and BCl_3 are harder than $AlCl_3$. BF_3 has F that is more electronegative than the Cl in BCl_3. BF_3 is harder than BCl_3.

b) $Al^{3+} > Tl^{3+} > Tl^{+}$ Al^{3+} and Tl^{3+} are harder than Tl^{+} because they have a greater charge. Al^{3+} is harder than Tl^{3+} because it is above Tl^{3+} in the periodic table.

c) $AlCl_3 > AlBr_3 > AlI_3$ The electronegativities of the anions increase Cl > Br > I. Therefore, the hardness of the compounds decrease: $AlCl_3 > AlBr_3 > AlI_3$

18.24 a) $NH_3 > PH_3 > SbH_3$ The softness of the molecules increases as the central atoms of the compounds move down a column of the periodic table.
b) $ClO_4^- > SO_4^{2-} > PO_4^{3-}$ The effective nuclear charge, Z_{eff}, increases moving across a row of the periodic table to the right. The central atom attracts the electrons of the surrounding atoms more strongly making the ions harder. Also, as the excess electrons (electrons greater than protons) increase for each ion, the electrons become more available and the ions become softer. The oxidation states are: Cl = +7, S = +6 and P = +5.
c) $O^{2-} > S^{2-} > Se^{2-}$ Moving down a column of the periodic table, the elements become larger and less electronegative and more polarizable. The ions become softer in the same direction.

18.25 SO_3 is a harder Lewis acid than SO_2 because the oxidation state of sulfur is +6 in SO_3 but only +4 in SO_2. The greater positive charge on S in SO_3 exerts a stronger pull on the electron cloud thus making it less polarizable and a harder Lewis acid.

18.26 Iodide ion has its valence electrons in larger orbitals (n = 5) than for chloride ion with smaller orbitals (n = 3). Chlorine is more electronegative than iodine. Therefore, iodide ion is more polarizable and a much softer base than chloride ion.

18.27 $(H_3C)_2N—PF_2$

$H_3C—\ddot{N}(CH_3)—\ddot{P}F—F$

BF_3 is a harder Lewis acid than BH_3; the phosphorus end of $(H_3C)_2N—PF_2$ is probably a harder Lewis base than the nitrogen end because the fluoride atoms are so electronegative. Therefore, the phosphorus end forms an adduct with BF_3 and the nitrogen end forms an adduct with BH_3.

BH_3—N(PF_2)(H_3C)(CH_3) and BF_3—P($N(CH_3)_2$)(F)(F)

18.28 a) chalcophiles: Pb, Hg, Ag and Au b) lithophiles: Ti, Ca and K
c) siderophiles: Ru, Rh, Pd, Os, Ir and Au

18.29 a) $NBr_3 + GaCl_3 \rightarrow Br_3N—GaCl_3$
b) $Al(CH_3)_3 + LiCH_3 \rightarrow$ no reaction
c) $SiF_4 + 2\ LiF \rightarrow Li_2[SiF_6]^{2-}$
d) $4LiCH_2CH_2CH_2CH_3 + SnCl_4 \rightarrow Sn(CH_2CH_2CH_2CH_3)_4 + 4\ LiCl$

18.30 The central Al atom of $AlCl_3$ has a vacant p orbital and can, therefore, act as a Lewis acid. The Cl atoms have lone pairs of electrons and can act as Lewis bases. One Cl atom from each of 2 $AlCl_3$ units acts as a Lewis base by sharing a lone pair of electrons with the Al atom of the other unit. This results in a pair of chlorine bridges being formed between the 2 Al atoms.

18.31 The bulkiness of the CH_3 groups hinders the ability of approaching Lewis bases to form bonds to the boron atom.

18.32 Arsenic trichloride is a Lewis base because of the lone pair of electrons on the arsenic(:$AsCl_3$). The As atom has vacant 4d orbitals that allow it to accept a pair of electrons and act as a Lewis acid.

As a base: $Cl_3As: + BCl_3 \rightarrow Cl_3As—BCl_3$

As an acid: $AsCl_3 + Cl^- \rightarrow AsCl_4^-$

18.33 $PCl_3 + (H_3C)_2NH \rightarrow Cl_3P—NH(CH_3)_2 \xrightarrow{-\,HCl} Cl_2P—N(CH_3)_2$

$\xrightarrow{+\,HN(CH_3)_2} ((H_3C)_2N)Cl_2P—NH(CH_3)_2 \xrightarrow{-\,HCl} ((H_3C)_2N)_2ClP$

$\xrightarrow{+\,HN(CH_3)_2} ((H_3C)_2N)_2ClP—NH(CH_3)_2 \xrightarrow{-\,HCl} P(N(CH_3)_2)_3$

18.34 a) $AlI_3 + 3\ NaCl \rightarrow AlCl_3 + 3\ NaI$
Al^{3+} is harder than Na^+. Cl^- is harder than I^-. Therefore, Al^{3+} couples with Cl^- and Na^+ couples with I^-.

b) $2\ TiCl_2 + TiI_4 \rightarrow TiCl_4 + 2\ TiI_2$
Ti^{4+} is harder than Ti^{2+} and Cl^- is harder than I^-.

c) $CaO + H_2S \rightarrow$ no reaction
Ca^{2+} is a hard acid and O^{2-} is a hard base.

d) $3\ CH_3Li + PCl_3 \rightarrow 3\ LiCl + P(CH_3)_3$
Li^+ is a harder acid than P and Cl is harder than CH_3.

e) $AgI + SiCl_4 \rightarrow$ no reaction
Ag^+ is a borderline to soft acid and I^- is a softer base than Cl^-.

18.35 Large atoms with low oxidation states, such as Hg^{2+}, are highly polarizable so they are soft Lewis acids. Zn, on the other hand, is a relatively smaller atom in the +2 oxidation state. As a result, Zn^{2+} is less polarizable and a harder Lewis acid than Hg^{2+}. Because Hg^{2+} is a soft Lewis acid it does not form compounds with hard Lewis bases. Zinc, a hard Lewis acid, forms compounds with hard Lewis bases such as oxides, carbonates, and silicates.

18.36 The polarizability of the halogen atoms increases in the order Cl < Br < I. As the polarizability of the electron clouds on the halogen atoms increases, the intermolecular forces increase. Therefore, the intermolecular forces increase: $BCl_3 < BBr_3 < BI_3$. Therefore, BCl_3 is a gas, BBr_3 is a liquid, and BI_3 is a solid.

18.37 Transition metals have vacant orbitals so they are Lewis acids. They will form bonds with ligands that have lone pairs of electrons (Lewis bases). Very few, if any, Lewis bases having lone pairs of electrons are cations; they are anions or neutral molecules.

18.38

a) ZnS (sphalerite)	chalcophile
b) SnO_2 (cassiterite)	lithophile
c) Na_3AlF_6 (cryolite)	lithophile
d) Ir	siderophile
e) $CaMg(CO_3)_2$ (dolomite)	lithophile
f) Li	lithophile
g) Ag_3SbS_3 (pyrargyrite)	chalcophile
h) Pd	siderophile

18.39

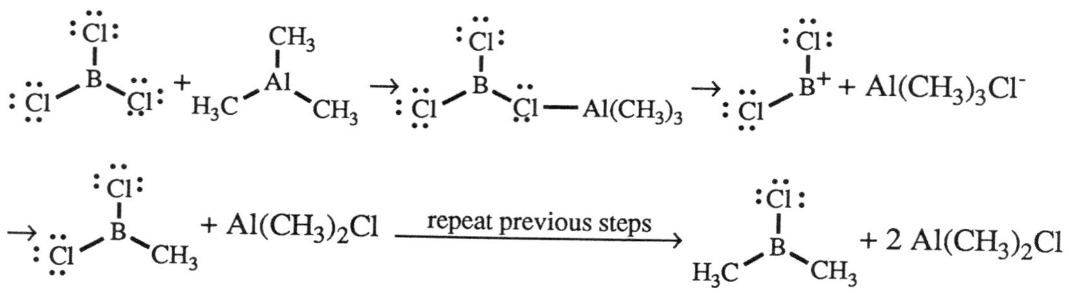

$$BCl_3 + Al(CH_3)_3 \rightarrow Cl_2B{-}Cl{-}Al(CH_3)_3 \rightarrow Cl_2B^+ + Al(CH_3)_3Cl^-$$

$$\rightarrow Cl_2B{-}CH_3 + Al(CH_3)_2Cl \xrightarrow{\text{repeat previous steps}} B(CH_3)_3 + 2\ Al(CH_3)_2Cl$$

a) The Lewis acid is $Al(CH_3)_3$ because it forms an adduct by reacting with an electron pair from a Cl of BCl_3. Further, it cannot be a Lewis base because it has no lone pair of electrons on the molecule.

b) As $Al(CH_3)_2Cl$ molecules are formed, they dimerize to form unreactive:

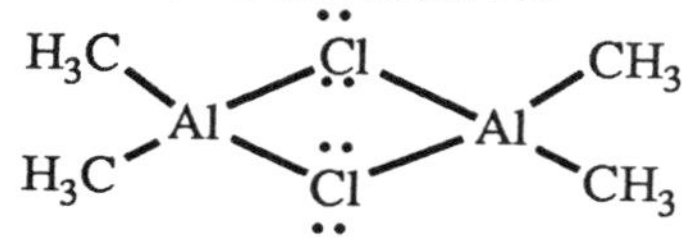

$$(H_3C)_2Al(\mu{-}Cl)_2Al(CH_3)_2$$

18.40 a) *cis*-tetraaquadichlorochromium(III) chloride
b) bromopentacarbonylmanganese(I)
c) *cis*-diamminedichloroplatinum(II)
d) tetraamminezinc sulfate
e) potassium hexacyanoferrate(III)

18.41 $CaO + SiO_2 \rightarrow CaSiO_3$

In this reaction, an O^{2-} anion adds to the SiO_2 structure to form an SiO_3^{2-} anion. Thus, O^{2-} acts as a Lewis base, donating a pair of electons to form an Si—O bond, and SiO_2 acts as a Lewis acid.

18.42 $PtCl_4 \cdot 2NH_3$

$[Pt(NH_3)_2Cl_4]$ diamminetetrachloroplatinum(IV) (*cis* and *trans* isomers)

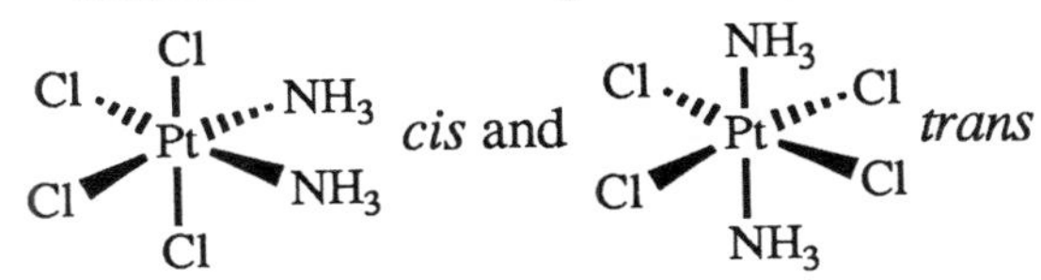

$PtCl_4 \cdot 3NH_3$

$[Pt(NH_3)_3Cl_3]Cl$ triamminetrichloroplatinum(IV) chloride

(*mer* and *fac* isomers)

Cl^- *mer* and Cl^- *fac*

$PtCl_4 \cdot 4NH_3$

$[Pt(NH_3)_4Cl_2]Cl_2$ tetraamminedichloroplatinum(IV) chloride

(*cis* and *trans* isomers)

2 Cl^- *cis* and 2 Cl^- *trans*

$PtCl_4$ $5NH_3$

$[Pt(NH_3)_5Cl]Cl_3$ pentaamminechloroplatinum(IV) chloride

3 Cl^-

$PtCl_4 \cdot 6NH_3$

$[Pt(NH_3)_6]Cl_4$ hexaammineplatinum(IV) chloride

4 Cl^-

18.43 $W(CO)_6$

e_g

$\uparrow\downarrow$ $\uparrow\downarrow$ $\uparrow\downarrow$ t_{2g}

Since the complex is colorless, it must be because the energy difference corresponds to photon energies in the ultraviolet.

18.44 Silver is a soft acid with a low affinity for hard oxygen atoms. It is more reactive toward the softer base, sulfide ion.

18.45 Sulfur is a soft Lewis base. Covalent sulfur–sulfur single bonds are a major determinant of protein structure. Enzymes are proteins. Sulfur is in many enzymes. If the soft metal ions (lead or mercury) are present, they bind with the sulfur atoms of the enzymes. This changes the enzymes' shape and the enzymes can no longer perform their function of catalyzing essential biochemical reactions in the body. This results in illness or even death.

18.46 The PCl_5 molecule has vacant 3d orbitals that allow it to function as a Lewis acid. The Cl atoms on the PCl_5 molecule have lone pairs of electrons so that the molecule can also function as a Lewis base. When PCl_5 is condensed from gas to a solid, the PCl_5 molecules are brought close to each other and 2 molecules react with 1 molecule acting as a Lewis base and the other as a Lewis acid.

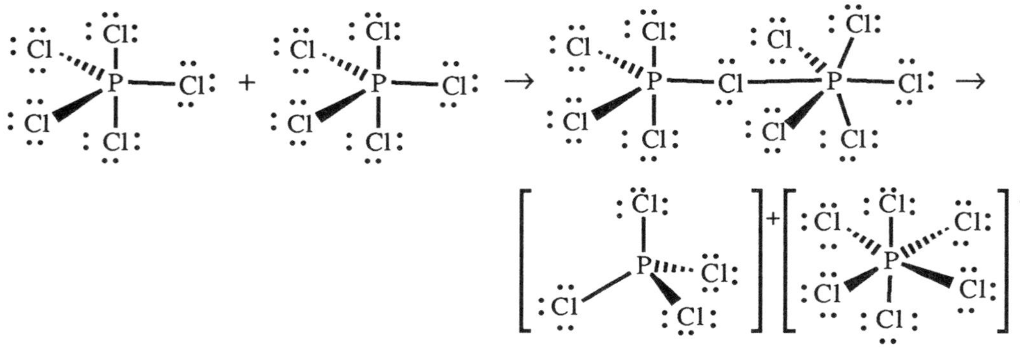

$$PCl_5 + PCl_5 \rightarrow Cl_4P{-}Cl{-}PCl_5 \rightarrow [PCl_4]^+[PCl_6]^-$$

The result is the tetrahedral cation $[PCl_4]^+$ and the octahedral anion $[PCl_6]^-$.

18.47 Strong adducts form when the donor and acceptor can approach each other closely. In dimethyltetrahydrofuran, the electron clouds around the two methyl groups repel an approaching BF_3 molecule, lengthening the O—B bond and making the adduct weaker.

18.48 a) tetraamminedibromocobalt(III) bromide $[Co(NH_3)_4Br_2]Br$

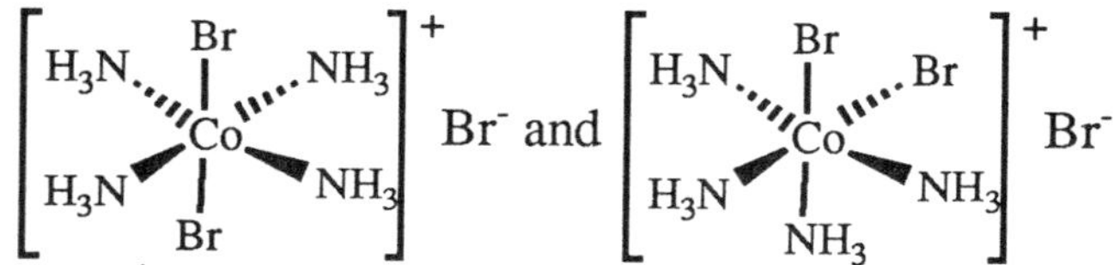

$[Co(NH_3)_4Br_2]^+ Br^-$ (*trans*) and $[Co(NH_3)_4Br_2]^+ Br^-$ (*cis*)

b) triamminetrichlorochromium(III) $[Cr(NH_3)_3Cl_3]$

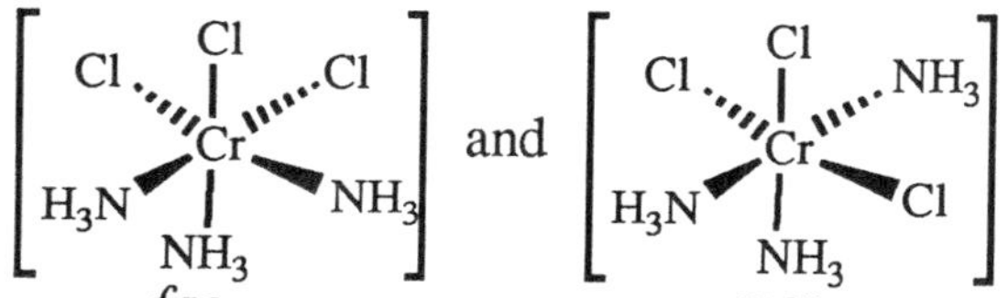

$[Cr(NH_3)_3Cl_3]$ (*fac*) and $[Cr(NH_3)_3Cl_3]$ (*mer*)

(continued)

(18.48 continued)

c) dicarbonylbis(trimethylphosphine)platinum(0) $[Pt(CO)_2(P(CH_3)_3)_2]$

This platinum complex is probably distorted square planar.

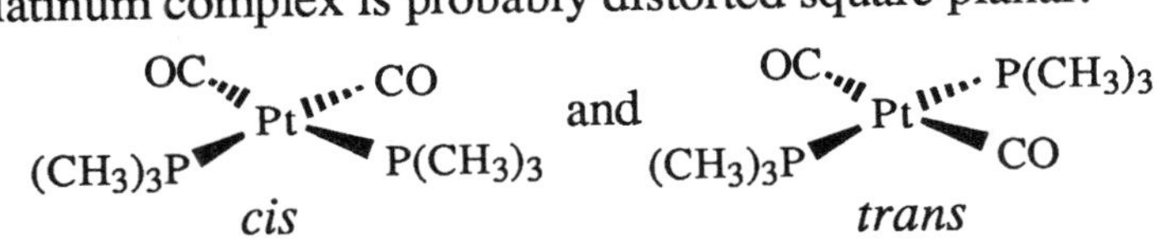

18.49 Because F^- is at the small splitting end of the spectrochemical series, $[FeF_6]^{3-}$ will absorb light of longer wavelengths than $[Fe(H_2O)_6]^{3+}$. This places its absorption in the infrared; so the complex is colorless.

18.50 In hemoglobin the Fe^{2+} is bound to the tetradentate porphyrin ligand. The d^6 Fe^{2+} ion is most stable in an octahedral environment, so two axial coordination sites are available to bind to other ligands. One axial site is occupied by a nitrogen donor atom from an amino acid side chain of the protein. For oxyhemoglobin the sixth site is occupied by the fairly strong ligand O_2. For deoxyhemoglobin the sixth site is occupied by the medium strength ligand H_2O. Because the octahedral environment of the Fe^{2+} ion is the same in each case except for the sixth site occupant, the energy level splitting will depend on this occupant of the sixth site. In each case the configuration for d^6 Fe^{2+} is $(t_{2g})^6$ and the color depends on the splitting between the t_{2g} and e_g. For deoxyhemoglobin the splitting depends on the strength of the ligand H_2O and results in a blue color ($\Delta \cong 200$ kJ/mol). The stronger ligand O_2 results in a red color ($\Delta \cong 240$ kJ/mol). One way to test this hypothesis would be to complex hemoglobin with other ligands. Weaker ligands like halides should result in green to colorless products. Intermediate strength ligands such as NH_3 would result in colors intermediate in the visible spectrum (violet or purple). Carbon monoxide is one of the very strongest ligands. Therefore, hemoglobin saturated with CO should have a greater splitting than H_2O or O_2 (Δ greater than 240 kJ/mol) and probably have a color somewhere between red-orange and yellow.

18.51 The $[Ni(CN)_4]^{2-}$ complex is diamagnetic meaning that all eight valence electrons of Ni^{2+} are paired. This can only occur if the complex is square planar (See Figure 18-16). The $[NiCl_4]^{2-}$ complex is paramagnetic because it has unpaired spins. This requires tetrahedral geometry (See Fig. 18-17).

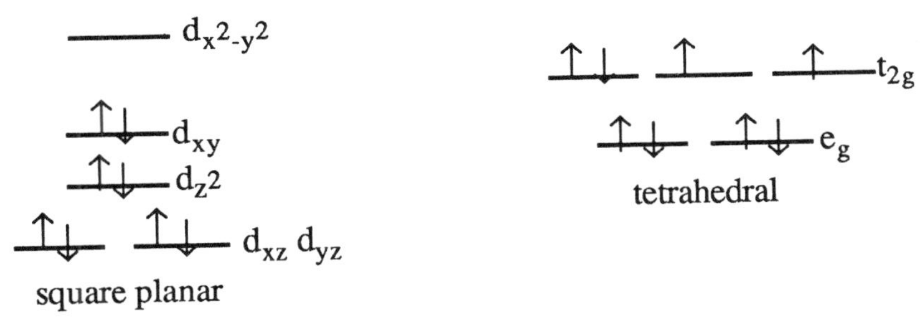

18.52 a) $[Fe(CN)_6]^{4-}$ The CN^- is a large energy level splitting ligand. Therefore, $\Delta > P$ and the complex is low spin. Fe^{2+} is $3d^6$.

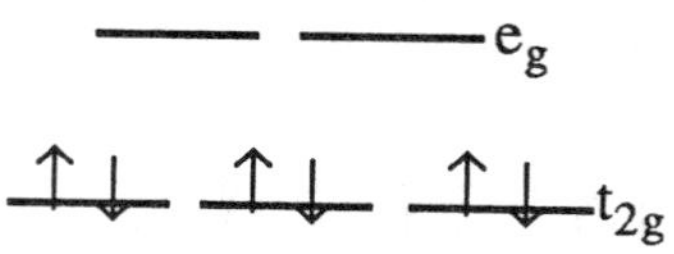

— — e_g

↑↓ ↑↓ ↑↓ t_{2g}

b) $[MnCl_4]^{2-}$ The oxidation state of Mn is +2 and Mn is in the fourth row. The Cl^- is a small energy level splitting ligand. Mn^{2+} is $3d^5$. Therefore, $P > \Delta$ and the complex is high spin.

↑ ↑ ↑ t_{2g}

↑ ↑ e_g

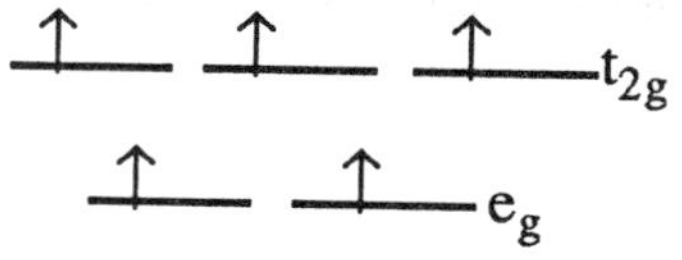

c) $[Rh(NH_3)_6]^{3+}$ The oxidation state of Rh is +3 and Rh is in the fifth row. Rh^{3+} is $4d^6$. The NH_3 is a middle to upper energy level splitting ligand. Therefore , $\Delta >$ P and the complex is low spin.

— — e_g

↑↓ ↑↓ ↑↓ t_{2g}

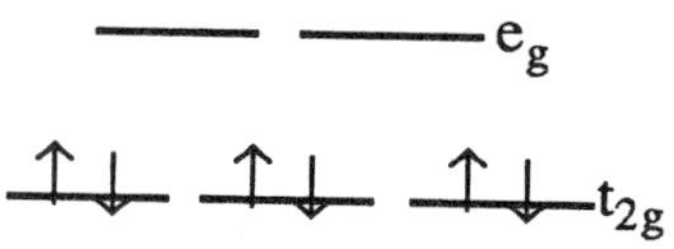

d) $[Co(H_2O)_6]^{2+}$ The oxidation state of Co is +2 and Co is in the 4th row. Co^{2+} is $3d^7$. The H_2O is a ligand of medium energy level splitting strength. Therefore, $P >$ Δ and the complex is high spin.

↑ ↑ e_g

↑↓ ↑↓ ↑ t_{2g}

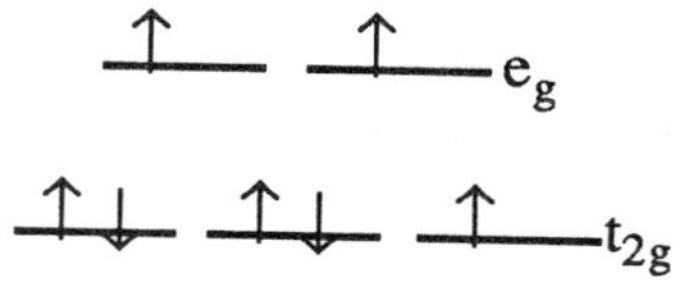

18.53 Rust removers containing oxalate anion take advantage of the chelate effect. The formation constant for the reaction of Fe^{2+} or Fe^{3+} with the bidentate oxalate must be larger than it is for monodentate O^{2-}. Thus, the iron cations in rust will preferentially react with oxalate. Once the iron is tied up with oxalate ligands, the water soluble complex can be washed away with water.

CHAPTER 19: NUCLEAR CHEMISTRY AND RADIOCHEMISTRY

19.1 a) $Z = 3, A = 6, N = 3$ b) $Z = 20, A = 43, N = 23$
c) $Z = 92, A = 238, N = 146$ d) $Z = 52, A = 130, N = 78$
e) $Z = 10, A = 20, N = 10$ f) $Z = 82, A = 205, N = 123$

19.2 a) $Z = 10, A = 22, N = 12$ b) $Z = 82, A = 202, N = 120$
c) $Z = 19, A = 41, N = 22$ d) $Z = 47, A = 109, N = 62$
e) $Z = 2, A = 3, N = 1$ f) $Z = 56, A = 126, N = 70$

19.3 a) ${}^{4}_{2}He$ b) ${}^{184}_{74}W$ c) ${}^{26}_{12}Mg$ d) ${}^{60}_{28}Ni$

19.4 a) ${}^{114}_{50}Sn$ b) ${}^{219}_{86}Rn$ c) ${}^{34}_{16}S$ d) ${}^{121}_{51}Sb$

19.5 a) unstable (neutron-poor, $Z > N$) b) unstable ($Z > 83$) c) stable
d) unstable (Z-N, odd-odd; Tc has no stable nuclides)

19.6 $\Delta m = 2(1.007276\ g/mol) = 2.014552\ g/mol$

$\Delta E = \Delta m(8.988 \times 10^{10}\ kJ/g) = (2.014552\ g/mol)(8.988 \times 10^{10}\ kJ/g)$
$= 1.811 \times 10^{11}\ kJ/mol$

$\Delta E = (1.811 \times 10^{11}\ kJ/mol) \div (6.022 \times 10^{23}\ events/mol)$
$= 3.007 \times 10^{-13}\ kJ/event$

19.7 $\Delta m = 1.00\ met\ ton = 1.00(1000\ kg) = 1.00(10^{3}\ kg)(10^{3}\ g/kg) = 1.00\ 10^{6}\ g$

$\Delta E = \Delta m(8.988 \times 10^{10}\ kJ/g) = (1.00 \times 10^{6}\ g)(8.988 \times 10^{10}\ kJ/g) = 8.99 \times 10^{16}\ kJ$

19.8 $MM_{Co} = 58.933\ g/mol\left({}^{59}_{27}Co\right)$

protons: $27(1.007276\ g/mol) = 27.196452\ g/mol$
neutrons: $32(1.008665\ g/mol) = 32.277280\ g/mol$
electrons: $27(0.0005486\ g/mol) = \underline{0.0148122\ g/mol}$
$59.488544\ g/mol$

$\Delta m = 58.933\ g/mol - 59.4885\ g/mol = -0.5555\ g/mol$

$\Delta E = \Delta m(8.988 \times 10^{10}\ kJ/g) = (-0.5555\ g/mol)(8.988 \times 10^{10}\ kJ/g)$
$= -4.99 \times 10^{10}\ kJ/mol$

$$\frac{-4.99 \times 10^{10}\ kJ/mol}{59\ nucleons} = -8.5 \times 10^{8}\ kJ/mol\ nucleons$$

19.9 $MM_{Cs} = 132.905\ g/mol \quad {}^{133}_{55}Cs$

protons: 55(1.007276 g/mol) = 55.400180 g/mol
neutrons: 78(1.008665 g/mol) = 78.675870 g/mol
electrons: 55(0.0005486 g/mol) = 0.0301730 g/mol
134.106223 g/mol

$\Delta m = 132.905\ g/mol - 134.106\ g/mol = -1.201\ g/mol$

$\Delta E = (-1.201\ g/mol)(8.988 \times 10^{10}\ kJ/g) = -1.079 \times 10^{11}\ kJ/mol$

$$\Delta E \text{ per nucleon} = \frac{-1.079 \times 10^{11}\ kJ/mol}{133\ nucleons} = -8.11 \times 10^{8}\ kJ/mol\ nucleons$$

19.10 Barrier for fusion of 2 deuterium nuclei:

$$E = \frac{(1.389 \times 10^{5}\ kJ\ pm/mol)(1)(1)}{(2.8 \times 10^{-3}\ pm)} = 5.0 \times 10^{7}\ kJ/mol$$

Barrier of fusion of 2 protons:
The charges would be the same as above, but the radii of the protons would be smaller than the radii of deuterium nuclei. Therefore, the energy barrier would be greater for the proton fusion.

The resulting ${}^{2}_{2}He$ is less stable than the reacting ${}^{1}_{1}H$ or the normal ${}^{4}_{2}He$ because the neutron-proton ratio is 0 for ${}^{2}_{2}He$ and 1 for ${}^{4}_{2}He$.

19.11

	symbol	name
a) a photon (high energy)	γ	Gamma-ray
b) positive particle with mass number 4	α	alpha-particle
c) positron:	β^{+}	positron

19.12

	charge number	mass number	symbol
a) an α - particle	2	4	α
b) a β-ray	-1	0	β
c) a neutron	0	1	n

19.13 a) ${}^{125}_{52}Te \rightarrow {}^{125}_{52}Te + \gamma$ b) ${}^{123}_{52}Te + {}^{0}_{-1}e \rightarrow {}^{123}_{51}Sb$

c) ${}^{127}_{52}Te \rightarrow {}^{0}_{-1}\beta + {}^{127m}_{53}I$

${}^{127m}_{53}I \rightarrow {}^{127}_{53}I + \gamma$

19.14 $$N = \#\text{ of nuclei} = \frac{(4.7 \times 10^{-6}\ g)(6.022 \times 10^{23}\ nuclei/mol)}{(94\ g/mol)} = 3.0 \times 10^{16}\ nuclei$$

$$t_{1/2} = \frac{N \ln 2}{(\Delta N/\Delta t)} = \frac{(3.0 \times 10^{16}\ nuclei)\ln 2}{(20\ nuclei/min)} = 1.0 \times 10^{15}\ min = 2.0 \times 10^{9}\ yr$$

19.15 $^{232}_{90}Th \rightarrow ^{4}_{2}\alpha + ^{228}_{88}Ra$

$^{228}_{88}Ra \rightarrow ^{0}_{-1}\beta + ^{228}_{89}Ac$

$^{228}_{89}Ac \rightarrow ^{0}_{-1}\beta + ^{228}_{90}Th$

$^{228}_{90}Th \rightarrow ^{4}_{2}\alpha + ^{224}_{88}Ra$

$^{224}_{88}Ra \rightarrow ^{4}_{2}\alpha + ^{220}_{86}Rn$

$^{220}_{86}Rn \rightarrow ^{4}_{2}\alpha + ^{216}_{84}Po$

$^{216}_{84}Po \rightarrow ^{0}_{-1}\beta + ^{216}_{85}At$

$^{216}_{85}At \rightarrow ^{4}_{2}\alpha + ^{212}_{83}Bi$

$^{212}_{83}Bi \rightarrow ^{0}_{-1}\beta + ^{212}_{84}Po$

$^{212}_{84}Po \rightarrow ^{4}_{2}\alpha + ^{208}_{82}Pb$

19.16 atoms Co = number $CoCl_2$

$$= \frac{(12.5 \text{ x } 10^{-3} \text{ g})(6.022 \text{ x } 10^{23} \text{ CoCl}_2 / \text{mol})\left(\frac{1 \text{ atom Co}}{1 \text{ CoCl}_2}\right)}{(129.84 \text{ g CoCl}_2 / \text{mol})}$$

$$= 5.80 \text{ x } 10^{19} \text{ atoms Co}$$

$$t_{1/2} = \frac{N \ln 2}{(\Delta N / \Delta t)}$$

$$N = \frac{t_{1/2}(\Delta N / \Delta t)}{\ln 2} = \frac{(5.26 \text{ yr})(350\ ^{60}\text{Co/min})}{0.693}$$

$$= \frac{(5.26 \text{ yr})(365 \text{ days / yr})(24 \text{ hr/ day})(60 \text{ min/hr})(350 \text{ min}^{-1})}{\ln 2}$$

$$= 1.40 \text{ x } 10^{9}\ ^{60}\text{Co}$$

$$\%\ ^{60}\text{Co} = \frac{1.40 \text{ x } 10^{9}\ ^{60}\text{Co}}{5.80 \text{ x } 10^{19} \text{ Co atoms}} \text{ x } 100\% = 2.41 \text{ x } 10^{-9}\%$$

19.17 a) $^{12}_{6}C + ^{1}_{0}n \rightarrow (^{13m}_{6}C) \rightarrow ^{12}_{5}B + ^{1}_{1}p$

b) $^{16}_{8}O + ^{4}_{2}\alpha \rightarrow (^{20m}_{10}Ne) \rightarrow ^{20}_{10}Ne + \gamma$

c) $^{247}_{96}Cm + ^{11}_{5}B \rightarrow (^{258m}_{101}Md) \rightarrow ^{255}_{101}Md + 3\ ^{1}_{0}n$

19.18 $^{14}_{7}N + ^{1}_{0}n \rightarrow ^{14}_{6}C + ^{1}_{1}p$

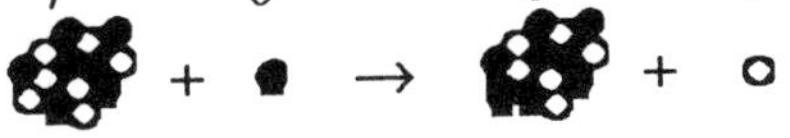

19.19 $^{14}_{7}N + ^{4}_{2}\alpha \rightarrow ^{18}_{8}O + ^{0}_{+1}\beta^{+}$

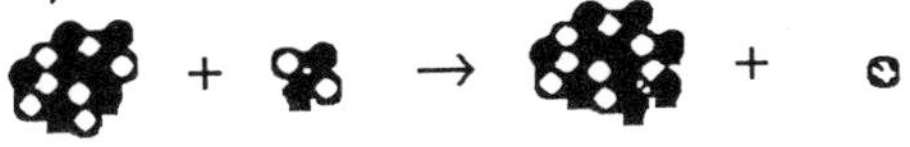

19.20 $$^{235}_{92}U + ^{1}_{0}n \rightarrow ^{100}_{42}Mo + ^{134}_{50}Sn + 2^{1}_{0}n$$

MM, g / mol: 235.0439 1.0087 99.9076 133.9125

Δm = [(99.9076 g/mol) + 133.9125 g/mol + 2(1.0087 g/mol)] - [235.0439 g/mol + 1.0087 g/mol] = -0.2151 g/mol

ΔE = (8.988 x 10^{10} kJ/g)(-0.2151 g/mol) = -1.933 x 10^{10} kJ/mol

It matches quite closely the result in Exercise 19.5.1 for 1 neutron (i.e., 1 net neutron).

19.21 The most likely products have A = 95 and A = 138.
Element with stable isotope with A = 95: Mo
Elements with stable isotopes with A = 138: Ba, La, Ce
Other elements with stable isotopes with A = 90 - 100: Zr, Nb, Ru, Rh
Other elements with stable isotopes with A = 132 - 143: Xe, Cs, Pr, Nd

19.22 The dangers of an accident in a coal-burning power plant consist of damages caused by the physical results of an explosion and fire, asphyxiation caused by high concentrations of gases that prevent victims from getting enough oxygen, or carbon monoxide poisoning. Surviving victims are very likely to recover with no serious long-term effects. An accident in a nuclear power plant has all the dangers above, but also has long-term dangers. If a victim inhales gases from a nuclear accident and initially survives, the radioactive contents of the gases may cause radioactive poisoning because of damage done to the lungs, etc. If one does not suffer from radioactive poisoning, the damage done to body cells by the radioactivity may result in increased chances for decades of cancer, leukemia and other disease related to cell abnormalities.

The dangers of the products of the coal-burning plant are short-lived. Gases quickly mix with the air and are diluted to safe concentrations. Solid and liquid particles settle to the ground and present little or no danger. The radioactive products of a nuclear accident may also be diluted, but they will still be dangerous. Even in dilute concentrations, long-term exposure can be dangerous. Solids and liquids contaminate the environment. They can be absorbed by plant life. Animals consuming these plants acquire and concentrate the radioactive contaminants. The animals and their products may be unsafe to use. The radioactive contaminants may last (have half-lifes) of thousands, millions or even billions of years.

The area of danger for the coal-burning plant would be much smaller. It would probably be measured in feet from the plant. Although people living within a mile or two might be evacuated as a precaution, once the situation was under control (within a day or two at most) they would be allowed to return. The danger area for a nuclear plant accident is likely to be measured in miles. People living within 20 miles might have to be evacuated permanently. Winds might carry radioactive substances hundreds of miles away to make other areas uninhabitable. All this assumes a worst-case scenario—that containment was broken and reaction products escaped. Even if containment is not breached, the nuclear accident would still be more serious. After a cool-down period, the coal-burning plant could be cleaned up and rebuilt. The cool-down period for the nuclear plant could be thousands or millions of years. Not only could the site not be cleaned up and reused, but it would have to be constantly monitored over this time to be sure there was not a breach of containment and an escape of radioactive material.

19.23 It is important that heat be transferred from the core to the turbines without transfer of matter because the core generates lethal radioactive products that cannot be allowed to escape or transfer radioactivity. Liquid sodium metal or high pressure water are the two most used substances for circulation around the core to absorb the heat produced by the nuclear fission. This hot fluid then passes through a steam generator, transfers its heat energy to water resulting in the water being vaporized to steam. This steam is used to produce electricity by driving a conventional steam turbine. Sodium metal or water are good choices to carry the heat from the core to the steam generator because if they were to form radioactive products while passing through the core, these products have short half-lifes, decay by β emission (which can be shielded) and produce stable products.

19.24

$$\begin{array}{ll} {}^{6}_{3}\text{Li} + {}^{1}_{0}\text{n} \rightarrow {}^{4}_{2}\text{He} + {}^{3}_{1}\text{H} & \Delta E = -4.6 \times 10^{8}\ \text{kJ/mol} \\ {}^{2}_{1}\text{H} + {}^{3}_{1}\text{H} \rightarrow {}^{4}_{2}\text{He} + {}^{1}_{0}\text{n} & \Delta E = -1.7 \times 10^{9}\ \text{kJ/mol} \\ \hline {}^{6}_{3}\text{Li} + {}^{2}_{1}\text{H} \rightarrow 2\,{}^{4}_{2}\text{He} & \Delta E = -2.16 \times 10^{9}\ \text{kJ/mol} \end{array}$$

$$(1.50\ \text{g}\ {}^{2}_{1}\text{H})\left(\frac{1\ \text{mol}\ {}^{2}_{1}\text{H}}{2.0141022\ \text{g}\ {}^{2}_{1}\text{H}}\right) = 0.7447\ \text{mol}\ {}^{2}_{1}\text{H}$$

$$(1.50\ \text{g}\ {}^{6}_{3}\text{Li})\left(\frac{1\ \text{mol}\ {}^{6}_{3}\text{Li}}{6.015121\ \text{g}\ {}^{6}_{3}\text{Li}}\right) = 0.2494\ \text{mol}\ {}^{6}_{3}\text{Li}$$

${}^{6}_{3}\text{Li}$ serves as the limiting reactant.

$$(0.2494\ \text{mol}\ {}^{6}_{3}\text{Li})(-2.16 \times 10^{9}\ \text{kJ/mol}) = -5.4 \times 10^{8}\ \text{kJ}$$

19.25 mass of ${}^{3}_{1}\text{H}$ = 3.01605 g/mol

Assuming ${}^{3}_{1}\text{H}$ and ${}^{2}_{1}\text{H}$ both have a radius of 1.4×10^{-3} pm

$$E = \frac{(1.389 \times 10^{5}\ \text{kJ pm/mol})(Z_1)(Z_2)}{d}$$

$$= \frac{(1.389 \times 10^{5}\ \text{kJ pm/mol})\ (1)(1)}{(2 \times 1.4 \times 10^{-3}\ \text{pm})} = 5.0 \times 10^{7}\ \text{kJ/mol}$$

$$E = 1/2mv^2$$

$$v = \sqrt{\frac{E}{1/2\ m}} = \sqrt{\frac{(5.0 \times 10^{7}\ \text{kJ/mol})(10^{3}\ \text{J/kJ})\left(\frac{\text{kg} \bullet \text{m}^2}{\text{s}^2 \bullet \text{J}}\right)}{\left(\frac{1}{2}\right)(3.01605\ \text{g/mol})\left(\frac{\text{kg}}{10^{3}\ \text{g}}\right)}} = 5.8 \times 10^{6}\ \text{m/s}$$

19.26 Fission Power

Advantages: Now commercially possible

Disadvantages: Possible for only a few, rare, heavy nuclides.
Large amounts of radioactive wastes to be disposed of.
Strong public opinion against it.

Fusion Power

Advantages: Possible for abundant light nuclides.
Some reactions release more energy per unit mass than fission.
Product nuclides usually stable; therefore, small amounts of radioactive wastes
Not as unacceptable to the public.

Disadvantages: Not now commercially possible
Feasibility uncertain.
Large developmental expenditures still required.
Must operate at very high temperatures that at present cannot be contained.

Obviously, fusion power plants are to be preferred and continued attempts should be made to develop them. Recent successes give new hope. The only reason to continue to develop fission power is if it might be required as a stopgap between fossil fuels and fusion power. It is not known when fusion power will be commercially feasible, or for that matter, if it will ever be feasible. We also do not know how much longer we will be able to use fossil fuels. While we can estimate how long it will be before they are depleted, there is the possibility that their use might have to be curtailed because of pollution from their combustion. To be safe, further development of fission power should continue in case it is needed.

19.27 Presently fusion can occur only on a very large scale. It occurs only at very high temperatures that require high-intensity power sources. Confinement of the reaction requires high-intensity fields that are beyond current technology as the temperatures are too high for physical containment. If a sustained fusion reaction is achieved, some way must be found to convert its energy output (consisting of photons) into electricity. This will require huge amounts of money.

19.28

First-generation star	Temperature	Composition
hydrogen-burning stage	4×10^7 K	${}^1_1H$, ${}^{0}_{-1}e$, ${}^4_2He$
helium-burning stage	10^8 K	${}^4_2He$, $({}^8_4He)$, ${}^{12}_6C$, ${}^{16}_8O$
carbon-burning stage	10^9 K	${}^{12}_6C$, ${}^{16}_8O$, ${}^{23}_{11}Na$, ${}^{20}_{10}Ne$, ${}^{24}_{12}Mg$, ${}^{31}_{15}P$, ${}^1_1H$, ${}^4_2He$, and other nuclides up to ${}^{56}_{26}Fe$

19.29

Second-generation star	Temperature	Composition
hydrogen-burning stage	probably lower than first generation as ^{12}C catalyzed	$^{1}_{1}H$, $^{12}_{6}C$, $^{4}_{2}He$, ($^{13}_{6}C$), $^{0}_{-1}e$, ($^{14}_{6}C$), ($^{15}_{7}N$) and other nuclides up to Z = 26
helium-burning stage	10^8 K	$^{13}_{6}C$, $^{4}_{2}He$, $^{16}_{8}O$, $^{1}_{0}n$, $^{56}_{26}Fe$, $^{57}_{27}Co$ and many more reactions because $^{1}_{0}n$ are generated; all possible stable nuclides can form
carbon-burning stage	10^9 K	May become unstable, explode, and produce still heavier nuclides

A second-generation star is a combination of interstellar hydrogen and matter from an exploded first-generation star. Therefore, a second-generation star will contain higher-Z nuclides because a first-generation star does not become unstable and explode (supernova) until it reaches the third stage at which time nuclides with Z and A values up to iron-56 are being produced. A greater range of reactions takes place in a second-generation star because the fusion of $^{4}_{2}He$ and $^{13}_{6}C$ generates neutrons which can be sequentially captured resulting in the production of all possible stable nuclides. An explosion of a second-generation star produces not only heavier nuclide debris, but also large numbers of neutrons that can be captured and for still more heavy nuclides.

19.30 Carbon fusion requires a higher temperature (10^9 K) than is present before the third stage of evolution.

19.31 Iron (Z = 26) is the most stable nuclide. Fusion reactions do not produce nuclides with Z greater than 26. Higher-Z nuclides can be produced by sequential capture of neutrons and β-emission, but first-generation stars do not have sources of neutrons and, therefore, do not produce nuclides with Z > 26.

19.32 $$(1.50\text{ mL})\left(\frac{1\text{ L}}{10^3\text{ mL}}\right)\left(\frac{2.50 \times 10^{-9}\text{ mol Tc}}{\text{L}}\right)(1.35 \times 10^7\text{ kJ / mol})$$

$$= 5.06 \times 10^{-4}\text{ kJ} = 0.506\text{ J}$$

19.33 $$N = (1.50\text{ mL})\left(\frac{1\text{ L}}{10^3\text{ mL}}\right)\left(\frac{2.50 \times 10^{-9}\text{ mol Tc}}{\text{L}}\right)\left(\frac{6.022 \times 10^{23}\text{ atoms Tc}}{\text{mol Tc}}\right)$$

$$= 2.26 \times 10^{13}\text{ atoms Tc}$$

$$\frac{\Delta N}{\Delta t} = \frac{-N \ln 2}{t_{1/2}} = \frac{-(2.26 \times 10^{13}\text{ Tc})(0.693)}{(6.0\text{ hr})(60\text{ min / hr})(60\text{ s / min})} = -7.3 \times 10^8\text{ Tc/s}$$

7.3×10^8 Tc decays/s = 7.3×10^8 γ-rays emitted per second.

19.34 Cells that divide most rapidly tend to be most easily damaged. Cells in the reproductive organs are included among these rapidly dividing cells. Also, radiation may cause genetic mutations which can be passed on to a future child.

19.35 After the foods are packaged, the package is exposed to enough gamma radiation to kill microorganisms. Moderate doses retard spoilage and high doses can prevent spoilage completely for as long as the package remains intact. There is fear that irradiation might generate unhealthy, or even carcinogenic, byproducts by causing chemical reactions.

19.36 $N_0 = N_U + N_{Pb}$

$$\frac{N_0}{N} = \frac{N_U + N_{Pb}}{N_U} = 1 + \frac{N_{Pb}}{N_U} = 1 + \left(\frac{1}{1.21}\right) = 1 + 0.826 = 1.826$$

$$t = \frac{t_{1/2}}{\ln 2}\ln\left(\frac{N_0}{N}\right) = \frac{(4.5 \text{ x } 10^9 \text{ yr})}{(0.693)}\ln(1.826) = 3.9 \text{ x } 10^9 \text{ yr (age of ore)}$$

19.37 $N_0 = N_{Sr} + N_{Rb}$

$$\frac{N_0}{N} = \frac{N_{Sr} + N_{Rb}}{N_{Rb}} = 1 + \frac{N_{Sr}}{N_{Rb}} = 1 + 0.0050 = 1.0050$$

$$t = \frac{t_{1/2}}{\ln 2}\ln\left(\frac{N_0}{N}\right) = \frac{(4.9 \text{ x } 10^{11} \text{ yr})}{(0.693)}\ln(1.0050) = 3.5 \text{ x } 10^9 \text{ yr}$$

19.38 $$\frac{N_0}{N} = \frac{18{,}400 \text{ counts/g 20 hr}}{1020 \text{ counts / (0.250 g)(24 hrs)}} = \frac{920 \text{ counts/g hr}}{170 \text{ counts/g hr}} = 5.4$$

$$t = \frac{t_{1/2}}{\ln 2}\ln\left(\frac{N_0}{N}\right) = \frac{(5730 \text{ yr})}{(0.693)}\ln(5.4) = 1.4 \text{ x } 10^4 \text{ yr}$$

(These numbers reflect changes planned for the second printing of the textbook.)

19.39 Advantageous properties of the ^{99m}Te for use as a medical imaging isotope are a half-life of 6 hours (long enough to use but short enough to minimize long-term effects); emits γ-rays of moderate energy (easy to detect but not highly damaging to living tissue); element can be bound to chemicals that the body recognizes and processes (can be used to image the thyroid gland, brain, lungs, heart, liver, stomach, kidneys and bones); and can readily be extracted in pure form from natural molybdenum that has been exposed to neutron bombardment.

19.40 a) $^{238}_{92}\text{U} \rightarrow {}^{4}_{2}\alpha + {}^{234}_{90}\text{Th}$

b) $^{60}_{28}\text{Ni} + {}^{1}_{0}\text{n} \rightarrow \left({}^{61m}_{28}\text{Ni}\right) \rightarrow {}^{60}_{27}\text{Co} + {}^{1}_{1}\text{p}$

c) $^{239}_{93}\text{Np} + {}^{12}_{6}\text{C} \rightarrow {}^{248}_{99}\text{Es} + 3\,{}^{1}_{0}\text{n}$

d) $^{35}_{17}\text{Cl} + {}^{1}_{1}\text{p} \rightarrow \left({}^{36m}_{18}\text{Ar}\right) \rightarrow {}^{4}_{2}\alpha + {}^{32}_{16}\text{S}$

e) $^{60}_{27}\text{Co} \rightarrow {}^{60}_{28}\text{Ni} + {}^{0}_{-1}\beta$

19.41 $^{209}_{83}Bi$

Protons:	83(1.007276 g/mol) =	83.603908 g/mol
Neutrons:	126(1.008665 g/mol) =	127.091790 g/mol
Electrons:	83(0.0005486 g/mol) =	0.0455338 g/mol
		210.741232 g/mol

Δm = 208.980 g/mol - 210.741232 g/mol = -1.7612 g/mol

ΔE = (-1.7612 g/mol)(8.988×10^{10} kJ/g) = -1.583×10^{11} kJ/mol

binding energy per mole nucleons =

$$\frac{\Delta E}{209 \text{ nucleons}} = \frac{(-1.583 \times 10^{11} \text{ kJ/mol})}{209 \text{ nucleons}} = -7.574 \times 10^{8} \text{ kJ/mol nucleons}$$

binding energy per nucleon =

$$\frac{(-7.574 \times 10^{8} \text{ kJ/mol nucleons})}{6.022 \times 10^{23} \text{ nucleons / mol nucleons}} = 1.258 \times 10^{-15} \text{ kJ / nucleon}$$

19.42 a) $^{241}_{95}Am$ no. protons = 95 no. neutrons = 146 $\frac{n}{p} = \frac{146}{95} = 1.537$

b) $$N = -\frac{\Delta N}{\Delta t}\frac{t_{1/2}}{\ln 2} = \left(\frac{5 \text{ atoms}}{\text{s}}\right)\left(\frac{458 \text{ yr}}{0.693}\right)\left(\frac{365 \text{ days}}{\text{yr}}\right)\left(\frac{24 \text{ hr}}{\text{day}}\right)\left(\frac{60 \text{ min}}{\text{hr}}\right)\left(\frac{60 \text{ s}}{\text{min}}\right)$$

$= 1.04 \times 10^{11}$ atoms Am

$$\text{mass Am} = \left(\frac{1.04 \times 10^{11} \text{ atoms Am}}{6.022 \times 10^{23} \text{ atom/mol}}\right)(241 \text{ g Am/mol})$$

$= 4.17 \times 10^{-11}$ g ^{241}Am

c) $$\ln\left(\frac{N_0}{N}\right) = \frac{t \ln 2}{t_{1/2}} \qquad \frac{N_0}{N} = \left(\frac{5/\text{s}}{3.5/\text{s}}\right) = 1.43$$

$$t = \frac{t_{1/2}}{\ln 2}\ln\left(\frac{N_0}{N}\right) = \frac{458 \text{ yr}}{0.693}\ln(1.43) = 236 \text{ yr}$$

d) Yes, you would be exposed to γ-ray radiation. The α-particles are stopped within approximately 10 cm of air, but the γ-rays travel many meters before losing their destructive power.

19.43 (10^5 neutrons/s)(30 s) = 3.0×10^6 neutrons

$$\ln\left(\frac{N_0}{N}\right) = \frac{t \ln 2}{t_{1/2}} = \ln\left(\frac{3.0 \times 10^6 \text{ neutrons}}{N}\right) = \left(\frac{(\ln 2)(1 \text{ hr})\left(\frac{3600 \text{ s}}{\text{hr}}\right)}{1100 \text{ s}}\right) = 2.268$$

(continued)

(19.43 continued)

$$\frac{3.0 \times 10^6 \text{ neutrons}}{N} = e^{2.268} = 9.66$$

$$N = \frac{3.0 \times 10^6 \text{ neutrons}}{9.66} = 3.1 \times 10^5 \text{ neutrons (if } t_{1/2} = 1100 \text{ s)}$$

$$\ln\left(\frac{3.0 \times 10^6 \text{ neutrons}}{N}\right) = \left(\frac{(\ln 2)(1 \text{ hr})\left(\frac{3600 \text{ s}}{\text{hr}}\right)}{876 \text{ s}}\right) = 2.849$$

$$N = \frac{3.0 \times 10^6 \text{ neutrons}}{e^{2.849}} = 1.7 \times 10^5 \text{ neutrons (if } t_{1/2} = 876 \text{ s)}$$

19.44 Yes, the precipitate will be radioactive. The ratio of sodium-24 to sodium-23 in the sodium nitrate precipitate will be the same as the ratio of sodium-24 to sodium-23 in the solution because the chemical properties of sodium-24 and sodium-23 are identical.

19.45 $^{1}_{0}n \rightarrow ^{1}_{1}p + ^{0}_{-1}\beta$ β-ray (an electron) is the other product.

See Sample Problem 19-2:

$\Delta m = (1.007276 \text{ g/mol}) + 0.0005486 \text{ g/mol} - 1.008665 \text{ g/mol} = -0.0008404 \text{ g/mol}$

$$\Delta E_{\text{per electron}} = \frac{(8.988 \times 10^{10} \text{ kJ / g})(-0.0008404 \text{ g / mol})}{(6.022 \times 10^{23} \text{ electrons / mol})}$$

$$= -1.25 \times 10^{-16} \text{ kJ / electron} = -1.25 \times 10^{-13} \text{ J / electron}$$

19.46 $^{30}_{14}Si + ^{1}_{0}n \rightarrow \left(^{31m}_{14}Si\right) \rightarrow ^{31}_{15}P + ^{0}_{-1}\beta$

19.47 $^{0}_{+1}\beta + ^{0}_{-1}e \rightarrow 2\ ^{0}_{0}\gamma$

$\Delta m = 0 - 2(0.0005486 \text{ g/mol}) = -0.0010972 \text{ g/mol}$

energy per mole of γ-rays =

$$\frac{\Delta E}{2} = \frac{(8.988 \times 10^{10} \text{ kJ / g})(-0.0010972 \text{ g / mol})}{2} = -4.931 \times 10^7 \text{ kJ / mol}$$

$$\text{energy per } \gamma\text{-ray} = \frac{-4.931 \times 10^7 \text{ kJ /mol}}{6.022 \times 10^{23} \text{ / mol}} = -8.188 \times 10^{-17} \text{ kJ or } -8.19 \times 10^{-14} \text{ J}$$

19.48 $\ln\left(\frac{N_0}{N}\right) = \frac{t \ln 2}{t_{1/2}}$ $\ln\left(\frac{5.0 \text{ mg } ^{210}\text{Po}}{N}\right) = \frac{(365 \text{ days})(\ln 2)}{(138.4 \text{ days})} = 1.828$

$\left(\frac{5.0 \text{ mg } ^{210}\text{Po}}{N}\right) = e^{1.828} = 6.22$ $N = 0.80 \text{ mg } ^{210}\text{Po}$ remain

$$\frac{\Delta N}{\Delta t} = \frac{N \ln 2}{t_{1/2}} = \frac{(0.80 \text{ mg})\left(\frac{10^{-3} \text{ g}}{\text{mg}}\right)\left(\frac{1 \text{ mol Po}}{210 \text{ g}}\right)(6.022 \times 10^{23} / \text{mol})(\ln 2)}{(138.4 \text{ days})(24 \text{ hr / day})(60 \text{ min / hr})(60 \text{ s / min})}$$

$= 1.3 \times 10^{11}$ emissions / s

19.49 $\ln\left(\frac{N_0}{N}\right) = \frac{t \ln 2}{t_{1/2}} = \frac{(1 \text{ yr})\ln 2}{103 \text{ yr}} = 0.00673$ $\frac{N_0}{N} = e^{0.00673} = 1.0068$ $\frac{N}{N_0} = 0.993$

Assume one mole of ^{209}Po and one mole of ^{210}Po at beginning:

for ^{209}Po: $\ln\left(\frac{1 \text{ mol}}{N}\right) = \frac{(10 \text{ yr})(\ln 2)}{103 \text{ yr}} = 0.0673$

$\left(\frac{1 \text{ mol}}{N}\right) = e^{0.0673} = 1.070$ $N = 0.935 \text{ mol } ^{209}\text{Po}$

for ^{210}Po: $\ln\left(\frac{1 \text{ mol}}{N}\right) = \frac{(10 \text{ yr})(\ln 2)}{(138.4 \text{ days})(1 \text{ yr / } 365.25 \text{ days})} = 18.29$

$\left(\frac{1 \text{ mol}}{N}\right) = e^{18.29} = 8.77 \times 10^7$ $N = 1.14 \times 10^{-8} \text{ mol } ^{210}\text{Po}$

$^{210}\text{Po}/^{209}\text{Po} = (1.14 \times 10^{-8} \text{ mol } ^{210}\text{Po})/(0.935 \text{ mol } ^{209}\text{Po}) = 1.2 \times 10^{-8}$

19.50 $E_{\text{coul.}} = \frac{(1.389 \times 10^5 \text{ kJ pm / mol})(Z_1)(Z_2)}{d} = \frac{(1.389 \times 10^5 \text{ kJ pm / mol})(6)(6)}{(6.0 \times 10^{-3} \text{ pm})}$

$= 8.3 \times 10^8 \text{ kJ/mol} = 8.3 \times 10^{11} \text{ J/mol}$

$\frac{8.3 \times 10^7 \text{ J /mol}}{6.022 \times 10^{23} C \text{ nuclei / mol}} = 1.4 \times 10^{-12} \text{ J /C nucleus}$

19.51 $^{11}_{6}\text{C}$ $N / Z = 5/6 = 0.833$

$^{15}_{8}\text{O}$ $N / Z = 7/8 = 0.875$

Both ^{11}C and ^{15}O are located below the "belt of stability" (N/Z < 1).

$^{11}_{6}\text{C} \rightarrow \ ^{0}_{+1}\beta + \ ^{11}_{5}\text{B}$

$^{15}_{8}\text{O} \rightarrow \ ^{0}_{+1}\beta + \ ^{15}_{7}\text{N}$

19.52 $^{13}_{7}N$ and $^{18}_{9}F$ ($^{17}_{9}F$ is also a $^{0}_{+1}\beta$ - emitter.)

$$^{13}_{7}N \rightarrow {}^{0}_{+1}\beta + {}^{13}_{6}C$$

$$^{18}_{9}F \rightarrow {}^{0}_{+1}\beta + {}^{18}_{8}O$$

19.53 $$\ln\left(\frac{15.3 \text{ counts / g min}}{0.03 \text{ counts / g min}}\right) = \frac{t \ln 2}{5730 \text{ yr}}$$

$$t = \ln\left(\frac{15.3}{0.03}\right)\frac{5730 \text{ yr}}{\ln 2} = 5.2 \times 10^{4} \text{ yrs}$$

19.54 a) $^{90}_{38}Sr \rightarrow {}^{90}_{39}Y + {}^{0}_{-1}\beta$

$^{90}_{39}Y \rightarrow {}^{90}_{40}Zr + {}^{0}_{-1}\beta$

b) The $^{90}_{38}Sr$ has the higher atomic mass because it has 52 neutrons and 38 protons vs. $^{90}_{40}Zr$ having 50 neutrons and 40 protons; neutrons have greater mass than protons plus electrons (1.008665 vs. 1.007276 + 0.0005486).

c) net reaction: $^{90}_{38}Sr \rightarrow {}^{90}_{40}Zr + 2\ {}^{0}_{-1}\beta$

Δm = 89.9043 g/mol + 2(0.0005486 g/mol) - 89.9073 g/mol = -0.0019 g/mol

ΔE = (8.988 x 10^{10} kJ/g)(-0.0019 g/mol) = -1.7 x 10^{8} kJ/mol

19.55 a) $^{64}_{29}Cu \rightarrow {}^{0}_{-1}\beta + {}^{64}_{30}Zn$ and $^{64}_{29}Cu \rightarrow {}^{0}_{+1}\beta + {}^{64}_{28}Ni$

b) ΔE = (8.988 x 10^{10} kJ/g)(Δm)

$$\Delta m = \frac{\Delta E}{(8.988 \times 10^{10} \text{ kJ / g})(10^{3} \text{ J / kJ})}$$

for $^{0}_{-1}\beta$: $$\Delta m = \frac{-(9.3 \times 10^{-14} \text{ J / atom})(6.022 \times 10^{23} \text{ atom / mol})}{(8.988 \times 10^{10} \text{ kJ / g})(10^{3} \text{ J / kJ})}$$

$$= -0.000623 \text{ g/mol}$$

for $^{0}_{+1}\beta$: $$\Delta m = \frac{-(1.04 \times 10^{-13} \text{ J / atom})(6.022 \times 10^{23} \text{ atom / mol})}{(8.988 \times 10^{10} \text{ kJ / g})(10^{3} \text{ J / kJ})}$$

$$= -0.000697 \text{ g / mol}$$

for $^{0}_{-1}\beta$: $\Delta m = -0.000623 \text{ g / mol} = m_e + m_{Zn} - m_{Cu}$

m_{Zn} = – 0.000623 g/mol + 63.92976 g/mol – 0.0005486 g/mol = 63.92859 g / mol

for $^{0}_{+1}\beta$: $\Delta m = -0.000697 \text{ g / mol} = m_e + m_{Ni} - m_{Cu}$

m_{Ni} = – 0.000697 g/mol – 0.0005486 g/mol + 63.92976 g/mol = 63.92851 g/mol

19.56 a) $^{26}_{14}Si \rightarrow {}^{0}_{+1}\beta + {}^{26}_{13}Al$

b) $^{82}_{38}Sr + {}^{0}_{-1}e \rightarrow {}^{82}_{37}Rb$

c) $^{210}_{84}Po \rightarrow {}^{4}_{2}\alpha + {}^{206}_{82}Pb$

d) $^{4}_{2}\alpha + {}^{9}_{4}Be \rightarrow \left({}^{13m}_{6}C\right) \rightarrow {}^{1}_{0}n + {}^{12}_{6}C$

e) $^{99m}_{43}Tc \rightarrow {}^{0}_{0}\gamma + {}^{99}_{43}Tc$

19.57 $^{232}_{90}Th$ $^{238}_{92}U$ $^{206,207,208}_{82}Pb$

α-emission decreases A by 4 units and Z by 2 units.

β-emission increases Z by 1 unit and leave A unchanged.

γ-emission changes neither A nor Z.

Difference of A for Th and A for each Pb isotope:
^{206}Pb: 232 - 206 = 26
^{207}Pb: 232 - 207 = 25
^{208}Pb: 232 - 208 = 24

Difference of A for U and A for each Pb isotope:
^{206}Pb: 238 - 206 = 32
^{207}Pb: 238 - 207 = 31
^{208}Pb: 238 - 208 = 30

For radioactive decay consisting entirely of α-, β- and γ-emissions, only α-emissions change the mass number (A) and then only by 4 units. Therefore, an isotope decaying by this scheme can only produce isotopes of new elements which differ in mass numbers by a multiple of 4.

Therefore, neither ^{232}Th nor ^{238}U can decay by this scheme and produce ^{207}Pb (differences of A are 25 and 31, respectively). The decay of ^{232}Th can produce ^{208}Pb (difference of A = 24 = 4 x 6) but not ^{206}Pb (difference of A = 26). The decay of ^{238}U can produce ^{206}Pb (difference of A = 32 = 4 x 8) but not ^{208}Pb (difference of A = 30).

The decay of $^{232}_{90}Th$ to $^{208}_{82}Pb$ requires 6 α-emissions and 4 β-emissions to counteract the Z change.

The decay of $^{238}_{92}U$ to $^{206}_{82}Pb$ requires 8 α-emissions and 6 β-emissions to counteract the Z change.

The lead from ^{238}U decay has a lower atomic mass than lead from ^{232}Th decay.

19.58 The Earth captures 3.4×10^{17} J/s which is $1/4.5 \times 10^{10}$ of the sun's total energy output. The sun's total output is $(3.4 \times 10^{17}\ J/s)(4.5 \times 10^{10}) = 1.53 \times 10^{28}$ J/s

a) $\Delta E = \Delta m(8.988 \times 10^{10}\ kJ/g)$

$$\Delta m = \frac{\Delta E}{8.988 \times 10^{10}\ kJ/g} = \frac{1.53 \times 10^{28}\ J/s}{(8.988 \times 10^{10}\ kJ/g)(10^{3}\ J/kJ)}$$

$$= 1.7 \times 10^{14}\ g/s \text{ or } 1.7 \times 10^{11}\ kJ/s$$

(continued)

(19.58 continued)

b) According to Section 19.7, p. 958, 2.5 x 10^9 kJ/mol of ^4He (or per 4 moles ^1_1H) are produced in the hydrogen-burning stage.

$$\frac{(1.53 \times 10^{28}\ J/s)}{(2.5 \times 10^9\ kJ/4\ mol\ ^1_1H)(10^3\ J/kJ)} = 2.4 \times 10^{16}\ mol\ ^1_1H/s$$

19.59 $^{10}_5B + ^1_0n \rightarrow ^4_2\alpha + ^7_3Li$

This does not present a health hazard because it is more difficult to shield for ^1_0n than for $^4_2\alpha$. The shielding necessary to protect against neutrons easily shields against $^4_2\alpha$.

19.60 $\ln\left(\frac{N_0}{N}\right) = \frac{t \ln 2}{t_{1/2}}$ $\qquad t = \frac{(t_{1/2})\ln(N_0/N)}{\ln 2}$

a) 2% of nuclide decayed; $N = 0.98\ N_0$

$$t = \frac{(150\ s)\ln(N_0/0.98\ N_0)}{\ln 2} = 4.4\ s$$

b) 99.5% of nuclide decayed; $N = 0.005\ N_0$

$$t = \frac{(150\ s)\ln(N_0/0.005\ N_0)}{\ln 2} = 1.1 \times 10^3\ s$$

19.61 1% of ^{226}Ra disappeared; $N = 0.99\ N_0$

$$t = \frac{(1622\ yr)\ln(N_0/0.99\ N_0)}{\ln 2} = 24\ yr$$

1% of ^{226}Ra remains; $N = 0.01\ N_0$

$$t = \frac{(1622\ yr)\ln(N_0/0.01\ N_0)}{\ln 2} = 1 \times 10^4\ yr$$

19.62 a) ^3_1H N/Z ratio = 2/1 = 2; too many neutrons; ratio too great

b) $^{238}_{92}U$; no stable nuclides with Z > 83

c) $^{40}_{19}K$; Z = 19 and N = 21 Only 5 stable nuclides with both Z and N odd with only 1 with Z > 7 and $^{40}_{19}K$ is not it.

d) $^8_4Be \rightarrow 2\ ^4_2He$ 2(4.00260 g/mol) - 8.005305 g/mol = -0.0000105 g/mol

The splitting of ^8_4Be into 2 ^4_2He is an exothermic process so ^8_4Be may not be stable.

19.63 $PV = nRT\ ;\ \ n = \frac{PV}{RT} = \frac{(1\ \text{atm})(6.0 \times 10^{-5}\ \text{cm}^3)(1\ \text{mL}/1\ \text{cm}^3)(\text{L}/10^3\ \text{mL})}{(0.0821\ \text{L atm / mol K})(298\ \text{K})}$

$= 2.45 \times 10^{-9}\ \text{mol}\ {}^4_2\text{He}$

$$n_U = \frac{1.3 \times 10^{-7}\ \text{g U}}{238\ \text{g U / mol U}} = 5.46 \times 10^{-10}\ \text{mol U};\ \ N = 5.46 \times 10^{-10}\ \text{mol U}$$

$$N_0 = 5.46 \times 10^{-10}\ \text{mol U} + (2.45 \times 10^{-9}\ \text{mol}\ {}^4_2\text{He})(1\ \text{mol U}/8\ \text{mol}\ {}^4_2\text{He})$$

$$= 8.52 \times 10^{-10}\ \text{mol U}$$

$$\ln\left(\frac{N_0}{N}\right) = \frac{t \ln 2}{t_{1/2}}$$

$$t = \frac{t_{1/2} \ln (N_0/N)}{\ln 2} = \frac{(4.51 \times 10^9\ \text{yr})\left(\ln\left(\frac{8.52 \times 10^{-10}\ \text{mol U}}{5.46 \times 10^{-10}\ \text{mol U}}\right)\right)}{0.6931} = 2.9 \times 10^9\ \text{yr}$$

19.64 $N = 0.15\ N_0;\ \ t = \frac{t_{1/2} \ln (N_0/N)}{\ln 2} = \frac{(14.6\ \text{days}) \ln(N_0/0.15\ N_0)}{\ln 2} = 40.0\ \text{days}$

19.65 $\ln(N_0/N) = \frac{(3250\ \text{yr}) \ln 2}{(5730\ \text{yr})} = 0.3931;\ \ \frac{N_0}{N} = e^{0.3931} = 1.482;\ \ N_0 = 1.482\ N;$

if $N = 1$, $N_0 = 1.48$, but if N_0 was 20% greater than thought, $N_0 = (1.20)(1.482) = 1.778$

$$\ln\left(\frac{1.78}{1}\right) = \frac{t \ln 2}{(5730\ \text{yr})};\ \ t = \frac{(5730\ \text{yr})(\ln(1.778/1))}{\ln\ 2} = 4.76 \times 10^3\ \text{yr}$$

19.66 ${}^{209}_{83}\text{Bi} + {}^{58}_{26}\text{Fe} \rightarrow {}^{267}_{109}\text{Une}$ Natural Bi is ${}^{209}\text{Bi}$.

19.67 ${}^{197}_{79}\text{Au}$; molar mass = 196.967 g/mol

79 protons: (79)(1.007276 g/mol) = 79.574804 g/mol
79 electrons: (79)(0.0005486 g/mol) = 0.0433394 g/mol
118 neutrons: (118)(1.007276 g/mol) = 119.022470 g/mol
198.640613 g/mol

Δm = 196.967 g/mol - 198.6406 g/mol = -1.6736 g/mol

$\Delta E = (8.988 \times 10^{10}\ \text{kJ/g})(-1.6736\ \text{g/mol}) = -1.504 \times 10^{11}\ \text{kJ/mol}$

$$\text{binding energy per nucleus of gold} = \frac{\Delta E}{6.023 \times 10^{23}\ \text{nuclei/mol}}$$

$$= \frac{-1.504 \times 10^{11}\ \text{kJ/mol}}{6.023 \times 10^{23}\ \text{nuclei/mol}} = -2.498 \times 10^{-13}\ \text{kJ / nucleus}$$

(continued)

(19.67 continued)

$$\text{binding energy per nucleon} = \frac{-2.498 \times 10^{-13}\ \text{kJ/nucleus}}{197\ \text{nucleons/nucleus}}$$

$$= -1.268 \times 10^{-15}\ \text{kJ/nucleon} = -1.268 \times 10^{-12}\ \text{J/nucleon}$$

19.68 0.5 mg = 0.5×10^{-3} g = $5. \times 10^{-4}$ g injected

$(5 \times 10^{-4}$ g$)(0.45) = 2.25 \times 10^{-4}$ g binds to thyroid gland.

$$\ln\left(\frac{N_0}{N}\right) = \frac{t \ln 2}{t_{1/2}}$$

$$t = \frac{t_{1/2} \ln (N_0/N)}{\ln 2} = \frac{(13.2\ \text{hr})\ln(2.25 \times 10^{-4}\ \text{g}/0.1 \times 10^{-6}\ \text{g})}{\ln 2} = 147\ \text{hr}$$

19.69 $^{27}_{13}\text{Al} + ^{4}_{2}\alpha \rightarrow \left(^{31m}_{15}\text{P}\right) \rightarrow ^{1}_{0}\text{n} + ^{30}_{15}\text{P}$

The $^{30}_{15}$P will not be stable because it has an odd number of protons and an odd number of neutrons and is not one of the five such isotopes that are stable. It has an equal number of protons and neutrons which gives a N/Z ratio of 1. Thus, it will probably decay by positron-emission that would increase N by 1 and decrease Z by 1 becoming $^{30}_{14}$Si. $^{30}_{14}$Si would have a N/Z ratio greater than 1 and is stable.

19.70 There are no nuclides of technetium that lie in the belt of stability; you can see the void that they would occupy if they were stable (See Figure 19-12, p. 927). If there were nuclides of Tc in the belt of stability, they might be:

$^{94}_{43}$Tc, $^{96}_{43}$Tc, $^{97}_{43}$Tc, $^{98}_{43}$Tc, $^{99}_{43}$Tc, $^{100}_{43}$Tc & $^{102}_{43}$Tc lying between $^{92}_{42}$Mo, $^{94}_{42}$Mo, $^{95}_{42}$Mo, $^{96}_{42}$Mo, $^{97}_{42}$Mo, $^{98}_{42}$Mo, $^{100}_{42}$Mo, & $^{96}_{44}$Ru, $^{98}_{44}$Ru, $^{99}_{44}$Ru, $^{100}_{44}$Ru, $^{101}_{44}$Ru, $^{102}_{44}$Ru, $^{104}_{44}$Ru.

All of the isotopes of technetium with an even mass number have odd numbers of protons and neutrons and have N/Z ratios of :1.19, 1.23, 1.28, 1.33, 1.37. These would decay by positron-emission or β-emission which makes the number of protons and number of neutrons even and changes the N/Z ratios slightly. The lower mass isotopes would be more likely to decay by positron-emission which increases the N/Z ratio. The higher mass isotopes would be more likely to decay by β-emission which decreases the N/Z ratio. Those isotopes with odd mass numbers could decay similarly.

19.71 $\ln\left(\frac{N_0}{N}\right) = \frac{t \ln 2}{t_{1/2}}$; $t = \frac{t_{1/2} \ln(N_0/N)}{\ln 2} = \frac{(15.0\ \text{hr})\ln(25\ \mu\text{g}/1\ \mu\text{g})}{\ln 2} = 70\ \text{hr}$

19.72 energy released by 20 kiloton bomb =

$(20 \times 10^3$ met ton TNT)(1000 kg/met ton)(2500 kJ/kg TNT) = 5.0×10^{10} kJ

energy released by mole of ^{235}U =

$(2.9 \times 10^{-11}$ J/nucleus)$(6.022 \times 10^{23}$ nuclei/mol) = 1.75×10^{13} J/mol

(continued)